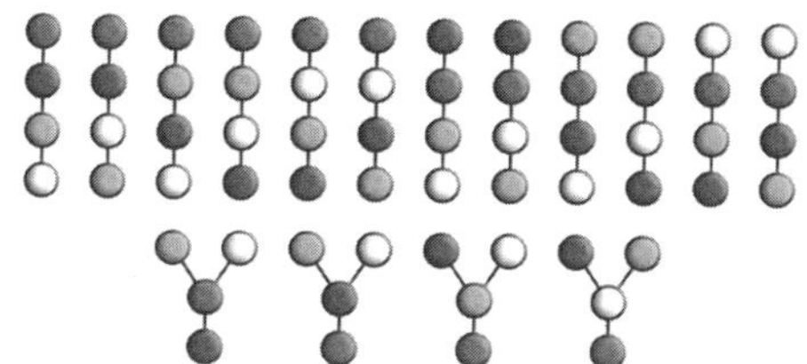

GIBBS MEASURES
ON CAYLEY TREES

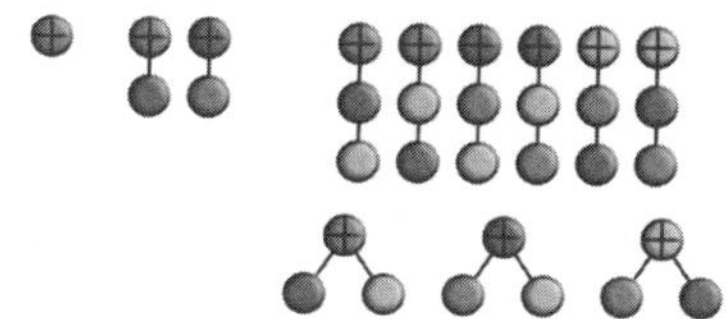

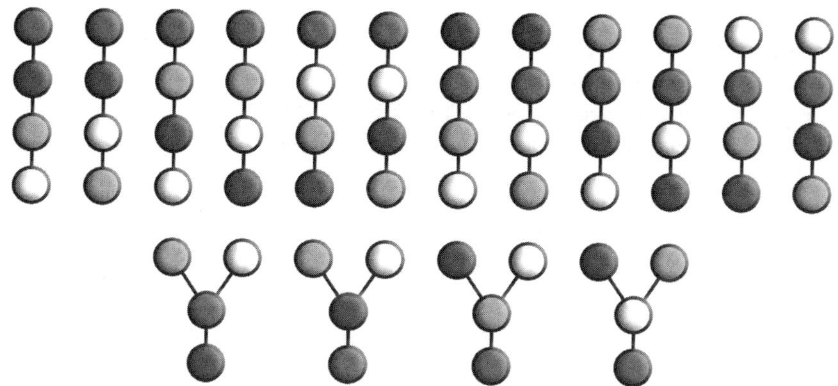

GIBBS MEASURES
ON CAYLEY TREES

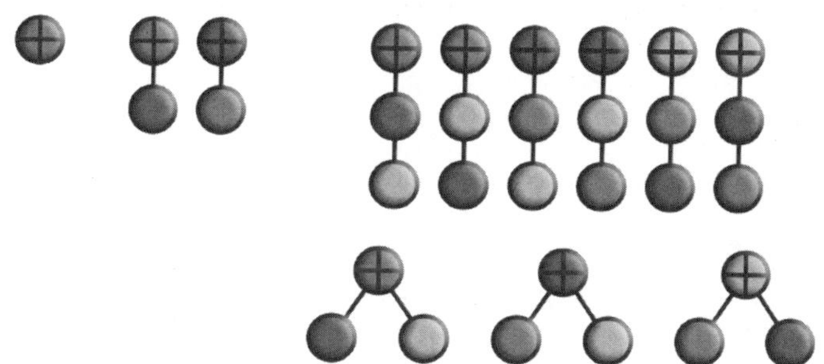

Utkir A Rozikov

Institute of Mathematics, Uzbekistan

World Scientific

NEW JERSEY • LONDON • SINGAPORE • BEIJING • SHANGHAI • HONG KONG • TAIPEI • CHENNAI

Published by

World Scientific Publishing Co. Pte. Ltd.

5 Toh Tuck Link, Singapore 596224

USA office: 27 Warren Street, Suite 401-402, Hackensack, NJ 07601

UK office: 57 Shelton Street, Covent Garden, London WC2H 9HE

Library of Congress Cataloging-in-Publication Data
Rozikov, Utkir A., 1970–
    Gibbs measures on Cayley trees / by Utkir A. Rozikov (Institute of Mathematics, Uzbekistan).
     pages cm
    Includes bibliographical references and index.
    ISBN 978-981-4513-37-1 (hardcover : alk. paper)
    1. Probability measures. 2. Distribution (Probability theory)   I. Title.
    QA273.6.R69 2013
    519.2--dc23

                   2013014066

British Library Cataloguing-in-Publication Data
A catalogue record for this book is available from the British Library.

Printed in Singapore by World Scientific Printers.

To the memory of my brothers
Hamza Rozikov (1965-1992) and Jamshid Rozikov (1972-2004)

Preface

The purpose of this book is to present systematically all known mathematical results on Gibbs measures on Cayley trees (Bethe lattices).

The Gibbs measure is a probability measure, which has been an important object in many problems of probability theory and statistical mechanics. It is the measure associated with the Hamiltonian of a physical system (a model) and generalizes the notion of a canonical ensemble. More importantly, when the Hamiltonian can be written as a sum of parts, the Gibbs measure has the Markov property (a certain kind of statistical independence), thus leading to its widespread appearance in many problems outside of physics, such as biology, Hopfield networks, Markov networks, and Markov logic networks. Moreover, the Gibbs measure is the unique measure that maximizes the entropy for a given expected energy.

Using MathSciNet and a Google search one can see that there are about 600 published works devoted to several models defined on Cayley trees. But most of them are written for physicists audience with little regard for mathematical rigor. There are about 150 papers which contain mathematically rigorous results about Gibbs measures on Cayley trees. This book is mainly based on these mathematical papers.

There are a few books written from a mathematical perspective which are devoted to spin systems and Gibbs measures. Some of these books do not have any result about Gibbs measures on trees; others have a small part devoted to Gibbs measures on trees but were published a long time ago. To the best of my knowledge, there is no mathematical book devoted to (recently) obtained results about Gibbs measures on trees.

The method used for the description of Gibbs measures on Cayley trees is the method of Markov random field theory and recurrent equations of this theory, but the modern theory of Gibbs measures on trees uses new tools such as group theory, information flows on trees, node-weighted random walks, contour methods on trees, non-linear analysis. This book discusses all the mentioned methods, which were developed recently.

In the bibliography of this book I have collected the possibly maximal list of references which are related to Gibbs measures on trees. Moreover, the last section of Chapter 11 contains a brief review and references on quantum models of statistical mechanics on trees, a review about p-adic Gibbs measures on Cayley trees, and a comparison between real-valued Gibbs measures and p-adic measures.

Thus, the book informs the reader about what has been (mathematically) done in the theory of Gibbs measures on trees and about where the corresponding results were published. The book is divided into 11 chapters; at the end of each chapter I give commentaries and references related to the chapter.

Chapter 1 is devoted to algebraic properties of the Cayley tree. We give a (non-commutative) group representation of a Cayley tree of order k, which is a free product of $k+1$ cyclic groups of the second order. We construct several subgroups of the group. Moreover, the partition structures of the Cayley tree with respect to normal subgroups of the group are studied. These results are applied in the following chapters to describe periodic and weakly periodic Gibbs measures on the Cayley tree.

In the first section of Chapter 2 we give general definitions of configuration space, Hamiltonian and Gibbs measures. This chapter contains all known results about Gibbs measures of the Ising model on Cayley trees. Some of the results were obtained very recently. We show that the Ising model may have up to three translation-invariant Gibbs measures. It may have *only* periodic measures with period two, which are a 'chess-board' periodic. To describe more 'rich' set of Gibbs measures we introduce the notion of weakly periodic Gibbs measures and show that the Ising model has at least seven such measures (under some conditions on parameters). The extremality criterion of the disordered Gibbs measure is proved. Moreover we give two constructions of uncountable sets of non-periodic Gibbs measures. We show that under some conditions on the temperature, one can construct for each known Gibbs measure on a Cayley tree of order k_0 a new Gibbs measure on the Cayley tree of order k, $k > k_0$. Some explicit formulae of the free energies (and entropies) according to these Gibbs measures are presented.

Chapter 3 is devoted to two Ising type models with competing interactions on Cayley trees. One of them is known as Vannimenus's model; the second is a model with four competing interactions (external field, nearest neighbor, second neighbors and triples of neighbors) on the Cayley tree of order two. We show non-uniqueness of the Gibbs measure for some parameter values of the models. Our second result gives a complete description of periodic Gibbs measures for the models. We also construct uncountably many non-periodic extreme Gibbs measures.

In Chapter 4 we consider a process on a tree T in which information is transmitted from the root of the tree to all the nodes of the tree. Each node inherits information from its parent with some probability of error. The transmission process is assumed to have identical distribution on all the edges, and different edges of the tree are assumed to act independently. The basic question of this chapter is: Does the configuration obtained at level n of T typically contain significant information on the root variable? This problem arose independently in biology, information theory and statistical physics. For models of statistical physics on trees, the problem is related to extremality of the disordered Gibbs measure. In this chapter, we give results and challenges related to this problem. In the following chapters we shall apply the results to extremality conditions of Gibbs measures.

Chapter 5 contains results related to Gibbs measures of the q-state Potts model on Cayley trees. The description of such measures is reduced to solution of a vector-valued functional equation. We show that under some conditions on the parameters there exist $q + 1$ distinct translation-invariant Gibbs measures. We apply the results of Chapter 4 to find the extremality conditions of the disordered Gibbs measure. Moreover using the Bleher-Ganikhodjaev construction we show the existence of an uncountable set of non-translation-invariant Gibbs measures. Compared to the Ising model, one can see that many problems related to the Potts model are open. For example, periodic and weakly periodic Gibbs measures have not been studied yet.

In Chapter 6 we consider a nearest-neighbor SOS (solid-on-solid) model, with several spin values $0, 1, \ldots, m$, $m \geq 2$, and zero external field, on a Cayley tree of order k. We mainly assume that $m = 2$ or $m = 3$ and study Gibbs measures.

For $m = 2$, in the anti-ferromagnetic case, we show that the translation-invariant Gibbs measure is unique for all temperatures. In the ferromagnetic case, for $m = 2$, the number of such measures varies with the temperature: this gives an interesting example of phase transition. Here we identify a critical inverse temperature, $\beta_{\mathrm{cr}}^1 \in (0, \infty)$ such that for all $0 \leq \beta \leq \beta_{\mathrm{cr}}^1$, there exists a unique translation-invariant measure and for all $\beta > \beta_{\mathrm{cr}}^1$ there are exactly three such measures. For $\beta > \beta_{\mathrm{cr}}^1$ we also construct a continuum of distinct, non-translation-invariant Gibbs measures. A complete description of the set of periodic Gibbs measures for the SOS model on a Cayley tree is given. A complete description of periodic Gibbs measures means a characterization of such measures with respect to any given normal subgroup of finite index in the representation group of the tree. We show that (i) for an ferromagnetic SOS model, for any normal subgroup of finite index, each periodic Gibbs measure is in fact translation-invariant. Further, (ii) for an anti-ferromagnetic SOS model, for any normal subgroup of finite index, each periodic Gibbs measure is either translation-invariant or has period two (i.e., is a chess-board Gibbs measure). For $m = 3$ similar results are obtained. But the case $m \geq 4$ is not studied yet.

In Chapter 7 we consider models which have "hard constraints". Such models are defined by the space $\mathrm{Hom}(\Gamma^k, H)$ of homomorphisms from a Cayley tree Γ^k to a fixed finite constraint graph H. For any assignment λ of positive real activities to the nodes of H, there is at least one Gibbs measure on $\mathrm{Hom}(\Gamma^k, H)$, but there may be more than one (phase transition). We mainly consider the case where graph H contains two vertices or three vertices. In such simple cases, a hard core model with two spin values and several hard core models with three spin values are discussed. In the last section of this chapter we give a model with two spin values (without hard constraints), but with interaction radius equal to two. We show that this model can be "transformed" to a nearest-neighbor interaction model with 8 spin values and with hard constrains on the Cayley tree. In each case we construct several kinds of Gibbs measures of these models.

Chapter 8 is devoted to a nearest-neighbor Potts model, with countable spin values $0, 1, \ldots$, and non-zero external field, on a Cayley tree of order k. We study translation-invariant 'splitting' Gibbs measures, which depend on k and a probability measure ν (with $\nu(i) > 0$ on the set of all non-negative integer numbers $\Phi = \{0, 1, ...\}$). This problem is reduced to the description of the solutions of an infinite system of equations. For any $k \geq 1$ and any fixed probability measure ν we show that the set of translation-invariant splitting Gibbs measures contains at most one point, independently on parameters of the Potts model with countable set of spin values on a Cayley tree. Also we give a description of the class of measures ν on Φ such that with respect to each element of this class the infinite system of equations has unique solution $\{a^i, i = 1, 2, ...\}$, where $a \in (0, 1)$.

In Chapter 9 we present very recently obtained results for models with nearest-neighbor interactions and with the set $[0, 1]$ of spin values, on a

Cayley tree of order $k \geq 1$. We reduce the problem of describing the "splitting Gibbs measures" of the model to the description of the solutions of some non-linear integral equation. For $k = 1$ we show that the integral equation has a unique solution. In case $k \geq 2$ some models (with the set $[0, 1]$ of spin values) which have a unique splitting Gibbs measure are constructed. Also for the Potts model with uncountable set of spin values it is proven that there is unique splitting Gibbs measure. For arbitrary $k \geq 2$ we find a sufficient condition under which the integral equation has unique solution; hence under this condition the corresponding model has unique splitting Gibbs measure. Finally, we construct several models with the set $[0, 1]$ of spin values and show that each of the constructed model has at least two translational-invariant Gibbs measures.

Chapter 10 is devoted to recently developed contour methods on Cayley trees. In the first section of this chapter we consider a one-dimensional model with nearest-neighbor interactions $I_n, n \in \mathbb{Z}$, and spin values ± 1. We show that under some conditions on parameters I_n the phase transition occurs for the model. We define a notion of "phase separation point" between two phases. We prove that the expectation value of the point is zero and its mean square fluctuation is bounded by a constant $C(\beta)$ which tends to $\frac{1}{4}$ if $\beta \to \infty$. Here $\beta = \frac{1}{T}$, $T > 0$-temperature. In other sections of this chapter we consider a q-component model and the Ising model with competing two-step interactions on a Cayley tree of order $k \geq 1$. We constructively describe (periodic and weakly periodic) ground states and verify the Peierls condition for these models. We define the notion of a contour for the models on the Cayley tree. Using a contour argument we show the existence of several different Gibbs measures. This chapter also contains a general contour argument for a finite range lattice models on Cayley tree with two basic properties: the existence of only a finite number of ground states and with Peierls type condition. We define a general contour for such models on the Cayley tree. By a contour argument we show the existence of s different (where s is the number of ground states) Gibbs measures.

Chapter 11 contains several models not discussed in previous chapters. In Sections 1-9 we give the definitions of these models and describe some known results for each model. The last section contains a review of other models. This section is divided into three subsections: the first is devoted to classical (real-valued) models, the second subsection to quantum models, and the third one to models with p-adic values. Moreover, we give a brief description of the differences of behavior between classical (real) models and p-adic models on Cayley trees.

The bibliography is possibly a maximal list of mathematical references related to Gibbs measures on trees.

The book is based on materials collected by the author during several visits to Statistical Laboratory, DPMMS, University of Cambridge and Newton Institute (Cambridge, UK); International Center for Theoretical Physics (Trieste, Italy); Physical department of "La Sapienza" University (Rome, Italy); Institut des Hautes Etudes Scientifiques (IHES, Bures-sur-Yvette, France); Universite du Sud Toulon Var and Centre de Physique Theorique (Marseille, France); University Santiago de Compostela (Santiago de Compostela, Spain); Institute for Applied Mathematics of University of Bonn, (Bonn, Germany).

Acknowledgements. I thank all the above-mentioned institutions for their warm hospitality and excellent working conditions. My special thanks go to IHES, where last chapters of the book were written.

I learned about Gibbs measures from N.N. Ganikhodjaev (my supervisor). I want to thank him for many fruitful conversations about Gibbs measures.

While working on Gibbs measures on trees and writing this book, I had numerous discussions with S. Albeverio, Sh.A. Ayupov, Ph. Blanchard, M. Cassandro, V.I. Chilin, A.C.D. van Enter, P. Falco, G. Gallavotti, D. Gandolfo, A. Giuliani, M. Gromov, B.M. Gurevich, Y. Higuchi, F. Hiroshima, M. Kontsevich, M. Ladra, J. Lörinczi, J.B. Martin, J.F.F. Mendes, I. Merola, R.A. Minlos, S. Miracle-Sole, F.M. Mukhamedov, P. Picco, D.Ruelle, J. Ruiz, Ya.G. Sinai, Yu.M. Suhov, S. Shlosman, V. Zagrebnov and many others. I am grateful to everyone for their help.

I thank World Scientific Publishing for the opportunity of publishing this book.

I am indebted to my parents (Abdullo Rozikov and Soliha Bozorova), wife (Dilfuza Gadoeva) and children (Azamat, Laziz, Sevinch) for their warm attitude to my work and patiently waiting for me when I am abroad.

Utkir A. Rozikov
Tashkent, Uzbekistan
March 2013

Contents

Chapter 1

Properties of a group representation of the Cayley tree

This chapter is devoted to algebraic properties of the Cayley tree. Here we shall give a group representation of the Cayley tree and construct several subgroups of the group. Moreover, the partition structures of the Cayley tree with respect to normal subgroups of the group are studied. These results will be applied in the following chapters to describe periodic and weakly periodic Gibbs measures on the Cayley tree.

1.1 Cayley tree

The Cayley tree (Bethe lattice [18]) Γ^k of order $k \geq 1$ is an infinite tree, i.e., a graph without cycles, such that exactly $k+1$ edges originate from each vertex (see Fig. 1.1). Let $\Gamma^k = (V, L)$ where V is the set of vertices and L the set of edges. Two vertices x and y are called *nearest neighbors* if there exists an edge $l \in L$ connecting them. We will use the notation $l = \langle x, y \rangle$. A collection of nearest neighbor pairs $\langle x, x_1 \rangle, \langle x_1, x_2 \rangle, ..., \langle x_{d-1}, y \rangle$ is called a *path* from x to y. The distance $d(x, y)$ on the Cayley tree is the number of edges of the shortest path from x to y.

For a fixed $x^0 \in V$, called the root, we set

$$W_n = \{x \in V \mid d(x, x^0) = n\}, \qquad V_n = \bigcup_{m=0}^{n} W_m$$

and denote

$$S(x) = \{y \in W_{n+1} : d(x, y) = 1\}, \quad x \in W_n,$$

the set of *direct successors* of x.

Now we give three well-known distinctions of Cayley tree and $\mathbb{Z}^d$:

(i) the group representation G_k (see the next section) of the Cayley tree is non-commutative, but $\mathbb{Z}^d$ is commutative group.

1

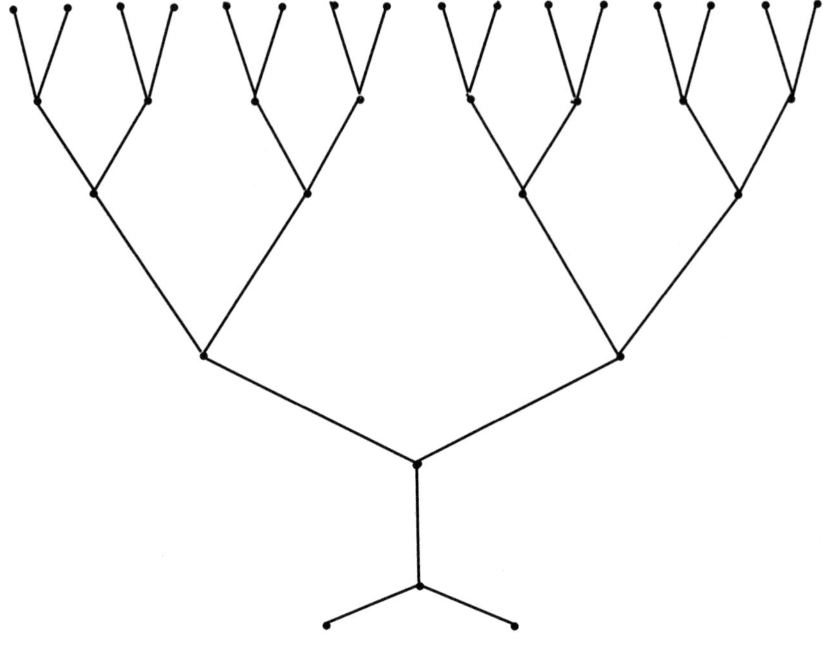

Fig. 1.1   Cayley tree of order two, i.e., $k = 2$.

(ii) The ratio of the number of boundary sites to the number of interior sites of the lattice $\mathbb{Z}^d$ becomes small in the thermodynamic limit of a large system. For the Cayley tree it does not, since both numbers grow exponentially like k^n. The Cayley tree is a non-amenable graph.

(iii) Consider any regular lattice $\mathbf{L}$. Let c_n be the number of sites within n steps of a given site of $\mathbf{L}$. For $\mathbf{L} = \mathbb{Z}^d$ it is easy to see that

$$\lim_{n \to \infty} \frac{\ln c_n}{\ln n} = d;$$

where d is the dimensionality of the lattice $\mathbb{Z}^d$. The above relation can be regarded as a definition of dimensionality d.

For the Cayley tree, $c_n = (k-1)^{-1}(k+1)(k^n - 1)$. Thus in this case we have $d = \infty$, so in this sense the Cayley tree is "infinite-dimensional".

1.2 A group representation of the Cayley tree

Let G_k be a free product of $k + 1$ cyclic groups of the second order with generators $a_1, a_2, \ldots, a_{k+1}$, respectively.

Proposition 1.1. *There exists a one-to-one correspondence between the set of vertices V of the Cayley tree Γ^k and the group G_k.*

Proof. Fix an arbitrary element $x_0 \in V$ and correspond it to the unit element e of the group G_k. Without loss of generality we assume that the Cayley tree is a planar graph. Using $a_1, \ldots, a_{k+1}$ we numerate nearest-neighbors of element e, moving by positive direction (see Fig. 1.2). Now we shall give numeration of the nearest-neighbors of each a_i, $i = 1, \ldots, k+1$ by $a_i a_j$, $j = 1, \ldots, k + 1$. Since all a_i have the common neighbor e we give to it $a_i a_i = a_i^2 = e$. Other neighbors are numerated starting from $a_i a_i$ by the positive direction. We numerate the set of all nearest-neighbors of each $a_i a_j$ by words $a_i a_j a_q$, $q = 1, \ldots, k + 1$, starting from $a_i a_j a_j = a_i$ by the positive direction. Iterating this argument one gets a one-to-one correspondence between the set of vertices V of the Cayley tree Γ^k and the group G_k. $\square$

The group representation given above is called *right* representation, since in this case if x and y are nearest-neighbors on tree and g, h corresponding elements of the group G_k, then $g = ha_i$ or $h = ga_j$ for some i or j. Similarly, one can define *left* representation.

In the group G_k, let us consider the left (right) shift transformations defined as follows. For $g \in G_k$, let us set

$$T_g(h) = gh, \quad (T_g(h) = hg), \quad \text{for all } \ h \in G_k. \tag{1.1}$$

The set of all left (right) shifts in G_k is isomorphic to the group G_k. By virtue of Proposition 1.1, any transformation S of the group G_k induces the transformation $\hat{S}$ of the set of vertices V of the Cayley tree.

The following theorem obviously holds.

Theorem 1.1. *The group of left (right) shifts on the right (left) representations of the Cayley tree is the group translations of the Cayley tree.*

In this book we use only right representations of the Cayley tree.

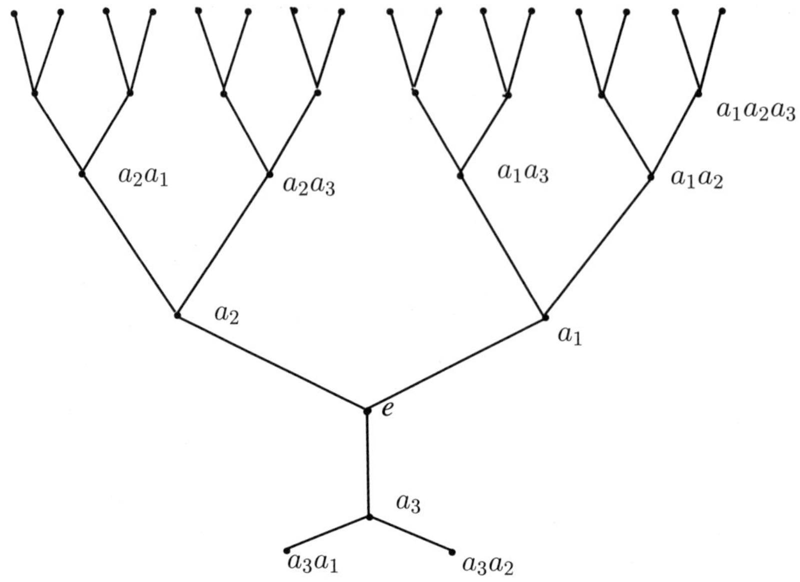

Fig. 1.2   Some elements of group G_2 on Cayley tree of order two.

1.3   Normal subgroups of finite index for the group representation of the Cayley tree

In this section we answer to the following question: Does G_k have subgroups (in particular, normal subgroups) of any finite index?

Any (minimal represented) element $x \in G_k$ has the following form:

$$x = a_{i_1} a_{i_2} \ldots a_{i_n}, \quad \text{where} \ \ 1 \leq i_m \leq k+1, \, m = 1, \ldots, n.$$

The number n is called the length of the word x and is denoted by $l(x)$. The number of letters a_i, $i = 1, \ldots, k+1$, that enter the non-contractible representation of the word x is denoted by $\omega_x(a_i)$.

The following proposition is well known in group theory:

Proposition 1.2. [121, 143]. *Let φ be a homomorphism of the group G_k with the kernel H. Then H is a normal subgroup of the group G_k and $\varphi(G_k) \simeq G_k/H$, (where G_k/H is a factor group) i.e., the index $|G_k : H|$ coincides with the order $|\varphi(G_k)|$ of the group $\varphi(G_k)$.*

By virtue of Proposition 1.2 in order to construct a normal subgroup of a finite index of the group G_k, one should construct a homomorphism of the group G_k into some finite group.

Definition 1.1. Let $M_1, M_2, \ldots, M_m$ be some sets and $M_i \neq M_j$, for $i \neq j$. We call the intersection $\cap_{i=1}^{m} M_i$ contractible if there exists i_0 $(1 \leq i_0 \leq m)$ such that

$$\cap_{i=1}^{m} M_i = \left(\cap_{i=1}^{i_0-1} M_i \right) \cap \left(\cap_{i=i_0+1}^{m} M_i \right).$$

Let $N_k = \{1, \ldots, k+1\}$. The following theorem describes several normal subgroups of G_k.

Theorem 1.2. *For any $\emptyset \neq A \subseteq N_k$, there exists a subgroup $H_A \subset G_k$ with the following properties:*

(a) H_A *is a normal subgroup and* $|G_k : H_A| = 2$;

(b) $H_A \neq H_B$ *for all* $A \neq B \subseteq N_k$;

(c) $|H_A \cap H_B| = \infty$ *and* $H_A \cap H_B \subset H_{A \triangle B}$, *for all* $A, B \subseteq N_k$, *where* $A \triangle B = (A \cup B) \setminus (A \cap B)$;

(d) *If* $A_1, A_2, \ldots, A_m \subseteq N_k$ *and* $A_i \cap A_j = \emptyset$ *for any* $i \neq j$, *then*

$$\bigcap_{i=1}^{m} H_{A_i} \subset H_{\cup_{i=1}^{m} A_i};$$

(e) *Let* $A_1, A_2, \ldots, A_m \subseteq N_k$. *If* $\cap_{i=1}^{m} H_{A_i}$ *is non-contractible, then it is a normal subgroup of index* 2^m;

(f) *For any* $m = 1, 2, \ldots, 2k$ *(where k is the order of the tree), there exist non-contractible intersections* $\cap_{i=1}^{m} H_{A_i}$.

Proof. (a) Let $\emptyset \neq A \subseteq N_k$. Define the mapping $f_A : G_k \to \{-1, 1\}$ as follows:

$$f_A(x) = \begin{cases} 1, & \text{if } \sum_{i \in A} \omega_x(a_i) \text{ is even,} \\ -1, & \text{if } \sum_{i \in A} \omega_x(a_i) \text{ is odd.} \end{cases}$$

We shall prove that f_A is a homomorphism, i.e., $f_A(xy) = f_A(x) f_A(y)$ for any $x, y \in G_k$.

We have

$$\sum_{i\in A}\omega_{xy}(a_i) = \begin{cases} \text{even,} \quad \text{if} & \left[\begin{array}{l} \sum_{i\in A}\omega_x(a_i),\ \sum_{i\in A}\omega_y(a_i) \ \text{ are even,} \\ \text{or} \\ \sum_{i\in A}\omega_x(a_i),\ \sum_{i\in A}\omega_y(a_i) \ \text{ are odd,} \end{array}\right. \\[2em] \text{odd,} \quad \text{if} & \left[\begin{array}{l} \sum_{i\in A}\omega_x(a_i) - \text{even},\ \sum_{i\in A}\omega_y(a_i) - \text{odd,} \\ \text{or} \\ \sum_{i\in A}\omega_x(a_i) - \text{odd},\ \sum_{i\in A}\omega_y(a_i) - \text{even.} \end{array}\right. \end{cases}$$

Then to the rule "even plus odd equal to odd" corresponds to the equality $1 \cdot (-1) = -1$ etc., i.e., f_A is a homomorphism. Then by Proposition 1.2 the set

$$H_A = \left\{ x \in G_k : \sum_{i\in A}\omega_x(a_i) - \text{even} \right\}$$

is a normal subgroup of index two.

(b) Since $A \neq B$, we can take $i_0 \in A$ (or $i_0 \in B$) such that $i_0 \notin B$ (resp. $i_0 \notin A$). Then it is easy to see that $a_{i_0} \notin H_A$, but $a_{i_0} \in H_B$. Hence $H_A \neq H_B$.

One can easily check that H_A satisfies assertions (c), (d), (e) of Theorem 1.2. Here we shall prove (f).

(f) For $m = 2k$ we construct a non-contractible intersection as follows. For a fixed i_0, where $1 \leq i_0 \leq k+1$ we define

$$H_{2k} = \bigcap_{\substack{i=1: \\ i\neq i_0}}^{k+1} \left(H_{\{i\}} \cap H_{\{i_0,i\}}\right). \tag{1.2}$$

By (c) and (d) one can see that H_{2k} is non-contractible. Now we shall show that there is no subgroup H_A with $H_{2k} \cap H_A$ non-contractible. Indeed, take $A \subseteq N_k$, if $i_0 \in A$ then by part (d) we have $\cap_{i\in A}H_{\{i\}} \subset H_A$, i.e., $H_{2k} \cap H_A$ is contractible. If $i_0 \in A$ then for a $j \in A \setminus \{i_0\}$ we have

$$\bigcap_{i\in A\setminus\{j,i_0\}} \left(H_{\{i\}} \cap H_{\{j,i_0\}}\right) \subset H_A.$$

Hence $H_{2k} \cap H_A$ is contractible. Thus we proved the existence of a non-contractible intersection with length $2k$. An intersection of length $2k - l$ can be obtained from (1.2) by omitting any l normal subgroups. $\qquad\square$

Let $x = a_{i_1} a_{i_2} \ldots a_{i_n}$, where $1 \leq i_m \leq k+1$, $m = 1, \ldots, n$. For $x \in G_k$ with $l(x) = n$, we set $\nu_x(a_j) = \{m \in N_{n-1} : i_m = j\}$. For instance if $x = a_1 a_4 a_8 a_4 a_1 a_2 a_4$, then $\nu_x(a_4) = \{2, 4, 7\}$.

Theorem 1.3. *Let $e \neq x_0 \in G_k$. Then there exists a normal finite-index subgroup H_{x_0}, which does not contain the element x_0.*

Proof. Let x_0 contains m, $m \leq k+1$, generators (from the set $\{a_1, \ldots, a_{k+1}\}$) we denote them by $a_1', a_2', \ldots, a_m'$. Then, x_0 reads

$$x_0 = a_{i_1}' a_{i_2}' \ldots a_{i_n}'.$$

In the symmetric group S_{n+1} that acts on the symbols $1, 2, \ldots, n+1$, we choose the permutations π_i, $i = 1, \ldots, m$ as follows: if $\nu_{x_0}(a_j) = \{j_1, j_2, \ldots, j_{m_1}\}$, where $j_{k_1} \in N_{n-1}$ and $k_1 = 1, \ldots, m_1$, then

$$\pi_j = \begin{pmatrix} 1 \ldots & j_1 & j_1 + 1 \ldots & j_{m_1} & j_{m_1} + 1 \ldots & n+1 \\ 1 \ldots & j_1 + 1 & j_1 & \ldots j_{m_1} + 1 & j_{m_1} & \ldots & n+1 \end{pmatrix},$$

$j = 1, \ldots, m$. Obviously, $\pi_i^2 = \pi_0$ for all $i = 1, \ldots, m$, where π_0 is the identical permutation.

Let us define the mappings $u : \{a_1, \ldots, a_{k+1}\} \to \{\pi_0, \ldots, \pi_m\}$ and $f_{x_0} : G_k \to S_{n+1}$ as follows:

$$u(x) = \begin{cases} \pi_0, & \text{if } x \notin \{a_1', \ldots, a_m'\}, \\ \pi_j, & \text{if } x = a_j', \quad j = 1, \ldots, m. \end{cases}$$

$$f_{x_0}(x) = u(a_{i_1}) u(a_{i_2}) \ldots u(a_{i_n}).$$

Since $\pi_i^2 = \pi_0$, the mapping f_{x_0} is a homomorphism. The kernel H_{x_0} of this homomorphism is a normal finite-index subgroup. Obviously, the element x_0 does not belong to this normal subgroup. $\square$

Proposition 1.3. *The following relations hold:*

(1) $H_{a_i} = H_{\{i\}}$, $i = 1, \ldots, k+1$;
(2) If x is generated by more than one generators, then $|G_k : H_x| \geq 4$;
(3) $|G_k : H_{a_i a_j}| = 6$, for any $i \neq j \in N_k$.

Proof. The proof follows from Theorems 1.2 and 1.3. $\square$

Theorem 1.4.

1. The group G_k does not have normal subgroups of odd index $(\neq 1)$.

2. The group G_k has a normal subgroups of arbitrary even index.

Proof. 1. The quotient group by any normal subgroup H contains a subgroup of order 2, since $a_i^2 = e$. By Lagrange's theorem, it follows that there is no normal subgroup of odd index.

2. Let n be an odd number (for an even n the proof is similar). Using notations of the proof of Theorem 1.3 we take x_0, with $l(x_0) = n$, which is generated only by two generators, say a_1, a_2. Then it is easy to see that the group $f_{x_0}(G_k)$ is generated by the permutations

$$\pi_0, \quad \pi_1 = (1,2)(3,4)(5,6)\ldots(n, n+1), \pi_2 = (2,3)(4,5)\ldots(n-1, n).$$

Note that $\pi_1\pi_2 = (1,3,5,\ldots,n,n+1,n-1,n-3,\ldots,4,2)$ this is a cycle with length $n+1$. Thus $< \pi_1\pi_2 >$ is a cyclic subgroup of S_{n+1}, with order $n+1$, i.e., $(\pi_1\pi_2)^{n+1} = \pi_0$. The last equality gives $(\pi_1\pi_2)^{n-s+1} = (\pi_2\pi_1)^s$, for any $s = 1,\ldots,n$. Hence, the group $f_{x_0}(G_k)$ contains the following distinct elements :

$$\pi_0, \pi_1, \pi_1\pi_2, \pi_1\pi_2\pi_1, (\pi_1\pi_2)^2, \ldots, (\pi_1\pi_2)^n, (\pi_1\pi_2)^n\pi_1 = \pi_2,$$

their number is $2(n+1)$. Consequently, index of the corresponding H_{x_0} is the even number, $2(n+1)$. $\qquad\qquad\square$

1.3.1   *Subgroups of infinite index*

There are normal subgroups of infinite index. Some of them can be constructed as follows. Fix $M \subseteq N_k$ such that $|M| > 1$. $|\bullet|$ is the cardinality of $\bullet$. Let the mapping $\pi_M : \{a_1, ..., a_{k+1}\} \longrightarrow \{a_i, \ i \in M\} \cup \{e\}$ be defined by

$$\pi_M(a_i) = \begin{cases} a_i, \text{ if } \ i \in M \\ e, \ \text{ if } \ i \notin M. \end{cases}$$

Denote by G_M the free product of cyclic groups $\{e, a_i\}$, $i \in M$. Consider

$$f_M(x) = f_M(a_{i_1} a_{i_2} ... a_{i_m}) = \pi_M(a_{i_1})\pi_M(a_{i_2})...\pi_M(a_{i_m}).$$

Then it is easy to see that f_M is a homomorphism and hence $H_M = \{x \in G_k : \ f_M(x) = e\}$ is a normal subgroup of infinite index.

1.4   Partition structures of the Cayley tree

Let $H_0 \subset G_k$ be an arbitrary normal subgroup of index n of the group G_k. Obviously, each normal subgroup of the group G_k is the kernel of some homomorphism φ of the group G_k into some group G^*. Introduce the following equivalence relation on the set G_k: $x \sim y$ if $xy^{-1} \in H_0$.

Proposition 1.4.

(1) $xy \sim xz$ if and only if $y \sim z$ and $x, y, z \in G_k$;
(2) $yx \sim zx$ if and only if $y \sim z$ and $x, y, z \in G_k$.

Proof. 1. *Necessity.* Let $xy \sim xz$, i.e., $xy(xz)^{-1} = xyz^{-1}x^{-1} \in H_0$. Hence, $\varphi(xyz^{-1}x^{-1}) = e \in G^*$ (e is the unit element of the group G^*). Since φ is a homomorphism, we obtain

$$\varphi(x)\varphi(yz^{-1})\varphi(x^{-1}) = e \qquad (1.3)$$

or $\varphi(yz^{-1}) = [\varphi(x)]^{-1}[\varphi(x^{-1})]^{-1} = \varphi(x^{-1})[\varphi(x^{-1})]^{-1} = e$, i.e., $y \sim z$.

2. *Sufficiency.* Let $y \sim z$, i.e.,

$$\varphi(yz^{-1}) = e. \qquad (1.4)$$

Consider the element $\varphi(xyz^{-1}x^{-1})$. As φ is a homomorphism, we have, by virtue of (1.4), $\varphi(xyz^{-1}x^{-1}) = e$, i.e., $xy \sim xz$.

Statement (1) is proved. Statement (2) can be proved analogously. $\square$

Corollary 1.1.

(1) $xa_i \sim xa_j$ if and only if $a_i \sim a_j$, where $a_i, a_j \in \{e, a_1, \ldots, a_{k+1}\}$ and $x \in G_k$;
(2) $xa_i \sim ya_i$ if and only if $x \sim y$, where $a_i \in \{e, a_1, \ldots, a_{k+1}\}$ and $x, y \in G_k$.

Denote by $S_1(x) = \{y \in G_k : \langle x, y \rangle\}$ the set of all nearest neighbors of the word $x \in G_k$. Let $G_k/H_0 = \{H_0, H_1, \ldots, H_{n-1}\}$ be the factor-group w.r.t. H_0. In addition, let $q_i(x) = |S_1(x) \cap H_i|$, $i = 0, \ldots, n-1$, and $Q(x) = (q_0(x), q_1(x), \ldots, q_{n-1}(x))$, $x \in G_k$.

Proposition 1.5. *If $x \sim y$, then $q_i(x) = q_i(y)$ for $i = 0, \ldots, n-1$.*

Proof. Let $x \sim y$. Then, by virtue of Corollary 1.1, $xa_i \sim ya_i$ for any $i = 0, \ldots, n-1$. Therefore, if $S_1(x) \cap H_i = \left\{ xa_{i_1}, xa_{i_2}, \ldots, xa_{i_{q_i(x)}} \right\}$, then $S_1(y) \cap H_i = \left\{ ya_{i_1}, ya_{i_2}, \ldots, ya_{i_{q_i(x)}} \right\}$ for any $i = 0, \ldots, n-1$, i.e., $|S_1(x) \cap H_i| = |S_1(y) \cap H_i|$. $\square$

Corollary 1.2. *If $x \sim y$, then $Q(x) = Q(y)$.*

Introduce the following notations: $q_i(H_0) = q_i(e) = |\{j : a_j \in H_i\}|$, $Q(H_0) = (q_0(H_0), \ldots, q_{n-1}(H_0))$ and $N(H_0) = |\{j : q_j(H_0) \neq 0\}|$.

Theorem 1.5. *For any $x \in G_k$, there exists a permutation π_x of the coordinates of the vector $Q(H_0)$ such that*

$$\pi_x Q(H_0) = Q(x). \tag{1.5}$$

Proof. Obviously, $S_1(x) = xS_1(e) = x\{a_1, a_2, \ldots, a_{k+1}\} = \{xa_1, xa_2, \ldots, xa_{k+1}\}$. By virtue of Corollary 1.1, for any $i = 0, \ldots, n-1$, there exists an index $j(i) \in \{0, 1, \ldots, n-1\}$ such that $q_i(H_0) = |\{j : a_j \in H_i\}| = |\{xa_m : xa_m \in H_{j(i)}\}| = q_{j(i)}(x)$. Set $\pi_x(i) = j(i)$. $\square$

Let $N(x) = |\{j : q_j(x) \neq 0\}|$.

Corollary 1.3. *For any $x \in G_k$ we have $N(x) = N(H_0)$.*

Consider the following example:

Example 1.1. Let $k = 2$, $H_0 = H_{\{1\}} \cap H_{\{2\}} = \{x \in G_2 : \omega_x(a_1) -$ even, $\omega_x(a_2) -$ even$\}$. Note that H_0 is a normal subgroup of index 4. The group $G_2/H_0 = \{H_0, H_1, H_2, H_3\}$ has the following elements:

$$H_1 = \{x \in G_2 : \omega_x(a_1) - \text{even}, \ \omega_x(a_2) - \text{odd}\},$$
$$H_2 = \{x \in G_2 : \omega_x(a_1) - \text{odd}, \ \omega_x(a_2) - \text{even}\},$$
$$H_3 = \{x \in G_2 : \omega_x(a_1) - \text{odd}, \ \omega_x(a_2) - \text{odd}\}.$$

We have

$$q_0(H_0) = |\{a_1, a_2, a_3\} \cap H_0| = |\{a_3\}| = 1,$$
$$q_1(H_0) = |\{a_1, a_2, a_3\} \cap H_1| = |\{a_2\}| = 1,$$
$$q_2(H_0) = |\{a_1, a_2, a_3\} \cap H_2| = |\{a_1\}| = 1,$$
$$q_3(H_0) = |\{a_1, a_2, a_3\} \cap H_3| = |\emptyset| = 0.$$

Hence $Q(H_0) = (1, 1, 1, 0)$.

Assume now $x \in H_1$ (cases $x \in H_2, H_3$ are similar), then $\omega_x(a_1)$-even and $\omega_x(a_2)$-odd, $\omega_{xa_1}(a_1)$-odd, $\omega_{xa_2}(a_1)$-even, $\omega_{xa_3}(a_1)$-even, $\omega_{xa_1}(a_2)$-odd, $\omega_{xa_2}(a_2)$-even, $\omega_{xa_3}(a_2)$-odd. Consequently,

$$q_0(x) = |\{xa_1, xa_2, xa_3\} \cap H_0| = |\{xa_2\}| = 1,$$
$$q_1(x) = |\{xa_1, xa_2, xa_3\} \cap H_1| = |\{xa_3\}| = 1,$$
$$q_2(x) = |\{xa_1, xa_2, xa_3\} \cap H_2| = |\emptyset| = 0,$$
$$q_3(x) = |\{xa_1, xa_2, xa_3\} \cap H_3| = |\{xa_1\}| = 1.$$

Hence $Q(x) = (1, 1, 0, 1)$. Note that the permutation π_x of coordinates of $Q(H_0)$ has the form $\pi_x = (34)$ and $N(x) = N(H_0) = 3$. In Fig. 1.3, the partitions of Γ^2 with respect to $H_0 = H_{\{1\}} \cap H_{\{2\}}$ is given. The elements of the class H_i, $i = 0, 1, 2, 3$ are denoted by i.

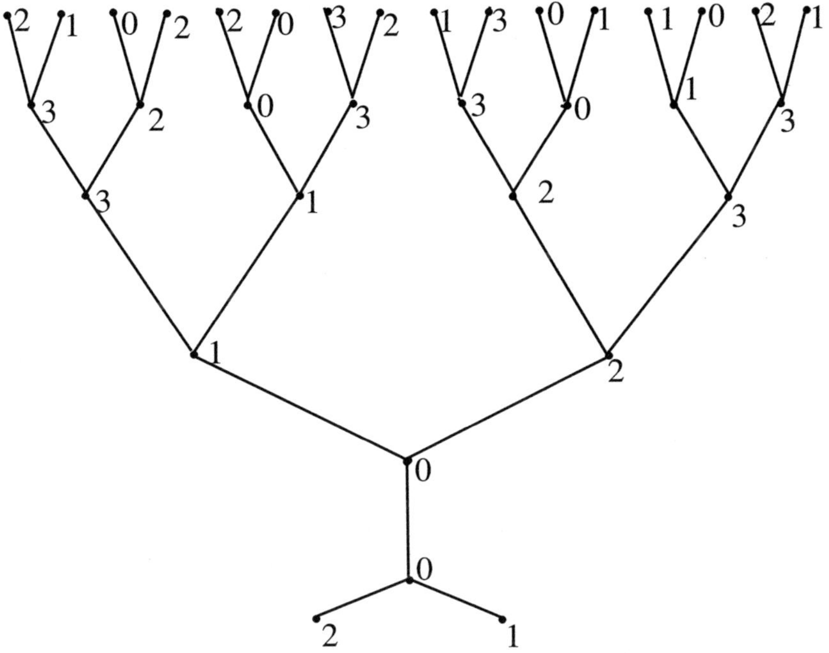

Fig. 1.3 Partition of Γ^2 by subgroup H_0 of index 4.

1.5 Density of edges in a ball

In this section we consider a group representation of a Cayley tree and its partition with respect to an arbitrary subgroup of index two. This partition gives a 2-vertex-coloring on the Cayley tree and induces a 4-edge-coloring. We give explicit formulae for the numbers of edges of each color in a ball V_n of radius n with the center at the root. We also compute for each color the limiting behavior of the ratio of these numbers with respect to V_n as $n \to \infty$. Results of this section will be used to compute free energies corresponding to weakly periodic Gibbs measures of the Ising model (see Chapter 2).

Consider a subgroup of index two:

$$H_A = \left\{ x \in G_k : \sum_{i \in A} \omega_x(a_i) \text{ is even} \right\}, \tag{1.6}$$

where $\emptyset \neq A \subseteq N_k = \{1, 2, \ldots, k+1\}$, and $\omega_x(a_i)$ is the number of a_i in a word $x \in G_k$.

Let $G_k/H_A = \{H_0, H_1\}$ be the factor group, where $H_0 = H_A, H_1 = G_k \setminus H_A$.

Denote

$$\mathcal{A}_n = |\{\langle x, y \rangle \in L_n : x \in H_0,\ y = x_\downarrow \in H_0\}|,$$

$$\mathcal{B}_n = |\{\langle x, y \rangle \in L_n : x \in H_0,\ y = x_\downarrow \in H_1\}|, \qquad (1.7)$$

$$\mathcal{C}_n = |\{\langle x, y \rangle \in L_n : x \in H_1,\ y = x_\downarrow \in H_0\}|,$$

$$\mathcal{D}_n = |\{\langle x, y \rangle \in L_n : x \in H_1,\ y = x_\downarrow \in H_1\}|,$$

where L_n is the set of edges in V_n and $x_\downarrow$ denotes the ancestor of x. Let

$$\alpha_n = \mathcal{A}_n - \mathcal{A}_{n-1}$$
$$\beta_n = \mathcal{B}_n - \mathcal{B}_{n-1}$$
$$\gamma_n = \mathcal{C}_n - \mathcal{C}_{n-1}$$
$$\delta_n = \mathcal{D}_n - \mathcal{D}_{n-1}$$

where $A = \{1, 2, 3, ..., j\}$ (with $1 \le j \le k + 1$).

Let M be the set of all unit balls with vertices in V and let $S_1(x)$ denotes the set of all nearest neighbors of x. For $b \in M$ the center of b is denoted by c_b.

Lemma 1.1. *If $c_b \in H_0$, then*

$$|\{x \in S_1(c_b) : x \in H_1\}| = j, \quad |\{x \in S_1(c_b) : x \in H_0\}| = k - j + 1.$$

If $c_b \in H_1$, then

$$|\{x \in S_1(c_b) : x \in H_1\}| = k - j + 1, \quad |\{x \in S_1(c_b) : x \in H_0\}| = j.$$

Proof. We have $S_1(c_b) = \{c_b a_p : p = 1, 2, \ldots, k + 1\}$.

If $c_b \in H_0$, (the case $c_b \in H_1$ is similar) then $c_b a_p \in H_1$ for $p = 1, 2, ..., j$ and $c_b a_p \in H_0$ for $p = j + 1, j + 2, ..., k + 1$, i.e., we have

$$|\{x \in S_1(c_b) : x \in H_1\}| = j, \quad |\{x \in S_1(c_b) : x \in H_0\}| = k - j + 1. \qquad \square$$

Consider $b = V_1 \in M$ with the center $x^0 = e \in H_0$, then in W_1 we have j vertices which belong to H_1, and $k - j + 1$ vertices which belong in H_0, consequently,

$$\alpha_1 = k - j + 1, \quad \beta_1 = 0, \quad \gamma_1 = j, \quad \delta_1 = 0.$$

Lemma 1.2. *For any $n \in N$ the following recurrence system holds*

$$\begin{cases} \alpha_{n+1} = (k-j)\alpha_n + (k-j+1)\beta_n \\[2mm] \beta_{n+1} = (j-1)\gamma_n + j\delta_n \\[2mm] \gamma_{n+1} = (j-1)\beta_n + j\alpha_n \\[2mm] \delta_{n+1} = (k-j)\delta_n + (k-j+1)\gamma_n, \end{cases} \tag{1.8}$$

with initial values $\alpha_1 = k - j + 1$, $\beta_1 = 0$, $\gamma_1 = j$, $\delta_1 = 0$.

Proof. By Lemma 1.1, an edge $\langle x, y \rangle \in L_n \setminus L_{n-1}$ with $x \in H_0$, $y = x_{\downarrow} \in H_0$ has $(k-j)$ neighbor edges $\langle z, x \rangle \in L_{n+1} \setminus L_n$ with $z \in H_0$, $x = z_{\downarrow} \in H_0$. An edge $\langle z, t \rangle \in L_n \setminus L_{n-1}$ with $z \in H_0$, $t = z_{\downarrow} \in H_1$ has $(k - j + 1)$ neighbor edges $\langle u, z \rangle \in L_{n+1} \setminus L_n$ with $u \in H_0$, $z = u_{\downarrow} \in H_0$. Moreover, it is easy to see that only α_n and β_n have contribution to α_{n+1}. Hence we have $\alpha_{n+1} = (k - j)\alpha_n + (k - j + 1)\beta_n$. Other equations of the system (1.8) can be obtained by a similar way. $\qquad\square$

Remark 1.1. For $j = k + 1$ by Lemmas 1.1 and 1.2 we get $\alpha_n = \delta_n = 0$, for any $n \geq 1$ and

$$\beta_n = \begin{cases} 0, & \text{if } n = 2m - 1 \\[2mm] (k + 1)k^{2m-1}, & \text{if } n = 2m \end{cases}, \quad m = 1, 2, \ldots$$

$$\gamma_n = \begin{cases} 0, & \text{if } n = 2m \\[2mm] (k + 1)k^{2(m-1)}, & \text{if } n = 2m - 1 \end{cases}, \quad m = 1, 2, \ldots$$

So in the sequel of this section we consider j as $1 \leq j \leq k$.

Lemma 1.3. *For α_n we have*

$$\alpha_{n+2} = j(k-j+1)|W_n| + (k-2j)\alpha_{n+1} - (2j-1)\alpha_n - k\alpha_{n-1}, \quad n \geq 2, \tag{1.9}$$

with initial values

$$\alpha_1 = k-j+1, \quad \alpha_2 = (k-j)(k-j+1), \quad \alpha_3 = \left((k-1)^2 + j(j-1)\right)(k-j+1). \tag{1.10}$$

Proof. The initial values follow from Lemma 1.2.

By definitions of α_n, β_n, γ_n, δ_n we have

$$\alpha_n + \beta_n + \gamma_n + \delta_n = |W_n| = k^{n-1}(k + 1), \quad n \geq 1. \tag{1.11}$$

From (1.8) we get

$$\beta_n = \frac{1}{k-j+1}(\alpha_{n+1} - (k - j)\alpha_n),$$

$$\gamma_n = \frac{j-1}{k-j+1}(\alpha_n - (k - j)\alpha_{n-1}) + j\alpha_{n-1}, \tag{1.12}$$

$$\delta_n = \frac{1}{j}(\beta_{n+1} - (j - 1)\gamma_n).$$

Substituting these values in (1.11) and then simplifying we get (1.9). $\qquad\square$

To find solution of (1.9) we denote

$$\alpha_n = q_n k^{n-2}. \tag{1.13}$$

From (1.9) we get

$$k^n q_{n+2} = k^{n-1}(k+1)(k-j+1)j + (k-2j)k^{n-1}q_{n+1} -$$

$$(2j-1)k^{n-2}q_n - k^{n-2}q_{n-1},$$

dividing by k^n we obtain

$$q_{n+2} = \frac{(k+1)(k-j+1)j}{k} + \frac{k-2j}{k}q_{n+1} - \frac{2j-1}{k^2}q_n - \frac{1}{k^2}q_{n-1}, \tag{1.14}$$

with initial values

$$q_1 = k(k-j+1),\ q_2 = (k-j)(k-j+1),\ q_3 = \frac{\left((k-1)^2 + j(j-1)\right)(k-j+1)}{k}. \tag{1.15}$$

In order to find solution to (1.14) first we rid the term $\frac{(k+1)(k-j+1)j}{k}$ by denoting

$$q_n = p_n + \frac{k(k-j+1)}{2}. \tag{1.16}$$

Substituting (1.16) into (1.14) we get

$$p_{n+2} = \frac{k-2j}{k}p_{n+1} - \frac{2j-1}{k^2}p_n - \frac{1}{k^2}p_{n-1}, \tag{1.17}$$

with

$$p_1 = \frac{1}{2}k(k-j+1),\ p_2 = \frac{(k-2j)(k-j+1)}{2},$$

$$p_3 = \frac{(k^2 - 4k + 2 + 2j^2 - 2j)(k-j+1)}{2k}. \tag{1.18}$$

The characteristic equation for (1.17) has the following form (setting $p_n = \lambda^n$):

$$\lambda^3 - \frac{k-2j}{k}\lambda^2 + \frac{2j-1}{k^2}\lambda + \frac{1}{k^2} = 0,$$

which has solutions

$$\lambda_1 = -\frac{1}{k},\ \lambda_{2,3} = \frac{k-2j+1 \pm \sqrt{(k-2j)^2 - 2(k+2j) + 1}}{2k}. \tag{1.19}$$

Then the general solution to (1.17) is

$$p_n = A_1\lambda_1^n + A_2\lambda_2^n + A_3\lambda_3^n, \tag{1.20}$$

where the coefficients A_1, A_2, A_3 are determined by the initial conditions (1.18).

Using (1.16) and (1.13) we get

$$\alpha_n = \frac{k-j+1}{2}k^{n-1} + A_1 \cdot \frac{(-1)^n}{k^2} + \frac{A_2}{k^2} \cdot (k\lambda_2)^n + \frac{A_3}{k^2} \cdot (k\lambda_3)^n. \quad (1.21)$$

Then using (1.12) and (1.21) one can find β_n, γ_n and δ_n.

We have

$$\mathcal{A}_n = \sum_{m=1}^{n} \alpha_m = \frac{k-j+1}{2(k-1)}\left(k^n - 1\right) + \frac{A_1}{2k^2}\left((-1)^n - 1\right) +$$

$$\frac{A_2\lambda_2 k}{k^2(\lambda_2 k - 1)}\left((\lambda_2 k)^n - 1\right) + \frac{A_3\lambda_3 k}{k^2(\lambda_3 k - 1)}\left((\lambda_3 k)^n - 1\right). \quad (1.22)$$

Proposition 1.6. *For any $j = 1,\ldots,k$ and any fixed $q = 0,1,2,\ldots$ we have*

$$\lim_{n\to\infty}\frac{\alpha_{n-q}}{|V_n|} = \lim_{n\to\infty}\frac{\delta_{n-q}}{|V_n|} = \frac{(k-1)(k-j+1)}{2(k+1)k^{q+1}}.$$

$$\lim_{n\to\infty}\frac{\beta_{n-q}}{|V_n|} = \lim_{n\to\infty}\frac{\gamma_{n-q}}{|V_n|} = \frac{(k-1)j}{2(k+1)k^{q+1}}.$$

$$\lim_{n\to\infty}\frac{\mathcal{A}_{n-q}}{|V_n|} = \lim_{n\to\infty}\frac{\mathcal{D}_{n-q}}{|V_n|} = \frac{k-j+1}{2(k+1)k^{q}}.$$

$$\lim_{n\to\infty}\frac{\mathcal{B}_{n-q}}{|V_n|} = \lim_{n\to\infty}\frac{\mathcal{C}_{n-q}}{|V_n|} = \frac{j}{2(k+1)k^{q}}.$$

Proof. It is easy to check that $|\lambda_2| < 1$ and $|\lambda_3| < 1$, i.e.,

$$\left|k - 2j + 1 \pm \sqrt{(k-2j)^2 - 2(k+2j) + 1}\right| < 2k, \quad \text{for any } 1 \le j \le k.$$

Using these inequalities, formulae (1.21) and (1.23) we get

$$\lim_{n\to\infty}\frac{\alpha_{n-q}}{|V_n|} = \frac{(k-1)(k-j+1)}{2(k+1)k^{q+1}}.$$

Now using this formula together with (1.12) we obtain

$$\lim_{n\to\infty}\frac{\beta_{n-q}}{|V_n|} = \frac{1}{k-j+1}\left(\lim_{n\to\infty}\frac{\alpha_{n+1-q}}{|V_n|} - (k-j)\lim_{n\to\infty}\frac{\alpha_{n-q}}{|V_n|}\right)$$

$$= \frac{j(k-1)}{2(k+1)k^{q+1}}.$$

The formulae involving γ_n and δ_n are obtained in a similar way.

By (1.22) and

$$|V_n| = \frac{(k+1)\cdot k^n - 2}{k-1} \qquad (1.23)$$

we get

$$\lim_{n\to\infty} \frac{\mathcal{A}_{n-q}}{|V_n|} = \frac{k-j+1}{2(k+1)k^q}.$$

From (1.12) we get

$$\mathcal{B}_n = \tfrac{1}{k-j+1}\left(\mathcal{A}_n - \alpha_1 + \alpha_{n+1} - (k-j)\mathcal{A}_n\right),$$

$$\mathcal{C}_n = \tfrac{j-1}{k-j+1}\left(\mathcal{A}_n - (k-j)\mathcal{A}_{n-1}\right) + j\mathcal{A}_{n-1}, \qquad (1.24)$$

$$\mathcal{D}_n = \tfrac{1}{j}\left(\mathcal{B}_n - \beta_1 + \beta_{n+1} - (j-1)\mathcal{C}_n\right),$$

that allows one to prove the remaining formulae. $\qquad\square$

Remark 1.2. By Proposition 1.6 it is clear that the values of A_1, A_2, A_3 do not give any contribution to the equalities of the proposition. This is why we did not compute A_1, A_2, A_3. But one can obtain the numbers by the initial conditions (1.18) for p_n. For example, in the case $k=4$, $j=1$ from (1.20) and (1.18) we have

$$p_n = A_1\left(-\frac{1}{4}\right)^n + A_2\left(\frac{3-\sqrt{7}i}{8}\right)^n + A_3\left(\frac{3+\sqrt{7}i}{8}\right)^n,$$

$$p_1 = 8, \quad p_2 = 4, \quad p_3 = 1.$$

The initial conditions give

$$\begin{cases} 8 = -\frac{A_1}{4} + A_2\frac{3-\sqrt{7}i}{8} + A_3\frac{3+\sqrt{7}i}{8} \\ 4 = \frac{A_1}{16} + A_2\frac{1-3\sqrt{7}i}{32} + A_3\frac{1+3\sqrt{7}i}{32} \\ 1 = -\frac{A_1}{64} + A_2\frac{-9-5\sqrt{7}i}{128} + A_3\frac{-9+5\sqrt{7}i}{128} \end{cases} \Rightarrow \begin{cases} A_1 = 0 \\ A_2 = 4 + \frac{20\sqrt{7}i}{7} \\ A_3 = 4 - \frac{20\sqrt{7}i}{7} \end{cases}.$$

Consequently, (for $k=4, j=1$) we have

$$\alpha_n = 2\cdot 4^{n-1} + \frac{1}{4}\left(1 - \frac{5\sqrt{7}i}{7}\right)\left(\frac{3+\sqrt{7}i}{2}\right)^n + \frac{1}{4}\left(1 + \frac{5\sqrt{7}i}{7}\right)\left(\frac{3-\sqrt{7}i}{2}\right)^n. \qquad (1.25)$$

Using (1.12) and (1.25) one can find β_n, γ_n and δ_n. Moreover, we have

$$\mathcal{A}_n = \sum_{m=1}^{n} \alpha_m = \frac{2(4^n - 1)}{3} + \frac{2\sqrt{7}i}{7}\cdot\frac{(3-\sqrt{7}i)^n - (3+\sqrt{7}i)^n}{2^n}.$$

Note that α_n and $\mathcal{A}_n$ are natural numbers for any $n \geq 1$.

Remark 1.3. In the case $j = k + 1$ by Remark 1.1 we get

$$\lim_{n \to \infty} \frac{\alpha_n}{|V_n|} = \lim_{n \to \infty} \frac{\delta_n}{|V_n|} = \lim_{n \to \infty} \frac{A_n}{|V_n|} = \lim_{n \to \infty} \frac{D_n}{|V_n|} = 0.$$

$$\lim_{m \to \infty} \frac{\beta_{2m-1}}{|V_{2m-1}|} = \lim_{m \to \infty} \frac{\gamma_{2m}}{|V_{2m}|} = 0.$$

$$\lim_{m \to \infty} \frac{\beta_{2m}}{|V_{2m}|} = \lim_{m \to \infty} \frac{\gamma_{2m-1}}{|V_{2m-1}|} = \frac{k-1}{k}.$$

$$\lim_{m \to \infty} \frac{\mathcal{B}_{2m}}{|V_{2m}|} = \lim_{m \to \infty} \frac{\mathcal{C}_{2m-1}}{|V_{2m-1}|} = \frac{k}{k+1}.$$

$$\lim_{m \to \infty} \frac{\mathcal{B}_{2m-1}}{|V_{2m-1}|} = \lim_{m \to \infty} \frac{\mathcal{C}_{2m}}{|V_{2m}|} = \frac{1}{k+1}.$$

Commentaries and references. Trees were reinvented a number of times. More than a century ago, Kirhchoff applied trees to the study of electric networks, and A. Cayley rediscovered trees in the enumeration problem for saturated carbo-hydrates and was the first one to explore their properties. At the same time C. Jordan introduced and studied trees as a purely mathematical object.

As mentioned in Chapter 1, results of the chapter will be applied in next chapters to describe periodic and weakly periodic Gibbs measures for models of statistical mechanics on the Cayley tree. But the algebraic properties of Cayley trees may be useful in the graph and group theory (see for example, [158], [242], [245]).

By Ganikhodjaev [87] results of Section 1.2 were obtained. It is known that a regular tree (Cayley tree) can be represented by a free group. In [216] it was proved that if a tree is representable as the free product of a finite set of cyclic groups of order two, then it is necessarily a Cayley tree. Moreover, only regular tree has a group representation.

In [89] the classes of finite-index subgroups of the group G_k are constructed. In [189] the assertion that the group G_k does not have normal subgroups of odd index ($\neq 1$) is proved.

In [244] normal subgroups of arbitrary even index of the group G_k are constructed. In [217] examples of normal subgroups of infinite index are constructed.

In [210] the structure of partitioning of the group G_k into adjacent classes w.r.t. a normal subgroup of finite index is described.

Note that the following problem still remains open: Assume the elements of the set $S_1(x)$ are colored by q colors (partition in q parts) is there a subgroup of the group G_k which parts the set $S_1(x)$ as a given colors. In [189] this problem is particularly solved. In [230] a general formulation of such a problem is given.

The densities of edges computed in the last section is due to [83].

Chapter 2

Ising model on the Cayley tree

In the first section we give general definitions. This chapter contains all known results about Gibbs measures of the Ising model on Cayley trees. Some of results were obtained very recently. We show that the Ising model may have up to three translational-invariant Gibbs measures. It may have only periodic measures with period two, which are a 'chess-board' periodic. To describe 'richer' set of Gibbs measures we introduce the notion of weakly periodic Gibbs measures and show that the Ising model has at least seven such measures (under some conditions on parameters). The extremality criterion of the disordered Gibbs measure is proved. Moreover, we give two constructions of uncountable sets of non-periodic Gibbs measures. We show that under some conditions on the temperature one can construct for each known Gibbs measure on a Cayley tree of order k_0 a new Gibbs measure on the Cayley tree of order k, $k > k_0$. Some explicit formulae of the free energies (and entropies) according to these Gibbs measures are presented.

2.1   Gibbs measure

2.1.1   *Configuration space*

For $A \subseteq V$ a spin *configuration* σ_A on A is defined as a function

$$x \in A \to \sigma_A(x) \in \Phi = \{1, 2, ..., q\}.$$

The set of all configurations coincides with $\Omega_A = \Phi^A$. We denote $\Omega = \Omega_V$ and $\sigma = \sigma_V$.

Let G_k^* be a subgroup of the group G_k. A configuration $\sigma \in \Omega$ is called G_k^*-periodic if $\sigma(yx) = \sigma(x)$ for any $x \in G_k$ and $y \in G_k^*$.

A configuration that is invariant with respect to all shifts is called *translation-invariant*.

2.1.2 *Hamiltonian*

The energy of the configuration $\sigma \in \Omega$ is given by the formal Hamiltonian

$$H(\sigma) = \sum_{\substack{A \subset V: \\ \mathrm{diam}(A) \leq r}} I(\sigma_A) \tag{2.1}$$

where $r \in N$, $\mathrm{diam}(A) = \max_{x,y \in A} d(x,y)$, $I(\sigma_A) : \Omega_A \to R$ is a given potential.

For a finite domain $D \subset V$ with the boundary condition φ_{D^c} given on its complement $D^c = V \setminus D$, the conditional Hamiltonian is

$$H(\sigma_D | \varphi_{D^c}) = \sum_{\substack{A \subset V: A \cap D \neq \emptyset \\ \mathrm{diam}(A) \leq r}} I(\sigma_A), \tag{2.2}$$

where

$$\sigma_A(x) = \begin{cases} \sigma(x) & \text{if} \quad x \in A \cap D \\ \varphi(x) & \text{if} \quad x \in A \cap D^c. \end{cases}$$

2.1.3 *The ground state*

A *ground state* of (2.1) is a configuration φ in Γ^k such that $H(\varphi) \leq H(\sigma)$ for all $\sigma \in \Omega$.

2.1.4 *Gibbs measure*

A probability measure μ on $(\Omega, \mathcal{B})$ (where $\mathcal{B}$ is σ-algebra generated by cylinder subsets of Ω) is called a *Gibbs measure* (with Hamiltonian H) if it satisfies the Dobrushin-Lanford-Ruelle (DLR) equation (see [58, 144]): for all finite $D \subset V$ and $\sigma_D \in \Omega_D$:

$$\mu\left(\{\omega \in \Omega : \omega|_D = \sigma_D\}\right) = \int_{\Omega} \mu(\mathrm{d}\varphi)\nu_{\varphi}^D(\sigma_D), \tag{2.3}$$

where ν_{φ}^D is the conditional probability:

$$\nu_{\varphi}^D(\sigma_D) = \frac{1}{Z_{D,\varphi}} \exp\left(-\beta H\left(\sigma_D | \varphi_{D^c}\right)\right). \tag{2.4}$$

Here $\beta = \frac{1}{T}, T > 0-$ temperature and $Z_{D,\varphi}$ stands for the partition function in D, with the boundary condition φ:

$$Z_{D,\varphi} = \sum_{\tilde{\sigma}_D \in \Omega_D} \exp\left(-\beta H\left(\tilde{\sigma}_D | \varphi_{D^c}\right)\right).$$

The main problem:

We are interested in the following two basic problems:

1) *To investigate when there exists at least one Gibbs measure for a given Hamiltonian.*

2) *To study the structure of the set $\mathcal{G}(H)$ of all Gibbs measures corresponding to a given Hamiltonian.*

Note that if H is a continuous Hamiltonian (see [204], p.40) then it is known that $\mathcal{G}(H)$ is a non-empty, compact convex subset of the set of all probability measures defined on $(\Omega, \mathcal{B})$ ([204], p.37).

A point $\mu \in \mathcal{G}(H)$ is called *extreme point* of $\mathcal{G}(H)$ if there does not exist ν_1, $\nu_2 \in \mathcal{G}(H)$ with $\nu_1 \neq \nu_2$ and $\mu = \frac{1}{2}(\nu_1 + \nu_2)$.

Hamiltonians which we shall consider in this book are continuous, hence the corresponding sets of Gibbs measures are non-empty, compact convex subset of the set of all probability measures defined on $(\Omega, \mathcal{B})$. The main purpose of this book is to study extreme points of the set of Gibbs measures for several concrete Hamiltonians.

2.2 A functional equation for the Ising model

There are several approaches to derive the equation solutions of which describes the limit Gibbs measures for lattice models on the Cayley tree. One approach is based on properties of Markov random fields on Cayley tree [204] and [250]. Another approach is based on recurrent equations for partition functions [94], [134].

Here we shall use the method of Markov random field theory.

2.2.1 *Hamiltonian of the Ising model*

Consider Ising model where the spin takes values in the set $\Phi := \{-1, 1\}$, and is assigned to the vertices of the tree. A configuration σ on V is then defined as a function $x \in V \mapsto \sigma(x) \in \Phi$; the set of all configurations is Φ^V.

The (formal) Hamiltonian of ferromagnetic Ising model is

$$H(\sigma) = -J \sum_{\langle x,y \rangle \in L} \sigma(x)\sigma(y), \qquad (2.5)$$

where $J > 0$ is a coupling constant and $\langle x, y \rangle$ stands for nearest neighbor vertices.

2.2.2   *Finite dimensional distributions*

Define a finite-dimensional distribution of a probability measure μ in the volume V_n as

$$\mu_n(\sigma_n) = Z_n^{-1} \exp\left\{ -\beta H_n(\sigma_n) + \sum_{x \in W_n} h_x \sigma(x) \right\}, \qquad (2.6)$$

where $\beta = 1/T$, $T > 0$–temperature, Z_n^{-1} is the normalizing factor, $\{h_x \in R, x \in V\}$ is a collection of real numbers and

$$H_n(\sigma_n) = -J \sum_{\langle x,y \rangle \in L_n} \sigma(x)\sigma(y).$$

We say that the probability distributions (2.6) are compatible if for all $n \geq 1$ and $\sigma_{n-1} \in \Phi^{V_{n-1}}$:

$$\sum_{\omega_n \in \Phi^{W_n}} \mu_n(\sigma_{n-1} \vee \omega_n) = \mu_{n-1}(\sigma_{n-1}). \qquad (2.7)$$

Here $\sigma_{n-1} \vee \omega_n$ is the concatenation of the configurations. In this case, according to the Kolmogorov theorem, (see, e.g. [243]), there exists a unique measure μ on Φ^V such that, for all n and $\sigma_n \in \Phi^{V_n}$,

$$\mu(\{\sigma|_{V_n} = \sigma_n\}) = \mu_n(\sigma_n).$$

Such a measure is called a *splitting Gibbs measure* corresponding to the Hamiltonian (2.5) and function $h_x, x \in V$.

The following statement describes conditions on h_x guaranteeing compatibility of $\mu_n(\sigma_n)$.

Theorem 2.1. *Probability distributions $\mu_n(\sigma_n)$, $n = 1, 2, \ldots$, in (2.6) are compatible iff for any $x \in V$ the following equation holds:*

$$h_x = \sum_{y \in S(x)} f(h_y, \theta). \qquad (2.8)$$

Here, $\theta = \tanh(J\beta)$, $f(h, \theta) = \operatorname{arctanh}(\theta \tanh h)$ and $S(x)$ is the set of direct successors of x on Cayley tree of order k.

Proof. Necessity. Suppose that (2.7) holds; we want to prove (2.8). Substituting (2.6) into (2.7), obtain that for any configurations σ_{n-1}: $x \in V_{n-1} \mapsto \sigma_{n-1}(x) \in \{-1, 1\}$:

$$\frac{Z_{n-1}}{Z_n} \sum_{\omega_n \in \Omega_{W_n}} \exp\left(\sum_{x \in W_{n-1}} \sum_{y \in S(x)} (J\beta\sigma_{n-1}(x)\omega_n(y) + h_y\omega_n(y)) \right) =$$

$$\exp\left(\sum_{x \in W_{n-1}} h_x \sigma_{n-1}(x) \right),$$

$$(2.9)$$

where $\omega_n \colon x \in W_n \mapsto \omega_n(x)$.

From (2.9) we get:

$$\frac{Z_{n-1}}{Z_n} \sum_{\omega_n \in \Omega_{W_n}} \prod_{x \in W_{n-1}} \prod_{y \in S(x)} \exp\left(J\beta\sigma_{n-1}(x)\omega_n(y) + h_y\omega_n(y)\right) =$$

$$\prod_{x \in W_{n-1}} \exp\left(h_x\sigma_{n-1}(x)\right). \tag{2.10}$$

Fix $x \in W_{n-1}$ and consider two configurations $\sigma_{n-1} = \overline{\sigma}_{n-1}$ and $\sigma_{n-1} = \tilde{\sigma}_{n-1}$ on W_{n-1} which coincide on $W_{n-1} \setminus \{x\}$, and rewrite now the equality (2.10) for $\overline{\sigma}_{n-1}(x) = 1$ and $\tilde{\sigma}_{n-1}(x) = -1$, then dividing first of them to the second one we get

$$\prod_{y \in S(x)} \frac{\sum_{u \in \{-1,1\}} \exp\left(J\beta u + h_y u\right)}{\sum_{u \in \{-1,1\}} \exp\left(-J\beta u + h_y u\right)} = \exp\left(2h_x\right),$$

which implies (2.8) where one has to use the following formula:

$$f(h,\theta) = \operatorname{arctanh}(\theta \tanh h) = \frac{1}{2} \ln \frac{(1+\theta)e^{2h} + (1-\theta)}{(1-\theta)e^{2h} + (1+\theta)}.$$

Sufficiency. Suppose that (2.8) holds. It is equivalent to the representations

$$\prod_{y \in S(x)} \sum_{u \in \{-1,1\}} \exp\left(J\beta tu + h_y u\right) = a(x) \exp\left(th_x\right), t \in \{-1,1\} \tag{2.11}$$

for some function $a(x) > 0, x \in V$. We have

$$\text{LHS of } (2.7) = \frac{1}{Z_n} \exp(-\beta H(\sigma_{n-1})) \times \tag{2.12}$$

$$\prod_{x \in W_{n-1}} \prod_{y \in S(x)} \sum_{u \in \{-1,1\}} \exp\left(J\beta\sigma_{n-1}(x)u + h_y u\right).$$

Substituting (2.11) into (2.12) and denoting $A_n(x) = \prod_{x \in W_{n-1}} a(x)$, we get

$$\text{RHS of } (2.12) = \frac{A_{n-1}}{Z_n} \exp(-\beta H(\sigma_{n-1})) \prod_{x \in W_{n-1}} \exp(h_x\sigma_{n-1}(x)).$$

$$\tag{2.13}$$

Since $\mu^{(n)}$, $n \geq 1$ is a probability, we should have

$$\sum_{\sigma_{n-1} \in \Omega_{V_{n-1}}} \sum_{\omega_n \in \Omega_{W_n}} \mu^{(n)}(\sigma_{n-1}, \omega_n) = 1.$$

Hence from (2.13) we get $Z_{n-1}A_{n-1} = Z_n$, and (2.7) holds. $\qquad\square$

From Theorem 2.1 it follows that for any $h = \{h_x, \ x \in V\}$ satisfying the functional equation (2.8) there exists a unique Gibbs measure μ and vice versa. However, the analysis of solutions to (2.8) is not easy.

In next sections we shall give several solutions to (2.8).

Remark 2.1. Note that if there is more than one solution to equation (2.8), then there is more than one Gibbs measure corresponding to these solutions. One says that a phase transition occurs for the Ising model, if equation (2.8) has more than one solution. The number of the solutions of equation (2.8) depends on the parameter $\beta = \frac{1}{T}$. The phase transition usually occurs for low temperature. If it is possible to find an exact value T^* of temperature such that a phase transition occurs for all $T < T^*$, then T^* is called a critical value of temperature.

Finding the exact value of the critical temperature for some models means to exactly solve the models.

2.3   Periodic Gibbs measures of the Ising model

Since the set of vertices V has the group representation G_k. Without loss of generality we identify V with G_k, i.e., we sometimes replace V with G_k.

In this section we study periodic solutions of (2.8).

Definition 2.1. Let K be a subgroup of $G_k, k \geq 1$. We say that a function $h = \{h_x \in R : x \in G_k\}$ is K-periodic if $h_{yx} = h_x$ for all $x \in G_k$ and $y \in K$. A G_k-periodic function h is called translation-invariant.

Definition 2.2. A Gibbs measure is called K-periodic if it corresponds to K-periodic function h.

Observe that a translation-invariant Gibbs measure is G_k-periodic.

2.3.1   *Translation-invariant measures of the Ising model*

2.3.1.1   *Ferromagnetic case*

In this subsection we shall find all translation-invariant solutions h_x to the functional equation (2.8) in case $J > 0$, i.e., ferromagnetic Ising model. Note that such solutions are constant functions, $h_x = h, \forall x \in G_k$. In this case from (2.8) we get

$$h = kf(h, \theta), \ \ 0 < \theta < 1. \tag{2.14}$$

The following properties of the function $f(h, \theta)$ are obvious:

1. $f(-x, \theta) = -f(x, \theta)$;

2. $\lim_{x \to \infty} f(x, \theta) = \operatorname{arctanh}(\theta)$;

3. $\frac{d}{dx} f(0, \theta) = \theta$, $\quad 0 < \frac{d}{dx} f(x, \theta) \le \theta$;

4. $\frac{d^2}{dx^2} f(x, \theta) < 0$, $\quad x > 0$.

From these properties it follows that equation (2.14) has unique solution $h = 0$, if $0 < \theta \le \theta_c = \frac{1}{k}$, and three solutions $h = 0, \pm h_*$, $h_* > 0$, if $\theta_c < \theta < 1$. We denote $h_* = 0$ for $0 < \theta \le \theta_c$. The critical temperature $T_{c,k}$ is found from equation $\theta_c = \tanh(\frac{1}{T_{c,k}} J) = \frac{1}{k}$.

Proposition 2.1. *Let* $J > 0$. *If* h_x *is a solution to (2.8) then*

$$-h_* \le h_x \le h_*, \quad \text{for any} \ \ x \in V. \tag{2.15}$$

Proof. By properties 1-3 of function f we have $-J\beta < f(h, \theta) < J\beta$. Using this inequality from (2.8) we get

$$-kJ\beta < h_x < kJ\beta.$$

Now we consider the function $f(h, \theta)$ on $[-kJ\beta, kJ\beta]$ and on this segment we get the estimations $-f(kJ\beta, \theta) < f(h, \theta) < f(kJ\beta, \theta)$. Then

$$-kf(kJ\beta, \theta) < h_x < kf(kJ\beta, \theta).$$

Iterating this argument we obtain

$$-g^n(kJ\beta, \theta) < h_x < g^n(kJ\beta, \theta),$$

where $g(h, \theta) = kf(h, \theta)$, and g^n is its nth iteration. Note that $h_* > 0$ is a fixed point of g. It is easy to see that $h_* \le g^n(kJ\beta, \theta)$ and the sequence $g^n(kJ\beta, \theta)$ monotone decreasing. Thus the sequence has a limit α, with $\alpha \ge h_*$. This limit point must be a fixed point for g. But since the function g has no fixed point in $(h_*, +\infty)$ we get that $\alpha = h_*$. $\qquad \square$

Theorem 2.2. *For the ferromagnetic Ising model on the Cayley tree of order* $k \ge 2$ *the following statements are true*

(1) If $T \ge T_{c,k}$ *then there is unique translation-invariant Gibbs measure* μ_0.

(2) If $T < T_{c,k}$ *then there are 3 translation-invariant Gibbs measures* μ_-, μ_0, μ_+. *(μ_0 is called disordered Gibbs measure.) Moreover,* μ_-, μ_+ *are extreme.*

Proof. The measures μ_-, μ_0, μ_+ correspond to solutions $h_x = -h_*$, h_0 and $h_x = h_*$ of (2.8). The extremality of measures μ_- and μ_+ can be deduced using the minimality and maximality of the corresponding values of $\mp h_*$. Assume that μ_+ is non-extreme, i.e., is decomposed:

$$\mu_+ = \int \mu(\omega)\nu(d\omega).$$

Then for any vertex $x \in V$ we have

$$h_* = \int h_x(\omega)\nu(d\omega). \tag{2.16}$$

By Proposition 2.1 h_* is an extreme point in the set $\{h_x : -h_* \le h_x \le h_*\}$, (2.16) holds if $h_x(\omega) = h_*$ for almost all ω. Hence, μ_+ is extreme. $\quad\square$

The extremeness of μ_0 is given later.

2.3.1.2 *Anti-ferromagnetic case*

In this subsection we shall consider (2.8) in case $J < 0$, i.e., anti-ferromagnetic Ising model. In this case we have $-1 < \theta < 0$ and the function $f(h, \theta)$ is a decreasing function. Therefore the equation (2.14) has unique solution $h = 0$. Hence anti-ferromagnetic Ising model has unique translation-invariant Gibbs measure.

2.3.2 *Periodic (non-translation-invariant) measures*

In this section we give a complete description of periodic Gibbs measures for the Ising model, i.e., a characterization of such measures with respect to any normal subgroup of finite index in G_k.

Let K be a subgroup of index r in G_k, and let $G_k/K = \{K_0, K_1, ..., K_{r-1}\}$ be the quotient group, with the coset $K_0 = K$.

Recall some notations from Chapter 1: Let $q_i(x) = |S_1(x) \cap K_i|$, $i = 0, 1, ..., r - 1$; $N(x) = |\{j : q_j(x) \ne 0\}|$, where $S_1(x) = \{y \in G_k : \langle x, y \rangle\}$, $x \in G_k$ and $|\cdot|$ is the number of elements in the set. Denote $Q(x) = (q_0(x), q_1(x), ..., q_{r-1}(x))$.

We note (see Theorem 1.5) that for every $x \in G_k$ there is a permutation π_x of the coordinates of the vector $Q(e)$ (where e is the identity of G_k) such that

$$\pi_x Q(e) = Q(x). \tag{2.17}$$

It follows from this equality that $N(x) = N(e)$ for all $x \in G_k$.

Each K-periodic collection is given by

$$\{h_x = h_i \ \text{ for } \ x \in K_i, \ \ i = 0, 1, ..., r - 1\}.$$

By Theorem 2.1 and (2.17), h_n, $n = 0, 1, ..., r - 1$, satisfies

$$h_n = \sum_{j=1}^{N(e)} q_{i_j}(e) f(h_{\pi_n(i_j)}, \theta) - f(h_{\pi_n(i_{j_0})}, \theta), \tag{2.18}$$

where $i_{j_0} = 1, \ldots, N(e)$, $N(e) = |\{i_1, \ldots, i_{N(e)}\}|$.

Proposition 2.2. $f(h, \theta) = f(u, \theta)$ *if and only if* $h = u$.

Proof. Follows from monotonicity of $f(h, \theta)$ with respect to h. $\qquad\square$

Let $G_k^{(2)}$ be the subgroup in G_k consisting of all words of even length. Clearly, $G_k^{(2)}$ is a subgroup of index 2.

Theorem 2.3. *Let K be a normal subgroup of finite index in G_k. Then each K-periodic Gibbs measure for the Ising model is either translation-invariant or $G_2^{(2)}$-periodic.*

Proof. We see from (2.18) that

$$f(h_{\pi_n(i)}, \theta) = f(h_{\pi_n(i')}, \theta), \tag{2.19}$$

for any $i, i' \in Q(e), n = 0, 1, ..., r - 1$. Hence by Proposition 2.2 we have

$$h_{\pi_n(i_1)} = h_{\pi_n(i_2)} = ... = h_{\pi_n(i_{N(e)})}.$$

Therefore,

$$h_x = h_y = h, \ \text{ if } \ x, y \in S_1(z), \ \ z \in G_k^{(2)};$$

$$h_x = h_y = l, \ \text{ if } \ x, y \in S_1(z), \ \ z \in G_k \setminus G_k^{(2)}.$$

Thus the measures are translation-invariant (if $h = l$) or $G_k^{(2)}$-periodic (if $h \neq l$). This completes the proof of the theorem. $\qquad\square$

Let K be a normal subgroup of finite index in G_k.

What condition on K will guarantee that each K-periodic Gibbs measure is translation-invariant? We put $I(K) = K \cap \{a_1, \ldots, a_{k+1}\}$, where a_i, $i = 1, \ldots, k + 1$ are generators of G_k.

Theorem 2.4. *If $I(K) \neq \emptyset$, then each K-periodic Gibbs measure for the Ising model is translation-invariant.*

Proof. Take $x \in K$. We note that the inclusion $xa_i \in K$ holds if and only if $a_i \in K$. Since $I(K) \neq \emptyset$, there is an element $a_i \in K$. Therefore K contains the subset $Ka_i = \{xa_i : x \in K\}$. By Theorem 2.3 we have $h_x = h$ and $h_{xa_i} = l$. Since x and xa_i belong to K, it follows that $h_x = h_{xa_i} = h = l$. Thus each K-periodic Gibbs measure is translation-invariant. $\quad\square$

Theorems 2.3 and 2.4 reduce the problem of describing K-periodic Gibbs measure with $I(K) \neq \emptyset$ to describing the fixed points of $kf(h,\theta)$ which describes translation-invariant Gibbs measures.

If $I(K) = \emptyset$, this problem is reduced to describing the solutions of the system:

$$\begin{cases} u = kf(v,\theta), \\ v = kf(u,\theta). \end{cases} \tag{2.20}$$

Evidently roots of the equation

$$u = g(u) = kf(kf(u,\theta),\theta), \tag{2.21}$$

describe the $G_k^{(2)}$-periodic Gibbs measures.

Using properties of function f one can easily note that, the system of equations (2.20) has a unique solution $h_0 = (0,0)$ if $-k^{-1} \leq \theta \leq k^{-1}$; three solutions $h_*^{(1)} = (-h_*, -h_*)$, $h_*^{(2)} = (0,0)$ and $h_*^{(3)} = (h_*, h_*)$, $(h_* > 0)$ for $k^{-1} < \theta < 1$; three solutions $h_*^{(\mp)} = (-h_*, h_*)$, $h_*^0 = (0,0)$ and $h_*^{(\pm)} = (h_*, -h_*)$, $(h_* > 0)$ for $-1 < \theta < -k^{-1}$. We denote by $\mu^{(1)}, \mu^{(2)}$, and μ_0 (resp. $\mu^{(\mp)}, \mu^{(\pm)}$) the Gibbs measures which correspond to these solutions. Note that the measures $\mu^{(1)}, \mu^{(2)}$, and μ_0 are translation-invariant.

Thus the following theorem is true:

Theorem 2.5.

1. *For the ferromagnetic $(J > 0)$ Ising model all periodic Gibbs measures are translation-invariant.*
2. *For anti-ferromagnetic $(J < 0)$ Ising model there are two extreme $G_k^{(2)}$-periodic Gibbs measures $\mu^{(\mp)}, \mu^{(\pm)}$.*

Proof. The proof is similar to the proof of Theorem 2.2. $\quad\square$

2.4 Weakly periodic Gibbs measures

For $x \in G_k$ we denote by $x_{\downarrow}$ the unique point of the set $\{y \in G_k : \langle x,y \rangle\} \setminus S(x)$.

Let $G_k/\widehat{G}_k = \{H_1, ..., H_r\}$ be a factor group, where $\widehat{G}_k$ is a normal subgroup of index $r \geq 1$.

Definition 2.3. A set of quantities $h = \{h_x, x \in G_k\}$ is called $\widehat{G}_k$-*weakly periodic*, if $h_x = h_{ij}$, for any $x \in H_i, x_\downarrow \in H_j$.

We note that the weakly periodic set of h coincides with an ordinary periodic one (see Definition 2.1) if the quantity h_x is independent of $x_\downarrow$.

Definition 2.4. A Gibbs measure μ is said to be $\widehat{G}_k$-weakly periodic if it corresponds to the $\widehat{G}_k$-weakly periodic set of h.

The aim of this section is to describe the set of weakly periodic Gibbs measures for the Ising model.

The level of difficulty in describing weakly periodic Gibbs measures is related to the structure and index of the normal subgroup relative to which the periodicity condition is imposed. From Chapter 1 we know that in the group G_k, there is no normal subgroup of odd index different from one. Therefore, we consider normal subgroups of even indices. Here, we restrict ourself to the cases of indices two and four.

2.4.1 *The case of index two*

We describe $\widehat{G}_k$-weakly periodic Gibbs measures for any normal subgroup $\widehat{G}_k$ of index two. We note (see Chapter 1) that any normal subgroup of index two of the group G_k has the form

$$H_A = \left\{ x \in G_k : \sum_{i \in A} \omega_x(a_i) - \text{even} \right\},$$

where $\emptyset \neq A \subseteq N_k = \{1, 2, \ldots, k+1\}$, and $\omega_x(a_i)-$ is the number of letters a_i in a word $x \in G_k$. Let $A \subseteq N_k$ and H_A be the corresponding normal subgroup of index two. We note that in the case $|A| = k + 1$, i.e., in the case $A = N_k$, weak periodicity coincides with ordinary periodicity. Therefore, we consider $A \subset N_k$ such that $A \neq N_k$. Then, in view of (2.8), the H_A-weakly periodic set of h has the form

$$h_x = \begin{cases} h_1, \ x \in H_A, \ x_\downarrow \in H_A, \\[2mm] h_2, \ x \in H_A, \ x_\downarrow \in G_k \backslash H_A, \\[2mm] h_3, \ x \in G_k \backslash H_A, \ x_\downarrow \in H_A, \\[2mm] h_4, \ x \in G_k \backslash H_A, x_\downarrow \in G_k \backslash H_A, \end{cases} \tag{2.22}$$

where $h_i, i = 1, 2, 3, 4$, satisfy the following equations:

$$\begin{cases} h_1 = |A|f(h_3, \theta) + (k - |A|)f(h_1, \theta), \\ h_2 = (|A| - 1)f(h_3, \theta) + (k + 1 - |A|)f(h_1, \theta), \\ h_3 = (|A| - 1)f(h_2, \theta) + (k + 1 - |A|)f(h_4, \theta), \\ h_4 = |A|f(h_2, \theta) + (k - |A|)f(h_4, \theta). \end{cases} \qquad (2.23)$$

Consider operator $W : R^4 \to R^4$, defined by

$$\begin{cases} h_1' = |A|f(h_3, \theta) + (k - |A|)f(h_1, \theta), \\ h_2' = (|A| - 1)f(h_3, \theta) + (k + 1 - |A|)f(h_1, \theta), \\ h_3' = (|A| - 1)f(h_2, \theta) + (k + 1 - |A|)f(h_4, \theta), \\ h_4' = |A|f(h_2, \theta) + (k - |A|)f(h_4, \theta). \end{cases} \qquad (2.24)$$

Note that the system of equations (2.23) is the equation $h = W(h)$.

It is obvious that the following sets are invariant with respect to operator W:

$$I_1 = \{h \in R^4 : h_1 = h_2 = h_3 = h_4\}, \quad I_2 = \{h \in R^4 : h_1 = h_4; h_2 = h_3\},$$

$$I_3 = \{h \in R^4 : h_1 = -h_4; h_2 = -h_3\}.$$

Set $\alpha = \frac{1-\theta}{1+\theta}$.

Theorem 2.6. *The following assertions hold:*

1) *For the Ising model, all H_A-weakly periodic Gibbs measures on I_1 and I_2 are translation invariant.*

2) *For $|A| = k$, all H_A-weakly periodic Gibbs measures are translation invariant.*

3) *For $|A| = 1$ and $k = 4$, there exists a critical value α_{cr} ($\approx 0,1569$) such that there exist five H_A-weakly periodic Gibbs measures $\mu_0, \mu_1^-, \mu_1^+, \mu_2^-, \mu_2^+$; for $0 < \alpha < \alpha_{\mathrm{cr}}$, three H_A-weakly periodic Gibbs measures μ_0, μ_1^-, μ_1^+, for $\alpha = \alpha_{\mathrm{cr}}$, and only one H_A-weakly periodic Gibbs measure μ_0 for $\alpha > \alpha_{\mathrm{cr}}$.*

4) *For $|A| = 1$, $k > 5$, and $\theta \in (\theta_1, \theta_2)$, where $\theta_{1,2} = \frac{k-1\pm\sqrt{k^2-6k+1}}{2k}$, there exist three H_A-weakly periodic Gibbs measures μ_0 and μ^+, μ^- on I_3.*

Proof. 1) It suffices to show that system of equations (2.23) has only one root of the form $h_1 = h_2 = h_3 = h_4$. The proof is obvious for the invariant set I_1. We prove this assertion for the invariant set I_2. Using the fact that

$$f(h, \theta) = \operatorname{arctanh}(\theta \tanh h) = \frac{1}{2} \ln \frac{(1+\theta)e^{2h} + (1-\theta)}{(1-\theta)e^{2h} + (1+\theta)},$$

and introducing the notation $z_i = e^{2h_i}$, $i = 1, 2, 3, 4$ we obtain the following system of equations instead of (2.23):

$$\begin{cases} z_1 - z_2 = A_1(z_3 - z_1), \\ z_1 - z_3 = A_2(z_1 - z_4) + B_2(z_3 - z_4) + C_2(z_3 - z_2), \\ z_1 - z_4 = A_3(z_1 - z_4) + B_3(z_3 - z_2), \\ z_2 - z_3 = A_4(z_3 - z_2) + B_4(z_1 - z_4), \\ z_2 - z_4 = A_5(z_3 - z_2) + B_5(z_1 - z_2) + C_5(z_1 - z_4), \\ z_3 - z_4 = A_6(z_4 - z_2), \end{cases} \tag{2.25}$$

where $A_i = (1 - \alpha^2)\tilde{A}_i(z_1, z_2, z_3, z_4)$, $B_i = (1 - \alpha^2)\tilde{B}_i(z_1, z_2, z_3, z_4)$, $C_i = (1 - \alpha^2)\tilde{C}_i(z_1, z_2, z_3, z_4)$ and $\tilde{A}_i$, $\tilde{B}_i$, $\tilde{C}_i$ are positive for all $i = 1, \ldots, 6$.

On the invariant set I_2 we have $h_2 = h_3$. As a result, for $\alpha < 1$ from $z_1 - z_2 = A_1(z_3 - z_1)$ we get $z_1 = z_2$.

In the anti-ferromagnetic case, i.e., $\alpha \in (1, +\infty)$ we obtain A_i, B_i, $C_i < 0$ for all i. On the invariant set I_2 we have $h_2 = h_3$. From (2.25) we get $z_2 - z_1 = -A_1(z_3 - z_1)$, consequently, $z_1 = z_2$. Hence, for all $\alpha \in (0, +\infty)$ we have $z_1 = z_2$, which implies $z_1 = z_2 = z_3 = z_4$ on I_2.

2) For $|A| = k$ from (2.23) we get

$$\begin{cases} h_2 = (k - 1)f(h_3, \theta) + f(kf(h_3, \theta), \theta), \\ h_3 = (k - 1)f(h_2, \theta) + f(kf(h_2, \theta), \theta). \end{cases} \tag{2.26}$$

We now prove that system (2.26) has only solutions of the form $h_2 = h_3$. We consider the case $h_2 > h_3$, then from (2.26) we get

$$h_2 - h_3 = (k - 1)(f(h_3, \theta) - f(h_2, \theta)) + f(kf(h_3, \theta), \theta) - f(kf(h_2, \theta), \theta). \tag{2.27}$$

It is easy to verify that the function f (for $J > 0$) is strictly increasing. Consequently, equality (2.27) cannot hold, because its left-hand side contains a positive quantity and its right-hand side contains a negative one. Equation (2.27) also does not hold in the case $h_2 < h_3$. Therefore, $h_2 = h_3$, which gives translation-invariant solutions of system (2.23).

3) The proof of the third assertion in the theorem follows from Lemma 2.3 below. In this case, it is necessary to analyze the solutions of the equation

$$h = 3f(h, \theta) - f(4f(h, \theta), \theta),$$

which is obtained by restricting the operator W to I_3.

4) Under conditions of theorem from (2.23) we obtain

$$h_1 = g(h_1, \theta, k), \qquad (2.28)$$

where $g(x) = g(x, \theta, k) = -f(kf(x, \theta); \theta) + (k-1)f(x, \theta)$, $x \in R$. We note that $g(0) = 0$ and g is an odd bounded function. It follows from these properties that if $g'(0) > 1$, then equation (2.28) has at least three solutions. It is easy to verify that the inequality $g'(0) > 1$ is equivalent to $\theta \in (\theta_1, \theta_2)$. In this case, equation (2.23) has three solutions:

$$(\pm h_1^*, \pm kf(h_1^*, \theta), \mp kf(h_1^*, \theta), \mp h_1^*), (0, 0, 0, 0).$$

$\square$

Remark 2.2. The measures $\mu^-, \mu^+, \mu_i^-, \mu_i^+, i = 1, 2$ are H_A-weakly periodic, and this provides new Gibbs measures for the Ising model. All the other measures constructed in Theorem 2.6 are translation invariant.

Remark 2.3. If $A \subset N_k$ is such that $|A| \neq 1$ or $|A| \neq k$, then it is difficult to obtain a solution of system of equations (2.23) outside the invariants I_1 and I_2. Even for the invariant I_3, system (2.23) is a system with two unknowns, which is difficult to solve.

2.4.2   *The case of index four*

Let $H_{\{a_1\}} = \{x \in G_k : \omega_x(a_1)\text{-even}\}$, $G_k^{(2)} = \{x \in G_k : l(x)\text{-even}\}$ and $G_k^{(4)} = H_{\{a_1\}} \bigcap G_k^{(2)}$ be the corresponding normal subgroup of index four.

Remark 2.4. Among all normal subgroups of index four, our chosen normal subgroup $G_k^{(4)}$ is convenient because we obtain a system of equations with eight unknowns from system (2.8) in this case, while the number of unknowns can reach 16 for an arbitrary normal subgroup of index four.

We consider a quotient group $G_k/G_k^{(4)} = \{H_0, H_1, H_2, H_3\}$, where

$$H_0 = \{x \in G_k : \omega_x(a_1)-\text{even}, \ l(x)-\text{even}\},$$

$$H_1 = \{x \in G_k : \omega_x(a_1)-\text{odd}, \ l(x)-\text{even}\},$$

$$H_2 = \{x \in G_k : \omega_x(a_1)-\text{even}, \ l(x)-\text{odd}\},$$

$$H_3 = \{x \in G_k : \omega_x(a_1)-\text{odd}, \ l(x)-\text{odd}\}.$$

Then by (2.8), the $G_k^{(4)}$-weakly periodic set of h has the form

$$h_x = \begin{cases} h_1, \ x \in H_3, \ x_\downarrow \in H_1 \\ h_2, \ x \in H_1, \ x_\downarrow \in H_3 \\ h_3, \ x \in H_3, \ x_\downarrow \in H_0 \\ h_4, \ x \in H_0, \ x_\downarrow \in H_3 \\ h_5, \ x \in H_1, \ x_\downarrow \in H_2 \\ h_6, \ x \in H_2, \ x_\downarrow \in H_1 \\ h_7, \ x \in H_2, \ x_\downarrow \in H_0 \\ h_8, \ x \in H_0, \ x_\downarrow \in H_2, \end{cases} \qquad (2.29)$$

where $h_i, i = 1, \ldots, 8$ satisfy the following system of equations

$$\begin{cases} h_1 = (k-1)f(h_2, \theta) + f(h_4, \theta) \\ h_2 = (k-1)f(h_1, \theta) + f(h_6, \theta) \\ h_3 = kf(h_2, \theta) \\ h_4 = kf(h_7, \theta) \\ h_5 = kf(h_1, \theta) \\ h_6 = kf(h_8, \theta) \\ h_7 = (k-1)f(h_8, \theta) + f(h_5, \theta) \\ h_8 = (k-1)f(h_7, \theta) + f(h_3, \theta). \end{cases} \qquad (2.30)$$

This system can be rewritten as $h = W(h)$, where the map $W : R^4 \to R^4$, is defined as $h' = W(h)$ if

$$\begin{cases} h_1' = (k-1)f(h_2, \theta) + f(kf(h_7, \theta), \theta) \\ h_2' = (k-1)f(h_1, \theta) + f(kf(h_8, \theta), \theta) \\ h_7' = (k-1)f(h_8, \theta) + f(kf(h_1, \theta), \theta) \\ h_8' = (k-1)f(h_7, \theta) + f(kf(h_2, \theta), \theta). \end{cases} \qquad (2.31)$$

It is easy to prove the following lemma.

Lemma 2.1. *The map W has the invariant sets*

$$I_1 = \{h \in R^4 : h_1 = h_2 = h_7 = h_8\}, \quad I_2 = \{h \in R^4 : h_1 = h_2; h_7 = h_8\},$$

$$I_3 = \{h \in R^4 : h_1 = -h_2; h_7 = -h_8\}, \quad I_4 = \{h \in R^4 : h_1 = h_2 = -h_7 = -h_8\},$$

$$I_5 = \{h \in R^4 : h_1 = h_7; h_2 = h_8\}, \quad I_6 = \{h \in R^4 : h_1 = -h_7; h_2 = -h_8\},$$

$$I_7 = \{h \in R^4 : h_1 = h_7 = -h_2 = -h_8\}, \quad I_8 = \{h \in R^4 : h_1 = h_8; h_2 = h_7\},$$

$$I_9 = \{h \in R^4 : h_1 = -h_8; h_2 = -h_7\}, \quad I_{10} = \{h \in R^4 : h_1 = h_8 = -h_2 = -h_7\}.$$

We note that restricting the operator W to I_1 yields translation-invariant measures studied previously. It is easy to verify that under certain additional conditions imposed on the variables, by restricting the operator W to the other sets I_i, $i = 2, \ldots, 10$, we can reduce the system of equations $W(h) = h$ to equations with one unknown having one of the forms

$$x = -(k-1)f(x, \theta) + f(kf(x, \theta), \theta), \tag{2.32}$$

$$x = (k-1)f(x, \theta) - f(kf(x, \theta), \theta). \tag{2.33}$$

Equation (2.32) reduces to the equation

$$(u^2 - 1) \cdot P_{2k-2}(u) = 0, \tag{2.34}$$

where $u = \frac{z + \alpha}{\alpha z + 1}$ and $P_{2k-2}(u)$ is a symmetric polynomial of degree $2k - 2$. It is well known that setting $u + 1/u = \xi$, we can decrease the degree of the equation $P_{2k-2}(u) = 0$ twofold, i.e., reduce this equation to the equation $P_{k-1}(\xi) = 0$, where $P_{k-1}(\xi)$ is a nonsymmetric polynomial of degree $k - 1$ in the general case. But for $k \geq 6$, the equation $P_{k-1}(\xi) = 0$ cannot be solved in radicals.

We consider the case $k = 4$, where equation (2.34) has the form

$$(u^2 - 1)\left(u^6 - \alpha u^5 + u^4 + (1 - \alpha)u^3 + u^2 - \alpha u + 1\right) = 0. \tag{2.35}$$

Lemma 2.2. *Equation (2.35) has three solutions $u_0 = 1, u_1 = u_*$, and $u_2 = \frac{1}{u_*}$ for $\alpha > \alpha_c = \frac{5}{3}$ and the unique solution $u_0 = 1$ for $0 < \alpha \leq \frac{5}{3}$.*

Proof. For equation (2.35), $u = 1$ is a solution. We assume that $u \neq 1$. Setting $\xi = u + \frac{1}{u} > 2$, we obtain the equation

$$\xi^3 - \alpha \xi^2 - 2\xi + \alpha + 1 = 0.$$

A detailed analysis of this equation shows that this lemma holds. $\square$

Similarly, for $k = 4$, from equation (2.33) we have

$$\alpha^2(u^8 - 1) - \alpha u(u^6 - 1) + u^3(u^2 - 1) = 0. \tag{2.36}$$

Lemma 2.3. *There exists a critical value $\alpha_{cr}(\approx 0.1569)$, such that equation (2.36) has*

1) five solutions $u_0 = 1$, $u_1 = u_*^{(1)}$, $u_2 = \frac{1}{u_*^{(1)}}$, $u_3 = u_*^{(2)}$, $u_4 = \frac{1}{u_*^{(2)}}$ for $0 < \alpha < \alpha_{cr}$.

2) three solutions $u_0 = 1$, $u_1 = u_*^{(1)}$, $u_2 = \frac{1}{u_*^{(1)}}$ for $\alpha = \alpha_{cr}$.

3) the unique solution $u_0 = 1$ for $\alpha > \alpha_{cr}$.

Proof. For any $\alpha > 0$, $u = 1$ is a solution of equation (2.36). Dividing equation (2.36) by $u^2 - 1$ and introducing the notation $\xi = u + \frac{1}{u}$, we reduce (2.36) to the equation

$$\varphi(\xi) = \alpha^2 \xi^3 - \alpha \xi^2 - 2\alpha^2 \xi + \alpha + 1 = 0.$$

The assertions in the lemma follow from the easily verified properties of the function $\varphi(\xi)$. Solving the equation $\varphi'(\xi) = 0$, we obtain $\xi = \xi_* = \frac{1+\sqrt{1+6\alpha^2}}{3\alpha}$. The values α_{cr} can be found from the equation $\varphi(\xi_*) = 0$, i.e., from the equation

$$f(\alpha) = 9\alpha^2 + 27\alpha - 2 - 2(\sqrt{6\alpha^2 + 1})^3 = 0, \quad 0 < \alpha < \frac{2}{5}.$$

We note that f increases in the interval $(0, 2/5)$. Because $f(0) = -4$ and $f(2/5) > 0$, the existence and uniqueness of a value α_{cr} such that $f(\alpha_{\mathrm{cr}}) = 0$ follow from the monotonicity of f. A computer calculation shows that $\alpha_{\mathrm{cr}} \approx 0.1569$. $\qquad\square$

Combining Lemmas 2.2 and 2.3, we obtain the following solutions of system of equations (2.30):

1. For $0 < \alpha < \alpha_{\mathrm{cr}}$ we have five solutions:

$$h_x = 0, \quad \pm h_x^{(i)} = \begin{cases} \pm h_*^{(i)}, & x \in H_3, \ x_\downarrow \in H_1 \\[4pt] \pm h_*^{(i)}, & x \in H_1, \ x_\downarrow \in H_3 \\[4pt] \pm 4f(h_*^{(i)}, \theta), & x \in H_3, \ x_\downarrow \in H_0 \\[4pt] \mp 4f(h_*^{(i)}, \theta), & x \in H_0, \ x_\downarrow \in H_3 \\[4pt] \pm 4f(h_*^{(i)}, \theta), & x \in H_1, \ x_\downarrow \in H_2 \\[4pt] \mp 4f(h_*^{(i)}, \theta), & x \in H_2, \ x_\downarrow \in H_1 \\[4pt] \mp h_*^{(i)}, & x \in H_2, \ x_\downarrow \in H_0 \\[4pt] \mp h_*^{(i)}, & x \in H_0, \ x_\downarrow \in H_2, \end{cases}$$

where $h_*^{(i)} = \frac{1}{2} \ln \frac{\alpha - u_*^{(i)}}{1 - \alpha u_*^{(i)}}$, $i = 1, 2$;

2. For $0 < \alpha = \alpha_{\mathrm{cr}}$ we have three solutions: $h_x = 0$ and $\pm h_x^{(1)}$.
3. For $\alpha_{\mathrm{cr}} < \alpha \leq \alpha_{\mathrm{c}}$ we have one solution $h_x = 0$;

4. For $\alpha > \alpha_c$ we have five solutions:

$$
h_x = 0, \quad \pm \overline{h}_x^{(1)} =
\begin{cases}
\pm \overline{h}_*^{(1)}, & x \in H_3,\ x_\downarrow \in H_1 \\[4pt]
\mp \overline{h}_*^{(1)}, & x \in H_1,\ x_\downarrow \in H_3 \\[4pt]
\mp 4 f(\overline{h}_*^{(1)}, \theta), & x \in H_3,\ x_\downarrow \in H_0 \\[4pt]
\pm 4 f(\overline{h}_*^{(1)}, \theta), & x \in H_0,\ x_\downarrow \in H_3 \\[4pt]
\pm 4 f(\overline{h}_*^{(1)}, \theta), & x \in H_1,\ x_\downarrow \in H_2 \\[4pt]
\mp 4 f(\overline{h}_*^{(1)}, \theta), & x \in H_2,\ x_\downarrow \in H_1 \\[4pt]
\pm \overline{h}_*^{(1)}, & x \in H_2,\ x_\downarrow \in H_0 \\[4pt]
\mp \overline{h}_*^{(1)}, & x \in H_0,\ x_\downarrow \in H_2,
\end{cases}
$$

on the invariant set I_7 and

$$
\pm \overline{h}_x^{(2)} =
\begin{cases}
\pm \overline{h}_*^{(2)}, & x \in H_3,\ x_\downarrow \in H_1 \\[4pt]
\mp \overline{h}_*^{(2)}, & x \in H_1,\ x_\downarrow \in H_3 \\[4pt]
\mp 4 f(\overline{h}_*^{(2)}, \theta), & x \in H_3,\ x_\downarrow \in H_0 \\[4pt]
\mp 4 f(\overline{h}_*^{(2)}, \theta), & x \in H_0,\ x_\downarrow \in H_3 \\[4pt]
\pm 4 f(\overline{h}_*^{(2)}, \theta), & x \in H_1,\ x_\downarrow \in H_2 \\[4pt]
\pm 4 f(\overline{h}_*^{(2)}, \theta), & x \in H_2,\ x_\downarrow \in H_1 \\[4pt]
\mp \overline{h}_*^{(2)}, & x \in H_2,\ x_\downarrow \in H_0 \\[4pt]
\pm \overline{h}_*^{(2)}, & x \in H_0,\ x_\downarrow \in H_2
\end{cases}
$$

on the invariant set I_{10}.

Lemma 2.1 can also be used to obtain the equations

$$\pm x = (k-1)f(x,\theta) + f(kf(x,\theta),\theta). \tag{2.37}$$

Because $f(x,-\theta) = -f(x,\theta)$, equations (2.37) can be reduced to equations (2.32), (2.33) by the change $\theta = -\theta$. The solutions of (2.37) are therefore obtained from the solutions of (2.32) and (2.33) by replacing θ with $-\theta$.

Thus the following theorem is proved:

Theorem 2.7. *For $k = 4$, there exist critical values $\alpha_{\mathrm{cr}}(\approx 0.1569)$ and $\alpha_c = 3/5$ such that*

1. *there exist at least seven $G_4^{(4)}$-weakly periodic Gibbs measures for $\alpha \in [0, \alpha_{\mathrm{cr}}) \cup (\alpha_{\mathrm{cr}}^{-1}, +\infty)$,*

2. *there exist at least five $G_4^{(4)}$-weakly periodic Gibbs measures for $\alpha = \alpha_{\mathrm{cr}}, \alpha_{\mathrm{cr}}^{-1}$,*

3. *there exist at least three $G_4^{(4)}$- weakly periodic Gibbs measures for $\alpha \in (\alpha_{\mathrm{cr}}, \alpha_{\mathrm{c}}) \cup (\alpha_{\mathrm{c}}^{-1}, \alpha_{\mathrm{cr}}^{-1})$, and*

4. *there exists at least one $G_4^{(4)}$-weakly periodic Gibbs measure for $\alpha \in (\alpha_{\mathrm{c}}, \alpha_{\mathrm{c}}^{-1})$.*

Remark 2.5. 1. In all cases in Theorem 2.7, one of the weakly periodic measures is translation-invariant, which corresponds to the solution $h_x = 0$. All the other measures are $G_4^{(4)}$-weakly periodic.

2. From $\alpha_{\mathrm{c}} = 3/5$, we obtain the critical value θ_{c} of the phase transition for the Ising model, i.e., $|\theta_{\mathrm{c}}| = 1/k$ for $k = 4$.

3. Five Gibbs measures among the seven in assertion 1 in Theorem 2.7 correspond to the stable solutions of equations (2.31) and (2.32). The above mentioned methods can therefore be used to prove that at least five of them are extreme (indecomposable).

The following conjecture is formulated based on computer calculations and Theorem 2.7.

Conjecture 1. 1. For the Ising model on the Cayley tree of order $k \geq 5$, the statements in Theorem 2.7 hold for the critical values $\alpha_{\mathrm{cr}} = \alpha_{\mathrm{cr}}(k)$ and $\alpha_{\mathrm{c}} = (k-1)/(k+1)$.

2. The estimate $\alpha_{\mathrm{cr}} = (1 - \theta_{\mathrm{cr}})/(1 + \theta_{\mathrm{cr}}) < (\sqrt{k} - 1)/(\sqrt{k} + 1)$ holds for α_{cr}, i.e., $|\theta_{\mathrm{cr}}| < |\theta_{\mathrm{c}}^{\mathrm{SG}}| = 1/\sqrt{k}$, where $\theta_{\mathrm{c}}^{\mathrm{SG}}$ is the critical value for the spin glass model and the second critical value for the Ising model, below which the measure corresponding to $h_x = 0$ is an extreme measure (see [25], [119]), which we shall discuss in the next section.

2.5 Extremality of the disordered Gibbs measure

In this section we give a proof that the Gibbs measure μ_0 (which is called disordered phase) which corresponds to the solution $h_x = 0$ of the functional equation (2.8) is extreme if and only if $T \geq T_{c,\sqrt{k}} = \dfrac{J}{\mathrm{arctanh}(1/\sqrt{k})}$, where $T_{c,\sqrt{k}}$ is the critical temperature of the spin-glass model (see [48]).

There are two distinct proofs of this statement, one in [25] another one in [119]. Here we shall present the proof of the paper [119].

If an arbitrary edge $\langle x^0, x^1 \rangle = l \in L$ is deleted from the Cayley tree Γ^k, it splits into two components– two semi-infinite trees Γ_0^k and Γ_1^k (see Fig. 2.1).

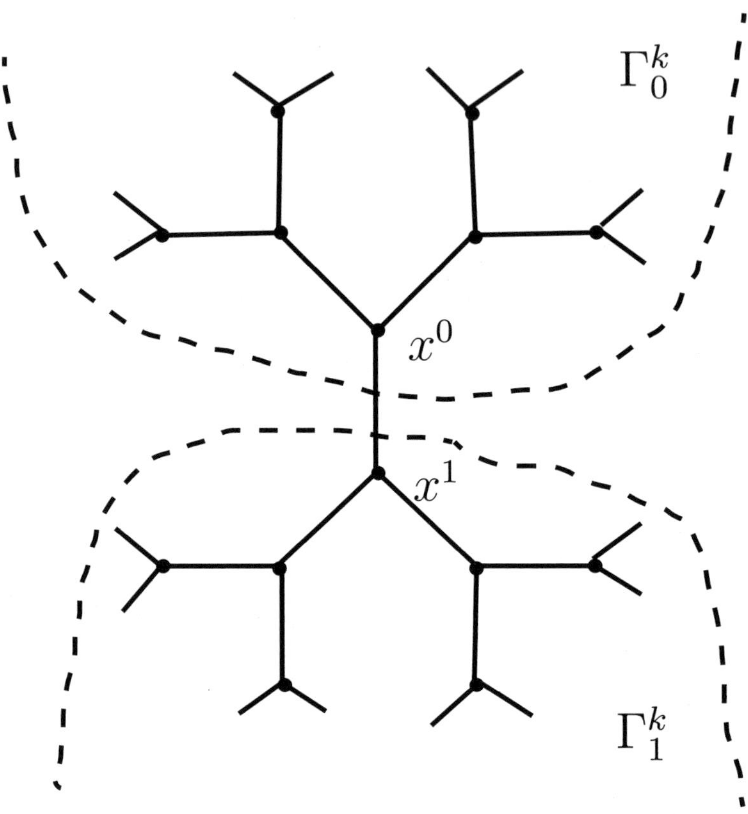

Fig. 2.1   Semi-infinite Cayley trees: Γ_0^2 and Γ_1^2.

Theorem 2.8. [107]. *A Gibbs measure μ on Γ^k, is an extreme measure if and only if there exist extreme Gibbs measures ν_0, ν_1 on Γ_0^k, Γ_1^k respectively, such that*

$$\mu = \nu_0 \nu_1 Z^{-1} \exp\{(J\beta)\sigma(x^0)\sigma(x^1)\}, \tag{2.38}$$

where $Z > 0$ is the normalizing constant.

This theorem reduces the description of extreme Gibbs measures on Γ^k to the semi-infinite tree Γ_0^k. Let V^0 be the set of vertices of Γ_0^k and L^0 its set of edges.

Without lost of generality we can assume $J = 1$ (i.e., denote $J\beta$ by β).

For $h_x = 0$ the restriction of the measure μ_0 on a finite subset $A \subset V^0$ has the following form:

$$\mu_{0,A} = Z_A^{-1} \exp\left\{ -\beta \sum_{\substack{\langle x,y\rangle: \\ x,y \in A}} \sigma(x)\sigma(y) \right\}. \tag{2.39}$$

Set $L_A = \{\langle x,y\rangle : x,y \in A\}$.

Consider the following representation of $\mu_{0,A}$.

Step 1. Consider an independent Bernoulli percolation on L_A, i.e., assign to each bond configuration $n_A \in \{0,1\}^{L_A}$ the probability

$$\mathbb{Q}_A^\theta(n_A) = \theta^{\sum_{b \in L_A} n_A(b)}(1 - \theta)^{|L_A| - \sum_{b \in L_A} n_A(b)}.$$

Step 2. Given a bond configuration n_A, two sites $x, y \in A$ are called connected if they are connected by the chain of open bonds in L_A, i.e., if $n_A = 1$ on all bonds from the (unique) chain leading from x to y. Thus, any (random) configuration n_A splits A into disjoint union of maximal connected components. In order to specify values of spins at various sites of A, paint independently each cluster of A into $+1$ or -1 with probability $1/2$ each.

After performing both steps above we end up with a probability distribution on $\Omega_A = \{-1, 1\}^A$. The important fact is that this measure happens to be precisely $\mu_{0,A}$.

The above two-step procedure can be, using some labeling algorithm to avoid ambiguities, equally applied to construct probability measures $\mu_{0,A}$ for infinite connected subsets $A \subseteq V^0$, in particular for V^0 itself. Thus, let $\mathbb{Q}^\theta$ denote the independent Bernoulli percolation measure on $\Omega = \{0,1\}^{L^0}$ and μ_0 denote the corresponding measure on Ω. Clearly, μ_0 satisfies the following

$$\mu_0(\sigma_A \in \bullet) = \mu_{0,A}(\bullet), \tag{2.40}$$

where σ_A is the restriction of the configuration $\sigma \in \Omega$ to A. Clearly μ_0 is the thermodynamic limit of the finite volume Gibbs measures (2.39).

Consequently, many questions about μ_0 and $\mu_{0,A}$ admit a natural percolation interpretation.

If $|A|$ is even, then one says that a configuration $n \in \{0,1\}^{L^0}$ splits A evenly, in case if there is even number of vertices of A in each maximal connected component of n.

For any finite subset $A \subset V^0$, set $\sigma^A = \prod_{x \in A} \sigma(x)$.

An example of how the percolation approach works is provided by the following proposition.

Proposition 2.3. *1. If $|A|$ is odd, then*

$$\langle \sigma^A \rangle_0 = 0, \tag{2.41}$$

where $\langle \cdot \rangle_0$ is the expectation with respect to μ_0.

2. If $|A|$ is even, then

$$\langle \sigma^A \rangle_0 = \mathbb{Q}^\theta(n : n \text{ splits } A \text{ evenly}). \tag{2.42}$$

Proof. Let $\Gamma^k(A) = (V(A), L(A))$ be the minimal connected subtree which spans A. Then, by (2.40), $\langle \sigma^A \rangle_0 = \langle \sigma^A \rangle_{0,V(A)}$, where $\langle \cdot \rangle_{0,V(A)}$ is the expectation with respect to $\mu_{0,V(A)}$. Simple combinatoric arguments, then, imply $\langle \sigma^A \rangle_{0,V(A)} = 0$, in the odd case, and $\langle \sigma^A \rangle_{0,V(A)} = \mathbb{Q}^\theta_{L(A)}(n : n$ splits A evenly) in the even case. Since $\mathbb{Q}^\theta_{L(A)}$ is the relativization of $\mathbb{Q}^\theta$, (2.42) follows. $\qquad\square$

Another example of how the percolation approach works is provided by the following proposition.

Proposition 2.4. *Let two disjoint finite subsets $A, B \subset V^0$ have edge disjoint minimal spanning trees, i.e., $L(A) \cap L(B) = \emptyset$.*

(a) If both $|A|$ and $|B|$ are even, then

$$\langle \sigma^A \sigma^B \rangle_0 = \langle \sigma^A \rangle_0 \langle \sigma^B \rangle_0. \tag{2.43}$$

(b) If both $|A|$ and $|B|$ are odd, then for any site x, which lies on the unique chain connecting $V(A)$ to $V(B)$,

$$\langle \sigma^A \sigma^B \rangle_0 = \langle \sigma^A \sigma(x) \rangle_0 \langle \sigma^B \sigma(x) \rangle_0. \tag{2.44}$$

Proof. Both statements are consequences of (2.41) and (2.42) above and independence relations for Bernoulli percolation. $\qquad\square$

Let $\sigma(x^0)$ be the value of the spin at the root x^0 of Γ^k_0. Also let B_N be the set of vertices at distance N from the root, where the distance $d(x,y)$ (on semi-infinite tree) between two vertices x and y is defined to be the number of edges in the unique path connecting these two vertices. Finally, define $\mathcal{F}_N$ to be the σ-algebra generated by the spin configurations from $\{-1,1\}^{B_N}$. Note that μ_0 is extremal, i.e., μ_0 has a trivial tail σ-field, if and only if,

$$\lim_{N\to\infty} \mathrm{Var}^\beta(\mathbb{E}^\beta(\sigma(x^0)|\mathcal{F}_N)) = 0, \qquad (2.45)$$

where all the expectations are computed with respect to μ_0. Since $\langle\sigma(x^0)\rangle_0 = 0$, we shall identify $\mathcal{F}_N$ with the Euclidean space of all μ_0-zero mean functions $f : \{-1,1\}^{B_N} \to R$ equipped with the scalar product $\langle\cdot\rangle_0$. Then $\mathcal{F}_N$ is spanned by the linearly independent family $\{\sigma_A\}_{A\subset B_N}$. To facilitate notations, let us use σ_N to denote the restriction of a configuration σ to $\{-1,1\}^{B_N}$ (instead of σ_{B_N} above).

Define

$$g_N(\sigma_N) = \mathbb{E}^\beta(\sigma(x^0)|\mathcal{F}_N) = \mathrm{proj}|_{\mathcal{F}_N}\sigma(x^0)$$

and set $\|g_N\| = \langle g_N, g_N\rangle_0$.

We have to show that $\lim_{N\to\infty}\|g_N\|^2 = 0$, as soon as $k\theta^2 \le 1$.

Let A_1, A_2 be two disjoint subsets of V^0, such that $V(A_1) \cap V(A_2) = \{x^0\}$, i.e., $\Gamma^k(A_1)$ and $\Gamma^k(A_2)$ are two edge disjoint trunks growing from the root x^0, and let $\mathcal{F}_1$, $\mathcal{F}_2$ and $\mathcal{F}$ be σ-algebras generated by the spins from A_1, A_2 and $A_1 \vee A_2$, respectively. Denote $\Omega_i = \{-1,1\}^{A_i}$, $i = 1,2$ and for fixed configurations $\xi_i \in \Omega_i$ set

$$g_i(\xi_i) = \mathbb{E}^\beta(\sigma(x^0)|\mathcal{F}_i), \quad g(\xi_1,\xi_2) = \mathbb{E}^\beta(\sigma(x^0)|\mathcal{F}).$$

Proposition 2.5. *There exists a positive constant $\alpha = \alpha(\theta, k)$ such that*

$$\|g\|^2 = \langle g, g\rangle_0 \le \|g_1\|^2 + \|g_2\|^2 - \alpha\|g_1\|^2\|g_2\|^2. \qquad (2.46)$$

Proof. We have

$$g = g_1 + g_2 - g_1 g_2 g. \qquad (2.47)$$

Indeed, let $\sigma_B \in \mathcal{F}$. Then $\sigma^B = \sigma^{B_1}\sigma^{B_2}$, where $\sigma^{B_i} \in \mathcal{F}_i$, $i = 1,2$. There are only two symmetric possibilities to consider: either $|B_1|$ is odd and $|B_2|$ is even or the other way around. So let us assume that $|B_1|$ is odd. Then, using (2.43), one obtains

$$\langle\sigma^B g\rangle_0 = \langle\sigma^{B_1}\sigma(x^0)\rangle_0\langle\sigma^{B_2}\rangle_0 = \langle\sigma^{B_1}g_1\rangle_0\langle\sigma^{B_2}\rangle_0 = \langle\sigma^B g_1\rangle_0.$$

Similarly, by (2.44) we get

$$\langle\sigma^B g_2\rangle_0 = \langle g_1\sigma^{B_1}\rangle_0\langle\sigma(x^0)g_2\sigma^{B_2}\rangle_0 = \langle g_1 g_2 g\sigma^B\rangle_0$$

and (2.47) follows.

Consequently, multiplying both sides of (2.47) by $\sigma(x^0)$, we obtain

$$\|g\|^2 = \|g_1\|^2 + \|g_2\|^2 - \langle g^2 g_1 g_2 \rangle_0.$$

Thus, it remain to show that for some $\alpha = \alpha(\theta, k) > 0$,

$$\langle g^2 g_1 g_2 \rangle_0 \geq \alpha \|g_1\|^2 \|g_2\|^2.$$

If the spin at the root is fixed, then the measures on Ω_1 and Ω_2 decouple, and, after summing out all the spins on $V(A_1)$ and $V(A_2)$, one obtains

$$g_i(\xi_i) = \tanh(H_i(\xi_1)), \; i = 1, 2$$

and

$$g(\xi_1, \xi_2) = \tanh(H_1(\xi_1) + H_2(\xi_2)), \tag{2.48}$$

where H_1, H_2 are the effective magnetic fields at x^0 given boundary conditions on A_1 and A_2, respectively. Note that H_i is an odd function of the boundary configuration ξ_i.

Define $\Omega_{i,+(-)}; i = 1, 2$, by

$$\Omega_{i,+,(-)} = \{\xi \in \Omega_i : H_i(\xi) > (<)0\}.$$

Using the symmetry of the model and (2.48), we obtain

$$\langle g^2 g_1 g_2 \rangle_0 = 2 \sum_{\xi_i \in \Omega_{i,+}} \left[\mu_0(\sigma_{A_1} = \xi_1, \sigma_{A_2} = \xi_2) \tanh^2(H_1(\xi_1) + H_2(\xi_2)) - \right.$$

$$\left. \mu_0(\sigma_{A_1} = \xi_1, \sigma_{A_2} = -\xi_2) \tanh^2(H_1(\xi_1) - H_2(\xi_2)) \right] g_1(\xi_1) g_2(\xi_2).$$

Now, it is easy to see that the probabilities above can be represented as follows: For any $\xi_i \in \Omega_i$, $i = 1, 2$

$$\mu_0(\sigma_{A_1} = \xi_1, \sigma_{A_2} = \xi_2) = \frac{1}{\mathbf{Z}} \mathbf{Z}_1(\xi_1) \mathbf{Z}_2(\xi_2) \cosh(H_1(\xi_1) + H_2(\xi_2)),$$

where $\mathbf{Z}_1$ and $\mathbf{Z}_2$ are positive even functions of boundary configurations and $\mathbf{Z}$ is the corresponding partition function. Therefore,

$$\langle g^2 g_1 g_2 \rangle_0 = 2 \sum_{\xi_i \in \Omega_{i,+}} g_1(\xi_1) g_2(\xi_2) \frac{1}{\mathbf{Z}} \mathbf{Z}_1 \mathbf{Z}_2 U(\xi_1, \xi_2),$$

where

$$U(\xi_1, \xi_2) = \frac{\sinh^2(H_1(\xi_1) + H_2(\xi_2))}{\cosh(H_1(\xi_1) + H_2(\xi_2))} - \frac{\sinh^2(H_1(\xi_1) - H_2(\xi_2))}{\cosh(H_1(\xi_1) - H_2(\xi_2))}.$$

Now bringing to the common denominator and regrouping terms, one gets

$$U(\xi_1, \xi_2) \geq \frac{\cosh(H_1(\xi_1) + H_2(\xi_2)) - \cosh(H_1(\xi_1) - H_2(\xi_2))}{\cosh(H_1(\xi_1) + H_2(\xi_2))\cosh(H_1(\xi_1) - H_2(\xi_2))}.$$

In view of (2.48), the denominator can be bounded above by a constant which depends only on θ and k and which we denote by $1/\alpha$. Thus,

$$\langle g^2 g_1 g_2 \rangle_0 \geq 2\alpha \sum_{\xi_i \in \Omega_{i,+}} g_1 g_2 \frac{1}{\mathbf{Z}} \mathbf{Z}_1 \mathbf{Z}_2 (\cosh(H_1 + H_2) - \cosh(H_1 - H_2)) =$$

$$\alpha \langle g_1, g_2 \rangle_0 = \alpha \|g_1\|^2 \|g_2\|^2. \qquad \square$$

The following theorem is the main result of this section:

Theorem 2.9. *For the Ising model on the Cayley tree of order $k \geq 2$, the Gibbs measure μ_0 is extreme if and only if $\theta \leq \theta_c^{\mathrm{SG}} = \frac{1}{\sqrt{k}}$ (i.e., $T \geq T_{c,\sqrt{k}}$).*

Proof. Because of the self-similarity of the measure μ_0, we may consider g_N to be just a function on $\{-1, 1\}^{k^N}$ and use it unambiguously to denote the projection of any spin to the σ-algebra generated by its Nth generation of descendants. Here one says that x belongs to the Nth generation of descendants of y, if $d(x, y) = N$ and y lies on the unique path leading from x to x^0.

There are k branches emanating from the root x^0. Let us denote by σ_N^l the restriction of a configuration $\sigma_N \in \{-1, 1\}^{B_N}$ to the lth branch and let $\mathcal{F}_{N,l}$ and $\mathcal{F}'_{N,l}$ to denote the σ-algebras, generated by the configurations $\{\sigma_N^l\}$ and $\{\sigma_N^m; m > l\}$, respectively. Finally, set $g'_{N,l} = \mathrm{proj}_{\mathcal{F}'_{N,l}} \sigma(x^0)$.

Note now that $\mathrm{proj}_{\mathcal{F}_{N,l}} \sigma(x^0) = \theta g_{N-1}(\sigma_N^l)$.

Indeed, as follows from Proposition 2.4,

$$\langle \sigma(x^0) \sigma^A \rangle_0 = \langle \sigma(x^0) \sigma_1^l \rangle_0 \langle \sigma_1^l \sigma^A \rangle_0 = \theta \langle g_{N-1}(\sigma_N^l) \sigma^A \rangle_0,$$

for any $A \in \mathcal{F}_{N,l}$. Hence, by Proposition 2.5, we have

$$\|g_N\|^2 \leq \theta^2 \|g_{N-1}(\sigma_N^l)\|^2 + \|g'_{N,1}\|^2 - \alpha \theta^2 \|g'_{N,1}\|^2 \|g_{N-1}\|^2 \leq$$

$$\theta^2 \|g_{N-1}\|^2 (1 - \alpha \|g_{N-1}\|^2) + \|g'_{N,1}\|^2.$$

Now using (2.46) once again we can decompose $\|g'_{N,1}\|^2$. Thus, after applying Proposition 2.5 $(k-1)$ times, we obtain

$$\|g_N\|^2 \leq k\theta^2 \|g_{N-1}\|^2 \left(1 - \frac{k-1}{k} \alpha \theta^2 \|g_{N-1}\|^2\right).$$

The above formula not only implies (2.45) and, hence, the assertion of the theorem, but also provides an estimate on the speed of convergence of $\|g_N\|^2$ to zero whenever $k\theta^2 \leq 1$. $\qquad \square$

2.6 Uncountable sets of non-periodic Gibbs measures

In this section we shall give some constructions of non-periodic extreme Gibbs measures for the Ising model on the Cayley tree.

2.6.1 *Bleher-Ganikhodjaev construction*

We shall use notations of previous sections. Consider semi-infinite trees Γ_0^k. Let V^0 be the set of vertices of Γ_0^k and L^0 its set of edges. For a fixed $x^0 \in V$ we set

$$W_n = \{x \in V^0 \,|\, d(x, x^0) = n\}, \qquad V_n = \bigcup_{m=1}^{n} W_m.$$

Denote

$$S_k(x) = \{y \in W_{n+1} : d(x, y) = 1\}, \quad x \in W_n,$$

this set is called a set of *direct successors* of x. On the tree Γ_0^k one can introduce a partial ordering, by saying that $y > x$ if there exists a path $x = x_0, x_1, ..., x_n = y$ from x to y that "goes upwards", i.e., such that $d(x_m, x^0) = d(x_{m-1}, x^0) + 1, m = 1, \ldots, n$. The set of vertices $V_x^0 = \{y \in V^0 | y \geq x\}$ and the edges connecting them from the semi-infinite tree Γ_x^k "growing" from the vertex $x \in V^0$.

Theorem 2.10. [107] *Let $n \geq 1$. In order for μ to be an extreme Gibbs measure on Γ_0^k, it is necessary and sufficient that there exist extreme Gibbs measures μ_x on Γ_x^k, $x \in W_n$ such that*

$$\mu = Z_n^{-1} \exp\{(J/T)H_n(\sigma)\} \prod_{x \in W_n} \mu_x, \qquad (2.49)$$

where $H_n(\sigma) = -J \sum_{\langle x,y \rangle : x,y \in V_n} \sigma(x)\sigma(y)$.

Denote by $\mathcal{E}$ the set of extreme Gibbs measures on Γ_0^k and by $\mathcal{F}$ the set of Gibbs measures μ on Γ_0^k such that there exist Gibbs measures μ_x on Γ_0^k, $x \in V^0$, such that for each $n \geq 0$ the factorization (2.49) is true. From Theorem 2.10 it follows that $\mathcal{E} \subset \mathcal{F}$.

We consider an arbitrary (finite or infinite) path $x^0 = x_0 < x_1 < x_2 < \ldots$ starting from the point x^0. We can represent the path by a sequence $i_1 i_2 i_3 \ldots$, where $i_n = 0, 1, \ldots, k-1$. Namely, we label by $l_0(x), l_1(x), \ldots, l_{k-1}(x)$ the edges going upward from the vertex $x \in V^0$. Then the path $x^0 = x_0 < x_1 < x_2 < \ldots$ can be assigned a sequence $i_1 i_2 \ldots$

such that $\langle x_{n-1}, x_n \rangle = l_{i_n}(x_{n-1})$, $n = 1, 2, \ldots$; the sequence $i_1 i_2 i_3 \ldots$ unambiguously determines the path $x^0 = x_0 < x_1 < x_2 < \ldots$.

A finite path $x^0 = x_0 < x_1 < x_2 < \cdots < x_n = x$ of length n determines a point $x \in W_n$. If the path $x^0 = x_0 < x_1 < x_2 < \cdots < x_n = x$ is represented by the sequence $i_1 i_2 \ldots i_n$, we assume that the vertex x is also represented by this sequence.

Let $x, y \in W_n$, and let x be represented by the sequence $i_1 i_2 \ldots i_n$ and y by $j_1 j_2 \ldots j_n$, where $i_1 = j_1$, $i_2 = j_2, \ldots$, $i_m = j_m$, but $i_{m+1} < j_{m+1}$ for some m. In this case, we write $x \prec y$.

Let $\pi = \{x^0 = x_0 < x_1 < \ldots\}$ be an infinite path, represented by the sequence $i_1 i_2 \ldots$. We assign the real number

$$t = t(\pi) = \sum_{n=1}^{\infty} \frac{i_n}{k^n}, \quad 0 \le t \le 1$$

to the path π. This assignment is a 1-1 correspondence everywhere except at those numbers t that can be decomposed into a finite sum,

$$t = \sum_{n=1}^{N} \frac{i_n}{k^n}, \quad i_N \ne 0.$$

Let Q_k denote the set of such fractions, i.e.,

$$Q_k = \{0 < t < 1 : t = \sum_{n=1}^{N} \frac{i_n}{k^n}, \quad i_N \ne 0\}.$$

If $t \in Q_k$, then t is known to be decomposed in two ways:

$$t = \sum_{n=1}^{N} \frac{i_n}{k^n}, \quad i_N \ne 0;$$

$$t = \sum_{n=1}^{N-1} \frac{i_n}{k^n} + \frac{(i_N - 1)}{k^N} + \sum_{n=N+1}^{\infty} \frac{k-1}{k^n}.$$

The first of these sequences is denoted by $\{i_n^{(1)}, n = 1, 2, \ldots\}$ and the second by $\{i_n^{(2)}, n = 1, 2, \ldots\}$, i.e.,

$$i_n^{(1)} = i_n^{(2)} = i_n, \quad n \le N - 1,$$

$$i_N^{(1)} = i_N, \, i_N^{(2)} = i_N - 1,$$

$$i_n^{(1)} = 0, \, i_n^{(2)} = k - 1, \quad n \ge N - 1.$$

Let the paths $\pi(t, 1)$ and $\pi(t, 2)$ be represented by the respective sequences $i_1^{(1)} i_2^{(1)} i_3^{(1)} \ldots$ and $i_1^{(2)} i_2^{(2)} i_3^{(2)} \ldots$. It is obvious that for $n > N + 1$, the path $\pi(t,1)$ constantly "goes right" and the path $\pi(t,2)$ constantly "goes left".

Let $\pi = \{x^0 = x_0 < x_1 < \dots\}$ be an infinite path. We assign the set of numbers $h^\pi = \{h^\pi_x, x \in V^0\}$ satisfying equation (2.8) to the path π. For $x \in W_n$, the set h^π is unambiguously defined by the conditions

$$h^\pi_x = \begin{cases} -h_*, & \text{if } x \prec x_n, \, x \in W_n, \\ h_*, & \text{if } x_n \prec x, \, x \in W_n, \end{cases} \tag{2.50}$$

$n = 1, 2, \dots$ (see Fig. 2.2).

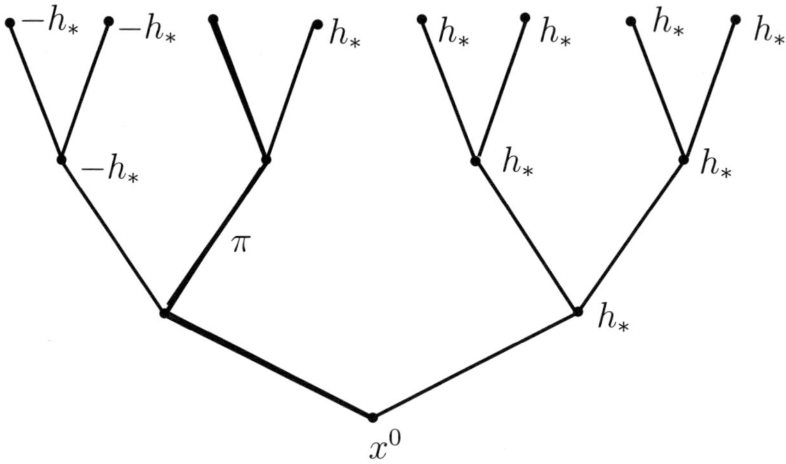

Fig. 2.2   Function h^π_x.

Theorem 2.11. *For any infinite path π, there exists a unique set of numbers $h^\pi = \{h^\pi_x, x \in V^0\}$ satisfying equations (2.8) and (2.50).*

Proof.   On W_n, we define the set

$$h^{(n)}_x = \begin{cases} -h_*, & \text{if } x \prec x_n, \, x \in W_n, \\ h_*, & \text{if } x_n \prec x, \, x \in W_n, \\ h^{(n)}_x, & \text{if } x = x_n, \end{cases} \tag{2.51}$$

where $h^{(n)}_{x_n} \in [-h_*, h_*]$ is an arbitrary number. We extend the definition of $h^{(n)}_x$ for all $x \in V_n = \cup^n_{m=0} W_m$ using recursion equations (2.8). We now prove that the limit

$$h_x = \lim_{n \to \infty} h_x^{(n)} \tag{2.52}$$

exists for every fixed $x \in V^0$ and is independent of the choice of $h_x^{(n)}$ for $x = x_n$. If $x \in W_{n-1}$ and $x \prec x_{n-1}$, then

$$h_x^{(n)} = \sum_{y \in W_n, y > x} f(h_y^{(n)}, \theta) = kf(-h_*, \theta) = -h_*.$$

Similarly, for $x \in W_{n-1}$ and $x_{n-1} \prec x$, we get $h_x^{(n)} = h_*$. Consequently, for any $x \in W_m$, $m \le n$ we have

$$h_x^{(n)} = \begin{cases} -h_*, & \text{if } x \prec x_m, \ x \in W_m, \\ h_*, & \text{if } x_m \prec x, \ x \in W_m. \end{cases} \tag{2.53}$$

This implies that limit (2.52) exists for $x \in W_m$ and $x \ne x_m$ and

$$h_x = \begin{cases} -h_*, & \text{if } x \prec x_m, \ x \in W_m, \\ h_*, & \text{if } x_m \prec x, \ x \in W_m. \end{cases}$$

Therefore, we only need to establish that limit (2.52) exists for $x = x_m$. Let $1 \le l \le n$. Then

$$h_{x_{l-1}}^{(n)} = \sum_{y \in W_l, y > x_{l-1}} f(h_y^{(n)}, \theta). \tag{2.54}$$

Consider two sets $\{\bar{h}_x^{(n)}, x \in V_n\}$ and $\{\tilde{h}_x^{(n)}, x \in V_n\}$ which correspond to two values $\bar{h}_x^{(n)}$ and $\tilde{h}_x^{(n)}$ for $x = x_n$, in (2.51), then from (2.54) we get

$$\tilde{h}_{x_{l-1}}^{(n)} - \bar{h}_{x_{l-1}}^{(n)} = \sum_{y \in W_l, y > x_{l-1}} \left[f(\tilde{h}_y^{(n)}, \theta) - f(\bar{h}_y^{(n)}, \theta) \right]. \tag{2.55}$$

Since $\tilde{h}_y^{(n)} = \bar{h}_y^{(n)}$ for any $y \ne x_l$, $y \in W_l$, in the RHS of (2.55) there is unique non-zero term at $y = x_l$, consequently,

$$\tilde{h}_{x_{l-1}}^{(n)} - \bar{h}_{x_{l-1}}^{(n)} = f(\tilde{h}_{x_l}^{(n)}, \theta) - f(\bar{h}_{x_l}^{(n)}, \theta). \tag{2.56}$$

Now by property $\left| \frac{df}{dx}(x, \theta) \right| \le \theta < 1$ we get

$$\left| \tilde{h}_{x_{l-1}}^{(n)} - \bar{h}_{x_{l-1}}^{(n)} \right| \le \theta \left| \tilde{h}_{x_l}^{(n)} - \bar{h}_{x_l}^{(n)} \right|. \tag{2.57}$$

Iterating this inequality we obtain

$$\left| \tilde{h}_{x_m}^{(n)} - \bar{h}_{x_m}^{(n)} \right| \le \theta^{n-m} \left| \tilde{h}_{x_n}^{(n)} - \bar{h}_{x_n}^{(n)} \right|. \tag{2.58}$$

For arbitrary $N, M > n$, we now consider the sets $\{h_x^{(N)}, x \in V_N\}$ and $\{h_x^{(M)}, x \in V_M\}$ determined by initial conditions of form (2.51) for $x \in W_N$ and $x \in W_M$ respectively and by recursion equations (2.8). We set $\bar{h}_{x_n}^{(n)} = h_{x_n}^{(N)}$, $\tilde{h}_{x_n}^{(n)} = h_{x_n}^{(M)}$. Then inequalities (2.58) imply

$$\left| h_{x_m}^{(N)} - h_{x_m}^{(M)} \right| \le \theta^{n-m} \left| h_{x_n}^{(N)} - h_{x_n}^{(M)} \right| \le 2h_* \theta^{n-m}.$$

This estimate implies that the sequence $h_{x_m}^{(n)}$ satisfies the Cauchy criterion as $n \to \infty$ for a fixed m; therefore, limit (2.52) exists and is independent of the choice of $h_{x_n}^{(n)}$ in (2.51). Because, by construction, the sets $\{h_x^{(n)}\}$ satisfy equation (2.8) before taking the limit, so does $\{h_x\}$. The uniqueness of $\{h_x\}$ obviously follows from estimate (2.58). $\qquad\square$

Lemma 2.4. *For every $t \in Q_k$, $t = \sum_{n=1}^{N} \frac{i_n}{k^n}$, the number sets $h^{\pi(t,1)}$ and $h^{\pi(t,2)}$ are identical, and $h_x^{\pi(t,1)} = h_*$ if $x = x_n$, $n \ge N + 1$.*

Proof. For $n > N$, we set

$$h_x = \begin{cases} -h_*, & \text{if } x \prec x_n, \\ h_*, & \text{if } x_n \preceq x, \end{cases}$$

and extend the definition of the set $\{h_x, x \in V^0\}$ for $x \in V_N$, using recursion equations (2.8). Then, as is easy to see, the set $\{h_x, x \in V^0\}$ satisfies these equations for all $x \in V^0$; further, in view of the uniqueness condition in Theorem 2.50, the thus constructed set $\{h_x, x \in V^0\}$ is precisely $\{h_x^{\pi(t,1)}, x \in V^0\}$. By the same token, the set $\{h_x, x \in V^0\}$ is also identical to $\{h_x^{\pi(t,2)}, x \in V^0\}$; therefore, $h^{\pi(t,1)} = h^{\pi(t,2)}$. $\qquad\square$

It follows from Lemma 2.4 that for any point $t \in [0, 1]$, the number set $h^{\pi(t)} = \{h_x^{\pi(t)}, x \in V^0\}$ is unambiguously defined. Let

$$h_0(t) = h_{x^0}^{\pi(t)}, \tag{2.59}$$

where x^0 is the root vertex of the graph Γ_0^k.

Lemma 2.5. *The function $h_0(t)$, $t \in [0, 1]$ is a strictly decreasing continuous function, and $h_0(0) = h_*$, $h_0(1) = -h_*$.*

Proof. Let

$$t = \sum_{n=1}^{\infty} \frac{j_n}{k^n}, \quad s = \sum_{n=1}^{\infty} \frac{i_n}{k^n}, \, t > s.$$

Then there exists N such that

$$k^N \sum_{n=1}^{N} \frac{j_n}{k^n} - k^N \sum_{n=1}^{N} \frac{i_n}{k^n} \geq 2 \qquad (2.60)$$

(otherwise $t = s$).

Let $\pi(t) = \{x^0 = x_0 < x_1 < x_2 < \dots\}$ and $\pi(s) = \{x^0 = y_0 < y_1 < y_2 < \dots\}$. Then (2.60) implies that $y_N \prec x_N$ and at least one more vertex z_N exists between x_N and y_N, i.e., $y_N \prec z_N \prec x_N$. By conditions (2.50), this leads to

$$h_x^{\pi(t)} \begin{cases} \leq h_x^{\pi(s)}, & x \in W_n, \\ < h_x^{\pi(s)}, & x = z_n. \end{cases} \qquad (2.61)$$

Setting $n = 0$, we obtain $h_{x^0}^{\pi(t)} < h_{x^0}^{\pi(s)}$, which proves that the function $h_0(t)$ is strictly monotonic. We now show it to be continuous. Let

$$t = \sum_{n=1}^{\infty} \frac{j_n}{k^n}, \quad s = \sum_{n=1}^{\infty} \frac{i_n}{k^n},$$

with

$$i_n = j_n, \quad \text{if} \quad n \leq N. \qquad (2.62)$$

Then paths $\pi(t)$, $\pi(s)$ coincide up to the level N: $x_n = y_n$, $n \leq N$. We set $\bar{h}_{x_N}^{(N)} = h_{x_N}^{\pi(t)}$, $\tilde{h}_{x_N}^{(N)} = h_{x_N}^{\pi(s)}$. Then, by (2.58) with $n = N$ and $m = 0$, we get

$$|h_0(t) - h_0(s)| \leq \theta^N 2h_*. \qquad (2.63)$$

Then it is obvious that

$$|h_0(t) - h_0(s)| \leq \theta^N 4h_*, \quad \text{if} \quad |t - s| \leq k^{-N-1}. \qquad (2.64)$$

This estimate leads to

$$|h_0(t) - h_0(s)| \leq C|t - s|^{\alpha},$$

where $\alpha = -\frac{\ln \theta}{\ln k}$, which proves the continuity of $h_0(t)$. Equalities $h_0(0) = h_*, h_0(1) = -h_*$ follow from Lemma 2.4. $\qquad \square$

Lemma 2.5 implies that the number sets $h^{\pi(t)}$ are distinct for different $t \in [0, 1]$. Let μ^t denote the Gibbs measure corresponding to $h^{\pi(t)}$. Because the measures μ^t are distinct for different $t \in [0, 1]$, we obtain a continuum of limiting Gibbs measures. Moreover, $\mu^{t=0} = \mu_+$ and $\mu^{t=1} = \mu_-$.

The main result of this subsection is the following theorem.

Theorem 2.12. *For any $t \in [0, 1]$, the Gibbs measure μ^t is extreme.*

Proof. Let us decompose μ^t into extreme Gibbs measures

$$\mu^t = \int \mu(\omega)\nu(d\omega).$$

By construction, $\mu^t \in \mathcal{F}$ and by Theorem 2.10 $\nu(\omega) \in \mathcal{F}$. We use the factorization (2.49)

$$Z_{n,t}^{-1} \exp\{-\beta H_n(\sigma)\} \prod_{x \in W_n} \mu_x^t =$$

$$\int Z_n^{-1}(\omega) \exp\{-\beta H_n(\sigma)\} \prod_{x \in W_n} \mu_x(\omega)\nu(d\omega),$$

from which it follows that

$$\prod_{x \in W_n} \mu_x^t = \int (Z_{n,t}/Z_n(\omega)) \prod_{x \in W_n} \mu_x(\omega)\nu(d\omega).$$

Integrating this equality with respect to $\sigma_z = \{\sigma(v), v \in V_z^0\}, z \neq x$, we obtain

$$\mu_x^t = \int L_{n,t}(\omega)\mu_x(\omega)\nu(d\omega), \tag{2.65}$$

where $L_{n,t}(\omega) > 0$. For $x \prec x_n$, where $\pi(t) = \{x^0 = x_0 < x_1 < x_2 < \dots\}$, we have $\mu_x^t = \mu_-$, (where μ_- corresponds to solution $h_x \equiv -h_*$ to (2.8)), by Theorem 2.2 μ_- is an extreme Gibbs measure, consequently $\mu_x(\omega) = \mu_-$ for almost all ω with respect to the measure $\nu(d\omega)$, if $x \prec x_n$. Hence $h_x(\omega) = h_x^{\pi(t)}$, if $x \neq x_n$, $n = 1, 2, \dots$, for almost all ω. By uniqueness mentioned in Theorem 2.11 we get that $h_x(\omega) = h_x^{\pi(t)}$, for any $x \in V^0$. Thus $\mu(\omega) = \mu^t$ for almost all ω with respect to the measure $\nu(d\omega)$, which means that μ^t is extreme. $\qquad\square$

Theorem 2.13. *For any collection $t_n = \{t(x) \in [0,1], x \in W_n\}$, the Gibbs measure*

$$\mu^{t_n} = Z_n^{-1} \exp\{-\beta H_n(\sigma)\} \prod_{x \in W_n} \mu^{t(x)}$$

is extreme.

Proof. This follows from Theorems 2.12 and 2.10. $\qquad\square$

Remark 2.6. The periodic and weakly periodic Gibbs measures described in Theorems 2.5, 2.6 and 2.7 allow describing (using Bleher-Ganikhodjaev construction) a continuum set of non-periodic Gibbs measures that differ from the above constructed ones.

2.6.2 *Zachary construction*

Assume $\theta > \theta_c$. We shall construct continuum distinct functions h^t, $t \in (-h_*, h_*)$, $h_* > 0$, which satisfy the functional equation (2.8).

Take $t \in (-h_*, h_*)$ with $t \neq 0$. Define the sequence $(t_n)_{n \geq 0}$ recursively by $t_0 = t$,

$$t_n = kf(t_{n+1}, \theta), \quad n \geq 0. \tag{2.66}$$

Since the function f maps the interval $(-h_*, h_*)$ bijectively onto itself, the definition of t_n given by (2.66) is correct.

Define the function h_x^t by $h_x^t = t_n$ for all $x \in W_n$. For any $x \in V$ there is $n \geq 0$ such that $x \in W_n$, consequently, we have

$$h_x^t = \sum_{y \in S_k(x)} f(h_y^t, \theta) = \sum_{y \in S_k(x)} f(t_{n+1}, \theta) = kf(t_{n+1}, \theta) = t_n,$$

i.e., the function h_x^t satisfies (2.8) for any t.

By the construction, distinct values of t between $-h_*$ and h_* define distinct functions $h^t = \{h_x^t, \; x \in V\}$. Denote by ν^t the Gibbs measure which corresponds to the function h^t.

Theorem 2.14. *For any $t \in (-h_*, h_*)$, the Gibbs measure ν^t is extreme.*

Proof. Follows from Corollary (12.18) and Theorem (7.26) of [107]. $\square$

2.7 New Gibbs measures

In this section we are going to describe new Gibbs measures of the Ising model on the Cayley tree of order $k \geq 3$.

The main result of this section is

Theorem 2.15. *For the ferromagnetic Ising model (i.e., $J > 0$) on the Cayley tree of order $k \geq 3$, if T such that $T_{c,\sqrt{k}} < T < T_{c,k_0}$, with $\sqrt{k} < k_0 < k$ then there is uncountable set $\mathcal{G}_{k_0,k}$ of extreme Gibbs measures which are different from the measures mentioned above in this Chapter.*

Proof. For given k_0 ($\sqrt{k} < k_0 < k$) denote by $\mathcal{G}_{k_0}$ the set of all known extreme Gibbs measures for the Ising model on Cayley tree of order k_0. By results of previous sections we know that $\mathcal{G}_{k_0}$ is an uncountable set. Now for any $\mu \in \mathcal{G}_{k_0}$, $\mu \neq \mu_0$ we shall construct a Gibbs measure $\nu = \nu(\mu)$ which is measure on the Cayley tree of order $k > k_0$. By Theorem 2.1 to each measure $\mu \in \mathcal{G}_{k_0}$ corresponds unique function $h_x = h_x(\mu)$ which satisfies

(2.8) on Γ^{k_0}. Construct function $\tilde{h}_x = \tilde{h}_x(\nu)$ on Γ^k as follows. Let V^k be the set of all vertices of the Cayley tree Γ^k. Since $k_0 < k$ one can consider V^{k_0} as a subset of V^k. Define the following function (see Fig. 2.3)

$$\tilde{h}_x = \begin{cases} h_x(\mu), & \text{if } x \in V^{k_0} \\ 0, & \text{if } x \in V^k \setminus V^{k_0}. \end{cases} \tag{2.67}$$

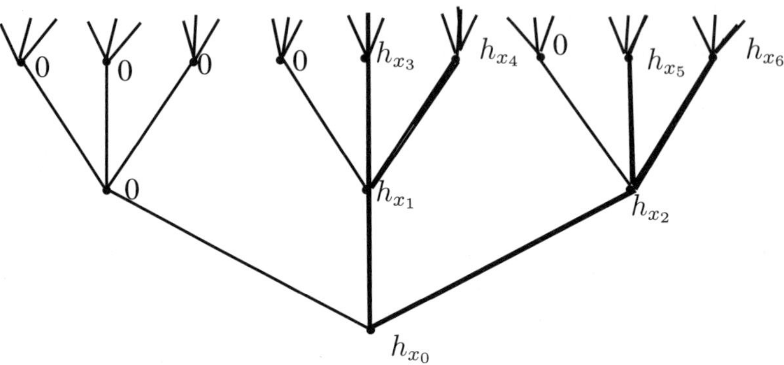

Fig. 2.3   The function $\tilde{h}_x$ on the Cayley tree of order $k = 3$ for $k_0 = 2$. The bold subtree has the set of vertices $V^{k_0} = V^2$.

Now we shall check that (2.67) satisfies (2.8) on Γ^k.

Let $x \in V^{k_0} \subset V^k$. We have

$$\tilde{h}_x = \sum_{y \in S_k(x)} f(\tilde{h}_y, \theta) = \sum_{y \in S_k(x) \cap V^{k_0}} f(\tilde{h}_y, \theta) +$$

$$\sum_{y \in S_k(x) \cap (V^k \setminus V^{k_0})} f(\tilde{h}_y, \theta) = \sum_{y \in S_{k_0}(x)} f(h_y, \theta) = h_x,$$

here we used $f(0, \theta) = 0$.

If $x \in V^k \setminus V^{k_0}$ then it is easy to see that $S_k(x) \subset V^k \setminus V^{k_0}$. Therefore we have

$$\tilde{h}_x = \sum_{y \in S_k(x)} f(\tilde{h}_y, \theta) = \sum_{y \in S_k(x)} f(0, \theta) = 0.$$

Thus $\tilde{h}_x, x \in V^k$ satisfies the functional equation (2.8) and we denote by $\nu = \nu(\mu)$ the Gibbs measure which by Theorem 2.1 corresponds to $\tilde{h}_x$. By the construction one can see that $\nu(\mu_1) \neq \nu(\mu_2)$ if $\mu_1 \neq \mu_2$ and the measure ν is different from measures mentioned in previous sections of this Chapter. Denote by $\mathcal{G}_{k_0,k}$ the set of all Gibbs measures defined on Γ^k which are constructed by Gibbs measures defined on Γ^{k_0}.

We shall prove that ν is an extreme Gibbs measure. Let us decompose ν into extreme Gibbs measures

$$\nu = \int \bar{\nu}(\omega)\lambda(d\omega).$$

By definition, $\nu \in \mathcal{F}$ and by Theorem 2.10 $\bar{\nu} \in \mathcal{F}$. We use the factorization (2.49)

$$Z_n^{-1}\exp\{-\beta H_n(\sigma)\}\prod_{x\in W_n}\nu_x =$$

$$\int Z_n^{-1}(\omega)\exp\{-\beta H_n(\sigma)\}\prod_{x\in W_n}\nu_x(\omega)\lambda(d\omega),$$

from which it follows that

$$\prod_{x\in W_n}\nu_x = \int (Z_n/Z_n(\omega))\prod_{x\in W_n}\nu_x(\omega)\lambda(d\omega).$$

Integrating this equality with respect to $\sigma_z = \{\sigma(t), t \in V_z^0\}, z \neq x$, we obtain

$$\nu_x = \int L_n(\omega)\nu_x(\omega)\lambda(d\omega), \tag{2.68}$$

where $L_n(\omega) > 0$. For $x \in V^k \setminus V^{k_0}$ we have $\nu_x = \mu_0$ (where μ_0 is the unordered Gibbs measure, which corresponds to solution $h_x \equiv 0$ to (2.8)), since $T_{c,\sqrt{k}} < T < T_{c,k_0}$, by Theorem 2.9 μ_0 is an extreme Gibbs measure, consequently $\nu_x(\omega) = \mu_0$ for almost all ω with respect to the measure $\lambda(d\omega)$. Hence if $h_y(\omega), y \in V_x^0$ corresponds to $\nu_x(\omega)$ then

$$h_y(\omega) = \tilde{h}_y(\mu) = 0, \quad \text{for all } \ y \in V^k \setminus V^{k_0} \ \text{ for almost all } \ \omega. \tag{2.69}$$

From (2.69) and (2.68) we get that measure ν_x is extreme for any $x \in V^k \setminus V^{k_0}$.

Now we shall prove that measure ν_x is extreme for any $x \in V^{k_0}$. Here we use the argument used in the proof of Theorem 3 of [115]. Let $|t|$ denote the distance between x and $\ t \in V_x^0 \cap V_n = V_{n,x}, \ \omega \in \Phi^{V_x^0}, \ s \in V_x^0$ and $n \geq |s|$ put:

$$H_{n,s}^{\omega}(\sigma) = -J\sum_{\substack{\langle x,y\rangle:\\x,y\in V_{n,s}}}\sigma(x)\sigma(y) - J\sum_{\substack{\langle x,y\rangle:\\x\in V_{n,s},\,y\in\partial V_{n,s}}}\sigma(x)\omega(y),$$

$$W^\omega_{n,s}(\varepsilon) = \sum_{\sigma \in \Phi^{V_{n,s}} : \sigma(s)=\varepsilon} \exp\left\{ -\beta H^\omega_{n,s}(\sigma) - J\beta\varepsilon\omega(t) \right\}, \quad \varepsilon = \pm 1,$$

$$R_{n,s}(\omega) = W^\omega_{n,s}(+1)/W^\omega_{n,s}(-1),$$

where t is the unique point such that $\langle s, t \rangle$. We shall use the following

Lemma 2.6. [115]. *Let P be a Gibbs measure of the Hamiltonian (2.5), then*

1) P is an extreme measure iff

$$\lim_{n\to\infty} R_{n,s}(\omega) = r_s \quad a.s.\,(P)$$

for each $s \in V^0_x$ with $|s| \geq N$ for some $N > 0$, where r_s is a constant depending only on s.

2) The limit $\lim_{n\to\infty} R_{n,s}(\omega) = r_s(\omega)$ exists almost surely with respect to P for every $s \in V^0_x$.

Denoting $G(x) = (1 + e^{2J\beta}x)/(e^{2J\beta} + x)$, we have (see formula (22) of [115])

$$r_t(\omega) = \lim_{n\to\infty} R_{n,s}(\omega) =$$

$$\prod_{u \in V^{k_0}:\langle t,u \rangle} G(r_u(\omega)) \prod_{v \in V^k \setminus V^{k_0}:\langle t,v \rangle} G(r_v(\omega)) = \prod_{u \in V^{k_0}:\langle t,u \rangle} G(r_u(\omega)), \quad (2.70)$$

here we used $r_v(\omega) = 1$ for all $v \in V^k \setminus V^{k_0}$ which follows from (2.69). Since on V^{k_0} the measure μ is extreme by definition of $\tilde{h}_x$ (see (2.67)) and Lemma 2.6 from (2.70) we get $r_t(\omega) = r_t$ a.s. (ν_x), $x \in V^{k_0}$.

Thus by Theorem 2.10 we conclude that the measure ν is extreme. $\square$

Corollary 2.1. *For ferromagnetic Ising model on the Cayley tree of order k, if $T_{c,\sqrt{k}} < T < T_{c,\sqrt{k}+1}$ then Gibbs measures of the set $\mathcal{G}_k = \bigcup_{k_0:\sqrt{k}<k_0<k} \mathcal{G}_{k_0,k}$ are extreme and different from measures described in previous sections of this Chapter.*

2.8   Free energies

On non-amenable graphs, not only Gibbs measures but also the free energy (and the entropy) depend on the boundary conditions.

The purpose of this section is to study this dependence for the Ising model on Cayley trees. For this model partition functions given by

$$Z_n \equiv Z_n(\beta, h) = \sum_{\sigma_n \in \Phi^{V_n}} \exp\left\{\beta J \sum_{\langle x,y \rangle \subset V_n} \sigma(x)\sigma(y) + \sum_{x \in W_n} h_x \sigma(x)\right\}.$$

$$(2.71)$$

Here the spin configurations σ_n belong to Φ^{V_n} and

$$h = \{h_x \in R, x \in V\}$$

is a collection of real numbers that stands for (generalized) boundary condition.

By the previous sections we know that for any boundary condition satisfying the functional equation (2.8) there exists a unique Gibbs measure, the correspondence being one-to-one.

We will be interested in the dependence with respect to boundary conditions of the free energy defined as the limit (when it exists):

$$F(\beta, h) = -\lim_{n \to \infty} \frac{1}{\beta |V_n|} \ln Z_n(\beta, h), \qquad (2.72)$$

where $|\cdot|$ denotes hereafter the cardinality of a set.

A boundary condition satisfying (2.8) will be in the sequel called *compatible*.

For compatible boundary conditions, the free energy is given by the formula

$$F(\beta, h) = -\lim_{n \to \infty} \frac{1}{|V_n|} \sum_{x \in V_n} a(x), \qquad (2.73)$$

where

$$a(x) = \frac{1}{2\beta} \ln[4 \cosh(h_x - \beta J) \cosh(h_x + \beta J)].$$

To see it, first notice that

$$Z_n(\beta, h) = A_{n-1} Z_{n-1}(\beta, h), \qquad (2.74)$$

where $A_n = \prod_{x \in W_n} b(x)$ with $b(x)$ satisfying

$$\prod_{y \in S(x)} \sum_{u = \pm 1} \exp(\beta J \varepsilon u + u h_y) = b(x) \exp(\varepsilon h_x), \quad \varepsilon = \pm 1. \qquad (2.75)$$

This formula used with both values of ε implies

$$b(x) = \prod_{y \in S(x)} (4 \cosh(h_y - \beta J) \cosh(h_y + \beta J))^{1/2} = \exp\left(\beta \sum_{y \in S(x)} a(y)\right).$$

It is then enough to insert this formula into the recursive equation (2.74) to get by iteration

$$Z_n(\beta, h) = \prod_{x \in V_{n-1}} b(x)$$

which gives (2.73).

Notice that

$$F(\beta, h) = F(\beta, -h), \qquad (2.76)$$

where $-h = \{-h_x, x \in V\}$.

The problem of convergence of (2.72) is an open problem. Let us give conditions on h that ensure the existence of the limit (2.73) and as a consequence of the corresponding free energy.

Note that $a(x)$ is bounded: $\beta^{-1} \ln 2 \leq a(x) \leq C_\beta$. Hence the limit (2.73) is also bounded. Let $\pi = \{x^0 = x_0, x_1, x_2, \dots\}$ be an infinite path. A function h_x on the path π is called monotone if $h_{x_i} \geq h_{x_{i+1}}$, $(h_{x_i} \leq h_{x_{i+1}})$, $i = 0, 1, 2, \dots$.

Proposition 2.6. *Let $h = \{h_x, x \in V\}$ be a compatible b.c. If on any infinite path starting at $x \in W_{n_0}$, $|h_x|$ is monotone, then the corresponding free energy $F(\beta, h)$ exists.*

Proof. Using Stolz-Cesáro theorem (see e.g. [142]) we get

$$\lim_{n \to \infty} \frac{1}{|V_n|} \sum_{x \in V_n} a(x) = \lim_{n \to \infty} A_n, \quad \text{with} \quad A_n = \frac{1}{|W_n|} \sum_{x \in W_n} a(x).$$

We shall show that A_n is monotone for $n > n_0$. We have

$$A_{n-1} - A_n = \frac{1}{|W_n|} \left(k \sum_{x \in W_{n-1}} a(x) - \sum_{x \in W_n} a(x) \right) =$$

$$\frac{1}{|W_n|} \sum_{x \in W_{n-1}} \left(ka(x) - \sum_{y \in S(x)} a(y) \right) = \frac{1}{|W_n|} \sum_{x \in W_{n-1}} \sum_{y \in S(x)} (a(x) - a(y)).$$

By monotonicity of $|h_x|$ and by evenness of $a(x)$ we notice that $a(x) - a(y)$ does not change sign for all x, y with $x < y$. Thus A_n is monotone and since it is a bounded sequence it has a limit. $\qquad \square$

Now we shall compute free energies of the Ising model which correspond to all known Gibbs measures (given in Sections 2.3-2.7).

Translation-invariant boundary conditions. (See Section 2.3.)

According to formula (2.73), the free energies of translation-invariant (TI) b.c. are given by:

$$F_{\text{TI}}(\beta, 0) = -\frac{1}{\beta} \ln(2\cosh(\beta J)).$$
(2.77)

$$F_{\text{TI}}(\beta, h_*) = F_{\text{TI}}(\beta, -h_*) = \frac{1}{2\beta} \ln[4\cosh(\beta J - h_*)\cosh(\beta J + h_*)].$$
(2.78)

Some particular plots are shown in Fig. 2.4.

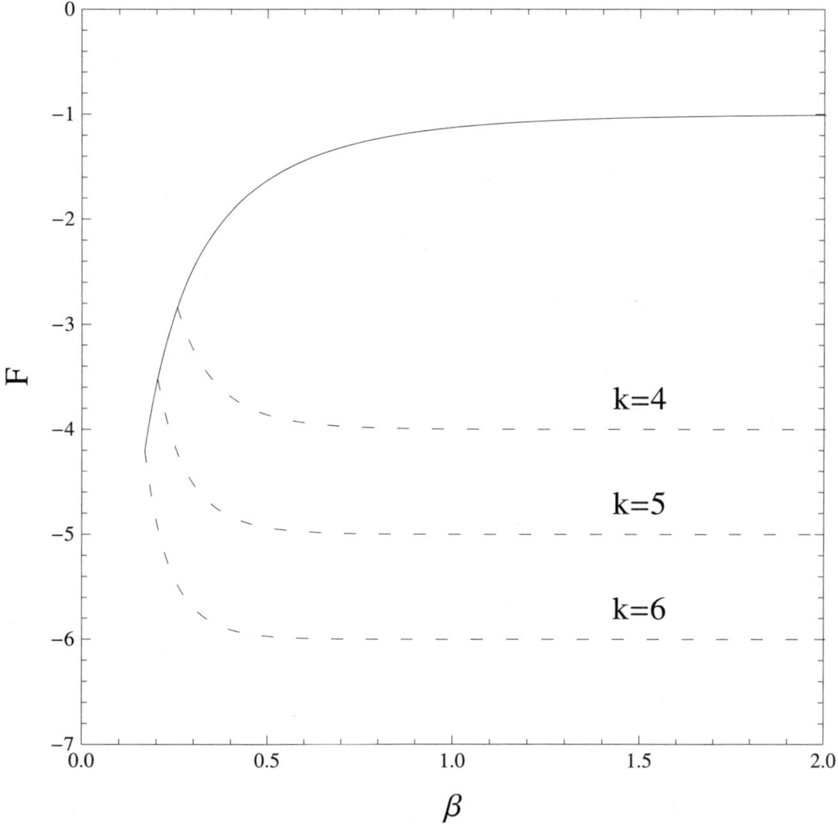

Fig. 2.4   The free energies $F_{\text{TI}}(\beta, 0)$ (solid line) and $F_{\text{TI}}(\beta, h_*)$ (dotted lines) for $\theta \geq \theta_c$ with $J = 1$ and $k = 4, 5, 6$.

In order to draw the free energy $F_{TI}(\beta, h)$ as a function of β, we notice that equation (2.14) gives

$$\beta(h) = \frac{1}{2J} \ln \frac{e^{(1+\frac{1}{k})2h} - 1}{e^{2h} - e^{\frac{2h}{k}}} \tag{2.79}$$

and use the parametric representation defined by the following mapping:

$$h \to (\beta(h), F(h)), \quad \text{for } h \geq 0.$$

The function $F(h)$ is defined by inserting (2.79) into (2.77) and (2.78).

Bleher-Ganikhodjaev construction.

Since the solution h^π (see subsection 2.6.1) can differ from h_* or $-h_*$ only on the path π, it is thus obvious that the values h_x^π, $x \in \pi$ do not contribute to the free energy. As a consequence, we get from the evenness (2.78), that the free energies of Bleher-Ganikhodjaev and translation-invariant boundary conditions coincide:

$$F_{\mathrm{BG}}(\beta, h^\pi) = F_{\mathrm{TI}}(\beta, h_*). \tag{2.80}$$

Zachary construction. Recall that this construction provides an (uncountable) set of distinct functions $h^{(t)}$ satisfying condition (2.8) and parametrized by $t \in (-h_*, h_*)$.

Take $t \neq 0$ and define the sequence $(t_n)_{n \geq 0}$ recursively as in subsection 2.6.2. One can see that $\lim_{n \to \infty} t_n = 0$ for each $t_0 = t$.

Consider the function $h_x^{(t)} = t_n$ for all $x \in W_n$. The corresponding free energies can be written as

$$F_{\mathrm{Zach}}(\beta, h^{(t)}) = -\frac{1}{2\beta} \lim_{n \to \infty} \frac{1}{|V_n|} \sum_{m=0}^{n} |W_m|\, \tilde{a}(t_m),$$

where $\tilde{a}(t) = \ln[4 \cosh(\beta J - t) \cosh(\beta J + t)]$.

By Stolz-Cesáro theorem applied to the sequences

$$a_n = \sum_{m=0}^{n} |W_m|\tilde{a}(t_m), \quad b_n = |V_n| \tag{2.81}$$

one has

$$\lim_{n \to \infty} \frac{1}{|V_n|} \sum_{m=0}^{n} |W_m|\, \tilde{a}(t_m) = \lim_{n \to \infty} \frac{\sum_{m=0}^{n+1} |W_m|\, \tilde{a}(t_m) - \sum_{m=0}^{n} |W_m|\, \tilde{a}(t_m)}{|V_{n+1}| - |V_n|}$$

$$= \lim_{n \to \infty} \frac{|W_{n+1}|\, \tilde{a}(t_{n+1})}{|W_{n+1}|} = \lim_{n \to \infty} \tilde{a}(t_{n+1}) = \tilde{a}(0). \tag{2.82}$$

As a consequence

$$F_{\text{Zach}}(\beta, h^{(t)}) = F_{\text{TI}}(\beta, 0). \tag{2.83}$$

ART construction. In this part we consider the measures (boundary conditions) constructed in Section 2.7.

For the corresponding free energy, we have

$$F_{\text{ART}}(\beta, \tilde{h}) = -\frac{1}{\beta} \lim_{n \to \infty} \frac{1}{|V_n^k|} \left\{ \left(|V_n^k| - |V_n^{k_0}|\right) \ln[2\cosh(\beta J)] + \beta \sum_{x \in V_n^{k_0}} a(x) \right\}. \tag{2.84}$$

Since

$$\lim_{n \to \infty} \frac{|V_n^{k_0}|}{|V_n^k|} = \frac{k-1}{k_0 - 1} \lim_{n \to \infty} \frac{(k_0 + 1)k_0^n - 2}{(k+1)k^n - 2} = 0,$$

by taking into account $0 \le a(x) \le C_\beta$, we get

$$0 \le \sum_{x \in V_n^{k_0}} a(x) \le |V_n^{k_0}| C_\beta.$$

As a consequence,

$$\lim_{n \to \infty} \frac{1}{|V_n^k|} \sum_{x \in V_n^{k_0}} a(x) = 0,$$

so that

$$F_{\text{ART}}(\beta, \tilde{h}) = -\frac{1}{\beta} \ln[2\cosh(\beta J)] = F_{\text{TI}}(\beta, 0). \tag{2.85}$$

Let us compute the entropy $S(\beta, h) = -\frac{dF(\beta, h)}{dT}$, which corresponds to the translation invariant b.c. To compute the entropy, we first notice that equation (2.79) gives

$$h'(\beta) = \frac{1}{\beta'(h)} = Jk \, \frac{\cosh(2h) - \cosh\left(\frac{2h}{k}\right)}{\sinh(2h) - k\sinh\left(\frac{2h}{k}\right)}. \tag{2.86}$$

As a result of easy computations, we get the formula

$$S(\beta, h) = \frac{1}{2} \ln\left[2\cosh(2h) + 2\cosh(2\beta J)\right]$$
$$- \beta J \, \frac{k\sinh(2h)\frac{\cosh(2h) - \cosh\left(\frac{2h}{k}\right)}{\sinh(2h) - k\sinh\left(\frac{2h}{k}\right)} + \sinh(2\beta J)}{\cosh(2h) + \cosh(2\beta J)} \tag{2.87}$$

and

$$S(\beta, 0) = \ln(2\cosh(\beta J)) - \beta J \tanh(\beta J). \tag{2.88}$$

Let us mention that in case $k = 2$, we get by solving equation (2.14) the following value of h_* as a function of the inverse temperature:

$$\pm h_* = \frac{1}{2} \ln\left[2^{-1}\left(e^{4\beta J} - 2e^{2\beta J} - 1 \pm (e^{2\beta J} - 1)\sqrt{(e^{2\beta J} + 1)(e^{2\beta J} - 3)}\right)\right].$$
$$(2.89)$$

The free energies and entropies then read

$$F(\beta, h_*) = -\frac{1}{\beta}\left(\ln(e^{2\beta J} - 1) + \frac{1}{2}\ln(e^{-2\beta J} + 1)\right), \qquad (2.90)$$

$$S(\beta, h_*) = \ln(2\sinh(\beta J)) + \frac{1}{2}\ln(2\cosh(\beta J)) - \beta J\frac{3\cosh(2\beta J) + 1}{2\sinh(2\beta J)}.$$

Periodic boundary conditions.

In this subsection, we consider periodic solutions of (2.8) (see subsection 2.3.2).

From subsection 2.3.2 we know that in the ferromagnetic case only translation invariant b.c. can be found. Let us also recall that for the antiferromagnetic Ising model:

(1) If $\theta \geq -1/k$, μ_0 is unique and extreme.
(2) If $\theta < -1/k$, there are two periodic measures $\mu^{(\pm)}$ and $\mu^{(\mp)}$, which are extreme.

For periodic measures, we have

$$F_{\text{Per}}(\beta, h_*^{(\pm)}) = -\frac{1}{2\beta}\ln[4\cosh(\beta J - h_*)\cosh(\beta J + h_*)] = F_{\text{TI}}(\beta, h_*).$$

Weakly periodic Gibbs measures. Here we compute free energies which correspond to the weakly periodic Gibbs measures constructed in subsection 2.4.1. For weakly periodic b.c. we have

$$F_{\text{WP}}(\beta, h) = -\frac{1}{2\beta}\lim_{n\to\infty}\frac{1}{|V_n|}\left\{(\mathcal{A}_n + \mathcal{D}_n)\ln[4\cosh(\beta J - h_1)\cosh(\beta J + h_1)]+\right.$$

$$\left.(\mathcal{B}_n + \mathcal{C}_n)\ln[4\cosh(\beta J - h_2)\cosh(\beta J + h_2)]\right\},$$

where $\mathcal{A}_n, \mathcal{B}_n, \mathcal{C}_n, \mathcal{D}_n$ are given in (1.7). By Proposition 1.6, for $k = 4$ and $j = 1$ we obtain

$$F_{\text{WP}}(\beta, h) = -\frac{1}{2\beta}\left\{\frac{4}{5}\ln[4\cosh(J\beta - h_1)\cosh(J\beta + h_1)]\right.$$

$$\left. + \frac{1}{5}\ln[4\cosh(J\beta - h_2)\cosh(J\beta + h_2)]\right\}. \qquad (2.91)$$

As it was shown in subsection 2.4.1 the equation

$$h_1 = 3f(h_1, \theta) - f(4f(h_1, \theta), \theta) \tag{2.92}$$

can be reduced to:

$$\alpha^2 \xi^3 - \alpha \xi^2 - 2\alpha^2 \xi + \alpha + 1 = 0, \tag{2.93}$$

which has two solutions ξ_1 and ξ_2 when $0 < \alpha < \alpha_{\mathrm{cr}}$.

The free energy then reads:

$$F_{\mathrm{WP}}(\beta, h) = -\frac{1}{10\beta} \times$$

$$\left\{ 4\ln\left[2\cosh(2\beta J) + \frac{2\xi\cosh(2\beta J) - 4}{2\cosh(2\beta J) - \xi} \right] + \ln[2\cosh(2\beta J) + \xi^4 - 4\xi^2 + 2] \right\}.$$

The solution ξ_1 is given by

$$\xi_1 = \frac{1}{3}\left(e^{2\beta} + \frac{2^{1/3}\left(6 + e^{4\beta}\right)}{U} + 2^{-1/3}U \right),$$

where

$$U = \left(-9e^{2\beta} - 27e^{4\beta} + 2e^{6\beta} + 3\sqrt{-96 - 39e^{4\beta} + 54e^{6\beta} + 69e^{8\beta} - 12e^{10\beta}} \right)^{1/3}.$$

The corresponding free energy reads

$$F_{\mathrm{WP}}(\beta) = -\frac{1}{10\beta}\ln \frac{\left(-e^{-2\beta} + e^{2\beta}\right)^8 \left(2 + e^{-2\beta} + e^{2\beta} - 4\left(e^{4\beta}\frac{W}{6}\right)^2 + \left(e^{4\beta}\frac{W}{6}\right)^4\right)}{\left(e^{-2\beta} + e^{2\beta} - e^{4\beta}\frac{W}{6}\right)^4},$$

where

$$W = \left(2e^{-2\beta} + \frac{V}{U} + 2^{2/3}e^{-4\beta}U \right) \quad \text{with} \quad V = 2^{4/3}e^{-4\beta}\left(6 + e^{4\beta}\right).$$

The plot is given in Fig. 2.5 from which we observe the strict inequalities

$$F_{\mathrm{TI}}(\beta, h_*) < F_{\mathrm{WP}}(\beta, h) < F_{\mathrm{TI}}(\beta, 0) \tag{2.94}$$

in the range $\alpha \le \alpha_{\mathrm{cr}}$.

2.9 Ising model with an external field

In this section we shall briefly give the generalizations of above mentioned results in case when the Ising model has an external field. This model is given by Hamiltonian:

$$H(\sigma) = -J \sum_{\langle x,y\rangle \in L} \sigma(x)\sigma(y) - \alpha \sum_{x \in V} \sigma(x), \tag{2.95}$$

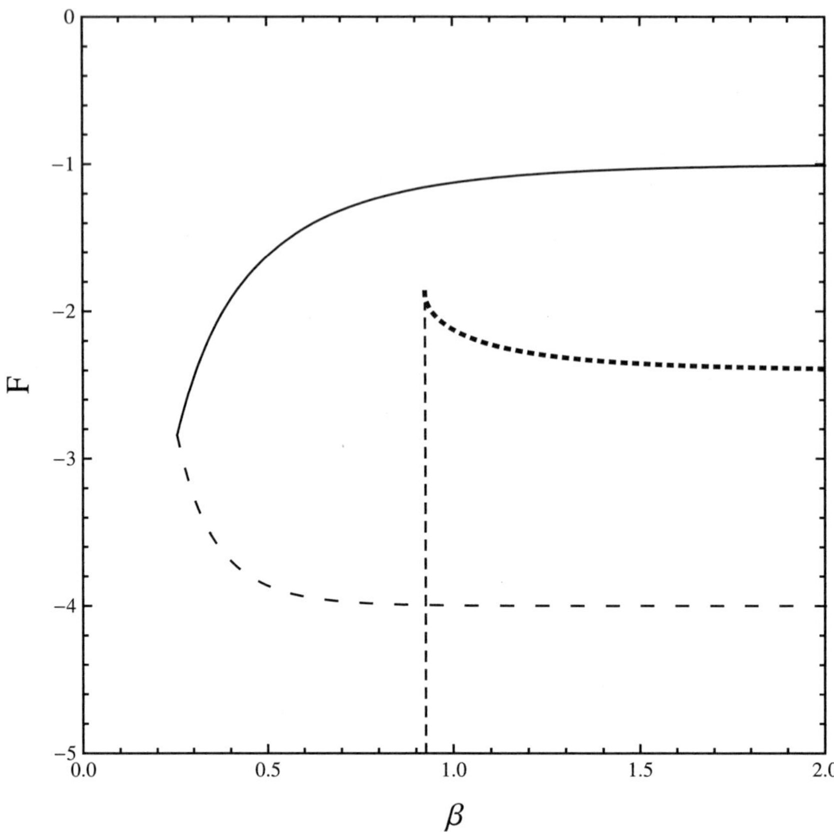

Fig. 2.5 The free energies $F_{\mathrm{WP}}(\beta)$ (dotted line) for $\alpha \leq \alpha_{\mathrm{cr}}$ together with the previous free energies $F_{\mathrm{TI}}(\beta,0)$ (solid line), and $F_{\mathrm{TI}}(\beta,h_*)$ (dashed line). Here $J=1$ and $k=4$.

where $J, \alpha \in R$.

In this case equation (2.8) has the form

$$h_x = \alpha\beta + \sum_{y \in S(x)} f(h_y, \theta). \tag{2.96}$$

By similar analysis as in the case $\alpha = 0$ one can prove the following

Theorem 2.16. [107], [23].

1) If $J > 0$ and $-(k-1)J < \alpha < (k-1)J$ then there is a critical temperature $T_c = T_c(\alpha)$, such that for $T < T_c$ there are two translation-invariant extreme Gibbs measures.

2) If $J < -\frac{1}{2}\ln\frac{k+1}{k-1}$ and $(k-1)J < \alpha < -(k-1)J$ then there are two $G_k^{(2)}$-periodic (non-translation-invariant) extreme Gibbs measures.

Using Bleher-Ganikhodjaev construction and Zachary construction one can describe continuum set of Gibbs measures for the Ising model with an external field. An analogue of Theorem 2.15 is not true for $\alpha \neq 0$, because 0 is not solution to (2.96). Also weakly periodic Gibbs measures are not studied for this model yet.

Commentaries and references. The most studied model of statistical mechanics is the Ising model, there are about 1700 papers devoted to the problems related to Ising model. In particular, this model plays a very special role in statistical mechanics and gives the simplest nontrivial example of a system undergoing phase transitions.

The one-dimensional Ising model (one-dimensional Cayley tree coincides with $\mathbb{Z}$) has no phase transition and was solved by Ising (1925) himself. The two-dimensional square lattice Ising model is much harder, and was given an analytic description much later, by Lars Onsager (1944).

The Ising model undergoes a phase transition between an ordered and a disordered phase in 2 dimensions or more. This was first proved by Rudolph Peierls in 1933, using what is now called a Peierls argument.

In [177] some physical motivations why the Ising model on a Cayley tree is interesting are given. In [123] and in [204] the existence of a phase transition for the Ising model on the Cayley tree for $k \geq 2$ is established. In [266] it was shown that either the set of Gibbs measures contains a single point or it contains infinitely many points. The paper [267] extends this result to general ferromagnetic or anti-ferromagnetic system. At the same time, in [115] for $J > 0$, $\alpha = 0$ it was shown that the boundary condition which is equal to $+1$ on a Cayley subtree and equal to -1 otherwise leads to a limiting Gibbs measure which is extreme and has a positive magnetization at the origin.

Theorem 2.1 is well known, in this book the author's proof is given.

In [107], [266] periodic Gibbs measures with period two of anti-ferromagnetic Ising model are constructed. The definition of a periodic Gibbs measure, using subgroups of the group representation of the Cayley tree, Theorems 2.3 and 2.4 were given by the author [214].

By Theorems 2.3 and 2.4 we know that the set of periodic Gibbs measures is "poor", it only contains specific periodic measures with period two, which are a "chess-board" periodic. Moreover, by these theorems we know that the periodic Gibbs measures, with period two, do not depend on structure of the subgroup of index ≥ 2.

To enlarge the set of Gibbs measures, which corresponds to some solutions of the functional equation (2.8), in [226], [228] the notion of a weakly periodic Gibbs measure is introduced. Theorem 2.7 shows that this set of Gibbs measure is "richer" than the case of periodic ones. Moreover, the set of weakly periodic Gibbs measures strongly depends on a fixed subgroup, with respect to which the weakly periodicity is required. We have only described such measures with respect to arbitrary subgroups of index two and some special subgroups of index 4 (according to Theorem 1.4 there is not any subgroup of odd index $\neq 1$). Note that there is no result about weakly periodic Gibbs measures, which are weakly periodic with respect to a subgroup of index ≥ 6. Even the case of index 4 is not completely studied.

In [25] and [119] the criterion of extremality of disordered Gibbs measure is found. Our proof is based on the paper [119]. The role played by the spin glass transition value $T_{\mathrm{cr}}^{\mathrm{SG}}$ was analyzed in details in the context of the Ising spin glass on the Cayley tree with $k = 2$ in [48].

Bleher-Ganikhodjaev construction is taken from [23] and Zachary's construction is taken from [107] and [266].

Theorem 2.15 is proven in [7].

In [83] the free energies are computed. The results for weakly periodic boundary conditions have to be compared with the inequalities given in [84]; one of weakly periodic boundary condition constructed in subsection 2.4.1 corresponds to the so-called dimer covering in [84]. There, the inequalities between free energies easier to catch with cluster expansion method in mind, the condition on the temperature is more restrictive, and the free energies cannot be expressed explicitly.

Results devoted to the Ising model with a non-zero external field are taken from [23] and [107].

In [211] a generalization of the Ising model is considered, some Gibbs measures of the model are constructed.

Of course, there are a lot of papers which are devoted to the Ising model on Cayley trees, but most of them are written in physical point of view. We did not include results of such papers in this mathematical book.

In [108] it is shown that the probability distribution for the infinite-volume, free-boundary- condition Ising ferromagnetic on the Cayley tree under zero external field is infinitely divisible with respect to the group operation of point wise multiplication of spin variables. For other results related to the Ising model on Cayley tree see [17].

It remains an open problem to give a correspondence between each solution of equation (2.8) and corresponding ground state of the Ising model. Another open problem is to know the decomposition of the disordered Gibbs measure into extremal states (a question of A. van Enter) [84].

Chapter 3

Ising type models with competing interactions

In this chapter we consider two Ising type models with competing interactions. One of them is known as Vannimenus's model; the second is a model with four competing interactions (external field, nearest neighbor, second neighbors and triples of neighbors) on the Cayley tree of order two. We show non-uniqueness of the Gibbs measure for some parameter values of the models. Our second result gives a complete description of periodic Gibbs measures for the models. We also construct uncountably many non-periodic extreme Gibbs measures.

3.1 Vannimenus model

3.1.1 *Definitions and equations*

The vertices x and y are called *next-nearest-neighbor* (NNN) which is denoted by $\rangle x, y \langle$, if there exists a vertex $z \in V$ such that x, z and y, z are NN. We will consider only *prolonged* NNN $\rangle x, y \langle$, for which there exist n such that $x \in W_n$ and $y \in W_{n+2}$.

We consider Ising model with competing NN and prolonged NNN interactions on a Cayley tree where the spin takes values in the set $\Phi := \{-1, 1\}$, and is assigned to the vertices of the tree. A configuration σ on V is then defined as a function $x \in V \mapsto \sigma(x) \in \Phi$; the set of all configurations is Φ^V. The (formal) Hamiltonian of the model is

$$H(\sigma) = -J_1 \sum_{\langle x,y \rangle \in L} \sigma(x)\sigma(y) - J_2 \sum_{\rangle x,y \langle} \sigma(x)\sigma(y), \qquad (3.1)$$

where $J_1, J_2 \in R$ are coupling constants and $\langle x, y \rangle$ stands for NN vertices and $\rangle x, y \langle$ stands for prolonged NNN.

67

The standard approach consists in writing down recurrence equation relating the partition function

$$Z_n = \sum_{\sigma_n \in \Phi^{V_n}} \exp\{-\beta H(\sigma_n)\},$$

of n-generation tree to the partition function Z_{n-1} of its subsystems containing $(n-1)$ generations. In [257] the partition function Z_n of the Hamiltonian (3.1) is given by

$$Z_n = \left(u_1^{(n)} + u_2^{(n)}\right)^2 + \left(u_3^{(n)} + u_4^{(n)}\right)^2, \quad n \geq 1. \tag{3.2}$$

Here $u^{(n)} = \left(u_1^{(n)}, u_2^{(n)}, u_3^{(n)}, u_4^{(n)}\right)$ satisfies the following recurrent equation

$$
\begin{aligned}
u_1^{(n+1)} &= a\left(bu_1^{(n)} + b^{-1}u_2^{(n)}\right)^2 \\[4pt]
u_2^{(n+1)} &= a^{-1}\left(bu_3^{(n)} + b^{-1}u_4^{(n)}\right)^2 \\[4pt]
u_3^{(n+1)} &= a^{-1}\left(b^{-1}u_1^{(n)} + bu_2^{(n)}\right)^2 \\[4pt]
u_4^{(n+1)} &= a\left(b^{-1}u_3^{(n)} + bu_4^{(n)}\right)^2,
\end{aligned}
\tag{3.3}
$$

where $a = \exp(J_1\beta)$, $b = \exp(J_2\beta)$.

Consider mapping $F : u = (u_1, u_2, u_3, u_4) \in R_+^4 \to F(u) = (u_1', u_2', u_3', u_4') \in R_+^4$ defined by

$$
\begin{aligned}
u_1' &= a\left(bu_1 + b^{-1}u_2\right)^2 \\[4pt]
u_2' &= a^{-1}\left(bu_3 + b^{-1}u_4\right)^2 \\[4pt]
u_3' &= a^{-1}\left(b^{-1}u_1 + bu_2\right)^2 \\[4pt]
u_4' &= a\left(b^{-1}u_3 + bu_4\right)^2.
\end{aligned}
\tag{3.4}
$$

Then the recurrent equation (3.3) can be written as $u^{(n+1)} = F(u^{(n)})$, $n \geq 0$ which in the theory of dynamical systems is called *trajectory* of the initial point $u^{(0)}$ under action of the mapping F. Thus asymptotic behavior of Z_n for $n \to \infty$ can be determined by values of $\lim u^{(n)}$, i.e., trajectory of $u^{(0)}$ under action of F. We study the trajectory (dynamical system) for a given initial point $u^{(0)} \in R_+^4$.

3.1.2   *Dynamics of F*

3.1.2.1   *Fixed points*

In this subsubsection we are going to define fixed points, i.e., solutions to $F(u) = u$.

Denote $\text{Fix}(F) = \{u : F(u) = u\}$.

We introduce the new variables $\alpha = \sqrt{a}$, $v_i = \sqrt{u_i}$, $i = 1, 2, 3, 4$. Then the equation $F(u) = u$ becomes

$$
\begin{aligned}
v_1 &= \alpha \left(bv_1^2 + b^{-1}v_2^2\right) \\
v_2 &= \alpha^{-1} \left(bv_3^2 + b^{-1}v_4^2\right) \\
v_3 &= \alpha^{-1} \left(b^{-1}v_1^2 + bv_2^2\right) \\
v_4 &= \alpha \left(b^{-1}v_3^2 + bv_4^2\right).
\end{aligned}
\tag{3.5}
$$

Lemma 3.1. *If a vector u is a fixed point of F then $u \in M_1 = \{u = (u_1, u_2, u_3, u_4) \in R_+^4 : u_1 = u_4, u_2 = u_3\}$ or $u \in M_2 = \{u = (u_1, u_2, u_3, u_4) \in R_+^4 : \sqrt{u_1} + \sqrt{u_4} = \varphi(\sqrt{u_2} + \sqrt{u_3})\}$, where $\varphi(x) = \frac{1 + a^{-1}bx}{ab + (b^2 - b^{-2})x}$.*

Proof. From (3.5) we get

$$
\begin{cases}
(v_1 - v_4)[\alpha b(v_1 + v_4) - 1] + (v_2 - v_3)[\alpha b^{-1}(v_2 + v_3)] = 0 \\
(v_1 - v_4)[(\alpha b)^{-1}(v_1 + v_4)] + (v_2 - v_3)[1 + \alpha^{-1}b(v_2 + v_3)] = 0.
\end{cases}
\tag{3.6}
$$

If $v_1 = v_4$ (resp. $v_2 = v_3$) from the second equation of (3.6) we get $v_2 = v_3$ (resp. $v_1 = v_4$). Thus $v_1 = v_4$ if and only if $v_2 = v_3$. Assume now $v_1 \neq v_4$ and $v_2 \neq v_3$ then system (3.6) can be reduced to the equation

$$
(b^2 - b^{-2})(v_1 + v_4)(v_2 + v_3) + \alpha b(v_1 + v_4) - \alpha^{-1}b(v_2 + v_3) - 1 = 0. \tag{3.7}
$$

Equation (3.7) gives $v_1 + v_4 = \varphi(v_2 + v_3)$. $\qquad\square$

Let us first study fixed points of F which belong in M_1: the condition $u_1 = u_4$, $u_2 = u_3$ reduces the equation $F(u) = u$ to the following equation

$$
x = f(x) \equiv a^2 \left(\frac{1 + b^2 x}{b^2 + x}\right)^2,
\tag{3.8}
$$

where $x = \frac{u_1}{u_2}$. Denote

$$
\tilde{a} = a^{-2}b^{-6}, \quad \tilde{b} = b^4, \quad y = b^2 x.
$$

The following lemma gives full description of solutions to (3.8).

Lemma 3.2. *(Cf. Lemma 10.7 in [204]). Equation (3.8) has a unique positive, stable solution if $\tilde{b} \leq 9$. If $\tilde{b} > 9$, then there exist $\nu_1(\tilde{b})$ and $\nu_2(\tilde{b})$ such that the conditions $0 < \nu_1(\tilde{b}) < \nu_2(\tilde{b})$ are satisfied and equation (3.8) has three solutions, $x_1^* < x_2^* < x_3^*$, x_1^* and x_3^* are stable and x_2^* is unstable,* *if $\nu_1(\tilde{b}) < \tilde{a} < \nu_2(\tilde{b})$ and has two solutions, x_1^*, x_2^*, x_1^* is unstable (saddle) and x_2^* is stable, if $\tilde{a} = \nu_1(\tilde{b})$ or $\tilde{a} = \nu_2(\tilde{b})$. In this case, we have*

$$\nu_i(\tilde{b}) = \frac{1}{y_i}\left(\frac{1+y_i}{\tilde{b}+y_i}\right)^2,$$

where y_1 and y_2 are the solutions of the equation $y^2 + (3 - \tilde{b})y + \tilde{b} = 0$.

Now we shall give some argument to find fixed points of F which belong in M_2. Again use variables v_i, $i = 1, 2, 3, 4$, assume $v_2 + v_3 = C$, with $C > \max\{0, \frac{\alpha b}{b^{-2}-b^2}\}$. Using Lemma 3.1 we get $v_3 = C - v_2$ and $v_4 = \varphi(C) - v_1$. Then from the first equation of (3.5) we obtain $v_2 = \sqrt{b(\alpha^{-1}v_1 - bv_1^2)}$ and from the second equation of (3.5) we have $P_4(v_1) = 0$ with a polynomial P_4 of degree 4, coefficients of which depend on parameters α, b and C. Thus a quartic equation can be obtained. Such an equation can be solved using known formulas (see http://mathworld.wolfram.com/QuarticEquation.html), since we will have some complicated formulas for the coefficients and the solutions, we do not present the solution here.

3.1.3　*Periodic points*

A point u in R_+^4 is called *periodic point* of F if there exists p so that $F^p(u) = u$ where F^p is the pth iterate of F. The smallest positive integer p satisfying $F^p(u) = u$ is called the *prime period* or least period of the point u. Denote by $\mathrm{Per}_p(F)$ the set of periodic points with prime period p.

Note that the set M_1 is invariant with respect to F, i.e., $F(M_1) \subset M_1$. In this subsection we shall describe some periodic points of F which belong in M_1.

Let us first describe periodic points with $p = 2$ on M_1, in this case the equation $F(F(u)) = u$ can be reduced to description of 2-periodic points of the function f defined in (3.8), i.e., to solution of the equation

$$f(f(x)) = x. \tag{3.9}$$

Note that the fixed points of f are solutions to (3.9), to find other solutions we consider the equation

$$\frac{f(f(x)) - x}{f(x) - x} = 0,$$

simple calculations show that the last equation is equivalent to the following

$$b^4(1+a^2\tilde{b}^2)^2 x^2 + a^2\{b^8 + 2(a^{-2}+a^2)b^6 + 4b^4 - 1\}x + b^4(a^2+b^2)^2 = 0. \tag{3.10}$$

The equation has two positive solutions iff $B < 0$ and $D > 0$ where

$$B = a^2\{b^8 + 2(a^{-2} + a^2)b^6 + 4b^4 - 1\}, \quad D = B^2 - (2b^4(a^2 + b^2)(1 + a^2b^2))^2.$$

We have

$$B = \begin{cases} > 0, & \text{if } b \geq \sqrt{\sqrt{2} - 1} \ \text{ or } \ b \leq \sqrt{\sqrt{2} - 1}, \ a^2 \in (0, \mathbf{b}^-) \cup (\mathbf{b}^+, +\infty) \\ 0, & \text{if } b \leq \sqrt{\sqrt{2} - 1}, \ a^2 = \mathbf{b}^-, \mathbf{b}^+ \\ < 0, & \text{if } b \leq \sqrt{\sqrt{2} - 1}, \ a^2 \in (\mathbf{b}^-, \mathbf{b}^+) \end{cases}$$

where

$$\mathbf{b}^{\pm} = \frac{1 - 4b^4 - b^8 \pm (1 - b^4)\sqrt{(b^4 - 1)^2 - 4b^4}}{4b^6};$$

$$D = -a^2(b^4 - 1)^2(4b^6 a^4 + (3b^8 + 6b^4 - 1)a^2 + 4b^6) =$$

$$\begin{cases} < 0, & \text{if } \sqrt{3^{-1}} < b, \ b \neq 1 \ \text{ or } \ b \leq \sqrt{3^{-1}}, a^2 \in (0, b_*^-) \cup (b_*^+, +\infty) \\ 0, & \text{if } b = 1 \ \text{ or } \ b \leq \sqrt{3^{-1}}, \ a^2 = b_*^-, b_*^+ \\ > 0, & \text{if } b \leq \sqrt{3^{-1}}, a^2 \in (b_*^-, b_*^+) \end{cases}$$

$$\tag{3.11}$$

where

$$b_*^{\pm} = \frac{1 - 3b^8 - 6b^4 \pm \sqrt{(b^4 - 1)^3(9b^4 - 1)}}{8b^6}.$$

One can check that $\sqrt{3^{-1}} < \sqrt{\sqrt{2} - 1}$, and for $b \leq \sqrt{3^{-1}}$ one has $\mathbf{b}^- \leq b_*^-$ and $b_*^+ \leq \mathbf{b}^+$. Thus we have proved the following

Lemma 3.3. *The solutions to (3.9) which are different from fixed points of f are varied as follows:*

1) *If $\sqrt{3^{-1}} < b, \ b \neq 1$ or $b \leq \sqrt{3^{-1}}, \ a^2 \in (0, b_*^-) \cup (b_*^+, +\infty)$ then equation (3.10) has no positive solution.*
2) *If $b = 1$ or $b \leq \sqrt{3^{-1}}$ and $a^2 = b_*^-, b_*^+$ then equation (3.10) has unique positive solution $x_1 = \frac{-B}{2b^4(1 + a^2b^2)^2}$.*
3) *If $b \leq \sqrt{3^{-1}}, \ a^2 \in (b_*^-, b_*^+)$ then there are two positive solutions $x_{\pm} = \frac{-B \pm \sqrt{D}}{2b^4(1 + a^2b^2)^2}$ to (3.10).*

The following lemma gives useful properties of the function f.

Lemma 3.4.

1) *If $b > 1$ then the sequence $x_n = f(x_{n-1})$, $n = 1, 2, \ldots$ converges for any initial point $x_0 > 0$, where f is defined in (3.8).*

2) If $b < 1$ then the sequence $y_n = g(y_{n-1})$, $n = 1, 2, \ldots$ converges for any initial point $y_0 > 0$, where $g(x) = f(f(x))$.

Proof. 1) For $b > 1$ we have $f'(x) = 2a^2(b^4 - 1)\frac{1+b^2 x}{(b^2+x)^3} > 0$, i.e., f is an increasing function. Here we consider the case when the function f has three fixed points x_i^*, $i = 1, 2, 3$ (see Lemma 3.2. This proof is more simple for cases when f has one or two fixed points). We have that the point x_2^* is repeller, i.e., $f'(x_2^*) > 1$ and the points x_1^*, x_3^* are attractive, i.e., $f'(x_1^*) < 1$, $f'(x_3^*) < 1$. Now we shall take arbitrary $x_0 > 0$ and prove that $x_n = f(x_{n-1})$, $n \geq 1$ converges as $n \to \infty$. Consider the following partition $(0, +\infty) = (0, x_1^*) \cup \{x_1^*\} \cup (x_1^*, x_2^*) \cup \{x_2^*\} \cup (x_2^*, x_3^*) \cup \{x_3^*\} \cup (x_3^*, +\infty)$. For any $x \in (0, x_1^*)$ we have $x < f(x) < x_1^*$, since f is an increasing function, from the last inequalities we get $x < f(x) < f^2(x) < f(x_1^*) = x_1^*$ iterating this argument we obtain $f^{n-1}(x) < f^n(x) < x_1^*$, which for any $x_0 \in (0, x_1^*)$ gives $x_{n-1} < x_n < x_1^*$, i.e., x_n converges and its limit is a fixed point of f, since f has unique fixed point x_1^* in $(0, x_1^*]$ we conclude that the limit is x_1^*. For $x \in (x_1^*, x_2^*)$ we have $x_2^* > x > f(x) > x_1^*$, consequently $x_n > x_{n+1}$, i.e., x_n converges and its limit is again x_1^*. Similarly, one can show that if $x_0 > x_2^*$ then $x_n \to x_3^*$ as $n \to \infty$.

2) For $b < 1$ we have f is decreasing and has unique fixed point x_1 which is repelling, but g is increasing since $g'(x) = f'(f(x))f'(x) > 0$. By Lemma 3.3 we have that g has at most three fixed points (including x_1). The point x_1 is repelling for g too, since $g'(x_1) = f'(f(x_1))f'(x_1) = (f'(x_1))^2 > 1$. But fixed points x_-, x_+ (see Lemma 3.3) of g are attractive. Hence one can repeat the same argument of the proof of part 1) for the increasing function g and complete the proof. $\qquad\square$

Lemma 3.3 shows that if $b > 1$, i.e., $J_2 > 0$ then there is no any 2-periodic trajectory of F on M_1. Moreover, the following lemma says that if $J_2 > 0$ then there is no any periodic trajectory on M_1.

Lemma 3.5.

1) If $J_2 > 0$ then for any $p \geq 2$ the equation $F^p(u) = u$ has no solution $u \in M_1 \setminus \mathrm{Fix}(F)$.

2) If $J_2 < 0$ then for any $p \geq 3$ the equation $F^p(u) = u$ has no solution $u \in M_1 \setminus (\mathrm{Fix}(F) \cup \mathrm{Per}_2(F))$.

Proof. 1) Assume there is a solution $u^{(0)} \in M_1 \setminus \mathrm{Fix}(F)$ then we get p-periodic trajectories $u_i^{(n+p)} = u_i^{(n)}$, $i = 1, 2$; $n = 0, 1, 2, \ldots$. Since the set

M_1 is invariant with respect to F, we obtain

$$x_{n+p} = \frac{u_1^{(n+p)}}{u_2^{(n+p)}} = \frac{u_1^{(n)}}{u_2^{(n)}} = x_n = f^n(x_0).$$

This is a contradiction, since by Lemma 3.4 we have x_n is not periodic.

2) Assume there is a solution $u^{(0)} \in M_1 \setminus (\mathrm{Fix}(F) \cup \mathrm{Per}_2(F))$ then we have

$$y_{n+p} = x_{2n+2p} = \frac{u_1^{(2n+2p)}}{u_2^{(2n+2p)}} = \frac{u_1^{(2n)}}{u_2^{(2n)}} = y_n = f^{2n}(y_0).$$

This is a contradiction, since by Lemma 3.4 we have y_n is not periodic. $\square$

3.1.4 *Exact values*

Starting from random initial condition (with $u_1^{(0)} \neq u_4^{(0)}$ and $u_2^{(0)} \neq u_3^{(0)}$), one iterates the recurrence equations (3.3) and observes their behavior after large number of iterations. In the simplest situation a fixed point $u^* = (u_1^*, u_2^*, u_3^*, u_4^*) \in R_+^4$ is reached. It corresponds to (see [257]):

a paramagnetic phase if $u^* \in M_1$;

a ferromagnetic phase if $u^* \in M_2$.

The above-mentioned phases have the following meaning (see [257]):

Ferromagnetic phase: the interior spins acquire a finite magnetization when the surface spins are slightly polarized;

Paramagnetic phase: the influence of symmetry-breaking boundary conditions dies out. In the high temperature paramagnetic phase, no symmetry is broken and the spins are as likely to point up as down, whatever the initial conditions: one then has $u_1 = u_4$, $u_2 = u_3$.

If the iterations give a cyclic (periodic), say with period p, sequence then the corresponding phase is called (p)-*commensurate phase*. Finally, the system may remain aperiodic, which corresponds to an *incommensurate phase*.

The condition $\tilde{b} > 9$ of Lemma 3.2 requires that $J_2 > 0$. Denote

$$T_c = \frac{2J_2}{\ln 3}, \quad J_2 > 0.$$

Lemma 3.2 gives the following

Theorem 3.1. *If $T \geq T_c$ then the model (3.1) has unique paramagnetic phase; If $T < T_c$ then there are exactly three (resp. two) paramagnetic phases if (J_1, J_2) is such that $b^3\sqrt{\nu_1} < a^{-1} < b^3\sqrt{\nu_2}$ (resp. $a^{-1} = b^3\sqrt{\nu_1}$ or $a^{-1} = b^3\sqrt{\nu_2}$).*

In a fixed point $u^* = (u_1^*, u_2^*, u_3^*, u_4^*) \in R_+^4$ is reached, then corresponding magnetization m is given by

$$m = \frac{2(1+x)(y_1 + y_2)}{(1+x)^2 + (y_1 + y_2)^2},$$

where

$$x = \frac{u_2^* + u_3^*}{u_1^* + u_4^*}, \quad y_1 = \frac{u_1^* - u_4^*}{u_1^* + u_4^*}, \quad y_2 = \frac{u_2^* - u_3^*}{u_1^* + u_4^*}.$$

Here the variable x is a measure of the frustration of the nearest-neighbor bonds [257]. Since for a paramagnetic phase we have $u_1^* = u_4^*$ and $u_2^* = u_3^*$, we get $y_1 = y_2 = 0$. Hence $m = 0$, but in case of coexistence of several paramagnetic phases their measure of the frustration (i.e., x) are different. These different values of x are the solutions to (3.8).

For the condition $b < \sqrt{3^{-1}}$ of Lemma 3.3 we need to condition $J_2 < 0$. In this case we have $T_c = \frac{-2J_2}{\ln 3}, \quad J_2 < 0$.

From Lemma 3.3 we get the following

Theorem 3.2. *If $T \geq T_c$ then the model (3.1) (on M_1) has unique 2-commensurate phase; If $T < T_c$ then there are exactly two (resp. one) 2-commensurate phases if $a^2 \in (b_*^-, b_*^+)$ (resp. $a^2 = b_*^-$ or $a^2 = b_*^+$).*

For a fixed temperature $T = \beta^{-1} < T_c$ we have two critical curves $a^2 = b_*^\pm$, i.e., on terms of J_1 and $J_2 < 0$ they are given by the following explicit relations

$$J_1 = \frac{1}{2\beta} \ln\left(8^{-1}\{1 - 3e^{8J_2\beta} - 6e^{4J_2\beta} \pm \sqrt{(e^{4J_2\beta} - 1)^3(9e^{4J_2\beta} - 1)}\}\right) - 3J_2.$$

Using Lemma 3.3 and formula (3.2) we can get explicit formulae for the sequence of periodic partition functions:

$$Z_n = Z_n(y) = 2a^{-2/3} \begin{cases} A, & \text{if } n = 2m \\ B, & \text{if } n = 2m+1, \end{cases}$$

where $m = 0, 1, 2, \ldots$ and

$$A = \left(\left(ab(b + \frac{1}{by})^2 + \frac{1}{ab}(\frac{1}{b} + \frac{b}{y})^2 \right)^{-\frac{2}{3}} + \right.$$

$$\left. a^{\frac{2}{3}} \left(\frac{a}{b}(by + \frac{1}{b})^2 + \frac{b}{a}(\frac{y}{b} + b)^2 \right)^{-\frac{2}{3}} \right)^2,$$

$$B = \left(\left(ab(b + \frac{1}{bf(y)})^2 + \frac{1}{ab}(\frac{1}{b} + \frac{b}{f(y)})^2\right)^{-\frac{2}{3}} + \right.$$

$$\left. a^{\frac{2}{3}}\left(\frac{a}{b}(bf(y) + \frac{1}{b})^2 + \frac{b}{a}(\frac{f(y)}{b} + b)^2\right)^{-\frac{2}{3}}\right)^2$$

with y is one of x_1, x_-, x_+ defined in Lemma 3.3 and function f is given in (3.8).

It is easy to see that if x is a fixed point of f then corresponding fixed point of F has the form $u^*(x) = (u_1^*(x), u_2^*(x), u_2^*(x), u_1^*(x))$ with $u_1^*(x) = a^{-1}(b + (bx)^{-1})^{-2}$ and $u_2^*(x) = a(b + b^{-1}x)^{-2}$. If y is a fixed point of g then the corresponding 2-periodic point of F has the form $u^{\mathrm{per}}(y) = (u_1^{\mathrm{per}}(y), u_2^{\mathrm{per}}(y), u_2^{\mathrm{per}}(y), u_1^{\mathrm{per}}(y))$ with

$$u_1^{\mathrm{per}}(y) = a^{-1/3}\left(ab(b + (by)^{-1})^2 + (ab)^{-1}(by^{-1} + b^{-1})^2\right)^{-2/3},$$

$$u_2^{\mathrm{per}}(y) = a^{1/3}\left(a^{-1}b(b + b^{-1}y)^2 + ab^{-1}(by + b^{-1})^2\right)^{-2/3}.$$

Lemma 3.5 gives

Theorem 3.3. *The model (3.1) (on M_1) has uncountable set S of incommensurate phases μ_u, where $u \in M_1 \setminus (\mathrm{Fix}(F) \cup \mathrm{Per}_2(F))$. Moreover the set of incommensurate phases can be classified to (uncountable) subsets*

$$S_x = \{\mu_u : \quad u \in M_1 \setminus (\mathrm{Fix}(F) \cup \mathrm{Per}_2(F)) \quad with \quad \lim_{n \to \infty} F^n(u) = u^*(x)\},$$

where x is an attractive fixed point of f and

$$S_y^{\mathrm{per}} = \{\mu_u : \quad u \in M_1 \setminus (\mathrm{Fix}(F) \cup \mathrm{Per}_2(F)) \quad with \quad \lim_{n \to \infty} F^{2n}(u) = u^{\mathrm{per}}(y)\},$$

where y is an attractive fixed point of g.

3.1.5 *Remarks*

Usually, to describe phases (Gibbs measures) of a given Hamiltonian on a Cayley tree one has correspondence between Gibbs measures and a collection of vectors (real numbers in some particular cases) $\{h_x, x \in V\}$, which satisfy a non-linear equation (as in Chapter 2). The recurrent equation (3.3) considered in this section (which was obtained in [257]) describes a vector function $\{u^{(n)}, n \in N\}$ which is a particular case of the above mentioned function h_x obtained as $h_x = u^{(n)}$ if $x \in W_n$, i.e., depends only on a number of the generation set which belongs to x but not on x itself. Thus the solutions to the recurrent equation (3.3) do not fully describe phases of the model (3.1). But deriving the functional equation for h_x corresponding to the Hamiltonian (3.1) is also difficult, since there is prolonged NNN interaction. Such model can also be studied by a contour argument (see [220], [222], [227]), but this argument does not give exact solutions, in general.

By a process of iteration, for the model (3.1) Vannimenus found a new modulated phase, in addition to the expected paramagnetic and ferromagnetic (fixed point) phases and a $(+ + - -)$ periodic (four cycle antiferromagnetic phase, which consisted of commensurate (periodic) and incommensurate (aperiodic) regions corresponding to the so-called "devil's staircase". In this section, using theory of dynamical systems we have analytically proved many above mentioned results, i.e., the following exact results are obtained:

Paramagnetic phase: The exact critical temperature and exact critical curves are found. It is proven that the number of the paramagnetic phases can be at most three (Theorem 3.1).

Ferromagnetic phase: The description of such phases reduced to solving a quadratic equation (i.e., solution of the equations on M_2). But we were not able to study the periodic solutions on M_2.

Commensurate phase: The exact critical temperature (which is obtained from the critical temperature of the paramagnetic phase by replacing J_2 with $-J_2$) and exact critical curves are found. On the set M_1 it is proven that the number of the 2-commensurate phases can be at most two and there is not p-commensurate phases if $p \geq 3$ (Lemma 3.5, Theorem 3.2). Also exact values of periodic partition functions are obtained.

Incommensurate phase: We proved that the model has uncountably many such phases. Moreover we classified them in two classes: the first class contains the phases which are "asymptotically fixed" (set S_x); the second class contains the phases which are "asymptotically periodic" (set S_y^{per}). Note that for the usual Ising model with external field on Cayley trees such infinitely many phases are known (see [107], p.250).

3.2 A model with four competing interactions

3.2.1 *The model*

We shall use notations of the previous chapters.

Let $\Gamma^k = (V, L)$ be a Cayley tree, where V is the set of vertices of Γ^k, L is the set of edges of Γ^k.

Fix $x^0 \in V$. A collection of the nearest-neighbor pairs $< x, x_1 >, ...,$ $< x_{d-1}, y >$ is called *a path* from x to y. We write $x < y$ if the path from x^0 to y goes through x.

The vertices x and y are called *second neighbor* which is denoted by $> x, y <$, if there exists a vertex $z \in V$ such that x, z and y, z are nearest-neighbors. We will consider only second neighbors $> x, y <$, for which there exist n such that $x, y \in W_n$. Three vertices x, y and z are called a *triple of neighbors* and they are denoted by $< x, y, z >$, if $< x, y >, < y, z >$ are nearest neighbors and $x, z \in W_n$, $y \in W_{n-1}$, for some $n \in N$. The fixed vertex x^0 is called the 0-th level and the vertices in W_n are called the *n-th level*.

In this chapter we consider the Ising model with four competing interactions on the Cayley tree which is defined by the following Hamiltonian

$$H(\sigma) = -J_3 \sum_{<x,y,z>} \sigma(x)\sigma(y)\sigma(z) - J \sum_{>x,y<} \sigma(x)\sigma(y) -$$

$$J_1 \sum_{<x,y>} \sigma(x)\sigma(y) - \alpha \sum_{x \in V} \sigma(x) \tag{3.12}$$

where the sum in the first term ranges all triples of neighbors, the second sum ranges all second neighbors, the third sum ranges all nearest neighbors and the spin variables $\sigma(x)$ have the values ± 1.

Remark 3.1. If one considers the Hamiltonian with all possible triple (without condition $x, z \in W_n$) and second neighbors (without condition $x, y \in W_n$) then the problem of describing of limit Gibbs measures becomes a difficult problem.

The various partial cases of this model have been investigated in numerous works, for example, in [175], [176] and [95] the case $J_3 = \alpha = 0$ is considered. In [95], the exact solution of an Ising model with competing restricted interactions with zero external field is presented. In [94], [176], and [181] the case $J = \alpha = 0$ is considered. In [94], the exact solution for the problem of phase transitions is found. In [181] it is proven that there are two translation-invariant and uncountable number of distinct non-translation-invariant extreme Gibbs measures. In [96] the phase transition problems for $\alpha = 0$, $J \cdot J_1 \cdot J_3 \neq 0$ and for $J_3 = 0$, $\alpha \cdot J \cdot J_1 \neq 0$ are solved.

In this section we shall consider the case $J \cdot J_1 \cdot J_2 \cdot \alpha \neq 0$.

Gibbs measures for the model. Let Λ be a finite subset of V. Assume $\Omega(\Lambda)$ is the set of all configurations on Λ, that is the functions $\{\sigma(x), x \in$

$\Lambda\}$. Let $\overline{\sigma}(V \setminus \Lambda)$ be a fixed boundary configuration. The total energy of configuration $\sigma(\Lambda) \in \Omega(\Lambda)$ under condition $\overline{\sigma}(V \setminus \Lambda)$ is defined as

$$H(\sigma(\Lambda)|\overline{\sigma}(V \setminus \Lambda)) = -J_3 \sum_{\substack{< x,\, y,\, z > \\ x,\, y,\, z \in \Lambda}} \sigma(x)\sigma(y)\sigma(z) - J \sum_{\substack{> x,\, y < \\ x,\, y \in \Lambda}} \sigma(x)\sigma(y) -$$

$$J_1 \sum_{\substack{< x,\, y > \\ x,\, y \in \Lambda}} \sigma(x)\sigma(y) - \alpha \sum_{x \in \Lambda} \sigma(x) - J_3 \sum_{\substack{< x,\, y,\, z > \\ x \in \Lambda,\, y \notin \Lambda,\, z \notin \Lambda \text{ or} \\ x \in \Lambda,\, y \in \Lambda,\, z \notin \Lambda}} \sigma(x)\sigma(y)\sigma(z) -$$

$$J \sum_{\substack{> x,\, y < \\ x \in \Lambda,\, y \notin \Lambda}} \sigma(x)\overline{\sigma}(y) - J_1 \sum_{\substack{< x,\, y > \\ x \in \Lambda,\, y \notin \Lambda}} \sigma(x)\overline{\sigma}(y)\,.$$

When all boundary points $\{\overline{\sigma}(y), y \in V \setminus \Lambda\}$ are fixed as $+1$, we have the positive boundary condition and when they are fixed as -1, we have negative boundary condition. The free boundary condition corresponds to the case when the last three sums in the above are absent, that is formally all boundary points are fixed as 0.

The partition function $Z_\Lambda(\overline{\sigma}(V \setminus \Lambda))$ in volume Λ under boundary condition $\overline{\sigma}(V \setminus \Lambda)$ is defined as

$$Z_\Lambda = \sum_{\sigma(\Lambda) \in \Omega(\Lambda)} \exp(-\beta H(\sigma(\Lambda))|\overline{\sigma}(V \setminus \Lambda)),$$

where $\beta = \frac{1}{kT}$ is the inverse temperature. Then the conditional Gibbs measure μ_Λ in volume Λ under boundary condition $\overline{\sigma}(V \setminus \Lambda)$ is defined as

$$\mu_\Lambda(\sigma(\Lambda)) = \frac{\exp(-\beta H(\sigma(\Lambda))|\overline{\sigma}(V \setminus \Lambda))}{Z_\Lambda}.$$

3.2.2   *The functional equation*

Let h_x be a real-valued function of $x \in V$. Given $n = 1, 2, ...,$ consider the probability measure $\mu^{(n)}$ on $\{-1, +1\}^{V_n}$ which defined by

$$\mu^{(n)}(\sigma_n) = Z_n^{-1} \exp\left\{ -\beta H(\sigma_n) + \sum_{x \in W_n} h_x \sigma(x) \right\}.$$

Here, as before, $\beta = \frac{1}{T}$ and $\sigma_n : x \in V_n \to \sigma_n(x)$ and Z_n is the corresponding partition function

$$Z_n = \sum_{\tilde{\sigma}_n \in \Omega(V_n)} \exp\left\{ -\beta H(\tilde{\sigma}_n) + \sum_{x \in W_n} h_x \tilde{\sigma}(x) \right\}.$$

The consistency condition for $\mu^{(n)}(\sigma_n)$, $n \geq 1$ is

$$\sum_{\sigma^{(n)}} \mu^{(n)}(\sigma_{n-1}, \sigma^{(n)}) = \mu^{(n-1)}(\sigma_{n-1}), \tag{3.13}$$

where $\sigma^{(n)} = \{\sigma(x), x \in W_n\}$.

Let $V_1 \subset V_2 \subset ..., \cup_{n=1}^{\infty} V_n = V$ and $\mu_1, \mu_2, ...$ be a sequence of the probability measures on $\Phi^{V_1}, \Phi^{V_2}, ...$ satisfying the consistency condition, where $\Phi = \{-1, +1\}$. Then, according to the Kolmogorov theorem, (see, e.g. [243]), there is a unique limit Gibbs measure μ_h on Ω such that for every $n = 1, 2, ...$ and $\sigma_n \in \Phi^{V_n}$ the following equality holds

$$\mu(\{\sigma|_{V_n} = \sigma_n\}) = \mu^{(n)}(\sigma_n).$$

The following statement describes the conditions on h_x which guarantee the consistency condition of measures $\mu^{(n)}(\sigma_n)$.

Proposition 3.1. *The measure $\mu^{(n)}(\sigma_n)$, $n = 1, 2, ...$ satisfies the consistency condition (3.13) if and only if for any $x \in V$ the following equation holds:*

$$h_x = \frac{1}{2} \log \left(\theta_4 \frac{\theta_1^2 \theta_2 \theta_3 e^{2(h_y + h_z)} + \theta_1(e^{2h_y} + e^{2h_z}) + \theta_2 \theta_3}{\theta_1^2 \theta_2 + \theta_1 \theta_3(e^{2h_y} + e^{2h_z}) + \theta_2 e^{2(h_y + h_z)}} \right), \tag{3.14}$$

here $S(x) = \{y, z\}$, $< y, x, z >$ is a ternary neighbor and $\theta_1 = e^{2\beta J_1}, \theta_2 = e^{2\beta J}, \theta_3 = e^{2\beta J_3}, \theta_4 = e^{2\beta \alpha}$.

Proof. *Necessity.* According to the consistency condition (3.13) we have

$$\frac{Z_{n-1}}{Z_n} \sum_{\sigma^{(n)}} exp\Big\{ -\beta H_{n-1}(\sigma_{n-1}) + \beta J_1 \sum_{\substack{x \in W_{n-1}, \\ y, z \in S(x)}} \sigma(x)(\sigma(y) + \sigma(z))$$

$$+ \beta J \sum_{\substack{x \in W_{n-1}, \\ y, z \in S(x)}} \sigma(y)\sigma(z) + \beta J_3 \sum_{\substack{x \in W_{n-1}, \\ y, z \in S(x)}} \sigma(x)\sigma(y)\sigma(z)$$

$$+ \sum_{\substack{x \in W_{n-1} \\ y \in S(x)}} \beta \alpha \sigma(y) + \sum_{x \in W_n} h_x \sigma(x) \Big\}$$

$$= exp\Big\{ -\beta H_{n-1}(\sigma_{n-1}) + \sum_{x \in W_{n-1}} h_x \sigma(x) \Big\}.$$

Consequently we have

$$\frac{Z_{n-1}}{Z_n} \sum_{\sigma^{(n)}} \prod_{x \in W_{n-1}} exp\{\beta J_1 \sigma(x)(\sigma(y) + \sigma(z)) + \beta J \sigma(y)\sigma(z)$$

$$+ \beta J_3 \sigma(x)\sigma(y)\sigma(z) + \beta \alpha(\sigma(y) + \sigma(z)) + h_y \sigma(y) + h_z \sigma(z)\}$$

$$= \prod_{x \in W_{n-1}} exp\{h_x \sigma(x)\}.$$

Assume $x \in W_{n-1}$ and $S(x) = \{y, z\}$, $\sigma_x^{(n)} = \{\sigma(y), \sigma(z)\}$. Since $\sigma^{(n)} = \cup_{x \in W_{n-1}} \sigma_x^{(n)}$, we get

$$\frac{Z_{n-1}}{Z_n} \prod_{x \in W_{n-1}} \sum_{\sigma_x^{(n)}} \exp\{\beta J_1 \sigma(x)(\sigma(y) + \sigma(z)) + \beta J \sigma(y)\sigma(z)$$

$$+\beta J_3 \sigma(x)\sigma(y)\sigma(z) + \beta\alpha(\sigma(y)+\sigma(z)) + h_y\sigma(y) + h_z\sigma(z)\} = \prod_{x \in W_n} \exp\{h_x\sigma(x)\}. \tag{3.15}$$

Now fix $x \in W_{n-1}$ and rewrite (3.15) for the cases $\sigma(x) = 1$ and $\sigma(x) = -1$. If $\sigma(x) = 1$, we have

$$N = \sum_{\sigma_x^{(n)}=\{\sigma(y),\sigma(z)\}} \exp\{\beta J_1(\sigma(y) + \sigma(z)) + \beta J \sigma(y)\sigma(z)$$

$$+\beta J_3\sigma(y)\sigma(z) + \beta\alpha(\sigma(y) + \sigma(z)) + h_y\sigma(y) + h_z\sigma(z)\}$$

$$= \exp\{h_x\};$$

and if $\sigma(x) = -1$, then

$$D = \sum_{\sigma_x^{(n)}=\{\sigma(y),\sigma(z)\}} \exp\{-\beta J_1(\sigma(y) + \sigma(z)) + \beta J\sigma(y)\sigma(z)\} -$$

$$\beta J_3\sigma(y)\sigma(z) - \beta\alpha(\sigma(y) + \sigma(z)) + h_y\sigma(y) + h_z\sigma(z)\}$$

$$= \exp\{-h_x\}.$$

So that

$$\frac{N}{D} = \exp\{2h_x\}. \tag{3.16}$$

The numerator N of the left-hand side is equal to

$$N = \exp(2\beta J_1 + \beta J + \beta J_3 + 2\beta\alpha + h_y + h_z) + \exp(-\beta J - \beta J_3 - h_y + h_z)$$

$$+ \exp(-\beta J - \beta J_3 + h_y - h_z) + \exp(-2\beta J_1 + \beta J + \beta J_3 - 2\beta\alpha - h_y - h_z)$$

while D is equal to

$$D = \exp(-2\beta J_1 + \beta J - \beta J_3 + 2\beta\alpha + h_y + h_z) + \exp(-\beta J + \beta J_3 - h_y + h_z)$$

$$+ \exp(-\beta J + \beta J_3 + h_y - h_z) + \exp(2\beta J_1 + \beta J - \beta J_3 - 2\beta\alpha - h_y - h_z).$$

For simplicity denote $h_x = \alpha + h_x$.

Then the equality $\frac{N}{D} = \exp\{2h_x\}$ implies (3.14).

Sufficiency. Assume that (3.14) is valid. From (3.16) we get

$$\sum_{\sigma_x^{(n)}=\{\sigma(y),\sigma(z)\}} \exp\{\beta J_1 \sigma(x)(\sigma(y) + \sigma(z)) + \beta J\sigma(y)\sigma(z)$$

$$+\beta J_3\sigma(x)\sigma(y)\sigma(z) + \beta\alpha(\sigma(y) + \sigma(z)) + h_y\sigma(y) + h_z\sigma(z)\} = a(x)\exp\{\sigma(x)h_x\},$$

where $\sigma(x) = \pm 1$. This equality implies

$$\prod_{x \in W_{n-1}} \sum_{\sigma_x^{(n)} = \{\sigma(y), \sigma(z)\}} \exp\{\beta J_1 \sigma(x)(\sigma(y) + \sigma(z)) +$$

$$\beta J \sigma(y)\sigma(z) + \beta J_3 \sigma(x)\sigma(y)\sigma(z) + \beta \alpha(\sigma(y) + \sigma(z)) + h_y\sigma(y) + h_z\sigma(z)\}$$

$$= \prod_{x \in W_{n-1}} a(x) \exp\{\sigma(x) h_x\}. \tag{3.17}$$

Denoting $A_n(x) = \prod_{x \in W_{n-1}} a(x)$, we have from (3.17)

$$Z_{n-1} A_n \mu^{(n-1)}(\sigma_{n-1}) = Z_n \sum_{\sigma^{(n)}} \mu^{(n)}(\sigma_{n-1}, \sigma^{(n)}).$$

As $\mu^{(n)}$, $n \geq 1$ is a probability, we have

$$\sum_{\sigma_{n-1}} \sum_{\sigma^{(n)}} \mu^{(n)}(\sigma_{n-1}, \sigma^{(n)}) = \sum_{\sigma_{n-1}} \mu^{(n-1)}(\sigma_{n-1}) = 1.$$

From these equalities we get $Z_{n-1} A_{n-1} = Z_n$, which means that (3.13) holds. $\qquad\square$

According to Proposition 3.1 the problem of describing the Gibbs measures is reduced to the description of the solutions of the functional equation (3.14).

3.2.3 *Translation-invariant Gibbs measures: phase transition*

The analysis of the solution of (3.14) is rather tricky. It is natural to begin with the translation-invariant solutions where $h_x = h$ is constant for all $x \in V$.

In this case from (3.14), we have

$$u = \theta_4 \frac{\theta_1^2 \theta_2 \theta_3 u^2 + 2\theta_1 u + \theta_2 \theta_3}{\theta_1^2 \theta_2 + 2\theta_1 \theta_3 u + \theta_2 u^2}, \tag{3.18}$$

where $u = e^{2h}$.

Note that if there is more than one positive solution for equation (3.18), then there is more than one translation-invariant Gibbs measure corresponding to these solutions.

Proposition 3.2. *If* $\theta_1^2 > 3$, $\theta_2 > \frac{2\theta_1}{\theta_1^2 - 3}$, *and*

$$\frac{\sqrt{\theta_2^2(\theta_1^4 + 2\theta_1^2 - 3) - 4\theta_1^2 - 8\theta_1\theta_2} - \sqrt{\theta_2^2(\theta_1^4 + 2\theta_1^2 - 3) - 4\theta_1^2 - 8\theta_1\theta_2 - 4\theta_1^2\theta_2^2}}{2\theta_1\theta_2}$$

$$< \theta_3 <$$

$$\frac{\sqrt{\theta_2^2(\theta_1^4 + 2\theta_1^2 - 3) - 4\theta_1^2 - 8\theta_1\theta_2} + \sqrt{\theta_2^2(\theta_1^4 + 2\theta_1^2 - 3) - 4\theta_1^2 - 8\theta_1\theta_2 - 4\theta_1^2\theta_2^2}}{2\theta_1\theta_2},$$

$$\eta_1(\theta_1, \theta_2, \theta_3) < \theta_4^2 < \eta_2(\theta_1, \theta_2, \theta_3)$$

then equation (3.18) has three positive roots $u_1^* < u_2^* < u_3^*$. *Here*

$$\eta_i(\theta_1, \theta_2, \theta_3) = \frac{1}{u_i} \frac{\theta_1^2\theta_2\theta_3 u_i^2 + 2\theta_1 u_i + \theta_2\theta_3}{\theta_1^2\theta_2 + 2\theta_1\theta_3 u_i + \theta_2 u_i^2}$$

where $u_i, i = 1, 2$ *are the solutions of*

$$\theta_1^2\theta_2^2\theta_3 u^4 + 4\theta_1\theta_2 u^3 + \theta_3(3\theta_2^2 - \theta_1^4\theta_2^2 + 4\theta_1^2)u^2 + 4\theta_1\theta_2\theta_3^2 u + \theta_1^2\theta_2^2\theta_3 = 0. \quad (3.19)$$

Proof.   Denote

$$f(u) = \frac{\theta_1^2\theta_2\theta_3 u^2 + 2\theta_1 u + \theta_2\theta_3}{\theta_1^2\theta_2 + 2\theta_1\theta_3 u + \theta_2 u^2}.$$

We have

$$f'(u) = 2\theta_2 \frac{\theta_1(\theta_1^2\theta_3^2 - 1)u^2 + \theta_2\theta_3(\theta_1^4 - 1)u + \theta_1(\theta_1^2 - \theta_3^2)}{(\theta_1^2\theta_2 + 2\theta_1\theta_3 u + \theta_2 u^2)^2},$$

$$f''(u) = 2\theta_2(\theta_2 u^2 + 2\theta_1\theta_3 u + \theta_1^2\theta_2)^{-3} \times$$

$$(-2\theta_1\theta_2(\theta_1^2\theta_3^2 - 1)u^3 - 3\theta_2^2\theta_3(\theta_1^4 - 1)u^2 + 6\theta_1\theta_2(\theta_3^2 - \theta_1^2)u + \theta_1^2\theta_3(\theta_2^2(\theta_1^4 - 1) - 4\theta_1^2 + 4\theta_3^2).$$

Denote

$$A = 2\theta_1\theta_2(\theta_1^2\theta_3^2 - 1); \quad B = 3\theta_2^2\theta_3(\theta_1^4 - 1);$$

$$C = 6\theta_1\theta_2(\theta_3^2 - \theta_1^2); \quad D = \theta_1^2\theta_3(\theta_2^2(\theta_1^4 - 1) - 4\theta_1^2 + 4\theta_3^2).$$

It is easy to see that under conditions of the proposition we have $A > 0, B > 0, C > 0, D > 0$. Equation $f''(u) = 0$ is equivalent to $Au^3 + Bu^2 - Cu - D = 0$, one can easily prove that the last equation has unique positive solution, say u_*. Thus f is convex for $u < u_*$ and concave for $u > u_*$. Consequently there are at most three solutions. On the other hand, it is easy to see that (3.18) has more than one solution if and only if there is more than one solution of the equation $uf'(u) = f(u)$ which is equivalent to equation (3.19).

Now consider (3.19), which can be rewritten as

$$\theta_1^2\theta_2^2\left(u + \frac{1}{u}\right)^2 + 4\theta_1\theta_2\left(\frac{u}{\theta_3} + \frac{\theta_3}{u}\right) + 3\theta_2^2 - \theta_1^4\theta_2^2 + 4\theta_1^2 - 2\theta_1^2\theta_2^2 = 0.$$

Denote

$$\varphi(u) = 4\theta_1\theta_2\left(\frac{u}{\theta_3} + \frac{\theta_3}{u}\right),$$

$$\psi(u) = \theta_2^2(\theta_1^4 + 2\theta_1^2 - 3) - \theta_1^2\theta_2^2\left(u + \frac{1}{u}\right)^2 - 4\theta_1^2.$$

A simple analysis of these functions show that under conditions of the proposition, equation (3.18) has three positive solutions. $\square$

Thus by Propositions 3.1 and 3.2 we can formulate the following

Theorem 3.4. *Assume the conditions of Proposition 3.2 are satisfied then for the model (3.12) there are three translation-invariant Gibbs measures μ_1, μ_2, μ_3, i.e., there is phase transition.*

Note that μ_1 (μ_3) corresponds to positive (resp. negative) boundary condition. The boundary condition corresponding to μ_2 is unclear.

The following Proposition 3.3 describes a useful property of general (non-translation-invariant) solutions h_x to (3.14).

Proposition 3.3. *Assume the conditions of Proposition 3.2 are satisfied and h_x is a solution to (3.14), with $u_x = e^{2h_x}$, then*

$$u_1^* \leq u_x \leq u_3^*, \quad x \in V \tag{3.20}$$

where $u_1^* < u_3^*$ are solutions of (3.18).

Proof. It is clear that $u_x > 0$, for any $x \in V$. For $u, v > 0$ denote

$$F(u, v) = \theta_4 \frac{\theta_1^2 \theta_2 \theta_3 uv + \theta_1(u + v) + \theta_2 \theta_3}{\theta_1^2 \theta_2 + \theta_1 \theta_3(u + v) + \theta_2 uv}.$$

Equation (3.14) can be rewritten as $u_x = F(u_y, u_z)$.

Observe that under conditions of Proposition 3.2 the function $F(u, v)$ is increasing with respect to u and v on $(0, \infty)$. Hence we conclude that

$$\frac{\theta_3 \theta_4}{\theta_1^2} < F(u, v) < \theta_1^2 \theta_3 \theta_4,$$

for all $u, v > 0$. Now we consider the function with $u, v \in (\frac{\theta_3 \theta_4}{\theta_1^2}, \theta_1^2 \theta_2 \theta_3)$. By similar reason as above we get

$$f\left(\frac{\theta_3 \theta_4}{\theta_1^2}\right) < F(u, v) < f(\theta_1^2 \theta_3 \theta_4),$$

where $f(u) = F(u, u)$. Repeating this argument one gets

$$f^{(n)}\left(\frac{\theta_3 \theta_4}{\theta_1^2}\right) < F(u, v) < f^{(n)}(\theta_1^2 \theta_3 \theta_4),$$

for all $n \geq 1$. Here $f^{(n)}$ is n-th iteration of the map $x \to f(x)$. The sequence $f^{(n)}(\theta_1^2 \theta_3 \theta_4)$ is decreasing and bounded from below by u_3^*. Its limit is a fixed point of f and thus equal to u_3^*. This proves that $u_x \leq u_3^*$. The lower bound for u_x is similar and gives u_1^*. $\qquad\square$

Using Proposition 3.13 by similar argument as in the proof of Theorem 2.2 one can prove the following

Theorem 3.5. *Assume conditions of Proposition 3.2 are satisfied then translation-invariant measures μ_1, μ_3 (see Theorem 3.4) are extreme.*

3.2.4   *Periodic Gibbs measures*

In this section we study periodic (see Definitions 2.1 and 2.2) solutions of (3.14).

For convenience of reader we shall recall some notions from Chapter 1. Let K be a subgroup of index r in G_k, and let $G_k/K = \{K_0, K_1, ..., K_{r-1}\}$ be the quotient group, with the coset $K_0 = K$. Let $q_i(x) = |S_1(x) \cap K_i|$, $i = 0, 1, ..., r - 1$; $N(x) = |\{j : q_j(x) \neq 0\}|$, where $S_1(x) = \{y \in G_k : \langle x, y \rangle\}$, $x \in G_k$ and $|\cdot|$ is the number of elements in the set. Denote $Q(x) = (q_0(x), q_1(x), ..., q_{r-1}(x))$.

For every $x \in G_k$ there is a permutation π_x of the coordinates of the vector $Q(e)$ (where e is the identity of G_k) such that

$$\pi_x Q(e) = Q(x). \tag{3.21}$$

It follows from (3.21) that $N(x) = N(e)$ for all $x \in G_k$.

Each K-periodic collection is given by

$$\{h_x = h_i \ \text{ for } \ x \in K_i, \ i = 0, 1, ..., r - 1\}.$$

For $k = 2$ by Proposition 3.1 and (3.21), h_n, $n = 0, 1, ..., r - 1$, satisfies

$$h_n = \frac{1}{2} \log F(e^{2h_{\pi_n(i)}}, e^{2h_{\pi_n(j)}}), \tag{3.22}$$

where $F(u, v)$ is defined in the proof of Proposition 3.3 and π_n is permutation of $Q(e)$ for $x \in K_n$, $i, j \in Q(e)$.

Proposition 3.4. *Suppose the conditions of Proposition 3.2 are satisfied then $F(u, v) = F(h, v)$ if and only if $u = h$ $(F(u, v) = F(u, h)$ if and only if $v = h)$.*

Proof.   Follows from monotonicity of F with respect to u (resp. v). $\square$

Let $G_2^{(2)}$ be the subgroup in G_2 consisting of all words of even length. Clearly, $G_2^{(2)}$ is a subgroup of index 2.

Theorem 3.6. *Let K be a normal subgroup of finite index in G_2. Then each K-periodic Gibbs measure for model (3.12) is either translation-invariant or $G_2^{(2)}$-periodic.*

Proof.   We see from (3.22) that

$$F(e^{h_{\pi_n(i)}}, e^{h_{\pi_n(j)}}) = F(e^{h_{\pi_n(i')}}, e^{h_{\pi_n(j')}}). \tag{3.23}$$

For any $i, j, i', j' \in Q(e), n = 0, 1, ..., r - 1$. Hence from Proposition 3.4 we have

$$h_{\pi_n(i_1)} = h_{\pi_n(i_2)} = ... = h_{\pi_n(i_{N(e)})}.$$

Therefore,

$$h_x = h_y = h, \quad \text{if } x, y \in S_1(z), \quad z \in G_2^{(2)};$$

$$h_x = h_y = l, \quad \text{if } x, y \in S_1(z), \quad z \in G_2 \setminus G_2^{(2)}.$$

Thus the measures are translation-invariant (if $h = l$) or $G_2^{(2)}$-periodic (if $h \neq l$). This completes the proof. $\square$

Let K be a normal subgroup of finite index in G_2. What condition on K will guarantee that each K-periodic Gibbs measure is translation-invariant? We put $I(K) = K \cap \{a_1, a_2, a_3\}$, where $a_i, \quad i = 1, 2, 3$ are generators of G_2.

Theorem 3.7. *If $I(K) \neq \emptyset$, then each K-periodic Gibbs measure for model (3.12) is translation-invariant.*

Proof. Take $x \in K$. We note that the inclusion $xa_i \in K$ holds if and only if $a_i \in K$. Since $I(K) \neq \emptyset$, there is an element $a_i \in K$. Therefore K contains the subset $Ka_i = \{xa_i : x \in K\}$. By Theorem 3.6 we have $h_x = h$ and $h_{xa_i} = l$. Since x and xa_i belong to K, it follows that $h_x = h_{xa_i} = h = l$. Thus each K-periodic Gibbs measure is translation-invariant. $\square$

Theorems 3.6 and 3.7 reduce the problem of describing K-periodic Gibbs measure with $I(K) \neq \emptyset$ to describing the fixed points of $f(u) = F(u, u)$ (see (3.18)) which describes translation-invariant Gibbs measures. If $I(K) = \emptyset$, this problem is reduced to describing the solutions of the system:

$$\begin{cases} u = f(v), \\ v = f(u) \end{cases} \tag{3.24}$$

with

$$f(u) = \theta_4 \frac{\theta_1^2 \theta_2 \theta_3 u^2 + 2\theta_1 u + \theta_2 \theta_3}{\theta_1^2 \theta_2 + 2\theta_1 \theta_3 u + \theta_2 u^2}.$$

Evidently the positive roots of the equation

$$\frac{f(f(u)) - u}{f(u) - u} = 0 \tag{3.25}$$

describe the periodic (non-translation-invariant) Gibbs measures.

Since we are looking for positive roots, equation (3.25) has the following form:

$$\theta_1^2 \theta_2 (\theta_1^2 \theta_2 \theta_3^2 \theta_4^2 + 2\theta_1 \theta_3^2 \theta_4 + \theta_2) u^2 + \theta_3 (\theta_1^4 \theta_2^2 \theta_4 + 2\theta_1^3 \theta_2 \theta_4^2 + 2\theta_1^2 \theta_2 + 4\theta_1^2 \theta_4 - \theta_2^2 \theta_4) u$$

$$+\theta_1^2\theta_2(\theta_2\theta_3^2\theta_4^2 + 2\theta_1\theta_4 + \theta_1^2\theta_2) = 0. \tag{3.26}$$

The discriminant Δ of (3.26) is equal to

$$\Delta = -4\theta_1^5\theta_2^3\theta_4^3(\theta_1\theta_2\theta_4 + 2)\theta_3^4 + A\theta_3^2 - 4\theta_1^5\theta_2^3(\theta_1\theta_2 + 2\theta_4),$$

where

$$A = -\theta_4^2(3\theta_1^8 + 6\theta_1^4 - 1)\theta_2^4 - 4\theta_1^3\theta_4(1 + \theta_4^2)(1 + \theta_1^4)\theta_2^3 +$$

$$4\theta_1^2(\theta_1^4\theta_4^4 + \theta_1^4 - 2\theta_4^2)\theta_2^2 + 16\theta_1^5\theta_4(1 + \theta_4^2)\theta_2 + 16\theta_1^4\theta_4^2.$$

Using simple analysis one can see that (3.26) has two positive solutions if

$$\theta_1 < 1, \quad \theta_2 > \frac{2\theta_1}{1 - \theta_1^2}, \quad \frac{1}{\theta_4^*} < \theta_4 < \theta_4^*, \tag{3.27}$$

where

$$\theta_4^* = \frac{\theta_2^2 - 4\theta_1^2 - \theta_1^4\theta_2^2 + \sqrt{(4\theta_1^2 + \theta_1^4\theta_2^2 - \theta_2^2)^2 - 16\theta_1^6\theta_2^2}}{4\theta_1^3\theta_2}$$

and

$$A^2 > 64\theta_1^{10}\theta_2^6\theta_4(\theta_1\theta_2\theta_4 + 2)(\theta_1\theta_2 + 2\theta_4), \tag{3.28}$$

$$\theta_3^- < \theta_3^2 < \theta_3^+, \tag{3.29}$$

where $\theta_3^{\mp}$ are solutions of $\Delta = 0$.

Therefore, the following theorem is proved:

Theorem 3.8. *Assume $(\theta_1, \theta_2, \theta_3, \theta_4)$ satisfies conditions (3.27)-(3.29) then for the model (3.12) there are two $G_2^{(2)}$-periodic Gibbs measures $\mu_1^{\mathrm{per}}, \mu_2^{\mathrm{per}}$.*

Remark 3.2. 1.   By the construction measures $\mu_1^{\mathrm{per}}, \mu_2^{\mathrm{per}}$ are non-translation-invariant, but periodic with period 2 (= index of normal subgroup).

2. For $\theta_4 = 1$ the condition (3.28) can be rewritten as (see [96])

$$(1 - 3\theta_1^2)(\theta_1^2 + 1)\left(\theta_2^2 - \frac{2\theta_1}{1 - 3\theta_1^2}\right)\left(\theta_2^2 - \frac{2\theta_1}{1 + \theta_1^2}\right)^2\left(\theta_2^2 + \frac{2\theta_1}{1 + \theta_1^2}\right) > 0.$$

This factorization gives more simple formulation of the conditions (3.27)-(3.29), i.e., for $\theta_4 = 1$ conditions (3.27)-(3.29) can be reduced to

$$0 < \theta_1 < \frac{1}{\sqrt{3}}, \quad \theta_2 > \frac{2\theta_1}{1 - 3\theta_1^2}, \quad \theta_3^- < \theta_3 < \theta_3^+.$$

3.2.5 *Non-periodic Gibbs measures*

In this section we consider the case of phase transition (i.e., assume that the conditions of Proposition 3.2 are satisfied). We show that functional equation (3.14) admits uncountably many non-periodic solutions. This section generalizes Bleher-Ganikhodjaev construction (see Chapter 2) for model (3.12). Take an arbitrary infinite path $\pi = \{x^0 = x_0, x_1, ...\}$ on the Cayley tree of order 2. There is one-to-one correspondence between such paths and real numbers $t \in [0; 1]$. We will map the path π to a function $h^\pi : x \in V \to h_x^\pi$ satisfying (3.14). Path π splits Cayley tree Γ^2 into two parts Γ_1^2 and Γ_2^2.

Function h^π is defined by

$$h_x^\pi = \begin{cases} \log u_1^*, & \text{if } x \in \Gamma_1^2 \\ \log u_3^*, & \text{if } x \in \Gamma_2^2. \end{cases} \tag{3.30}$$

Denote

$$\Phi(x, y) = \frac{1}{2} \log \left(\theta_4 \frac{\theta_1^2 \theta_2 \theta_3 e^{2(x+y)} + \theta_1(e^{2x} + e^{2y}) + \theta_2 \theta_3}{\theta_1^2 \theta_2 + \theta_1 \theta_3 (e^{2x} + e^{2y}) + \theta_2 e^{2(x+y)}} \right).$$

Proposition 3.5. *The following inequality holds:*

$$|\Phi(x_1, y) - \Phi(x_2, y)| \leq \gamma(\theta_1, \theta_2, \theta_3)|x_1 - x_2|,$$

where

$$\gamma(\theta_1, \theta_2, \theta_3) =$$

$$\max_{t \in [u_1^*, u_3^*]} \frac{|\sqrt{(\theta_1 t + \theta_2 \theta_3)(\theta_2 t + \theta_1 \theta_3)} - \theta_1 \sqrt{(\theta_1 \theta_2 \theta_3 t + 1)(\theta_3 t + \theta_1 \theta_2)}|}{\sqrt{(\theta_1 t + \theta_2 \theta_3)(\theta_2 t + \theta_1 \theta_3)} + \theta_1 \sqrt{(\theta_1 \theta_2 \theta_3 t + 1)(\theta_3 t + \theta_1 \theta_2)}} < 1.$$

Proof. The function $\Phi(x, y)$ can be rewritten as

$$\Phi(x, y) = \frac{1}{2} \log \theta_4 + \frac{1}{2} \log \frac{Ae^{2x} + B}{Ce^{2x} + D},$$

where A, B, C, D depend on $\theta_1, \theta_2, \theta_3$ and y. It is easy to see that

$$|\Phi_x'(x, y)| \leq \frac{|\sqrt{AD} - \sqrt{BC}|}{\sqrt{AD} + \sqrt{BC}}.$$

This completes the proof. $\qquad\qquad\square$

With the help of Proposition 3.5 it is easy to prove the following theorem, similar to Theorem 2.11.

Theorem 3.9. *For any infinite path π, there exists a unique function h^π satisfying (3.14) and (3.30).*

As in the Bleher-Ganikhodjaev construction one can prove that functions $h^{\pi(t)}$ are different for different $t \in [0; 1]$.

Now let $\mu(t)$ denote the Gibbs measure corresponding to function $h^{\pi(t)}$, $t \in [0; 1]$.

Using Theorem 3.5, similar to Theorem 2.12 we obtain the following

Theorem 3.10. *For any* $t \in [0; 1]$, *there exists a unique extreme Gibbs measure* $\mu(t)$. *Moreover, the above Gibbs measures* μ_i, $i = 1, 3$, *are specified as* $\mu(0) = \mu_3$, $\mu(1) = \mu_1$.

Because measures $\mu(t)$ are different for different $t \in [0; 1]$ we obtain a continuum of distinct extreme Gibbs measures.

Commentaries and references. In [257] the Vannimenus model was introduced.

It was then generalized in many directions:

In [118] a model with the competing NN and NNN interactions Ising model on a Cayley tree was considered. But in their case it is allowed for all interbranch NNN interactions on the coordination number three which was discussed earlier in [123] and it was obtained in addition to the expected paramagnetic, ferromagnetic and antiferromagnetic phases, an intermediate range of $J_2/J_1 < 0$ values where the local magnetization has chaotic oscillatory glass-like behavior.

Another generalization is due to Mariz et al. [152] these authors studied the phase diagram for the Ising model on a Cayley tree with competing NN interactions J_1 and NNN interactions J_2 and J_3 in the presence of an external magnetic field. At vanishing temperature, the phase diagram is fully determined, for all values and signs of J_2/J_1 and J_3/J_2; in particular, it was verified that values of J_3/J_2 high enough favor the paramagnetic phase. At finite temperatures, several interesting features for typical values of J_2/J_1 and J_3/J_2 (evolution of reentrances, separation of the modulated region into two disconnected pieces, etc.) are exhibited.

In [191] the next generalization is considered, where the lattice spin model with Q-component discrete spin variables restricted to having orientations orthogonal to the faces of Q-dimensional hypercube on the Cayley tree is considered. The partition function of the model with dipole-dipole and quadrupole-quadrupole interaction in terms of double graph expansions is presented. By analyzing the regions of stability of different types of fixed points of the system of recurrent relations (which is generalization of the Vannimenus's equations), the phase diagrams of the model are plotted.

For $Q \leq 2$ the phase diagram of the model is found to have three tricritical points.

In [99], [101] and [104] the next generalizations are considered. These authors have studied the phase diagram for Potts model on a Cayley tree with competing NN interactions J_1, prolonged NNN interactions J_p and one level NNN interactions J_o. In [104] the Potts model with $J_o \neq 0$ is considered. It is shown that for some values of J_o the multicritical Lifshitz point be at non-zero temperature and proven that as soon as the same-level interaction J_o is non-zero, the paramagnetic phase found at high temperatures for $J_o = 0$ disappears, while Ising model does not obtain such property.

But most results of the above mentioned works are obtained numerically.

Second section of this chapter is based on paper [102]. In [175], [176] and [95] the various partial cases of this model have been investigated. In [95], the exact solution of an Ising model with competing restricted interactions with zero external field is presented. In [94], [176], and [181] the case $J = \alpha = 0$ is considered. In [94], the exact solution for the problem of phase transitions is found. In [181] it is proven that there are two translation-invariant and uncountable number of distinct non-translation-invariant extreme Gibbs measures. In [96] the phase transition problems for $\alpha = 0$, $J \cdot J_1 \cdot J_3 \neq 0$ and $J_3 = 0$, $\alpha \cdot J \cdot J_1 \neq 0$ are solved.

For some other results related to models with competing interactions see [88], [105], [106], [185], [219].

Since the models considered in this chapter are more complicated than the usual Ising model, many results known for Ising model are not obtained for models with competing interactions. For example, models of this chapter are not studied in case when the order k of the Cayley tree is ≥ 3.

Chapter 4

Information flow on trees

In this chapter we consider a process on a tree T in which information is transmitted from the root of the tree to all the nodes of the tree. Each node inherits information from its parent with some probability of error. The transmission process is assumed to have identical distribution on all the edges, and different edges of the tree are assumed to act independently. The basic question of this chapter is: Does the configuration obtained at level n of T typically contain significant information on the root variable? This problem arose independently in biology, information theory and statistical physics. For models of statistical physics on trees, the problem is related to extremality of the disordered Gibbs measure. In this chapter, we give results and challenges related to this problem. In the following chapters we shall apply the results to extremality conditions of Gibbs measures. The theory of finite Markov chains implies that if the underlying Markov chain is ergodic (i.e., irreducible and aperiodic), then the variable at a single node at level n and the variable at the root are asymptotically independent as $n \to \infty$. However, for the tree process, information is duplicated, so it is conceivable that the configuration at level n contains a significant amount of information on the root variable.

4.1 Definitions and their equivalency

This chapter is based on papers [169], [170], [171]. We are interested on results of this chapter, because their applicability to find regions of extremality of disordered phases (Gibbs measures) for models of statistical mechanics.

Let $T = (V; E)$ be a tree. The information flow on each edge is given by a channel on a finite alphabet $\mathcal{A} = \{1, \ldots, k\}$. Let $\mathbf{M}_{i,j}$ be the transition probability from i to j; M be the random function (or channel) which satisfies for all i and j that $\mathbf{P}[M(i) = j] = \mathbf{M}_{i,j}$, and $\lambda_2(M)$ be the eigenvalue of $\mathbf{M}$ which has the second largest absolute value. We assume that M defines an ergodic Markov chain (irreducible and aperiodic). At the root ρ one of the symbols of A is chosen according to some initial distribution. We denote this (random) symbol by σ_ρ. This symbol is then propagated in the tree in the following way. Given that the parent of v, denoted v', has value $\sigma_{v'}$, the probability that σ_v is j is given by $\mathbf{M}_{\sigma_{v'},j}$. Precisely, for each vertex v having v' as a parent, we let $\sigma_v = M_{v',v}(\sigma_{v'})$, where $\{M_{v',v}\}$ are independent copies of M. Equivalently, for a vertex v, let v' be the parent of v, and let $\Gamma(v)$ be the set of all vertices which are connected to ρ through paths which do not contain v. Then the process satisfies:

$$\mathbf{P}[\sigma_v = j \,|\, (\sigma_w)_{w \in \Gamma(v)}] = \mathbf{P}[\sigma_v = j \,|\, \sigma_{v'}] = \mathbf{M}_{\sigma_{v'},j}.$$

Let $d(\cdot, \cdot)$ be the graph-metric distance on T, and $L_n = \{v \in V : d(\rho, v) = n\}$ be the nth level of the tree. For $v \in V$ and $e = (v, w) \in E$ we denote $|v| = d(\rho, v)$ and $|e| = \max\{|v|, |w|\}$. Denote by $\sigma_n = (\sigma(v))_{v \in L_n}$ the symbols (configuration) at the n-th level of the tree. We let $c_n = (c_1(1), \ldots, c_n(k))$, where

$$c_n(i) = |\{v \in L_n : \sigma(v) = i\}|.$$

In other words, c_n is the census of the n-th level. Note that both $(\sigma_n)_{n=1}^\infty$ and $(c_n)_{n=1}^\infty$ are Markov chains.

Definition 4.1. The reconstruction problem for T and M is solvable if there exist $i, j \in \mathcal{A}$ for which

$$\lim_{n \to \infty} |\mathbf{P}_n^i - \mathbf{P}_n^j| > 0, \tag{4.1}$$

where $\mathbf{P}_n^l$ denotes the conditional distribution of σ_n given that $\sigma_\rho = l$ and $|\cdot|$ denotes the total variation norm.

We define census solvability similarly, where the measures $\mathbf{P}_n^l$ are replaced by measures $\tilde{\mathbf{P}}_n^l$ which are conditional distributions of c_n given that $\sigma_\rho = l$.

4.1.1   *Equivalent definitions*

Note that if the reconstruction problem is solvable, then σ_n contains significant information on the root variable. This may be expressed in several equivalent ways. Assume that the variable at the root, σ_ρ, is chosen according to some initial distribution $(\pi_i)_{i \in \mathcal{A}}$, and let $\mathbf{P}^\pi$ denote the corresponding probability measure. The maximum-likelihood algorithm, which is the optimal reconstruction algorithm of σ_ρ given σ_n, is successful with probability

$$\Delta_n(\pi) = \sum_\sigma \mathbf{P}^\pi[\sigma_n = \sigma] \max_{i \in \mathcal{A}} \mathbf{P}^\pi[\sigma_\rho = i \,|\, \sigma_n = \sigma] \geq$$

$$\max_{i \in \mathcal{A}} \sum_\sigma \mathbf{P}^\pi[\sigma_n = \sigma] \mathbf{P}^\pi[\sigma_\rho = i \,|\, \sigma_n = \sigma] = \max_{i \in \mathcal{A}} \pi_i.$$

One can reconstruct σ_ρ with probability $\max_i \pi_i$ even when σ_n is unknown (using the algorithm which always reconstructs the i which maximizes π_i). It is natural to consider $\Delta_n(\pi) - \max_i \pi_i$ as a measurement for the dependency between σ_n and σ_ρ. Let H be the entropy function, and let $I(X, Y) = H(X) + H(Y) - H(X, Y)$ be the mutual-information operator (see e.g. [50] for definitions and basic properties). For a sequence of random variables X_n defined on the same probability space, let F_n be the σ-algebra defined by $(X_m)_{m \geq n}$, i.e., F_n is the minimal σ-algebra such that all the variables $(X_m)_{m \geq n}$ are measurable with respect to F_n. Let $F_\infty = \cap_{n=1}^\infty F_n$. We say that the sequence X_n has a trivial tail, if all the measurable sets with respect to F_∞ have probability either 0 or 1. Otherwise, we say the the sequence has a non-trivial tail.

In the theory of Markov random fields the notion of tail triviality is closely related to the extremality of the measure, see e.g. [107], [119].

The following proposition gives several equivalence to the solvability of reconstruction problem.

Proposition 4.1. *Let T be an infinite tree, M be a channel and π denotes initial distribution for σ_ρ. Then the following conditions are equivalent:*

(1) The reconstruction problem is solvable, i.e., there exist $i, j \in \mathcal{A}$ for which

$$\lim_{n \to \infty} |\mathbf{P}_n^i - \mathbf{P}_n^j| > 0. \tag{4.2}$$

(2) There exists a π for which

$$\lim_{n \to \infty} I(\sigma_\rho, \sigma_n) > 0. \tag{4.3}$$

(3) If π is the uniform distribution on $\mathcal{A}$, then

$$\lim_{n \to \infty} I(\sigma_\rho, \sigma_n) > 0. \tag{4.4}$$

(4) For any distribution π with $\min_i \pi_i > 0$, it holds that

$$\lim_{n \to \infty} I(\sigma_\rho, \sigma_n) > 0. \tag{4.5}$$

(5) There exists a π for which

$$\lim_{n \to \infty} \inf \Delta_n(\pi) > \max_i \pi_i. \tag{4.6}$$

(6) If π is the uniform distribution on $\mathcal{A}$, then

$$\lim_{n \to \infty} \inf \Delta_n(\pi) > \frac{1}{|\mathcal{A}|}. \tag{4.7}$$

The analogous 6 conditions are equivalent for c_n.

Proof. See [50] for some standard facts in information theory which we will use in the sequence. By the data processing lemma ([50], p.32), it follows that $I(\sigma_\rho, \sigma_n)$ is a decreasing sequence, so the limits (4.3), (4.4) and (4.5) exist. Similarly, using the coupling between $\mathbf{P}_i^n$ and $\mathbf{P}_j^n$, we see that the sequence in (4.2) is decreasing, so that the limit in (4.2) exists.

(4.2)≡(4.6)≡(4.7): Denote by $\mathbf{P}^\sigma$ the distribution $\mathbf{P}[\sigma_\rho = i \,|\, \sigma_n = \sigma]$, and by $\mathbf{P}_n$ the distribution $\sum_i \pi_i \mathbf{P}_n^i$ and the reconstruction probability given σ_n by $\Delta_n(\pi)$. Since given that $\sigma_n = \sigma$, the optimal algorithm will reconstruct a symbol j such that $\mathbf{P}^\sigma[j] = \max_i \mathbf{P}^\sigma[i]$, we have

$$\Delta_n(\pi) = \sum_\sigma \mathbf{P}_n[\sigma] \max_i \mathbf{P}^\sigma[i]. \tag{4.8}$$

By (4.8) we get

$$\Delta_n(\pi) - \max_i \pi_i = \sum_\sigma \mathbf{P}_n[\sigma] \left(\max_i \mathbf{P}^\sigma[i] - \max_i \pi_i \right) \leq \tag{4.9}$$

$$\sum_\sigma \mathbf{P}_n[\sigma] \left(\sum_i |\mathbf{P}^\sigma[i] - \pi_i| \right) =$$

$$\sum_\sigma \sum_i \pi_i \left| \mathbf{P}_n^i[\sigma] - \mathbf{P}_n[\sigma] \right| = \sum_i \pi_i \left| \mathbf{P}_n^i - \mathbf{P}_n \right| \leq$$

$$\sum_i \pi_i \max_{j,j'} \left| \mathbf{P}_n^j - \mathbf{P}_n^{j'} \right| = \max_{i,j} \left| \mathbf{P}_n^i - \mathbf{P}_n^j \right|,$$

where the inequality in (4.9) follows from the fact that $\mathbf{P}_n$ is an average of $\mathbf{P}_n^i$. Moreover, if π is the uniform distribution, then we have

$$\Delta_n(\pi) - \frac{1}{|\mathcal{A}|} = \sum_\sigma \mathbf{P}_n[\sigma] \left(\max_i \mathbf{P}^\sigma[i] - \frac{1}{|\mathcal{A}|} \right) = \tag{4.10}$$

$$\frac{1}{|\mathcal{A}|} \sum_\sigma \max_i \left(\mathbf{P}_n^i[\sigma] - \mathbf{P}_n[\sigma]\right) \geq$$

$$\frac{1}{|\mathcal{A}|^2} \sum_\sigma \max_{i,j} \left|\mathbf{P}_n^i[\sigma] - \mathbf{P}_n^j[\sigma]\right| \geq$$

$$\frac{1}{|\mathcal{A}|^2} \max_{i,j} \left|\mathbf{P}_n^i - \mathbf{P}_n^j\right|,$$

where the first inequality follows from the fact that, for a sequence $(a_i)_{i=1}^k$, we have

$$\max_i a_i - \frac{1}{k} \sum_i a_i \geq \frac{1}{k} \max_{i,j} |a_i - a_j|.$$

By (4.9) we have that (4.6) implies (4.2), and by (4.10) we have that (4.2) implies (4.7), which implies (4.6).

(4.7)$\equiv$(4.4): Recall the Kullback-Leibler distance (KL-distance) for $p = (p_1, \ldots, p_k)$ and $q = (q_1, \ldots, q_k)$ probability distributions:

$$D(p\|q) = \sum_{i=1}^k p_i \ln \left(\frac{p_i}{q_i}\right),$$

this is the average of the logarithmic difference between the probabilities p and q, where the average is taken using the probability p. The KL-distance is only defined if $q_i > 0$ for any i such that $p_i > 0$. If the quantity $0 \ln 0$ appears in the formula, it is interpreted as zero. Denote $G(x) = (2 \ln 2)^{-1} x^2$, $F(x) = -x \ln(x/k)$ for $0 \leq x \leq 1/2$ and $F(x) = \ln k$ otherwise. The following inequalities are known ([50]):

$$D(p\|q) \geq G(|p - q|), \tag{4.11}$$

$$|H(p) - H(q)| \leq F(|p - q|). \tag{4.12}$$

For an initial distribution π, using inequality (4.12) we get

$$I(\sigma_\rho, \sigma_n) = H(\pi) - \sum_\sigma \mathbf{P}_n[\sigma] H(\mathbf{P}^\sigma) \leq \tag{4.13}$$

$$\sum_\sigma \mathbf{P}_n[\sigma]|H(\pi) - H(\mathbf{P}^\sigma)| \leq \sum_\sigma \mathbf{P}_n[\sigma] F(|\pi - \mathbf{P}^\sigma|),$$

and by inequality (4.11) we obtain

$$I(\sigma_\rho, \sigma_n) = \sum_\sigma \mathbf{P}_n[\sigma] D(\mathbf{P}^\sigma \| \pi) \geq \sum_\sigma \mathbf{P}_n[\sigma] G(|\pi - \mathbf{P}^\sigma|). \tag{4.14}$$

For π the uniform distribution, we have

$$\Delta_n(\pi) - \frac{1}{|\mathcal{A}|} = \sum_\sigma \mathbf{P}_n[\sigma] \max_i (\mathbf{P}^\sigma[i] - \frac{1}{|\mathcal{A}|}) \geq \frac{1}{|\mathcal{A}|} \sum_\sigma \mathbf{P}_n[\sigma] |\mathbf{P}^\sigma - \pi| \quad (4.15)$$

and

$$\Delta_n(\pi) - \frac{1}{|\mathcal{A}|} = \sum_\sigma \mathbf{P}_n[\sigma] \max_i (\mathbf{P}^\sigma[i] - \frac{1}{|\mathcal{A}|}) \leq \sum_\sigma \mathbf{P}_n[\sigma] |\mathbf{P}^\sigma - \pi|. \quad (4.16)$$

By (4.13), (4.14), (4.15) and (4.16), it follows that

$$I(\sigma_\rho, \sigma_n) \to 0, \quad \text{iff} \quad \Delta_n(\pi) - \frac{1}{|\mathcal{A}|} \to 0.$$

(4.3)$\equiv$(4.4)$\equiv$(4.5): Note that if we write $p(x) = \mathbf{P}[\sigma = x]$ and $p(y|x) = \mathbf{P}[\tau = y|\sigma = x]$, then for fixed $p(y, x)$ the function $I(\sigma, \tau)$ is a concave function of $p(x)$ ([50], p.31). Suppose that $\lim_{n\to\infty} I(\sigma_\rho, \sigma_n) > 0$, where σ_ρ has density π. If σ'_ρ has the uniform distribution π', we can write

$$\pi' = \alpha\pi + (1 - \alpha)\pi((1 - \alpha)^{-1}(\pi' - \alpha\pi))$$

as a convex sum of distribution vectors, where $\alpha = (|\mathcal{A}| \max_i p_i)^{-1}$. Consequently, we obtain that

$$\lim_{n\to\infty} I(\sigma'_\rho, \sigma_n) \geq \alpha \lim_{n\to\infty} I(\sigma_\rho, \sigma_n) \geq |\mathcal{A}|^{-1} \lim_{n\to\infty} I(\sigma_\rho, \sigma_n) > 0. \quad (4.17)$$

Similarly, if σ'_ρ is a uniform variable on $\mathcal{A}$ and σ_ρ has distribution π, then we obtain

$$\lim_{n\to\infty} I(\sigma_\rho, \sigma_n) \geq |\mathcal{A}|^{-1}(\min_i \pi_i) \lim_{n\to\infty} I(\sigma'_\rho, \sigma_n) > 0. \quad (4.18)$$

By (4.17) we have that (4.3) implies (4.4), and by (4.18) we have that (4.4) implies (4.5) which implies (4.3). $\qquad\square$

Proposition 4.2. *Assume that $X_n = \sigma_n$ or $X_n = c_n$, and π is some distribution which satisfies $\min_i \pi_i > 0$. Then the conditions in Proposition 4.1 are all equivalent to the fact that the sequence $\{X_i\}_{i=1}^\infty$ has a nontrivial tail.*

Proof. If the conditions of Proposition 4.1 hold, then we have that

$$\lim_{n\to\infty} I(X_0, X_n) > 0.$$

In particular, the variable X_0 is not independent of the tail sigma field.

We shall prove the other direction (see [69] for a similar argument). For a set U, let $\sigma_U = (\sigma_u)_{u\in U}$. Fix a level n. For each $m \geq n$ and $w \in L_n$, let

$L(w,m)$ be the set of vertices in T which connect to ρ through w. Since the variables $\sigma_{L(w,m)}$ are conditionally independent given σ_{L_n}, we have

$$I(\sigma_{L_n}, \sigma_{L_m}) \leq \sum_{w \in L_n} I(\sigma_{L_n}, \sigma_{L(w,m)}) = \sum_{w \in L_n} I(\sigma_{L_w}, \sigma_{L(w,m)}),$$

and the right-hand side goes to 0 as $n \to \infty$. It follows that if for the sequence σ_n the conditions in Proposition 4.1 do not hold, the σ-tail is trivial.

Similar proof can be used for the sequence c_{L_n}. Let $c_{L(w,m)}$ be the count of the vertices of L_m which connect to ρ through w, and we get

$$I(c_{L_n}, c_{L_m}) \leq I(\sigma_{L_n}, c_{L_m}) \leq I(\sigma_{L_n}, (c_{L(w,m)})_{w \in L_n}) \leq$$

$$\sum_{w \in L_n} I(\sigma_{L_n}, c_{L(w,m)}) = \sum_{w \in L_n} I(\sigma_{L_w}, c_{L(w,m)}),$$

as before. For the sequence σ_{L_n}, it is easy to see that the claim remains true even without the assumption that $\min_i \pi_i > 0$. Indeed, if the assumptions of Proposition 4.1 hold, then we choose a vertex v with $\min_i \mathbf{P}[\sigma_v = i] > 0$. Then the variable σ_v is not independent of the tail σ-field. $\qquad \square$

4.2 Symmetric binary channels: the Ising model

The only family of channels for which the problem is well understood is the family of symmetric binary channels

$$\mathbf{M} = \begin{pmatrix} 1 - \epsilon & \epsilon \\ \epsilon & 1 - \epsilon \end{pmatrix}, \tag{4.19}$$

where $\lambda_2(M)$, the second largest (in absolute value) eigenvalue of $\mathbf{M}$, satisfies $\lambda_2(M) = 1 - 2\epsilon$. Channel (4.19) corresponds to the Ising model on the tree.

The free measure (disordered Gibbs measure) for the Ising model on a finite tree is the probability measure on configurations σ of ± 1, given by

$$\mathbf{P}[\sigma] = \frac{1}{Z} \exp\left(\sum_{\langle v,w \rangle} \sigma_v \sigma_w \right), \tag{4.20}$$

where Z is a normalizing constant (see Chapter 2). The correspondence between (4.19) and (4.20) is given by $\epsilon = \frac{\exp(-\beta)}{\exp(-\beta) + \exp(\beta)}$, or equivalently, $\lambda_2(M) = \tanh \beta$.

Theorem 4.1. *The reconstruction problem is solvable for the binary symmetric channel with error probability ϵ (4.19), and the b-ary tree T_b, if and only if $b\lambda_2^2(M) > 1$. If $b\lambda_2^2(M) > 1$, then the reconstruction problem is also census solvable.*

Proof. Consider spin values as ± 1. Let $\lambda = 1 - 2\epsilon$, and let S_n be the sum of the $\pm$ variables at level n of the tree. Given that the spin at the root is $+$, the expected value of S_n satisfies

$$\mathbf{E}^+[S_n] = \sum_{v \in L_n} \mathbf{E}^+[\sigma_v] = b^n \lambda^n. \tag{4.21}$$

Similarly, $\mathbf{E}^-[S_n] = -b^n \lambda^n$. The second moment of S_n satisfies the following

$$\mathbf{E}^+[S_n^2] = \mathbf{E}^-[S_n^2] = \sum_{v,w \in L_n} \mathbf{E}[\sigma_v \sigma_w] \tag{4.22}$$

$$= b^n \left(1 + \sum_{j=1}^{n} (b^j - b^{j-1})\lambda^{2j} \right) = \Theta(b^{2n}\lambda^{2n}),$$

where the last equality follows from the fact that $b\lambda^2 > 1$. By Cauchy-Schwartz,

$$\mathbf{E}^+[S_n] - \mathbf{E}^-[S_n] = \sum_{\sigma}(\mathbf{P}^+[\sigma] - \mathbf{P}^-[\sigma])S_n(\sigma) \leq$$

$$\sqrt{\sum_{\sigma} \frac{(\mathbf{P}^+[\sigma] - \mathbf{P}^-[\sigma])^2}{\mathbf{P}^+[\sigma] + \mathbf{P}^-[\sigma]}} \sqrt{\sum_{\sigma} S_n^2(\sigma)(\mathbf{P}^+[\sigma] + \mathbf{P}^-[\sigma])}.$$

By (4.21) and (4.22) for $b\lambda^2 > 1$ we get

$$\sum_{\sigma} \frac{(\mathbf{P}^+[\sigma] - \mathbf{P}^-[\sigma])^2}{\mathbf{P}^+[\sigma] + \mathbf{P}^-[\sigma]} = \Theta(1),$$

which implies that $|\mathbf{P}^+ - \mathbf{P}^-| = \Theta(1)$. $\qquad\square$

The proofs of the non-reconstruction result when $b\lambda_2^2(M) \leq 1$ are harder, and do not generalize to other channels. There are 4 different proofs for this result:

1. The first proof [25], is based on recursive analysis of the disordered Gibbs measure.

2. A proof of non-reconstruction which is based on information inequalities is given in [69] where it is shown that the mutual information between the variable at the root of the tree and the level n variables satisfy

$$I(\sigma_\rho, \sigma_n) \leq \sum_{v \in L_n} I(\sigma_\rho, \sigma_v),$$

   as in the case of conditionally independent variables.

3. In [119] recursive analysis is used in order to show $\mathbf{E}[\mathbf{E}^2[\sigma_\rho|\sigma_n]]$ tends to zero as $n \to \infty$ for the n-level tree, when $b\lambda_2^2(M) \leq 1$ (see Chapter 2).

4. Glauber dynamics is the following reversible Monte-Carlo method for sampling configurations σ according to the distribution (4.19) or (4.20). Given the current configuration σ, a vertex v is picked uniformly at random at rate 1, in which case the variable σ_v is replaced by a random variable σ'_v chosen according to the conditional distribution on the rest of the configuration, $(\sigma_w)_{w \neq v}$. In [19] it is shown that Glauber dynamics have spectral gap which is bounded away from zero when $b\lambda_2^2(M) < 1$. Then using a general principle (which is proven in a much more general context) it was obtained that the reconstruction problem is unsolvable when $b\lambda_2^2(M) < 1$.

4.2.1 *Reconstruction algorithms*

(Taken from [171]) Theorem 4.1 reveals a surprising phenomenon: reconstruction by global majority vote has the same threshold for success as maximum likelihood reconstruction (which is the optimal reconstruction strategy).

The parsimony method is popular in biology. Given a bicoloring of the boundary of a tree T, a parsimonious coloring of the internal nodes is any assignment of the two colors to these nodes that minimizes the total number of bicolored edges. A way of finding a parsimonious coloring is the following: Starting from the parents of the boundary nodes, assign recursively to each internal node the color of the majority of its ± 1-colored children. In case of a tie, assign the non-color "?". Then scan the tree from the root downwards and assign all vertices labeled by ? the same label as their parent.

On a fixed finite tree, when $\epsilon > 0$ is small, the maximum likelihood algorithm will reconstruct the same root value as one of the parsimonious colorings given the boundary.

However, this is not the case when ϵ is larger. For the binary tree, it is shown in [248] that the parsimony reconstruction algorithm has success probability bounded away from $1/2$ as $n \to 1$ if and only if $\epsilon \geq 1/8$. Thus when $\lambda_2(M) = 1 - 2\epsilon \in (1/\sqrt{2}, 3/4]$ on the binary tree, majority (and maximum likelihood) will have success probability bounded away from $1/2$, while the parsimony success probability tends to $1/2$.

4.2.2   *Census solvability*

The threshold $b\lambda_2^2(M) = 1$ which appeared as the threshold both for reconstruction solvability and for census solvability for the Ising model, turns out to be in general the threshold for census solvability.

Theorem 4.2. *Let M be a channel corresponding to an ergodic Markov chain. Let T_b be the b-ary tree. The reconstruction problem is census-solvable if $b\lambda_2^2(M) > 1$, and is not census solvable if $b\lambda_2^2(M) < 1$. For general trees, the reconstruction problem is solvable when $\mathrm{br}(T)\lambda_2^2(M) > 1$, where $\mathrm{br}(T)$ is the branching number of the tree.*

Proof.   *Census solvability:* In [125] a limit theorem for the variables c_n is proven. In particular it is shown that if $b\lambda_2^2(M) > 1$, then the distribution of the limiting variable depends on the initial variable at the root. This implies that the problem is census solvable. A more elementary proof is given in [170]. The proof follows the lines of the proof for the Ising model (Theorem 4.1), where S_n is replaced by the scalar product of c_n with any non-zero vector v satisfying $Mv = \lambda_2(M)v$. Note that this proof generalizes the proof for the Ising model, since for the Ising model, $v = \begin{pmatrix} 1 \\ -1 \end{pmatrix}$. This proof also generalizes to general trees.

    No census solvability: The CLT in [125] implies that if $b\lambda_2^2(M) \leq 1$ then the normalized value of c_n ($c_n/b^{n/2}$ if $b\lambda_2^2(M) < 1$) converges to a non-zero random variable which is independent of the variable of the root. However, this result does not imply that the reconstruction problem is not census solvable. Presumably, it may be the case that the first coordinate of c_n is more likely to be even for some value of the root variable than for others. This dependency between the root variable and c_n would not manifest itself in the limiting normalized variables. In [170] combining the results of [125] with the local central limit theorem, it was demonstrated that this could not happen. The idea of the proof is to use [125] in order to couple c_n^i, the value of c_n given that the root variable is i, and c_n^j, the value of c_n given that the root variable is j in such a way that the variables are close (i.e., $|c_n^i - c_n^j|_\infty < \epsilon b^{n/2}$). Then use the local central limit theorem in order to achieve a coupling of c_{n+l}^i and c_{n+l}^j with high probability. $\qquad\qquad\square$

Conjecture 1. The reconstruction problem is not census solvable when $b\lambda_2^2(M) = 1$:

In [170] this conjecture is formulated and verified for Potts models and asymmetric Ising models.

4.3 q-ary symmetric channels: the Potts model

For Gibbs measures of the Potts model on a Cayley tree, see the next chapter. Two of the natural generalizations of binary symmetric channels are asymmetric binary channels (which correspond to Ising models with external field, see Chapter 2), and q-ary symmetric channels (which correspond to Potts models with no external field, see the next chapter):

(a) Asymmetric binary channels have the state space $\{0, 1\}$ and the matrices:

$$\mathbf{M} = \begin{pmatrix} 1 - \delta_1 & \delta_1 \\ 1 - \delta_2 & \delta_2 \end{pmatrix}, \tag{4.23}$$

where $\lambda_2(M)$, the second largest (in absolute value) eigenvalue of $\mathbf{M}$, satisfies $\lambda_2(M) = \delta_2 - \delta_1$.

(b) Symmetric channels on q symbols have the state space $\{1, \ldots, q\}$ and $q \times q$-matrices:

$$\mathbf{M} = \begin{pmatrix} 1 - (q-1)\delta & \delta & \delta \ldots & \delta \\ \delta & 1 - (q-1)\delta & \delta \ldots & \delta \\ \ldots & \vdots & \ldots \vdots & \ldots \\ \delta & \delta & \delta \ldots & 1 - (q-1)\delta \end{pmatrix}, \tag{4.24}$$

with $\lambda_2(M) = 1 - q\delta$.

Depending on the sign of $\lambda_2(M)$ we distinguish between ferromagnetic Potts models where $\lambda_2(M) > 0$, and anti-ferromagnetic models where $\lambda_2(M) < 0$. In case $1 - (q-1)\delta = 0$ we obtain the model of proper colorings of the tree:

$$
\mathbf{M} = \begin{pmatrix}
0 & (q-1)^{-1} & (q-1)^{-1} & \ldots & (q-1)^{-1} \\
(q-1)^{-1} & 0 & (q-1)^{-1} & \ldots & (q-1)^{-1} \\
\ldots & \vdots & \ldots & \vdots & \ldots \\
(q-1)^{-1} & (q-1)^{-1} & (q-1)^{-1} & \ldots & 0
\end{pmatrix}.
\tag{4.25}
$$

Problem 1. For the 3 symbols Potts model (4.24) find the values for which the reconstruction problem is solvable on the b-ary tree.

It may be easier to solve the analogous problem for the Ising model with external field. The analogous problem for colorings was stated in [37]. Applying standard coupon-collector estimates recursively, it is easy to see that if $b \geq (1 + \delta)q \ln q$ and q is large, then the reconstruction problem is solvable for the coloring model.

Problem 2. For colorings, for which b and q is the reconstruction problem solvable on the b-ary tree?

Below we discuss several bounds for the reconstruction problem for Potts models.

Proposition 4.3. *If* $b\lambda_2^2(M) > 1$ *then the reconstruction problem is solvable.*

Proof. This follows from Theorem 4.2, and from the fact that census solvability implies solvability. $\qquad\square$

Proposition 4.4. *If* $b|\lambda_2(M)| \leq 1$, *then the reconstruction problem is unsolvable.*

Proof. Assume first that $\mathbf{M}$ is a ferromagnetic Potts model, i.e., $\lambda_2(M) > 0$. Consider two measures on the tree, one with i as the root variable and one with j as the root variable. We couple these measures in the following way: starting at the root if the two measures agree on the variable at v, then we couple in such a way that the measures also agree for all the children of v. If they do not agree at v, then for each of the children of v, use the optimal coupling in order to couple the measures. For each of the children, this has success probability $q\delta$. Thus the non-coupled vertices are a branching process with parameter $1 - q\delta = \lambda_2(M)$. When $b\lambda_2(M) \leq 1$

this process will eventually die; this means that for large n all the vertices at level n will have the same variables with probability going to 1 as $n \to \infty$, as needed. When **M** is anti-ferromagnetic, the coupling probability is $(q-2)\delta + 2(1 - (q-1)\delta) = 2 - q\delta$, therefore the branching process parameter is $1 - (2 - q\delta) = -\lambda_2(M) = |\lambda_2(M)|$. Similar arguments apply for Ising models with external fields. $\square$

Proposition 4.5. *If $b\lambda_2(M) > 1$ and q is sufficiently large, then the reconstruction problem is solvable.*

Proof. The detailed proof is given in [169]. It implies in particular that $b\lambda_2^2(M) = 1$ is not the threshold for the reconstruction problem for Potts models, as it is sometimes possible to reconstruct even when $b\lambda_2^2(M) < 1$. An analogous result is proven for the asymmetric binary channel. The idea behind the proof is the following. Channel (4.24) may be thought of in the following way: at each step the output is identical to the input with probability $\lambda_2(M)$, otherwise, the output is chosen uniformly among the q symbols. In particular if $\lambda_2(M) > 0$ is fixed and q is very large, then if two of the children of a vertex in the b-ary tree T_b have the same label, then with overwhelming probability, this is also the label of their parent. Now suppose that q is large and there exists a copy of $T_2 \subset T_b$ such that all the vertices of T_2 are labeled by i. Using a recursive argument we see that given this event, with large probability, the variable at the root is i. Moreover, it can be shown that when $b\lambda_2(M) > C$ for some constant $C > 1$, such a tree exists with positive probability. Therefore, it is possible to reconstruct the root variable based on the existence of such a unicolored tree. In order to obtain the result for $C = 1$, we replace the unicolored T_2 by a diluted unicolored T_2. $\square$

Proposition 4.6. *If $b\frac{(1-q\delta)^2}{1-(q-2)\delta} \leq 1$ then the reconstruction problem is unsolvable.*

This and the analogous result for asymmetric binary channels are proven in [170], we sketch the main idea of the proof following [171]:

Proof. To show that the reconstruction problem is unsolvable, it suffices to show that given the root value is 0 or 1 with probability $1/2$ each, it is asymptotically impossible to conclude from the variables at level n, if the root is 0 or 1. Suppose that in addition to the variables at level n, we are also given all the variables at all levels of the tree having variable j with $j \neq 0, 1$. Since we are given more information, it is easier to reconstruct.

The model where we are given this extra information is nothing but the symmetric binary channel with matrix

$$\mathbf{M} = \begin{pmatrix} \frac{1-(q-1)\delta}{1-(q-2)\delta} & \frac{\delta}{1-(q-2)\delta} \\[2mm] \frac{\delta}{1-(q-2)\delta} & \frac{1-(q-1)\delta}{1-(q-2)\delta} \end{pmatrix},$$

on the random tree which is obtained from the original tree by independently deleting an edge with probability $(q-2)\delta$ and retaining it with probability $1-(q-2)\delta$. The results of [69] imply that the binary symmetric channel on a general tree T, the reconstruction problem is unsolvable if $\mathrm{br}(T)\lambda_2^2(M) < 1$. For the branching process on the regular tree T_b, the branching number is "typically" $b(1-(q-2)\delta)$. The non-reconstruction criterion $\mathrm{br}(T)\lambda_2^2(M) < 1$, now is $b\frac{(1-q\delta)^2}{1-(q-2)\delta} \leq 1$. $\qquad\square$

Commentaries and references. As it was mentioned above, this chapter is due to papers [169]-[171]. Since the main object of this book is Gibbs measure, in this chapter we included results of the theory (of information flows on trees) which are applied for extremality problem of disordered Gibbs measures of Ising and Potts models. For example, we did not add in this chapter some results devoted to general trees and to general channels. We refer the reader to [169]-[174] and to corresponding references therein for more information about information flows on trees.

It is known that such an information flow represents propagation of a genetic property from ancestor to its descendants, it was studied in genetics, see e.g. [46], [249]. In communication theory, this process represents a communication network on the tree where information is transmitted from the root of the tree. Moreover, the process was studied in statistical physics, see e.g. [25], [115], [250]. More precisely, in statistical physics a non-uniqueness of the Gibbs measure means that for all n there exists σ_n such that the distribution of σ_ρ given σ_n has total variation distance at least $\delta > 0$ from uniform. This is a weaker condition than reconstruction solvability and it was studied in statistical physics for Ising and Potts models (see Chapter 2, Chapter 5 and [107], [113], [204]).

Reconstruction (and census) solvability when $b\lambda_2^2(M) > 1$ was first proved in [115], the paper [134] is earlier and does much more, but is formulated in the language of multi-type branching processes.

The crucial role of the reconstruction problem in Phylogeny was demonstrated in [173] and [174] extending the results of [172].

In the following chapters we shall apply results of this chapter for Potts and Hard-Core models on Cayley trees.

Chapter 5

The Potts model

This chapter contains results related to Gibbs measures of the q-state Potts model on Cayley trees. The description of such measures is reduced to a solution of a vector-valued functional equation. We show that under some conditions on the parameters there exist $q + 1$ distinct translation-invariant Gibbs measures. We apply the results of the previous chapter to find the extremality conditions of the disordered Gibbs measure. Moreover using the Bleher-Ganikhodjaev construction we show the existence of an uncountable set of non-translation-invariant Gibbs measures. Compared to the Ising model, one can see that many problems related to the Potts model are open. For example, periodic and weakly periodic Gibbs measures have not been studied yet.

5.1 The Hamiltonian and vector-valued functional equation

In this chapter we consider Potts model on a Cayley tree, where the spin takes values in the set $\Phi := \{1, 2, \ldots, q\}$, and is assigned to the vertices of the tree. A configuration σ on V is then defined as a function $x \in V \mapsto \sigma(x) \in \Phi$; the set of all configurations is Φ^V.

The (formal) Hamiltonian of Potts model is

$$H(\sigma) = -J \sum_{\langle x,y \rangle \in L} \delta_{\sigma(x)\sigma(y)} - \alpha \sum_{x \in V} \delta_{1\sigma(x)}, \qquad (5.1)$$

where $J \in R$ is a coupling constant, $\alpha \in R$ is an external field, $\langle x, y \rangle$ stands for nearest neighbor vertices and δ_{ij} is the Kroneker's symbol:

$$\delta_{ij} = \begin{cases} 0, & \text{if } i \neq j \\ 1, & \text{if } i = j. \end{cases}$$

105

Define a finite-dimensional distribution of a probability measure μ in the volume V_n as

$$\mu_n(\sigma_n) = Z_n^{-1} \exp\left\{ -\beta H_n(\sigma_n) + \sum_{x \in W_n} h_{\sigma(x),x} \right\}, \qquad (5.2)$$

where $\beta = 1/T$, $T > 0$-temperature, Z_n^{-1} is the normalizing factor and $\{h_x = (h_{1,x}, \ldots, h_{q,x}) \in R^q, x \in V\}$ is a collection of vectors and

$$H_n(\sigma_n) = -J \sum_{\langle x,y \rangle \in L_n} \delta_{\sigma(x)\sigma(y)} - \alpha \sum_{x \in V_n} \delta_{1\sigma(x)}.$$

We say that the probability distributions (5.2) are compatible if for all $n \geq 1$ and $\sigma_{n-1} \in \Phi^{V_{n-1}}$:

$$\sum_{\omega_n \in \Phi^{W_n}} \mu_n(\sigma_{n-1} \vee \omega_n) = \mu_{n-1}(\sigma_{n-1}). \qquad (5.3)$$

Here $\sigma_{n-1} \vee \omega_n$ is the concatenation of the configurations. In this case, there exists a unique measure μ on Φ^V such that, for all n and $\sigma_n \in \Phi^{V_n}$,

$$\mu(\{\sigma|_{V_n} = \sigma_n\}) = \mu_n(\sigma_n).$$

Such a measure is called a *splitting Gibbs measure* corresponding to the Hamiltonian (5.1) and vector-valued function $h_x, x \in V$.

The following statement describes conditions on h_x guaranteeing compatibility of $\mu_n(\sigma_n)$.

Theorem 5.1. *Probability distributions $\mu_n(\sigma_n)$, $n = 1, 2, \ldots$, in (5.2) are compatible iff for any $x \in V$ the following equation holds:*

$$h_x = \sum_{y \in S(x)} F(h_y, \theta, \alpha), \qquad (5.4)$$

where $F : h = (h_1, \ldots, h_{q-1}) \in R^{q-1} \to F(h, \theta, \alpha) = (F_1, \ldots, F_{q-1}) \in R^{q-1}$ is defined as

$$F_i = \alpha\beta\delta_{1i} + \ln\left(\frac{(\theta - 1)e^{h_i} + \sum_{j=1}^{q-1} e^{h_j} + 1}{\theta + \sum_{j=1}^{q-1} e^{h_j}} \right),$$

and $\theta = \exp(J\beta)$, $S(x)$ is the set of direct successors of x.

Proof. Necessity. Suppose that (5.3) holds, we want to prove (5.4). Substituting (5.2) into (5.3), obtain that for any configurations σ_{n-1}: $x \in V_{n-1} \mapsto \sigma_{n-1}(x) \in \{1, \ldots, q\}$:

$$\frac{Z_{n-1}}{Z_n} \sum_{\omega_n \in \Omega_{W_n}} \exp\left(\sum_{x \in W_{n-1}} \sum_{y \in S(x)} (J\beta\delta_{\sigma_{n-1}(x)\omega_n(y)} + \qquad (5.5) \right.$$

$$\alpha\beta\delta_{1\omega_n(y)} + h_{\omega_n(y),y}\Big) = \exp\left(\sum_{x\in W_{n-1}} h_{\sigma_{n-1}(x),x}\right),$$

where $\omega_n\colon x \in W_n \mapsto \omega_n(x)$.

From (5.5) we get:

$$\frac{Z_{n-1}}{Z_n} \sum_{\omega_n\in\Omega_{W_n}} \prod_{x\in W_{n-1}} \prod_{y\in S(x)} \exp\left(J\beta\delta_{\sigma_{n-1}(x)\omega_n(y)} + \alpha\beta\delta_{1\omega_n(y)} + h_{\omega_n(y),y}\right) =$$

$$\prod_{x\in W_{n-1}} \exp\left(h_{\sigma_{n-1}(x),x}\right).$$

Fix $x \in W_{n-1}$ and rewrite the last equality for $\sigma_{n-1}(x) = i$, $i = 1,\ldots,q-1$ and $\sigma_{n-1}(x) = q$, then dividing each of them to the last one we get

$$\prod_{y\in S(x)} \frac{\sum_{u=1}^{q} \exp\left(J\beta\delta_{iu} + \alpha\beta\delta_{1u} + h_{u,y}\right)}{\sum_{u=1}^{q} \exp\left(J\beta\delta_{qu} + \alpha\beta\delta_{1u} + h_{u,y}\right)} = \exp\left(h_{i,x} - h_{q,x}\right),\ i = 1,\ldots,q-1.$$

$$(5.6)$$

Now we change the variables as follows

$$\begin{cases} \alpha\beta + h_{1,x} - h_{q,x} \to h_{1,x}, \\ h_{i,x} - h_{q,x} \to h_{i,x}, \quad i = 2,\ldots,q-1. \end{cases}$$

Then (5.6) implies (5.4).

Sufficiency. Suppose that (5.4) holds. It is equivalent to the representations

$$\prod_{y\in S(x)} \sum_{u=1}^{q} \exp\left(J\beta\delta_{iu} + \alpha\beta\delta_{1u} + h_{u,y}\right) = a(x)\exp\left(h_{i,x}\right), \quad i = 1,\ldots,q-1$$

$$(5.7)$$

for some function $a(x) > 0, x \in V$. We have

$$\text{LHS of } (5.3) = \frac{1}{Z_n} \exp(-\beta H(\sigma_{n-1}))\times$$

$$\prod_{x\in W_{n-1}} \prod_{y\in S(x)} \sum_{u=1}^{q} \exp\left(J\beta\delta_{\sigma_{n-1}(x)u} + \alpha\beta\delta_{1u} + h_{u,y}\right). \qquad (5.8)$$

Substituting (5.7) into (5.8) and denoting $A_n(x) = \prod_{x\in W_{n-1}} a(x)$, we get

$$\text{RHS of } (5.8) = \frac{A_{n-1}}{Z_n} \exp(-\beta H(\sigma_{n-1})) \prod_{x\in W_{n-1}} \exp(h_{\sigma_{n-1}(x),x}). \qquad (5.9)$$

Since $\mu^{(n)}$, $n \geq 1$ is a probability, we should have

$$\sum_{\sigma_{n-1}\in\Omega_{V_{n-1}}} \sum_{\omega_n\in\Omega_{W_n}} \mu^{(n)}(\sigma_{n-1},\omega_n) = 1.$$

Hence from (5.9) we get $Z_{n-1}A_{n-1} = Z_n$, and (5.3) holds. $\qquad\square$

Hence, from Theorem 5.1 it follows that for any $h = \{h_x, \quad x \in V\}$ satisfying the functional equation (5.4) there exists a unique Gibbs measure μ and vice versa. However, the analysis of solutions to (2.8) is not easy.

5.2   Translation-invariant Gibbs measures

In this section, we consider Gibbs measures which are translation-invariant, i.e., we assume $h_x = h = (h_1, \ldots, h_{q-1}) \in R^{q-1}$ for all $x \in V$. Then from equation (5.4) we get $h = kF(h, \theta, \alpha)$, i.e.,

$$h_i = \alpha\beta\delta_{1i} + k\ln\left(\frac{(\theta-1)e^{h_i} + \sum_{j=1}^{q-1}e^{h_j} + 1}{\theta + \sum_{j=1}^{q-1}e^{h_j}}\right), \quad i = 1, \ldots, q-1. \quad (5.10)$$

Denoting $z_i = \exp(h_i), i = 1, \ldots, q-1$, we get from (5.10)

$$z_i = \exp(\alpha\beta\delta_{1i})\left(\frac{(\theta-1)z_i + \sum_{j=1}^{q-1}z_j + 1}{\theta + \sum_{j=1}^{q-1}z_j}\right)^k, \quad i = 1, \ldots, q-1. \quad (5.11)$$

5.2.1   *Anti-ferromagnetic case*

In this subsection consider the anti-ferromagnetic case: $J < 0$.

Lemma 5.1. *If $\theta < 1$ ($J < 0$) then the system (5.11) has unique solution.*

Proof.   For $i = 2, \ldots, q-1$ from (5.11) we get

$$(z_i - 1)A^k = B^k - A^k,$$

where

$$A = \theta + \sum_{j=1}^{q-1}z_j > 0, \quad B = 1 + (\theta-1)z_i + \sum_{j=1}^{q-1}z_j > 0.$$

Consequently,

$$(z_i - 1)\left\{A^k + (1-\theta)(B^{k-1} + \cdots + A^{k-1})\right\} = 0, \, i = 2, \ldots, q-1. \quad (5.12)$$

Since $0 < \theta < 1$ from (5.12) we get unique solution $z_i = 1$ for all $i = 2, \ldots, q-1$.

Substituting $z_i = 1, i = 2, \ldots, q-1$ into the first equation of the system of equations (5.11) we get

$$z_1 = \exp(\alpha\beta)\left(\frac{\theta z_1 + q - 1}{z_1 + \theta + q - 2}\right)^k. \quad (5.13)$$

Set

$$a = \theta^{-k} \exp(-\alpha\beta); \quad b = \frac{\theta(\theta + q - 2)}{q - 1}; \quad x = \frac{\theta}{q - 1} z_1.$$

Then equation (5.13) can be rewritten as

$$ax = \left(\frac{1 + x}{b + x}\right)^k. \tag{5.14}$$

The detailed analysis of solutions to equation (5.14) is given in [204], Proposition 10.7, which is the following:

Proposition 5.1. *Equation (5.14) with $x \geq 0$, $k \geq 1$, $a, b > 0$ has a unique solution if either $k = 1$ or $b \leq (\frac{k+1}{k-1})^2$. If $k > 1$ and $b > (\frac{k+1}{k-1})^2$ then there exist $\nu_1(b, k)$, $\nu_2(b, k)$, with $0 < \nu_1(b, k) < \nu_2(b, k)$, such that the equation has three solutions if $\nu_1(b, k) < a < \nu_2(b, k)$ and has two if either $a = \nu_1(b, k)$ or $a = \nu_2(b, k)$. In fact:*

$$\nu_i(b, k) = \frac{1}{x_i}\left(\frac{1 + x_i}{b + x_i}\right)^k,$$

where x_1, x_2 are the solutions of

$$x^2 + [2 - (b - 1)(k - 1)]x + b = 0.$$

By Proposition 5.1 the equation (5.13) has unique solution if $k = 1$ or $k \neq 1$ and

$$\frac{\theta(\theta + q - 2)}{q - 1} \leq \left(\frac{k + 1}{k - 1}\right)^2. \tag{5.15}$$

Since $0 < \theta < 1$ we have $\frac{\theta(\theta+q-2)}{q-1} < 1$, but $\left(\frac{k+1}{k-1}\right)^2 > 1$. Hence the inequality (5.15) is true for any $k > 1$, $q > 1$ and $0 < \theta < 1$. Thus equation (5.13) has unique solution $z_1 = z_*$. Consequently, system (5.11) has unique solution $z^* = (z_*, 1, 1, \ldots, 1)$. $\qquad\square$

Note that $h_x = h^* = (\ln z_*, 0, 0, \ldots, 0)$ is a solution to (5.4) which corresponds to z^*. We denote by μ_* the Gibbs measure which corresponds to h^*. Thus we have proved the following

Theorem 5.2. *For any $k \geq 1$, $q > 1$, $J < 0$, $\alpha \in R$ the anti-ferromagnetic Potts model (5.1) has unique translation-invariant Gibbs measure.*

5.2.2 Ferromagnetic case

Consider the ferromagnetic case: $J > 0$ and $\alpha = 0$.

5.2.2.1   *Case: $k = 2$, $q = 3$*

For simplicity of calculations let us first consider the case $k = 2$, and $q = 3$. In this case the mapping $F : R^2 \to R^2$ has the following form

$$
\begin{aligned}
F_1 &= F_1(h, \theta) = \ln\left(\frac{\theta \exp(h_1) + \exp(h_2) + 1}{\exp(h_1) + \exp(h_2) + \theta} \right), \\
F_2 &= F_2(h, \theta) = \ln\left(\frac{\exp(h_1) + \theta \exp(h_2) + 1}{\exp(h_1) + \exp(h_2) + \theta} \right).
\end{aligned}
\tag{5.16}
$$

The following proposition is obvious.

Proposition 5.2. *Let S be a rotation of the plane R^2 through angle $2\pi/3$. Then the transformations F and S commute with each other.*

For the transformation $F(h, \theta)$ of the plane R^2 into itself there exist three invariant lines $l_i = \{h \in R^2 : h_i = 0\}$, $i = 1, 2$; $l_3 = \{h \in R^2 : h_1 = h_2\}$. Denote by f_i the restriction of the mapping F to the invariant line l_i, $i = 1, 2, 3$.

Proposition 5.3. *For any $i = 1, 2, 3$, the function $f_i(h, \theta)$, $h \in R$, has the following properties:*

1) $f_i(h, \theta)$ is bounded;

2) $\frac{d}{dh} f_i(0, \theta) = \frac{\theta - 1}{\theta + 2}$; $0 < \frac{d}{dh} f_i(h, \theta) < \alpha = \alpha(\theta)$, $\alpha < 1$.

Proof.   Can be proved by simple calculations. $\square$

The invariant lines l_i, $i = 1, 2, 3$, divide the plane R^2 into six sectors, each of which is obviously invariant with respect to the transformation $F(h, \theta)$, so that it is sufficient to study the behavior of the trajectory $F(h, \theta)$ in one of the sectors.

As mentioned above, to describe phase transitions or translation-invariant Gibbs measures in the ferromagnetic Potts model with three states on the Cayley tree Γ^2, it is necessary to describe the fixed points of the mapping $2F$, i.e., to solve

$$
h = 2F(h, \theta). \tag{5.17}
$$

By the substitution $z_i = \exp(h_i)$, $i = 1, 2$, equation (5.17) is reduced to the system of equations

$$
z_1 = \left(\frac{\theta z_1 + z_2 + 1}{z_1 + z_2 + \theta} \right)^2, \quad z_2 = \left(\frac{z_1 + \theta z_2 + 1}{z_1 + z_2 + \theta} \right)^2. \tag{5.18}
$$

Simple calculations show that the system of equations (5.18) has the unique solution $z_1 = z_2 = 1$ for $1 < \theta < \theta'_c = 1 + 2\sqrt{2}$; it has seven solutions for $\theta'_c < \theta \neq \theta_c = 4$, and four solutions for $\theta = \theta'_c$ or $\theta = \theta_c$. It is easy to show that all the solutions of (5.17) lie on the invariant lines l_i, $i = 1, 2, 3$, described above. By virtue of Proposition 5.2, it is sufficient to study the fixed points that lie on one of the invariant lines.

We consider the line $l_3 : h_1 = h_2$. Then the restriction of the mapping $h' = 2F(h, \theta)$ to l_3 has the form

$$h' = 2 \ln \left(\frac{(\theta + 1) \exp(h) + 1}{2 \exp(h) + \theta} \right), \tag{5.19}$$

where $h \in R$, or, making the substitution $z = \exp(h)$, we obtain

$$\varphi(z, \theta) = \left(\frac{(\theta + 1)z + 1}{2z + \theta} \right)^2.$$

Simple arguments from analysis show that the function $\varphi(z, \theta)$ increases monotonically, is bounded, and satisfies $\frac{d}{dz} \varphi(1, \theta) < 1$ for $1 < \theta < \theta_c = 4$. By virtue of these properties, the fixed points of (5.17) on l_3 can be described as follows:

a) for $1 < \theta < \theta'_c$, the unique fixed point $\mathbf{h} = 0 \in R^2$ is stable;

b) for $\theta'_c < \theta < \theta_c$, there exist three fixed points: $\mathbf{h}^1 = 0$, $\mathbf{h}^2$, and $\mathbf{h}^3$, which are such that $h_1^2 = h_2^2 < 0$, $h_1^3 = h_2^3 < 0$, the fixed point $\mathbf{h}^1$ is stable while the other two are unstable;

c) for $\theta > 4$, there exist three fixed points: $\mathbf{h}^1 = 0$, $\mathbf{h}^2$, and $\mathbf{h}^3$, which are such that $h_1^2 = h_2^2 < 0$, $h_1^3 = h_2^3 > 0$, with the fixed point $\mathbf{h}^2$ stable and the other two unstable.

Thus, for $\theta \leq 4$ equation (5.17) has a unique stable solution $\mathbf{h}^1 = 0 \in R^2$ while for $\theta > 4$ it has three stable solutions $\mathbf{h}^i_*$, $i = 1, 2, 3$, which lie on the invariant lines l_i, i.e., have the following form: $\mathbf{h}^1_* = (0, h_*)$, $\mathbf{h}^2_* = (h_*, 0)$, and $\mathbf{h}^3_* = (-h_*, -h_*)$, where $h_* > 0$.

By Theorem 5.1 we know that Gibbs measures are uniquely determined by the vectors $\{h_x, x \in V\}$ namely, we have the following theorem.

Let $T_c > 0$ be determined from the equation $\exp(J/T_c) = \theta_c = 4$. It follows from Theorem 5.1 that for $T \geq T_c$, i.e., $1 < \theta \leq 4$ there exists a unique Gibbs measure.

Let $T < T_c$. Consider the set of vectors $\{h_x = \mathbf{h}^i_*, x \in V\}$, $i = 1, 2, 3$. Since $\mathbf{h}^i_*$, is a solution of equation (5.17), $i = 1, 2, 3$, this set satisfies equation (5.4). We denote by μ^i the Gibbs measure corresponding to the set of vectors $\{h_x = \mathbf{h}^i_*, x \in V\}$, $i = 1, 2, 3$.

Theorem 5.3. *The Gibbs measures μ^i, $i = 1, 2, 3$, are extreme.*

To prove the theorem, we shall prove some lemmas first.

Let P be the sector between the rays

$$l_1^- = \{h \in R^2 : h_1 = 0, \, h_2 \leq 0\}, \quad l_3^- = \{h \in R^2 : h_1 = h_2 < 0\}.$$

On the plane R^2 we consider with respect to the usual basis (e_1, e_2) the norm defined as follows: For $h = (h_1, h_2) \in R^2$, set $\|h\| = |h_1| + |h_2|$.

Let $S_{2a}(0) = \{h \in R^2 : \|h\| = 2a\}$ be the sphere of radius $2a$ with center at the origin and $a > 0$. With arbitrary point $A \in S_{2a} \cap P$ we associate a pair of points A_0 and A_1, setting $A_0 = S_{2a}(0) \cap l_3^-$ and $A_1 = S_{2a}(0) \cap l_1^-$.

The following lemma is important.

Lemma 5.2. *For arbitrary points $A, B \in P$, the following inequalities hold:*

a) $\|F(A_1)\| \leq \|F(A)\| \leq \|F(A_0)\|$;
b) $\|F(A) - F(B)\| \leq \max\{\|F(A_0) - F(B_0)\|, \|F(A_1) - F(B_1)\|\}$.

Proof.   It is easy to see that $A \in S_{2a}(0) \cap P$ for $a = \|A\|/2$. An arbitrary point of the segment $S_{2a}(0) \cap P$ can be represented in the form $A_\lambda(-(1 - \lambda)a, -(1 + \lambda)a)$, where $0 \leq \lambda \leq 1$, and A_0 and A_1 are precisely the points that are associated with A_λ for any $\lambda \in [0, 1]$. Let $F(A_\lambda) \in S_{2\mu(\lambda)}(0)$. We shall show that $\mu(\lambda)$ is a monotonically decreasing function. By the change of variables $z_i = \exp(h_i)$, $i = 1, 2$, the coordinates of the point A_λ can be represented in the form

$$z_1 = \exp(-(1 - \lambda)a), \quad z_2 = \exp(-(1 + \lambda)a),$$

while the coordinates of the point $F(A_\lambda)$ have the form

$$z_1' = \frac{\theta z_1 + z_2 + 1}{z_1 + z_2 + \theta}, \quad z_2' = \frac{z_1 + \theta z_2 + 1}{z_1 + z_2 + \theta},$$

and at the same time $z_1 z_2 = e^{-2a}$ and $z_1' z_2' = e^{-2\mu(\lambda)}$. We have

$$z_1' z_2' = \frac{\theta(z_1 + z_2)^2 + (\theta + 1)(z_1 + z_2) + (\theta - 1)^2 e^{-2a} + 1}{(z_1 + z_2 + \theta)^2},$$

where $z_1 + z_2 = e^{-a}(e^{\lambda a} + e^{-\lambda a})$. It must be verified that $z_1' z_2'$ as a function of the variable λ attains its minimal value at a point that satisfies the condition $z_1 + z_2 = \frac{2(\theta - 1)e^{-2a} - \theta - 2}{2\theta + 1}$ and to the right of it increases monotonically.

We shall show that for any $\lambda \in [0, 1]$

$$e^{-a}(e^{\lambda a} + e^{-\lambda a}) > \frac{2(\theta - 1)e^{-2a} - \theta - 2}{2\theta + 1},$$

for which it is sufficient to verify that

$$\frac{2(\theta - 1)e^{-a} - (\theta + 2)e^a}{2\theta + 1} < 2.$$

The validity of this inequality follows from the fact that the quadratic inequality

$$(\theta + 2)t^2 + 2(2\theta + 1)t - 2(\theta - 1) > 0$$

holds for all positive t. Since $z_1' z_2' = e^{-2\mu(\lambda)}$, it follows that $\mu(\lambda)$ is a monotonically decreasing function, and, in turn, that

$$\|F(A_1)\| \le \|F(A)\| \le \|F(A_0)\|$$

the equality signs holding when $\lambda = 0$ or $\lambda = 1$.

b) Assume, as in the previous part, that $A_\lambda(-(1 - \lambda)a, -(1 + \lambda)a)$ is an arbitrary point of the segment $S_{2a}(0) \cap P$, and that B is a fixed point. Then the coordinates of the point $F(A_\lambda) - F(B)$ have the form

$$\tilde{h}_1 = \ln\left(\frac{\theta \exp(-(1 - \lambda)a) + \exp(-(1 + \lambda)a) + 1}{\exp(-(1 - \lambda)a) + \exp(-(1 + \lambda)a) + \theta}\right) + (1 - \mu)\tilde{b},$$

$$\tilde{h}_2 = \ln\left(\frac{\exp(-(1 - \lambda)a) + \theta \exp(-(1 + \lambda)a) + 1}{\exp(-(1 - \lambda)a) + \exp(-(1 + \lambda)a) + \theta}\right) + (1 + \mu)\tilde{b},$$

where $\tilde{b} = \|F(B)\|/2$, and μ is fixed.

If $\tilde{h}_1 < 0$, $\tilde{h}_2 < 0$, then after the change of variables $z_i = \exp(h_i)$, $i = 1, 2$, we get

$$\tilde{z}_1 \tilde{z}_2 = \frac{\theta(z_1 + z_2)^2 + (\theta + 1)(z_1 + z_2) + (\theta - 1)^2 e^{-2a} + 1}{(z_1 + z_2 + \theta)^2} e^{2\tilde{b}},$$

from which, by virtue of a), we obtain

$$\|F(A_\lambda) - F(B)\| \le \|F(A_0) - F(B)\|.$$

If $\tilde{h}_1 > 0$, $\tilde{h}_2 > 0$, then

$$\tilde{z}_1 \tilde{z}_2 = \frac{(z_1 + z_2 + \theta)^2}{\theta(z_1 + z_2)^2 + (\theta + 1)(z_1 + z_2) + (\theta - 1)^2 e^{-2a} + 1} e^{2\tilde{b}},$$

from which it follows

$$\|F(A_\lambda) - F(B)\| \le \|F(A_1) - F(B)\|.$$

Consequently,

$$\|F(A_\lambda) - F(B)\| \le \max\{\|F(A_0) - F(B)\|, \|F(A_1) - F(B)\|\}.$$

Similarly, it can be proved that

$$\|F(A_0) - F(B)\| \le \|F(A_0) - F(B_0)\|, \quad \|F(A_1) - F(B)\| \le \|F(A_1) - F(B_1)\|,$$

and this completes the proof of the lemma. $\qquad\square$

Lemma 5.3. *There exists a bounded closed set $K \subset R^2$ that is invariant with respect to the transformation S and such that if the set of vectors $\{h_x, x \in V\}$ satisfies equation (5.4) then $h_x \in K$ for any $x \in V$. Moreover, the stable fixed points $\mathbf{h}_*^i$, $i = 1, 2, 3$, determined above for $\theta > 4$ are extreme points of the set K.*

Proof. In the notation of Lemma 5.2, let $K_P = S_{2h_*}(0) \cap P$, where $(-h_*, -h_*) = \mathbf{h}_*^3$, is a stable fixed point on l_3, $\tilde{K}_P$ is the set symmetric to K_P with respect to the invariant line l_3, $\tilde{K} = K_P \cup \tilde{K}_P$, and K be the set formed from the set $\tilde{K}$ and its rotations through $\pm 2\pi/3$. It is easy to show that by virtue of Lemma 5.2 the set K satisfies all the conditions of Lemma 5.3. $\qquad\square$

Proof. (of Theorem 5.3) By virtue of Lemma 5.3, if $\mathbf{h}_x \in K$ and $\|\mathbf{h}_x\| = 2h_*$, then $\mathbf{h}_x = (-h_*, -h_*)$. Suppose that μ^3 is not extreme and expand it with respect to extreme Gibbs measures:

$$\mu^3 = \int_\Omega \mu(\omega)\nu(d\omega).$$

Then for any vertex $x \in V$

$$\mathbf{h}_*^3 = \int_\Omega \mathbf{h}_x(\omega)\nu(d\omega), \tag{5.20}$$

and since $\mathbf{h}_*^3$ is an extreme point of the set K, (5.20) is satisfied if $\mathbf{h}_x(\omega) = \mathbf{h}_*^3$ for almost all ω, from which the assertion of theorem follows. $\qquad\square$

5.2.2.2   *The general case: $k \geq 2$, $q \geq 2$*

In this subsubsection we shall briefly give generalizations of the above mentioned results. Here we shall consider equation (5.10) for $J > 0$ and $\alpha = 0$.

Consider the line $l_q : h_1 = h_2 = \cdots = h_{q-1}$ in R^{q-1}, which is invariant with respect to $kF(h, \theta, q)$. The restriction of this mapping on l_q has the form

$$h' = k \ln \left(\frac{(\theta - q - 2)\exp(h) + 1}{(q - 1)\exp(h) + \theta} \right), \tag{5.21}$$

where $h \in R$, by the change $z = e^h$ we get

$$z = \varphi_k(z, \theta, q) = \left(\frac{(\theta - q - 2)z + 1}{(q - 1)z + \theta} \right)^k.$$

It is easy to check that the function $\varphi_k(z, \theta, q)$ is increasing, bounded and

$$\left. \frac{d\varphi_k(z, \theta, q)}{dz} \right|_{z=1} < 1, \quad \text{if } 1 < \theta < \theta_c = \frac{k + q - 1}{k - 1}.$$

Proposition 5.4. *For the mapping (5.21) there are $\theta'_c < \theta_c$ such that*

a) for $1 < \theta \leq \theta'_c$, the unique fixed point $\mathbf{h} = 0 \in R^{q-1}$ is stable;

b) for $\theta'_c < \theta \leq \theta_c$, there exist three fixed points: $\mathbf{h}^1 = 0$, $\mathbf{h}^2$, and $\mathbf{h}^3$, the fixed point $\mathbf{h}^1$ is stable while the other two are unstable;

c) for $\theta > \theta_c$, there exist three fixed points: $\mathbf{h}^1 = 0$, $\mathbf{h}^2$, and $\mathbf{h}^3$, the fixed point $\mathbf{h}^2$ is stable and the other two unstable.

Set

$$l_i = \{h \in R^{q-1} : h_j = 0, \; j \neq i, \; j = 1, \ldots, q-1\}, \quad i = 1, \ldots, q-1.$$

Obviously, the lines l_i are invariant with respect to mapping $kF(h, \theta, q)$. An analogue of Proposition 5.4 can be proved for restrictions of the mapping $kF(h, \theta, q)$ on each l_i, $i = 1, \ldots, q-1$.

Proposition 5.5. *Let $\theta > \theta_c$. The stable fixed points which lie on invariant sets l_i, $i = 1, \ldots, q$ are the only stable fixed points of $kF(h, \theta, q)$.*

Thus for $\theta > \theta_c$ the mapping $kF(h, \theta, q)$ has q stable fixed points $\mathbf{h}^i_*$, and the set of vectors $\{h_x = \mathbf{h}^i_*, \; x \in V\}$, $i = 1, \ldots, q$ satisfies equation (5.4). We denote by μ^i the Gibbs measure corresponding to the set of vectors $\{h_x = \mathbf{h}^i_*, \; x \in V\}$, $i = 1, \ldots, q$.

Theorem 5.4. *The Gibbs measures μ^i, $i = 1, \ldots, q$ are extreme.*

From $\theta_c = \exp(J/T_c) = \frac{k+q-1}{k-1}$ we find the critical temperature of phase transitions for q state Potts model:

$$T_c(k, q) = \frac{J}{\ln(1 + \frac{q}{k-1})}.$$

Note that $T_c(k, q) \to 0$ as $q \to \infty$. Thus the following theorem is true.

Theorem 5.5. *For the Potts model with countable many states on the Cayley tree, the translation-invariant Gibbs measure is unique.*

5.3 Extremality of the disordered Gibbs measure: The reconstruction solvability

Results presented in this section are based on methods developed in [169]-[171] (Chapter 4). In the previous section we proved that for $T < T_c(k, q) = J/\ln(1 + q/(k-1))$ equation (5.4) has q non-zero solutions, which for the Potts model correspond to q extreme translation-invariant Gibbs measures. Moreover, it is obvious that for any k, q, θ the functional equation (5.4) has a solution $h_x = h_0 = (0, 0, \ldots, 0)$. The Gibbs measure μ_0 corresponding to the solution h_0 is called the *disordered phase* or *disordered Gibbs measure*.

The following theorem gives some extremality properties of the measure μ_0.

Theorem 5.6. *For $k \geq 2$,*

(i) *If $J > 0$ and one of the following conditions is satisfied*

    (i.1)

$$T < T_c(\sqrt{k}, q) = \frac{J}{\ln(1 + \frac{q}{\sqrt{k}-1})} \tag{5.22}$$

    (i.2) *q is sufficiently large and*

$$T < T_c(k, q) = \frac{J}{\ln(1 + \frac{q}{k-1})} \tag{5.23}$$

then the disordered Gibbs measure μ_0 is not extreme.

(ii) *If one of the following conditions is satisfied*

    (ii.1)

$$T \geq T_c(k, q) = \frac{|J|}{\ln(1 + \frac{q}{k-1})} \tag{5.24}$$

    (ii.2) *If $J > 0$ and*

$$J \ln^{-1}\left(\frac{2k + q + \sqrt{(q-2)^2 + 8kq}}{2(k-1)} \right) \leq T \leq$$

$$J \ln^{-1}\left(\frac{2k + q - \sqrt{(q-2)^2 + 8kq}}{2(k-1)} \right) \tag{5.25}$$

then the disordered Gibbs measure μ_0 is extreme.

Proof. We apply Propositions 4.3-4.6. Consider a two-parameter family of Markov chains with states $\{1, 2, \ldots, q\}$ and transition probabilities

$$\mathbf{M}_{i,j} = \frac{\exp(J\beta\delta_{ij})}{\sum_{l=1}^{q} \exp(J\beta\delta_{il})}. \tag{5.26}$$

The matrix $\mathbf{M} = (\mathbf{M}_{i,j})$ coincides with the matrix (4.24) for $\delta = (q - 1 + \exp(J\beta))^{-1}$.

By Proposition 4.3 a sufficient condition for non-extremality of μ_0 is that $k\lambda_2^2(M) > 1$, where $\lambda_2(M) = 1 - q\delta$. Simple calculations show that the last condition is equivalent to the condition (5.22). Similarly, other statements of theorem can be proved using Propositions 4.4-4.6. $\qquad\square$

Remark 5.1. For symmetric models the disordered Gibbs measure corresponds to the zero solution of the recurrent equations (equation (2.8) for the Ising model, and (5.4) for the Potts model). But if the model is not symmetric, then the corresponding recurrent equation has no zero solution. In this case the disordered measure is defined as a translation-invariant Gibbs measure which corresponds to the free boundary condition. This is why, in some literatures (see for example [169]) the disordered measure is called free measure.

5.4 A construction of an uncountable set of Gibbs measures

Now we are going to constructively describe an uncountable set of non-translational invariant Gibbs measures for the three state ferromagnetic Potts model with zero external field. We shall use notations of Bleher-Ganikhodjaev construction of Chapter 2.

Consider an arbitrary (finite or infinite) path $x^0 = x_0 < x_1 < x_2 < \ldots$ starting from the point x^0. We assign the real number $t = t(\pi) \in [0,1]$ to the path π.

Let $\pi = \{x^0 = x_0 < x_1 < \ldots\}$ be an infinite path. We associate the path π with a set of vectors $\mathbf{h}^\pi = \{h_x^\pi, x \in V^0\}$ satisfying equation (5.4) for $q = 3$.

For $x \in W_n$, the set $\mathbf{h}^\pi$ is uniquely defined by the conditions

$$h_x^\pi = \begin{cases} \mathbf{h}_*^1, & \text{if } x \prec x_n, \ x \in W_n, \\ \mathbf{h}_*^2, & \text{if } x_n \prec x, \ x \in W_n, \end{cases} \tag{5.27}$$

$n = 1, 2, \ldots$, where $\mathbf{h}_*^i$, $i = 1, 2$ stable fixed points of $kF(h, \theta)$. Here one can also use another fixed points.

Theorem 5.7. *For any infinite path π, there exists a unique set of vectors $\mathbf{h}^\pi = \{h_x^\pi, x \in V^0\}$ satisfying equations (5.4) and (5.27).*

Proof. Let $\pi = \{x^0 = x_0 < x_1 < \ldots\}$ be an arbitrary infinite path. On W_n, we define the set

$$h_x^{(n)} = \begin{cases} \mathbf{h}_*^1, & \text{if } x \prec x_n, \ x \in W_n, \\ \mathbf{h}_*^2, & \text{if } x_n \prec x, \ x \in W_n, \\ h_x^{(n)} \in K, & \text{if } x = x_n, \end{cases} \tag{5.28}$$

where $h_{x_n}^{(n)} \in K$ is an arbitrary vector. We extend the definition of $h_x^{(n)}$ for all $x \in V_n = \cup_{m=0}^n W_m$ using recursion equations (5.4). We prove that the limit

$$h_x = \lim_{n \to \infty} h_x^{(n)} \tag{5.29}$$

exists for any fixed $x \in V^0$ and is independent of the choice of $h_x^{(n)}$ for $x = x_n$. If $x \in W_{n-1}$ and $x \prec x_{n-1}$, then

$$h_x^{(n)} = \sum_{y \in W_n, y > x} F(h_y^{(n)}, \theta) = kF(\mathbf{h}_*^1, \theta) = \mathbf{h}_*^1.$$

Similarly, for $x \in W_{n-1}$ and $x_{n-1} \prec x$, we get $h_x^{(n)} = \mathbf{h}_*^2$. Consequently, for any $x \in W_m$, $m \le n$ we have

$$h_x^{(n)} = \begin{cases} \mathbf{h}_*^1, & \text{if } x \prec x_m, \ x \in W_m, \\ \mathbf{h}_*^2, & \text{if } x_m \prec x, \ x \in W_m. \end{cases}$$

This implies that limit (5.29) exists for $x \in W_m$ and $x \ne x_m$ and

$$h_x = \begin{cases} \mathbf{h}_*^1, & \text{if } x \prec x_m, \ x \in W_m, \\ \mathbf{h}_*^2, & \text{if } x_m \prec x, \ x \in W_m. \end{cases}$$

Thus, we only need to establish that limit (5.29) exists for $x = x_m$. Let $1 \le l \le n$. Then

$$h_{x_{l-1}}^{(n)} = \sum_{y \in W_l, y > x_{l-1}} F(h_y^{(n)}, \theta). \tag{5.30}$$

We consider two sets of vectors $\{\bar{h}_x^{(n)}, x \in V_n\}$ and $\{\tilde{h}_x^{(n)}, x \in V_n\}$ correspond to two values $\bar{h}_x^{(n)}$ and $\tilde{h}_x^{(n)}$ for $x = x_n$, in (5.28), then from (5.30) we get

$$\tilde{h}_{x_{l-1}}^{(n)} - \bar{h}_{x_{l-1}}^{(n)} = \sum_{y \in W_l, y > x_{l-1}} \left[F(\tilde{h}_y^{(n)}, \theta) - F(\bar{h}_y^{(n)}, \theta) \right].$$

Since $\tilde{h}_y^{(n)} = \bar{h}_y^{(n)}$ for any $y \ne x_l$, $y \in W_l$, the unique non-vanishing term on the RHS of the last equation corresponds to $y = x_l$, and therefore

$$\tilde{h}_{x_{l-1}}^{(n)} - \bar{h}_{x_{l-1}}^{(n)} = F(\tilde{h}_{x_l}^{(n)}, \theta) - F(\bar{h}_{x_l}^{(n)}, \theta). \tag{5.31}$$

By virtue of Lemma 5.2 and Proposition 5.3 we get

$$\left| \tilde{h}_{x_{l-1}}^{(n)} - \bar{h}_{x_{l-1}}^{(n)} \right| \le \alpha \max \left(\| (\tilde{h}_{x_l}^{(n)})_0 - (\bar{h}_{x_l}^{(n)})_0 \|, \| (\tilde{h}_{x_l}^{(n)})_1 - (\bar{h}_{x_l}^{(n)})_1 \| \right). \tag{5.32}$$

This estimate establishes the contractive property for the vectors $h_{x_l}^{(n)}$. Iterating this inequality we obtain

$$\left| \tilde{h}_{x_m}^{(n)} - \bar{h}_{x_m}^{(n)} \right| \le \alpha^{n-m} \max \left(\| (\tilde{h}_{x_n}^{(n)})_0 - (\bar{h}_{x_n}^{(n)})_0 \|, \| (\tilde{h}_{x_n}^{(n)})_1 - (\bar{h}_{x_n}^{(n)})_1 \| \right). \tag{5.33}$$

For arbitrary $N, M > n$, we now consider the sets $\{h_x^{(N)}, x \in V_N\}$ and $\{h_x^{(M)}, x \in V_M\}$ determined by initial conditions of form (5.28) for $x \in W_N$ and $x \in W_M$ respectively and by recursion equations (5.4). We set

$$\bar{h}_{x_n}^{(n)} = h_{x_n}^{(N)}, \quad \tilde{h}_{x_n}^{(n)} = h_{x_n}^{(M)}.$$

Then from (5.33) we get

$$\left| h_{x_m}^{(N)} - h_{x_m}^{(M)} \right| \leq 4 h_* \alpha^{n-m}.$$

It follows from this estimate that the sequence of vectors $h_{x_m}^{(n)}$ satisfies the Cauchy criterion as $n \to \infty$ for a fixed m; therefore, limit (5.29) exists and is independent of the choice of $h_{x_n}^{(n)}$ in (5.28). By construction, the sets $\{h_x^{(n)}\}$ satisfy equation (5.4) before taking the limit, so does $\{h_x\}$. The uniqueness of $\{h_x\}$ obviously follows from estimate (5.33). $\qquad\square$

With an arbitrary path π (finite or infinite) emanating from the point x^0 we can in the standard manner associate a point $t \in [0, 1]$ (see Chapter 2). It is easy to show that for any $t \in [0, 1]$ there is uniquely determined a set of vectors $\mathbf{h}^{\pi(t)}$, and $\mathbf{h}^{\pi(t)}$ are different for all t. By Theorem 5.1, every set $\mathbf{h}^{\pi(t)}$ can be associated with a Gibbs measure, which we denote by μ^t.

The following theorem is an analogue of Theorem 2.12.

Theorem 5.8. *For any $t \in [0, 1]$, the Gibbs measure μ^t is extreme.*

Commentaries and references. The Potts model is a generalization of the Ising model, but the Potts model on a Cayley tree is not well studied, compared to the Ising model. For example, there is no any result about periodic and weakly periodic Gibbs measures of the Potts model on a Cayley tree.

In [85], for $q = 3$ and $k = 2$, Theorem 5.1 is proved. In [86], for the general case ($q \geq 2$, $k \geq 1$) Theorem 5.1 is announced. The proof presented in [85] has the following two steps:

Step 1. The set of states $\Phi = \{1, \ldots, q\}$ replaced with the set of vectors $\tilde{\Phi} = \{\sigma_1, \ldots, \sigma_q\} \subset R^{q-1}$, with

$$\sigma_i \sigma_j = \begin{cases} 1, & i = j, \\ -\frac{1}{q-1}, & i \neq j. \end{cases}$$

Then it is easy to see that

$$\delta_{\sigma(x)\sigma(y)} = \frac{q-1}{q} \left(\sigma(x)\sigma(y) + \frac{1}{q-1} \right),$$

for any $x, y \in V$. Using this formula the Hamiltonian of the Potts model is transformed to the Hamiltonian of the Ising model, but with q spin values from the set $\tilde{\Phi}$.

Step 2. The conditions for consistency of the vectors $\{h_x, x \in V\}$ are described. Step 1 allows now to write the potential of a point x of boundary as $\sigma(x)h_x$, where $\sigma(x)$ is the vector-spin value on x, and h_x is a vector interpreted as an external field on x. Note that the same formula of the potential (of x) is used for Ising model. So the driving of the equation is now similar to the case of Ising model.

We note that the transformation of Hamiltonian mentioned in Step 1 will give some transformation on the formula of critical value, i.e., the critical temperature $T_c(k, q) = \frac{J}{\ln(1 + \frac{q}{k-1})}$ is transformed to $\frac{q}{q-1} T_c(k, q)$ (see [86]).

The proof of Theorem 5.1 presented in this chapter is due to the author. We did not transform Step 1, the crucial point in our proof is that we took the potential of a boundary point x, with integer-spin value $\sigma(x)$, as $h_{\sigma(x),x}$, i.e., the variables i and x of $h_{i,x}$ are not separable.

In [213] Theorem 5.2 is proved. Theorems 5.3, 5.4, 5.5 and the construction of an uncountable extreme Gibbs measures are due to Ganikhodjaev [85], [86].

The extremality conditions (Theorem 5.6) is due to the author (is not published). In [92], using the idea of reductions to recursive estimates on second moments, inherited from the papers [48], [25], a rough estimate for critical temperature of the extremality is obtained.

For other results related to the Potts model on Cayley tree see [5], [6], [16], [195]-[200], [259].

Chapter 6

The Solid-on-Solid model

In this chapter we consider a nearest-neighbor SOS (solid-on-solid) model, with several spin values $0, 1, \ldots, m$, $m \geq 2$, and zero external field, on a Cayley tree of order k. We mainly assume that $m = 2$ or $m = 3$ and study Gibbs measures.

For $m = 2$, in the anti-ferromagnetic case, we show that the translation-invariant Gibbs measure is unique for all temperatures. In the ferromagnetic case, for $m = 2$, the number of such measures varies with the temperature: this gives an interesting example of phase transition. Here we identify a critical inverse temperature, $\beta_{\mathrm{cr}}^1 \in (0, \infty)$ such that $\forall\, 0 \leq \beta \leq \beta_{\mathrm{cr}}^1$, there exists a unique translation-invariant measure μ^* and $\forall\, \beta > \beta_{\mathrm{cr}}^1$ there are exactly three such measures: μ_+^*, μ_{m}^* and μ_-^*. For $\beta > \beta_{\mathrm{cr}}^1$ we also construct a continuum of distinct, non-translation-invariant Gibbs measures.

Our second result gives complete description of the set of periodic Gibbs measures for the SOS model on a Cayley tree. A complete description of periodic Gibbs measures means a characterization of such measures with respect to any given normal subgroup of finite index in the representation group of the tree. We show that (i) for a ferromagnetic SOS model, for any normal subgroup of finite index, each periodic Gibbs measure is in fact translation-invariant. Further, (ii) for an anti-ferromagnetic SOS model, for any normal subgroup of finite index, each periodic Gibbs measure is either translation-invariant or has period two (i.e., is a chess-board Gibbs measure).

For $m = 3$ similar results are obtained. But the case $m \geq 4$ is not studied yet.

6.1 The model and a system of vector-valued functional equations

We consider models where the spin takes values in the set $\Phi :=$ $\{0, 1, \ldots, m\}$, $m \geq 2$, and is assigned to the vertices of the tree. A configuration σ on V is then defined as a function $x \in V \mapsto \sigma(x) \in \Phi$; the set of all configurations is Φ^V. The (formal) Hamiltonian is of an SOS form:

$$H(\sigma) = -J \sum_{\langle x,y \rangle \in L} |\sigma(x) - \sigma(y)|, \tag{6.1}$$

where $J \in R$ is a coupling constant. As usual, $\langle x, y \rangle$ stands for nearest neighbor vertices.

The SOS model of this type can be considered as a generalization of the Ising model (which arises when $m = 1$). Here, $J < 0$ gives a ferromagnetic and $J > 0$ an anti-ferromagnetic model. In the ferromagnetic case the ground states are 'flat' configurations, with $\sigma(x) \equiv j \in \Phi$ (there are $m + 1$ of them), in the anti-ferromagnetic two 'contrasting' checker-board configurations where $|\sigma(x) - \sigma(y)| = m \; \forall \; \langle x, y \rangle$. Compared with the Potts model, the SOS has 'less symmetry' and therefore more diverse structure of phases. For example, in the ferromagnetic case it is intuitively plausible that the ground states corresponding to 'middle-level' surfaces will be 'dominant'. This observation was made formal in [156] for the model on a cubic lattice.

We use a standard definition of a Gibbs measure, a translation-invariant (TI) measure. Also, call measure μ symmetric if it is preserved under the simultaneous change $j \mapsto m - j$ at each vertex $x \in V$.

Now we shall drive a system of functional equations, solutions of which correspond to Gibbs measures of SOS model on the Cayley tree.

Let $h : \; x \mapsto h_x = (h_{0,x}, h_{1,x}, \ldots, h_{m,x}) \in R^{m+1}$ be a real vector-valued function of $x \in V \setminus \{x^0\}$. Given $n = 1, 2, \ldots$, consider the probability distribution μ_n on Φ^{V_n} defined by

$$\mu^{(n)}(\sigma_n) = Z_n^{-1} \exp\left(-\beta H(\sigma_n) + \sum_{x \in W_n} h_{\sigma(x),x}\right). \tag{6.2}$$

Here, $\sigma_n : x \in V_n \mapsto \sigma(x)$ and Z_n is the corresponding partition function:

$$Z_n = \sum_{\widetilde{\sigma}_n \in \Phi^{V_n}} \exp\left(-\beta H(\widetilde{\sigma}_n) + \sum_{x \in W_n} h_{\widetilde{\sigma}(x),x}\right). \tag{6.3}$$

We say that the probability distributions $\mu^{(n)}$ are compatible if $\forall\, n \geq 1$ and $\sigma_{n-1} \in \Phi^{V_{n-1}}$:

$$\sum_{\omega_n \in \Phi^{W_n}} \mu^{(n)}(\sigma_{n-1} \vee \omega_n) = \mu^{(n-1)}(\sigma_{n-1}). \tag{6.4}$$

Here $\sigma_{n-1} \vee \omega_n \in \Phi^{V_n}$ is the concatenation of σ_{n-1} and ω_n. In this case there exists a unique measure μ on Φ^V such that, $\forall\, n$ and $\sigma_n \in \Phi^{V_n}$,
$\mu\left(\left\{\sigma\big|_{V_n} = \sigma_n\right\}\right) = \mu^{(n)}(\sigma_n)$. Such a measure is called a *splitting Gibbs measure* (SGM) corresponding to Hamiltonian H and function $x \mapsto h_x$, $x \neq x^0$.

The following statement describes conditions on h_x guaranteeing compatibility of distributions $\mu^{(n)}(\sigma_n)$.

Proposition 6.1. *Probability distributions $\mu^{(n)}(\sigma_n)$, $n = 1, 2, \ldots$, in (6.2) are compatible iff for any $x \in V \setminus \{x^0\}$ the following equation holds:*

$$h_x^* = \sum_{y \in S(x)} F(h_y^*, m, \theta). \tag{6.5}$$

Here,

$$\theta = \exp(J\beta), \tag{6.6}$$

h_x^* stands for the vector $(h_{0,x} - h_{m,x}, h_{1,x} - h_{m,x}, \ldots, h_{m-1,x} - h_{m,x})$ and the vector function $F(\ \cdot\ , m, \theta)\ :\ R^m \to R^m$ is $F(h, m, \theta) = (F_0(h, m, \theta), \ldots, F_{m-1}(h, m, \theta))$, with

$$F_i(h, m, \theta) = \ln \frac{\sum_{j=0}^{m-1} \theta^{|i-j|} \exp(h_j) + \theta^{m-i}}{\sum_{j=0}^{m-1} \theta^{m-j} \exp(h_j) + 1}, \tag{6.7}$$

$h = (h_0, h_1, \ldots, h_{m-1}), i = 0, \ldots, m-1$.

Proof. *Necessity.* Suppose that (6.4) holds; we want to prove (6.5). Substituting (6.2) in (6.4), obtain, $\forall$ configurations $\sigma_{n-1}\colon x \in V_{n-1} \mapsto \sigma_{n-1}(x) \in \Phi$:

$$\frac{Z_{n-1}}{Z_n} \sum_{\omega_n \in \Phi^{W_n}} \exp\left(\sum_{x \in W_{n-1}} \sum_{y \in S(x)} (J\beta|\sigma_{n-1}(x) - \omega_n(y)| + h_{\omega_n(y),y})\right) = \tag{6.8}$$

$$\exp\left(\sum_{x \in W_{n-1}} h_{\sigma_{n-1}(x),x}\right),$$

where $\omega_n \colon x \in W_n \mapsto \omega_n(x)$.

From (6.8) we get:

$$\frac{Z_{n-1}}{Z_n} \sum_{\omega_n \in \Phi^{W_n}} \prod_{x \in W_{n-1}} \prod_{y \in S(x)} \exp\left(J\beta|\sigma_{n-1}(x) - \omega_n(y)| + h_{\omega_n(y),y}\right) = \quad (6.9)$$

$$\prod_{x \in W_{n-1}} \exp\left(h_{\sigma_{n-1}(x),x}\right).$$

Consequently, $\forall\, i \in \Phi$,

$$\prod_{y \in S(x)} \frac{\sum_{j \in \Phi} \exp\left(J\beta|i - j| + h_{j,y}\right)}{\sum_{j \in \Phi} \exp\left(J\beta|m - j| + h_{j,y}\right)} = \exp\left(h_{i,x} - h_{m,x}\right). \qquad (6.10)$$

Introducing θ as in (6.6) and denoting $h_{i,x}^* = h_{i,x} - h_{m,x}$, we get (6.5) from (6.10).

Sufficiency. From (6.5) we obtain (6.10), (6.9) and (6.8), i.e., (6.4). $\quad\square$

The following proposition is straightforward.

Proposition 6.2. *1) Any measure μ with local distributions $\mu^{(n)}$ satisfying (6.2), (6.4) is an SGM.*

2) An SGM μ is TI iff $h_{j,x}$ does not depend on x: $h_{j,x} \equiv h_j$, $x \in V$, $j \in \Phi$, and symmetric TI iff $h_j = h_{m-j}$, $j \in \Phi$.

6.2   Three-state SOS model

From Proposition 6.2 it follows that for any $h = \{h_x, \quad x \in V\}$ satisfying (6.5) there exists a unique Gibbs measure μ (with restrictions $\mu^{(n)}$ as in (6.2)) and vice versa. However, the analysis of solutions to (6.5) for an arbitrary m is not easy. We now suppose that the number of spin values $m+1$ is 3, i.e., $m = 2$ and $\Phi = \{0, 1, 2\}$. We assume that $h_{2,x} \equiv 0$ ($h_{m,x} \equiv 0$ for general m).

6.2.1   *The critical value β_{cr}^1*

It is natural to begin with TI solutions where $h_x = h \in R^m$ is constant. Unless otherwise stated, we concentrate on the simplest case where $m = 2$, i.e., spin values are 0, 1 and 2. In this case we obtain from (6.5), (6.6):

$$
\begin{aligned}
h_{0,x} &= \sum_{y \in S(x)} \ln \frac{\exp(h_{0,y}) + \theta \exp(h_{1,y}) + \theta^2}{\theta^2 \exp(h_{0,y}) + \theta \exp(h_{1,y}) + 1}, \\
h_{1,x} &= \sum_{y \in S(x)} \ln \frac{\theta \exp(h_{0,y}) + \exp(h_{1,y}) + \theta}{\theta^2 \exp(h_{0,y}) + \theta \exp(h_{1,y}) + 1}.
\end{aligned}
\qquad (6.11)
$$

Set $z_0 = \exp(h_{0,x})$, $z_1 = \exp(h_{1,x})$ (and $z_2 = 1$), $x \in V$. From (6.11) we have

$$z_0 = \left(\frac{z_0 + \theta z_1 + \theta^2}{\theta^2 z_0 + \theta z_1 + 1} \right)^k, \tag{6.12}$$

$$z_1 = \left(\frac{\theta z_0 + z_1 + \theta}{\theta^2 z_0 + \theta z_1 + 1} \right)^k. \tag{6.13}$$

Observe that $z_0 = 1$ satisfies equation (6.12) independently of k, θ and z_1. Substituting $z_0 = 1$ into (6.13), we obtain

$$z_1 = \left(\frac{2\theta + z_1}{\theta^2 + \theta z_1 + 1} \right)^k. \tag{6.14}$$

Set:

$$a = 2\theta^{k+1}, \quad b = \frac{1+\theta^2}{2\theta^2}, \quad x = \frac{z_1}{2\theta}. \tag{6.15}$$

Then from (6.14):

$$ax = \left(\frac{1+x}{b+x} \right)^k. \tag{6.16}$$

Now for this equation we can use Proposition 5.1.

Proposition 6.3. *If $J \geq 0$ then the system of equations (6.12), (6.13) has a unique solution.*

Proof. Let $A = z_0 + \theta z_1 + \theta^2$, $B = \theta^2 z_0 + \theta z_1 + 1$, then from (6.12) we have:

$$(z_0 - 1)[B^k + (\theta^2 - 1)(A^{k-1} + ... + B^{k-1})] = 0. \tag{6.17}$$

Since $\theta \geq 1$ ($J \geq 0$), we deduce from (6.17) that $z_0 = 1$ is the only solution. Then $b = \frac{1+\theta^2}{2\theta^2} \leq 1 < (\frac{k+1}{k-1})^2$. By Proposition 5.1, equation (6.14) has a unique solution. Thus we have proved that system (6.12), (6.13) has a unique solution. $\square$

Proposition 6.4. *If $J < 0$ then for $\beta \leq \frac{1}{2J} \ln \frac{(k-1)^2}{k^2+6k+1}$, the system of equations (6.12), (6.13) has a unique solution of the form $(1, z^*)$ (i.e., a unique solution (z_0^*, z_1^*) with $z_0^* = 1$) and for $\beta > \frac{1}{2J} \ln \frac{(k-1)^2}{k^2+6k+1}$, precisely three such solutions, $(1, z_{1,-}^*)$, $(1, z_{1,\mathrm{m}}^*)$, $(1, z_{1,+}^*)$, with $0 < z_{1,-}^* < z_{1,\mathrm{m}}^* < z_{1,+}^*$ and $z_{1,i}^* = e^{h_{1,i}^*}$, $i = -, \mathrm{m}, +$, (see (6.11)).*

Proof. The value $\frac{1}{2J} \ln \frac{(k-1)^2}{k^2+6k+1}$ is the solution of equation $b = (\frac{k+1}{k-1})^2$. Other statements of Proposition 6.4 are simple consequences of Proposition 5.1. $\qquad\qquad\qquad\qquad\qquad\qquad\qquad\qquad\qquad\qquad\qquad\qquad\quad\square$

For brevity we say that z^* and z^*_-, z^*_{m}, z^*_+ give symmetric solutions to (6.12) and (6.13).

In the ferromagnetic case, set:

$$\beta^1_{\mathrm{cr}} = \frac{1}{2J} \ln \frac{(k-1)^2}{k^2+6k+1} > 0. \tag{6.18}$$

Going back to (6.6), summarize:

Theorem 6.1.

(i) For the anti-ferromagnetic SOS model, with $J > 0$ and $m = 2$, the TISGM exists and is unique $\forall\ \beta \geq 0$. In fact, it is a symmetric measure.

(ii) For the ferromagnetic SOS model, with $J < 0$ and $m = 2$:

    (ii.1) If $k \geq 2$ and $0 \leq \beta \leq \beta^1_{\mathrm{cr}}$ then there exists a unique symmetric TISGM, μ^*.

    (ii.2) If $k \geq 2$ and $\beta > \beta^1_{\mathrm{cr}}$ then there exist precisely three symmetric TISGMs μ^*_-, μ^*_{m}, μ^*_+ corresponding to $h^*_i = \ln z^*_i$, $i = -, \mathrm{m}, +$.

Remark 6.1. Note that in the anti-ferromagnetic case, the phase transition is manifested in the break of the TI property. More precisely, it is expected that for β small there exists a unique translation-periodic SGM (this is TI) while for β large there are several such measures.

In the ferromagnetic case, observe that values z^*_i, $i = -, \mathrm{m}, +$, vary with β. It is easy to show that as $\beta \to \infty$, $z^*_- \to 0$, $z^*_{\mathrm{m}} \to 1$ and $z^*_+ \to \infty$. Correspondingly, we make a

Conjecture 1. For $m = 2$, $k \geq 2$ and $J < 0$, as $\beta \to \infty$, measure μ^*_- tends to the half-sum $\frac{1}{2}\left(\delta_{\omega^0} + \delta_{\omega^2}\right)$, μ^*_{m} to the mean $\frac{1}{3}\left(\delta_{\omega^0} + \delta_{\omega^1} + \delta_{\omega^2}\right)$ and μ^*_+ to δ_{ω^1}. Here δ_ω stands for the Dirac delta-measure sitting on configuration $\omega \in \Phi^V$ and ω^i has $\omega^i(x) \equiv i$, $i = 0, 1, 2$.

On the other hand, we can say that for $\beta \leq \beta^1_{\mathrm{cr}}$, all three measures coincide and in the limit $\beta \to 0$ give a Bernoulli measure, with iid and equiprobable values $\sigma(x) = 0, 1, 2$, $x \in V$.

Remark 6.2. Note that β^1_{cr} may not be the first critical value of the inverse temperature for the ferromagnetic model. Namely, there exists β^0_{cr} $(= \beta^{\mathrm{TIGM}}_{\mathrm{c}}) \in (0, \beta^1_{\mathrm{cr}}]$ such that (i) for $0 \leq \beta \leq \beta^0_{\mathrm{cr}}$, a minimal GM, μ_-, and a maximal, μ_+, coincide, and the whole set of GMs is reduced to a unique measure which is therefore extreme (and coincides with symmetric TISGM μ^*), (ii) for $\beta > \beta_{\mathrm{cr}}^0$, μ_- and μ_+ are distinct (they are always extreme TIS-GMs, but not symmetric). Thus, for $\beta \geq \beta_{\mathrm{cr}}^1$, there are five such SGMs (in a natural order: $\mu_- \leq \mu_-^* \leq \mu_{\mathrm{m}}^* \leq \mu_+^* \leq \mu_+$) three of which are symmetric. It is not known whether $\beta_{\mathrm{cr}}^0 = \beta_{\mathrm{cr}}^1$ (it is our **Conjecture 2**).

The following Proposition 6.5 describes a useful property of general (non-TI) solutions $h_x = (h_{0,x}; h_{1,x})$ to (6.11) with $h_{0,x} \equiv 0$ (or $z_0^* \equiv 1$). As before, $h_{0,x}$ gives a solution to the first equation in (6.11), regardless of $h_{1,x}$ and θ.

Proposition 6.5. *For $J < 0$, $k \geq 2$ and $\beta > \beta_{\mathrm{cr}}^1$, if $h_x = (0; h_{1,x})$ is a solution of (6.11) then, with $h_{1,x} = \ln z_{1,x}$,*

$$z_-^* \leq z_{1,x} \leq z_+^*, \quad x \in V \tag{6.19}$$

where $z_-^* < z_+^*$ are the symmetric solutions of (6.12), (6.13), i.e., the solutions of (6.14).

Proof. Denote $z_x = \exp(h_{1,x})$. Then from (6.11) we get

$$z_x = \prod_{i=1}^{k} \frac{2\theta + z_{x_i}}{1 + \theta^2 + \theta z_{x_i}}, \quad z_{x_j} > 0, \quad j = 1, \ldots, k,$$

where x_j, $j = 1, \ldots, k$ are direct successors of x. Denote $\varphi(x, \theta) = \frac{2\theta + x}{1 + \theta^2 + \theta x}$. Consider

$$G(x_1, ..., x_k) = \prod_{i=1}^{k} \varphi(x_i, \theta), \quad x_i > 0, \quad i = 1, \ldots, k.$$

Set the map $x \mapsto \psi(x, \theta, k) = (\varphi(x, \theta))^k$. Clearly, $\psi(0, \theta, k) \leq G(x_1, ..., x_k) \leq \psi(\infty, \theta, k)$. Thus for z_x we get $\psi(0, \theta, k) \leq z_x \leq \psi(\infty, \theta, k)$. Now consider $G(x_1, ..., x_k)$ with $\psi(0, \theta, k) \leq x_j \leq \psi(\infty, \theta, k)$. Here we have

$$\psi(\psi(0, \theta, k), \theta, k) \leq z_x \leq \psi(\psi(\infty, \theta, k), \theta, k).$$

Repeating this argument, we see that for the nth iteration $\psi^{(n)}$ of ψ:

$$\psi^{(n)}(0, \theta, k) \leq z_x \leq \psi^{(n)}(\infty, \theta, k),$$

for all $n \geq 1$ and $x \in V \setminus \{x^0\}$. The sequence $\psi^{(n)}(\infty, \theta, k)$ is decreasing and bounded from below by z_+^*. Its limit is a fixed point for ψ and thus equal to z_+^*. The lower bound for z_x is similar and gives z_-^*. $\qquad\square$

Proposition 6.6. *For $J < 0$ and $\beta \le \beta^1_{cr}$, measure μ^* is the only SGM such that $z_{0,x} = 0$, $x \in V \setminus \{x^0\}$ (regardless of whether it is TI or not). Thus, μ^* is the only symmetric SGM.*

Proof. In this case equation (6.11) with $h_{0,x} = 0$ has a unique solution $h_x = (0, \ln z^*)$. $\qquad\qquad\qquad\qquad\qquad\qquad\qquad\qquad\qquad\qquad\qquad\quad \square$

Conjecture 3. In the case $m = 2$, $J < 0$ and $\beta \le \beta^1_{cr}$, μ^* is the unique GM and hence extreme.

Conjecture 4. In the case $k \ge 2$, $J < 0$ and $\beta > \beta^1_{cr}$, the boundary condition for the top symmetric TISGM μ^*_+ is $\omega^{(n)}(x) \equiv 1$.

The boundary conditions for the bottom and middle symmetric TISGM, μ^*_- and μ_m, are unclear. In the case of a general m, we also have two conjectures.

Conjecture 5. For all $m, k \ge 2$ and $J < 0$, there exist symmetric solutions $h = (h_0, h_1, ..., h_{m-1})$ to (6.7), with $h_0 = 0$ and $h_i = h_{m-i}$, $i = 1, 2, ..., m - 1$.

Conjecture 6. $\forall\, m, k \ge 2$ and $J > 0$, $\forall\, \beta \ge 0$ the TISGM is unique and is a symmetric measure.

6.2.2   *Periodic SGMs*

In this section we study a periodic solutions of system (6.11).

Periodic Gibbs measures of the SOS model can be defined similarly as in Chapter 2, by using the group representation G_k of the Cayley tree Γ^k.

We give a complete description of periodic Gibbs measures, i.e., a characterization of such measures with respect to any normal subgroup of finite index in G_k.

For convenience of the reader we recall some necessary notations: Let K be a subgroup of index r in G_k, and let $G_k/K = \{K_0, K_1, ..., K_{r-1}\}$ be the quotient group, with the coset $K_0 = K$. Let $q_i(x) = |S_1(x) \cap K_i|$, $i = 0, 1, ..., r - 1$; $N(x) = |\{j : q_j(x) \ne 0\}|$, where $S_1(x) = \{y \in G_k : \langle x, y \rangle\}$, $x \in G_k$ and $|\cdot|$ is the number of elements in the set. Denote $Q(x) = (q_0(x), q_1(x), ..., q_{r-1}(x))$.

For every $x \in G_k$ there is a permutation π_x of the coordinates of the vector $Q(e)$ (where e is the identity of G_k) such that

$$\pi_x Q(e) = Q(x).$$

Each K-periodic collection is given by

$$\{h_x = h_i \text{ for } x \in K_i, \quad i = 0, 1, ..., r-1\}.$$

By Proposition 6.2 (for $m = 2$), vector h_n, $n = 0, 1, ..., r-1$, satisfies the system

$$h_n = \sum_{j=1}^{N(e)} q_{i_j}(e) F(h_{\pi_n(i_j)}; \theta) - F(h_{\pi_n(i_{j_0})}; \theta), \tag{6.20}$$

where $j_0 = 1, ..., N(e)$, and function $h \mapsto F(h, m, \theta)$ defined in Proposition 6.2 now takes the form $h \mapsto F(h) = (F_0(h, \theta), F_1(h, \theta))$ where

$$F_0(h, \theta) = \ln \frac{\exp(h_0) + \theta \exp(h_1) + \theta^2}{\theta^2 \exp(h_0) + \theta \exp(h_1) + 1},$$

$$F_1(h, \theta) = \ln \frac{\theta \exp(h_0) + \exp(h_1) + \theta}{\theta^2 \exp(h_0) + \theta \exp(h_1) + 1}. \tag{6.21}$$

Recall, θ has been defined in (6.6).

Proposition 6.7. *If $\theta \neq 1$, then $F(h) = F(l)$ if and only if $h = l$.*

Proof. *Necessity.* From $F(h) = F(l)$ we get the system of equations

$$\begin{cases} \theta\Big(\exp(h_0 + l_1) - \exp(h_1 + l_0) \Big) + \\ (1 + \theta^2)\Big(\exp(h_0) - \exp(l_0) \Big) + \theta\Big(\exp(h_1) - \exp(l_1) \Big) = 0, \\ \theta\Big(\exp(h_0) - \exp(l_0) \Big) + \exp(h_1) - \exp(l_1) = 0, \end{cases} \tag{6.22}$$

where $h = (h_0, h_1)$, $l = (l_0, l_1)$. Using the fact that

$$\exp(h_0 + l_1) - \exp(h_1 + l_0) =$$

$$\exp(l_1)\Big(\exp(h_0) - \exp(l_0) \Big) - \exp(l_0)\Big(\exp(h_1) - \exp(l_1) \Big),$$

we obtain

$$\begin{cases} \Big(1 + \theta^2 + \theta \exp(l_1)\Big)\Big(\exp(h_0) - \exp(l_0) \Big) + \\ \theta\Big(1 - \exp(l_0)\Big)\Big(\exp(h_1) - \exp(l_1) \Big) = 0, \\ \theta\Big(\exp(h_0) - \exp(l_0) \Big) + \exp(h_1) - \exp(l_1) = 0. \end{cases} \tag{6.23}$$

From (6.23) we get

$$\Big(1 + \theta^2 \exp(l_0) + \theta \exp(l_1)\Big)\Big(\exp(h_0) - \exp(l_0) \Big) = 0. \tag{6.24}$$

It follows from (6.24) that $h_0 = l_0$. Consequently, from second equation in (6.23) we have $h_1 = l_1$.

Sufficiency. Straightforward. $\qquad\square$

Let $G_k^{(2)}$ be the subgroup in G_k consisting of all words of even length.

Theorem 6.2. *Let K be a normal subgroup of finite index in G_k. Then each K-periodic GM for SOS model is either TI or $G_k^{(2)}$-periodic.*

Proof. Is similar to that of Theorem 2.3. $\square$

Let K be a normal subgroup of finite index in G_k. What condition on K will guarantee that each K-periodic GM is TI? We put $I(K) = K \cap \{a_1, ..., a_{k+1}\}$, where a_i, $i = 1, ..., k+1$ are generators of G_k.

Theorem 6.3. *If $I(K) \neq \emptyset$, then each K-periodic GM for SOS model is TI.*

Proof. Is similar to that of Theorem 2.4. $\square$

Theorems 6.2 and 6.3 reduce the problem of describing $K-$ periodic GM with $I(K) \neq \emptyset$ to describing the fixed points of $kF(h; \theta)$ (see (6.12), (6.13)) which describes TIGM. If $I(K) = \emptyset$, this problem is reduced to describing the solutions of the system:

$$\begin{cases} h = kF(l; \theta), \\ l = kF(h; \theta). \end{cases} \tag{6.25}$$

Denote $z_i = \exp(h_i)$, $t_i = \exp(l_i)$, $i = 0, 1$. Then from (6.25) we get

$$\begin{cases} z_0 = \left(\dfrac{t_0 + \theta t_1 + \theta^2}{\theta^2 t_0 + \theta t_1 + 1} \right)^k, \\[2mm] z_1 = \left(\dfrac{\theta t_0 + t_1 + \theta}{\theta^2 t_0 + \theta t_1 + 1} \right)^k, \\[2mm] t_0 = \left(\dfrac{z_0 + \theta z_1 + \theta^2}{\theta^2 z_0 + \theta z_1 + 1} \right)^k, \\[2mm] t_1 = \left(\dfrac{\theta z_0 + z_1 + \theta}{\theta^2 z_0 + \theta z_1 + 1} \right)^k. \end{cases} \tag{6.26}$$

Proposition 6.8. *For a ferromagnetic SOS model, with $J < 0$ $(\theta < 1)$ (and even for $J = 0$), the system of equations (6.26) has solutions with $z_0 = t_0$ and $z_1 = t_1$ only.*

Proof. Denote $u_i = z_i^{1/k}$, $v_i = t_i^{1/k}$, $i = 0, 1$. Then from (6.26) we have

$$u_0 - v_0 = \frac{(1 - \theta^2)[\theta(u_1^k v_0^k - u_0^k v_1^k) + (\theta^2 + 1)(v_0^k - u_0^k) + \theta(v_1^k - u_1^k)]}{(\theta^2 v_0^k + \theta v_1^k + 1)(\theta^2 u_0^k + \theta u_1^k + 1)}$$

$$\tag{6.27}$$

and

$$u_1 - v_1 = \frac{(1-\theta^2)[\theta(v_0^k - u_0^k) + (v_1^k - u_1^k)]}{(\theta^2 v_0^k + \theta v_1^k + 1)(\theta^2 u_0^k + \theta u_1^k + 1)}. \tag{6.28}$$

Using the fact that

$$u_1^k v_0^k - u_0^k v_1^k = u_1^k(v_0^k - u_0^k) + u_0^k(u_1^k - v_1^k)$$

we obtain

$$\begin{cases} [A + (1-\theta^2)(\theta u_1^k + \theta^2 + 1)B_0](u_0 - v_0) + \\ \theta(1-\theta^2)(1 - u_0^k)B_1(u_1 - v_1) = 0, \\ \\ \theta(1-\theta^2)B_0(u_0 - v_0) + [A + (1-\theta^2)B_1](u_1 - v_1) = 0, \end{cases} \tag{6.29}$$

where

$$A = (\theta^2 v_0^k + \theta v_1^k + 1)(\theta^2 u_0^k + \theta u_1^k + 1) > 0,$$

$$B_i = u_i^{k-1} + u_i^{k-2} v_i + \ldots + v_i^{k-1} > 0, \quad i = 0, 1.$$

From (6.29) we get

$$\left[A^2 + (1-\theta^2)\left(B_1 + (\theta u_1^k + \theta^2 + 1)B_0 \right)A + \right. \tag{6.30}$$

$$\left. (1-\theta^2)^2(\theta u_1^k + \theta^2 u_0^k + 1)B_0 B_1 \right](u_0 - v_0) = 0.$$

Since $\theta \leq 1$ $(J \leq 0)$, we deduce from (6.30) that $u_0 = v_0$. Then from second equation of (6.29) we have $u_1 = v_1$. $\qquad\square$

Now consider anti-ferromagnetic case, with $J > 0$ $(\theta > 1)$. By Proposition 6.3 we know that if $J > 0$ then the system of equations (6.26) has a unique solution with $z_0 = t_0$, $z_1 = t_1$. Moreover, $z_0 = 1$. For $z_0 = t_0 = 1$ from (6.26) we have

$$\begin{cases} z_1 = \left(\frac{2\theta + t_1}{\theta^2 + \theta t_1 + 1} \right)^k, \\ \\ t_1 = \left(\frac{2\theta + z_1}{\theta^2 + \theta z_1 + 1} \right)^k. \end{cases} \tag{6.31}$$

The following proposition gives a condition under which (6.26) has solutions with $z_0 = t_0 = 1$ and $z_1 \neq t_1$.

Proposition 6.9. *Let (z_*, z_*) be the unique solution of (6.31). If*

$$\frac{kz_*(\theta^2 - 1)}{(2\theta + z_*)(1 + \theta^2 + \theta z_*)} > 1, \tag{6.32}$$

then the system of equations (6.31) has at least three solutions $(z_-^*, z_+^*), (z_*, z_*), (z_+^*, z_-^*)$, where $z_-^* = \psi(z_+^*, \theta, k)$ and

$$\psi(x, \theta, k) = \left(\frac{2\theta + x}{1 + \theta^2 + \theta x} \right)^k.$$

Proof. Under (6.32) z_* is unstable fixed point of the map $z > 0 \to \psi(z, \theta, k)$. For any $z \geq 1$, iterates $\psi^{(2n)}(z, \theta, k)$ remain $> z_*$ monotonically decrease and hence converge to a limit, $z_+^* \geq z_*$ which solves

$$z = \psi(\psi(z, \theta, k), \theta, k). \tag{6.33}$$

However, $z_+^* > z_*$ as z_* is unstable. Then $z_-^* = \psi(z_+^*, \theta, k)$ is $< z_*$ and also solves (6.33). This completes the proof. $\qquad\qquad\square$

Summarizing, we obtain the following

Theorem 6.4. *For the SOS model with respect to any normal subgroup $K \subset G_k$ of finite index the following assertions hold:*

(i) In the ferromagnetic case ($J < 0$), and for $J = 0$ (no interaction), the K-periodic GMs coincide with TIGMs.

(ii) In the anti-ferromagnetic case ($J > 0$):

> *(a) if $I(K) \neq \emptyset$ then K-periodic GMs coincide with TIGMs;*
> *(b) if (6.32) holds and $I(K) = \emptyset$ then there are three K-periodic GMs μ_{12}, μ_{21} and μ_*. Moreover, measure μ_* is TI and measures μ_{12} and μ_{21} are G_k^*-periodic.*

6.2.3 Non-periodic SGMs

In this subsection we consider the case $J < 0$, $m = 2$, $\beta > \beta_{\mathrm{cr}}^1$. We use measures μ_i^*, $i = -, \mathrm{m}, +$, to show that system (6.11) admits uncountably many non-periodic solutions.

Take an arbitrary infinite path $\pi = \{x_0, x_1, ...\}$ on the Cayley tree Γ^k starting at the origin x^0: $x_0 = x^0$. We will establish a 1-1 correspondence between such paths and real numbers $t \in [0; 1]$ (see Chapter 2). In fact, let $\pi_1 = \{x_0, x_1, ...\}$ and $\pi_2 = \{y_0, y_1, ...\}$ be two such paths, with $x_0 = y_0 = x^0$. We will map the pair (π_1, π_2) to a vector-function $h^{\pi_1 \pi_2} : x \in V \mapsto h_x^{\pi_1 \pi_2}$ satisfying (6.11). Paths π_1 and π_2 split Γ^k into three components Γ_1^k, Γ_2^k and Γ_3^k when π_1, π_2 are distinct and into two components Γ_1^k and Γ_3^k when π_1, π_2 coincide. Vector-function $h^{\pi_1 \pi_2}$ is then defined by

$$h_x^{\pi_1 \pi_2} = \begin{cases} h_-^*, & \text{if } x \in \Gamma_1^k, \\[4pt] h_\mathrm{m}^*, & \text{if } x \in \Gamma_2^k, \\[4pt] h_+^*, & \text{if } x \in \Gamma_3^k, \end{cases} \tag{6.34}$$

where vectors $h_i^* = (0, \ln z_{1,i}^*)$, $i = -, \mathrm{m}, +$, are solutions of (6.11).

Let $h = (h_0, h_1) \in R^2$. Denote

$$\|h\| = \max\{|h_0|, |h_1|\}.$$

Let function $h \mapsto F(h) = F(h, \theta)$ be defined by (6.21).

Proposition 6.10. *For any* $h = (h_0, h_1) \in R^2$ *the following inequalities hold:*

a)

$$\left| \frac{\partial F_i}{\partial h_j} \right| \leq \frac{|\theta^2 - 1|}{\theta^2}, \quad i, j = 0, 1,$$

b)

$$\|F(h, \theta) - F(l, \theta)\| \leq 2\frac{|\theta^2 - 1|}{\theta^2}\|h - l\|, \quad h, l \in R^2,$$

c) *for any* $h = (0, h_1)$ *and* $l = (0, l_1)$:

$$\|F(h) - F(l)\| \leq \frac{|\theta^2 - 1|}{1 + 3\theta^2 + 2\theta\sqrt{2(\theta^2 + 1)}}\|h - l\|, \quad h, l \in R^2.$$

d)

$$|F_0(h)| \leq \frac{|\theta^2 - 1|}{\theta^2 + 1}|h_0|, \quad h = (h_0, h_1) \in R^2.$$

Proof.

a) Write:

$$\frac{\partial F_0}{\partial h_0} = \frac{(1 - \theta^2)e^{h_0}(\theta e^{h_1} + \theta^2 + 1)}{(e^{h_0} + \theta e^{h_1} + \theta^2)(\theta^2 e^{h_0} + \theta e^{h_1} + 1)}.$$

To assess the derivative $\frac{\partial F_0}{\partial h_0}$, consider two cases:

Case 1: $h_0 \geq 0$. Then

$$\frac{\theta e^{h_1} + \theta^2 + 1}{e^{h_0} + \theta e^{h_1} + \theta^2} \leq 1, \quad \frac{e^{h_0}}{\theta^2 e^{h_0} + \theta e^{h_1} + 1} < \frac{1}{\theta^2}.$$

Case 2: $h_0 \leq 0$. Then

$$\frac{1}{e^{h_0} + \theta e^{h_1} + \theta^2} \leq \frac{1}{\theta^2}, \quad \frac{(\theta e^{h_1} + \theta^2 + 1)e^{h_0}}{\theta^2 e^{h_0} + \theta e^{h_1} + 1} \leq 1.$$

Hence, $|\frac{\partial F_0}{\partial h_0}| \leq \frac{|1 - \theta^2|}{\theta^2}$.

To assess $\frac{\partial F_0}{\partial h_1}$, we again consider two cases:

Case 3: $h_0 \geq 0$. Then

$$\left| \frac{dF_0}{dh_1} \right| = |\theta^2 - 1|\theta \frac{e^{h_1}}{e^{h_0} + \theta e^{h_1} + \theta^2} \frac{e^{h_0} - 1}{\theta^2 e^{h_0} + \theta e^{h_1} + 1} \leq \frac{|1 - \theta^2|}{\theta^2}.$$

Case 4: $h_0 < 0$. Then

$$\left|\frac{dF_0}{dh_1}\right| = |\theta^2 - 1|\theta \, \frac{e^{h_1}}{\theta^2 e^{h_0} + \theta e^{h_1} + 1} \, \frac{1 - e^{h_0}}{e^{h_0} + \theta e^{h_1} + \theta^2} \leq \frac{|1 - \theta^2|}{\theta^2}.$$

Finally, to assess the derivatives of F_1, write:

$$\left|\frac{\partial F_1}{\partial h_0}\right| = \left|\theta(\theta^2 - 1)\frac{1}{\theta e^{h_0} + e^{h_1} + \theta} \, \frac{e^{h_0}}{\theta^2 e^{h_0} + \theta e^{h_1} + 1}\right| \leq \frac{|1 - \theta^2|}{\theta^2}$$

and

$$\left|\frac{\partial F_1}{\partial h_1}\right| = \left|(\theta^2 - 1)\frac{1}{\theta e^{h_0} + e^{h_1} + \theta} \, \frac{e^{h_1}}{\theta^2 e^{h_0} + \theta e^{h_1} + 1}\right| \leq \frac{|1 - \theta^2|}{\theta^2}.$$

b) Write:

$$\|F(h) - F(l)\| = \max\{|F_0(h) - F_0(l)|, |F_1(h) - F_1(l)|\} \leq$$

$$\max_{i=0,1}\{|(F_i)'_{h_0}|\,|h_0 - l_0| + |(F_i)'_{h_1}|\,|h_1 - l_1|\} \leq 2\frac{|\theta^2 - 1|}{\theta^2}\|h - l\|.$$

In cases c) and d) the inequalities are straightforward. $\qquad\square$

With the help of Proposition 6.10 it is easy to prove the following Theorem 6.5, similar to arguments of the Bleher-Ganikhodjaev construction mentioned in Chapter 2:

Theorem 6.5. *For any two infinite paths π_1, π_2, there exists a unique vector-function $h^{\pi_1\pi_2}$ satisfying (6.11) and (6.34).*

Next, we map (π_1, π_2) to a pair $(t, s) \in [0, 1] \times [0, 1]$. In the standard way (see Chapter 2) one can prove that functions $h^{\pi_1(t)\pi_2(s)}$ are different for different pairs $(t, s) \in D$ where $D = \{(u, v) \in [0, 1]^2 : u \leq v\}$.

Now let $\mu(t, s)$ denote the SGM corresponding to function $h^{\pi_1(t)\pi_2(s)}$, $(t, s) \in D$. We obtain the following

Theorem 6.6. *For any pair $(t, s) \in D$, there exists a unique extreme SGM $\mu(t, s)$. Moreover, the above GMs μ_i^*, $i = -, m, +$, are specified as $\mu(0, 0) = \mu_+^*$, $\mu(0, \frac{k+1}{k}) = \mu_m^*$, $\mu(\frac{k+1}{k}, \frac{k+1}{k}) = \mu_-^*$.*

Because measures $\mu(t, s)$ are different for different $(t, s) \in D$ we obtain a continuum of distinct extreme SGMs.

6.3 Four-state SOS model

As the number of TISGMs for the SOS model on $\mathbb{Z}^d$ depends on the parity of m (see [156]), a natural question arises about how the number of TISGMs depends on the parity of m in the case of the SOS model on a Cayley tree. The situation with an even m, $m = 2$, was considered in previous section, and in this section we consider the case $m = 3$. Therefore, the results in this section allow comparing the number of TISGMs on a Cayley tree for different values of m.

6.3.1 *Translation-invariant measures*

For $m = 3$ from (6.5) we get:

$$h_{0,x} = \sum_{y\in S(x)} \ln \frac{\exp(h_{0,y})+\theta\exp(h_{1,y})+\theta^2\exp(h_{2,y})+\theta^3}{\theta^3\exp(h_{0,y})+\theta^2\exp(h_{1,y})+\theta\exp(h_{2,y})+1},$$

$$h_{1,x} = \sum_{y\in S(x)} \ln \frac{\theta\exp(h_{0,y})+\exp(h_{1,y})+\theta\exp(h_{2,y})+\theta^2}{\theta^3\exp(h_{0,y})+\theta^2\exp(h_{1,y})+\theta\exp(h_{2,y})+1}, \qquad (6.35)$$

$$h_{2,x} = \sum_{y\in S(x)} \ln \frac{\theta^2\exp(h_{0,y})+\theta\exp(h_{1,y})+\exp(h_{2,y})+\theta}{\theta^3\exp(h_{0,y})+\theta^2\exp(h_{1,y})+\theta\exp(h_{2,y})+1}.$$

It is natural to begin with translation-invariant solution (6.35), i.e., $=$to assume that $h_x = h \in R^3$. We set $z_i = e^{h_i}$, $i = 0,1,2$. From (6.35), we obtain

$$z_0 = \left(\frac{z_0 + \theta z_1 + \theta^2 z_2 + \theta^3}{\theta^3 z_0 + \theta^2 z_1 + \theta z_2 + 1}\right)^k, \qquad (6.36)$$

$$z_1 = \left(\frac{\theta z_0 + z_1 + \theta z_2 + \theta^2}{\theta^3 z_0 + \theta^2 z_1 + \theta z_2 + 1}\right)^k, \qquad (6.37)$$

$$z_2 = \left(\frac{\theta^2 z_0 + \theta z_1 + z_2 + \theta}{\theta^3 z_0 + \theta^2 z_1 + \theta z_2 + 1}\right)^k. \qquad (6.38)$$

We note that $z_0 = 1$ and $z_1 = z_2 = z$ satisfy equation (6.36) independently of k, θ.

Substituting $z_0 = 1$ and $z_1 = z_2 = z$ in (6.37) and (6.38) gives

$$z = \left(\frac{\theta^2 + (\theta + 1)z + \theta}{\theta^3 + (\theta^2 + \theta)z + 1}\right)^k. \qquad (6.39)$$

Denote

$$a = \theta^{k+1}, \quad b = \frac{1 - \theta + \theta^2}{\theta^2}, \quad x = \frac{z}{\theta}. \qquad (6.40)$$

Then from (6.39) we get an equation like (6.16) so here we can also use Proposition 5.1.

Lemma 6.1. *If $J \geq 0$ then the system of equations (6.36)-(6.38) has a unique solution $(1, z_*, z_*)$.*

Proof. Let

$$A = z_0 + \theta z_1 + \theta^2 z_2 + \theta^3, \quad B = \theta^3 z_0 + \theta^2 z_1 + \theta z_2 + 1,$$

$$A_1 = \theta z_0 + z_1 + \theta z_2 + \theta^2, \quad B_1 = \theta^2 z_0 + \theta z_1 + z_2 + \theta.$$

From (6.36)-(6.38), we obtain

$$\begin{cases} [(1 - \theta^3)M - 1](z_0 - 1) + (\theta - \theta^2)M(z_1 - z_2) = 0, \\ (\theta - \theta^2)M_1(z_0 - 1) + [(1 - \theta)M_1 - 1](z_1 - z_2) = 0, \end{cases} \qquad (6.41)$$

where $M = (A^{k-1} + \cdots + B^{k-1})/B^k > 0$ and $M_1 = (A_1^{k-1} + \cdots + B_1^{k-1})/B_1^k > 0$.

We note that, for $\theta \geq 1$ ($J \geq 0$), the determinant of homogeneous system (6.41) is positive. Consequently, the system has a unique solution, $z_0 = 1$, $z_1 = z_2$. We have thus proved that system (6.36)-(6.38) has only solutions of the form $(1, z, z)$, for $J \geq 0$. Since $b = \frac{1 - \theta + \theta^2}{\theta^2} \leq 1 < (\frac{k+1}{k-1})^2$, by Proposition 5.1, system (6.36)-(6.38) has a unique solution $(1, z_*, z_*)$. $\square$

For $\theta \neq 1$ we obtain the following corollary from (6.41).

Corollary 6.1. *We have $z_0 = 1$ if and only if $z_1 = z_2$.*

Let

$$\beta_{\mathrm{C}} = \frac{1}{J} \ln \frac{2(k - 1)}{k - 1 + \sqrt{k^2 + 14k + 1}}.$$

Lemma 6.2. *If $J < 0$, then system of equations (6.36)-(6.38) has a unique solution $z^* = (1, z_*, z_*)$ for $\beta \leq \beta_{\mathrm{C}}$, and it has exactly three solutions $z_-^* = (1, z_{1,-}^*, z_{1,-}^*)$, $z_0^* = (1, z_{1,0}^*, z_{1,0}^*)$ and $z_+^* = (1, z_{1,+}^*, z_{1,+}^*)$ for $\beta > \beta_{\mathrm{C}}$, with $0 < z_{1,-}^* < z_{1,0}^* < z_{1,+}^*$ and $z_{1,i}^* = e^{h_{1,i}^*}$, $i = -, 0, +$.*

Proof. The value of β_{C} is the solution of equation $b = (\frac{k+1}{k-1})^2$. Other statements of lemma are consequences of Proposition 5.1. $\square$

We note that z^*, z^*_-, z^*_0, z^*_+ are (mirror) symmetric solution of system (6.36)-(6.38). Consequently, we have the following theorem.

Theorem 6.7. *For the SOS model with $J > 0$ and $m = 3$, a TISGM exists and is unique for every $\beta > 0$. For $J < 0$ and $m = 3$, if $k \geq 2$ and $0 < \beta \leq \beta_C$, then there is a unique TISGM μ^*, and if $k \geq 2$ and $\beta > \beta_C$, then there are three TISGMs μ^*_-, μ^*_0, and μ^*_+ corresponding to $h^*_i = \ln z^*_i$, $i = -, 0, +$.*

The following Lemma can be proved as Proposition 6.5.

Lemma 6.3. *For $J < 0$, $k \geq 2$, and $\beta > \beta_C$, if $h_x = (0, h_{1,x}, h_{1,x})$ is a solution of (6.35) then, with $h_{1,x} = \ln z_{1,x}$, we have*

$$z^*_{1,-} \leq z_{1,x} \leq z^*_{1,+}, \quad x \in V. \tag{6.42}$$

Lemma 6.4. *For $J < 0$ and $\beta \leq \beta_C$, the measure μ^* is a unique SGM.*

Proof. Under condition $z^*_{1,-} = z^*_{1,+} = z^*_1$, we conclude from (6.42) that $h_x = (0, \ln z^*_1, \ln z^*_1)$ is a unique solution. $\square$

6.3.2 Construction of periodic SGMs

In this section we study periodic solutions of system (6.35).

Each K-periodic collection is given by

$$\{h_x = h_i \in R^3 \text{ for } x \in K_i, \quad i = 0, 1, ..., r - 1\}.$$

By Proposition 6.2 (for $m = 3$), vector h_n, $n = 0, 1, ..., r - 1$, satisfies the system

$$h_n = \sum_{j=1}^{N(e)} q_{i_j}(e) F(h_{\pi_n(i_j)}, \theta) - F(h_{\pi_n(i_{j_0})}, \theta), \tag{6.43}$$

where $j_0 = 1, ..., N(e)$.

Here

$$F : h = (h_0, h_1, h_2) \in R^3 \mapsto F(h, \theta) = (F_0(h, \theta), F_1(h, \theta), F_2(h, \theta)) \in R^3$$

defined as

$$F_0(h, \theta) = \ln \frac{e^{h_0} + \theta e^{h_1} + \theta^2 e^{h_2} + \theta^3}{\theta^3 e^{h_0} + \theta^2 e^{h_1} + \theta e^{h_2} + 1},$$

$$F_1(h, \theta) = \ln \frac{\theta e^{h_0} + e^{h_1} + \theta e^{h_2} + \theta^2}{\theta^3 e^{h_0} + \theta^2 e^{h_1} + \theta e^{h_2} + 1}, \tag{6.44}$$

$$F_2(h, \theta) = \ln \frac{\theta^2 e^{h_0} + \theta e^{h_1} + e^{h_2} + \theta}{\theta^3 e^{h_0} + \theta^2 e^{h_1} + \theta e^{h_2} + 1}.$$

Lemma 6.5. *If $\theta \neq 1$, then $F(h, \theta) = F(l, \theta)$ if and only if $h = l$.*

Proof. *Necessity.* From $F(h, \theta) = F(l, \theta)$, we get the system of equations

$$
\begin{cases}
[1 + \theta^2 + \theta^4 + \theta^2 t_1 + \theta(1 + \theta^2)t_2](z_0 - t_0) + \\[4pt]
[\theta^2(t_2 - t_0) + \theta(1 + \theta^2)](z_1 - t_1) + \\[4pt]
\theta[\theta - (1 + \theta^2)t_0 - \theta t_1](z_2 - t_2) = 0, \\[10pt]
[\theta^2 t_2 + \theta(1 + \theta^2)](z_0 - t_0) + [1 + \theta^2 + \theta t_2](z_1 - t_1) + \\[4pt]
\theta[1 - t_1 - \theta t_2](z_2 - t_2) = 0. \\[10pt]
\theta^2(z_2 - t_0) + \theta(z_1 - t_1) + (z_2 - t_2) = 0,
\end{cases}
\tag{6.45}
$$

where $z_i = e^{h_i}$, $t_i = e^{l_i}$, $i = 0, 1, 2$.

We note that the determinant of homogeneous system (6.45) is positive for $\theta \neq 1$, $\Delta = (1 + \theta^3 t_0 + \theta^2 t_1 + \theta t_2)^2 > 0$. Consequently, system (6.45) has a unique solution $z_i - t_i = 0$, $i = 0, 1, 2$.

The *sufficiency* is obvious. $\qquad\qquad\qquad\qquad\qquad\qquad\qquad\qquad\square$

Using Lemma 6.5 one can prove analogues of Theorems 6.2, 6.3. Then to describe periodic Gibbs measures of the SOS model with four values of spin, $0, 1, 2, 3$, one has to solve the following system:

$$
z_0 = \left(\frac{t_0 + \theta t_1 + \theta^2 t_2 + \theta^3}{\theta^3 t_0 + \theta^2 t_1 + \theta t_2 + 1} \right)^k, \quad
t_0 = \left(\frac{z_0 + \theta z_1 + \theta^2 z_2 + \theta^3}{\theta^3 z_0 + \theta^2 z_1 + \theta z_2 + 1} \right)^k,
$$

$$
z_1 = \left(\frac{\theta t_0 + t_1 + \theta t_2 + \theta^2}{\theta^3 t_0 + \theta^2 t_1 + \theta t_2 + 1} \right)^k, \quad
t_1 = \left(\frac{\theta z_0 + z_1 + \theta z_2 + \theta^2}{\theta^3 z_0 + \theta^2 z_1 + \theta z_2 + 1} \right)^k, \tag{6.46}
$$

$$
z_2 = \left(\frac{\theta^2 t_0 + \theta t_1 + t_2 + \theta}{\theta^3 t_0 + \theta^2 t_1 + \theta t_2 + 1} \right)^k, \quad
t_2 = \left(\frac{\theta^2 z_0 + \theta z_1 + z_2 + \theta}{\theta^3 z_0 + \theta^2 z_1 + \theta z_2 + 1} \right)^k.
$$

Lemma 6.6. *For $J < 0$, $(\theta < 1)$, the system of equations (6.46) has solutions with $z_i = t_i$, $i = 0, 1, 2$ only.*

Proof. Denote $u_i = z_i^{1/k}$ and $v_i = t_i^{1/k}$, $i = 0, 1, 2$. Then from (6.46) we get

$$
\begin{cases}
[A + B_0(1 + \theta^2 + \theta^4 + \theta^2 u_1^k + \theta(1 + \theta^2)u_2^k)](u_0 - v_0) + \\
[B_1(\theta^2(u_2^k - u_0^k) + \theta^3 + \theta)](u_1 - v_1) + \\
[B_2(\theta^2 - (\theta^3 + \theta)u_0^k - \theta^2 u_1^k)](u_2 - v_2) = 0, \\[2mm]
[B_0(\theta^2 u_2^k + \theta^3 + \theta)](u_0 - v_0) + [A + B_1(\theta u_2^k + \theta^2 + 1)](u_1 - v_1) + \\
[B_2(\theta - \theta u_1^k - \theta^2 u_0^k)](u_2 - v_2) = 0, \\[2mm]
\theta^2 B_0(u_0 - v_0) + \theta B_1(u_1 - v_1) + (A + B_2)(u_2 - v_2) = 0,
\end{cases}
$$

$$(6.47)$$

where

$$
A = (\theta^3 v_0^k + \theta^2 v_1^k + \theta v_2^k + 1)(\theta^3 u_0^k + \theta^2 u_1^k + \theta u_2^k + 1)(1 - \theta^2)^{-1} > 0,
$$

$$
B_i = u_i^{k-1} + u_i^{k-2} v_i + \ldots + v_i^{k-1} > 0, \quad i = 0, 1, 2.
$$

A simple, but rather cumbersome, analysis shows that the determinant of system (6.47) is positive, and consequently $z_i = t_i$, $i = 0, 1, 2$. $\qquad\square$

We now consider the anti-ferromagnetic case $J > 0$. In this situation, by Lemma 6.1, system (6.46) has a unique solution $(1, z_1^*, z_1^*, 1, t_1^*, t_1^*)$. Let $z_0 = t_0 = 1$, $z_1 = z_2 = z$, and $t_1 = t_2 = t$. Then (6.46) implies

$$
\begin{cases}
z = \left(\dfrac{\theta + t}{1 - \theta + \theta^2 + \theta t} \right)^k, \\[3mm]
t = \left(\dfrac{\theta + z}{1 - \theta + \theta^2 + \theta z} \right)^k.
\end{cases}
$$

$$(6.48)$$

Lemma 6.7. *Let (z_*, z_*) be the unique solution of (6.48). If*

$$
\frac{k z_*(\theta - 1)}{(\theta + z_*)(1 - \theta + \theta^2 + \theta z_*)} > 1,
$$

$$(6.49)$$

then the system of equations (6.41) has at least three solutions $(z_-^*, z_+^*), (z_*, z_*)$ and (z_+^*, z_-^*), where $z_-^* = \psi(z_+^*, \theta, k)$ and

$$
\psi(x, \theta, k) = \left(\frac{\theta + x}{1 - \theta + \theta^2 + \theta x} \right)^k.
$$

Proof. Similar to the proof of Proposition 6.9. $\qquad\square$

We have thus proved the following theorem.

Theorem 6.8. *For the SOS model with $m = 3$, the following assertions are true for the Cayley tree relative to every normal divisor of finite index K:*

1. In the ferromagnetic case $(J \leq 0)$, the K-periodic GM coincides with a TIGM.

2. In the anti-ferromagnetic case $(J > 0)$,

a. if $I(K) \neq \emptyset$, then the K-periodic GM coincides with a TIGM, and

b. if inequality (6.49) holds and if $I(K) = \emptyset$, then there are three K-periodic GMs μ_{12}, μ_* and μ_{21}.

In this situation, μ_* is a TIGM, and μ_{12} and μ_{21} are periodic measures with period two, i.e., they are $G_k^{(2)}$-periodic measures.

6.3.3 *Uncountable set non-periodic SGMs*

We consider the case $J < 0$, $m = 3$, $\beta > \beta_C$ and shall use an argument similar to the Bleher-Ganikhodjaev construction. To do this we need the following lemma.

We introduce the notation

$$f(x, y, z) \equiv f(x, y, z, a_1, a_2, a_3, a_4, b_1, b_2, b_3, b_4) = \ln \frac{a_1 e^x + a_2 e^y + a_3 e^z + a_4}{b_1 e^x + b_2 e^y + b_3 e^z + b_4},$$
$$(6.50)$$

where

$$a_i \geq 0, \quad b_i \geq 0, \quad i = 1, 2, 3, 4, \quad a_1 + a_2 + a_3 + a_4 > 0, \quad b_1 + b_2 + b_3 + b_4 > 0.$$
$$(6.51)$$

Lemma 6.8. *The following estimates hold:*

(a) If $a_1 a_2 a_3 a_4 b_1 b_2 b_3 b_4 > 0$, then

$$\left| \frac{\partial f}{\partial x} \right| \leq \max \left\{ \frac{|a_1 b_2 - b_1 a_2|}{(\sqrt{a_1 b_2} + \sqrt{b_1 a_2})^2}, \frac{|a_1 b_3 - b_1 a_3|}{(\sqrt{a_1 b_3} + \sqrt{b_1 a_3})^2}, \frac{|a_1 b_4 - b_1 a_4|}{(\sqrt{a_1 b_4} + \sqrt{b_1 a_4})^2} \right\},$$

$$\left| \frac{\partial f}{\partial y} \right| \leq \max \left\{ \frac{|a_1 b_2 - b_1 a_2|}{(\sqrt{a_1 b_2} + \sqrt{b_1 a_2})^2}, \frac{|a_2 b_3 - b_2 a_3|}{(\sqrt{a_2 b_3} + \sqrt{b_2 a_3})^2}, \frac{|a_2 b_4 - b_2 a_4|}{(\sqrt{a_2 b_4} + \sqrt{b_2 a_4})^2} \right\},$$

$$\left| \frac{\partial f}{\partial z} \right| \leq \max \left\{ \frac{|a_1 b_3 - b_1 a_3|}{(\sqrt{a_1 b_3} + \sqrt{b_1 a_3})^2}, \frac{|a_2 b_3 - b_2 a_3|}{(\sqrt{a_2 b_3} + \sqrt{b_2 a_3})^2}, \frac{|a_3 b_4 - b_3 a_4|}{(\sqrt{a_3 b_4} + \sqrt{b_3 a_4})^2} \right\}.$$

(b) If $a_1 a_2 a_3 a_4 b_1 b_2 b_3 b_4 = 0$, then

$$\left|\frac{\partial f}{\partial x}\right| \le 1, \quad \left|\frac{\partial f}{\partial y}\right| \le 1, \quad \left|\frac{\partial f}{\partial z}\right| \le 1.$$

Proof.   Let $a, b, c, d \ge 0$, $a + b > 0$, and $c + d > 0$. We introduce the notation

$$f_1(x) = \ln \frac{ae^x + b}{ce^x + d}.$$

It is easy to verify that

$$|f_1'(x)| \le \begin{cases} \frac{|ad - bc|}{(\sqrt{ad} + \sqrt{bc})^2}, & \text{for} \quad abcd > 0, \\ 1, & \text{for} \quad abcd = 0. \end{cases} \tag{6.52}$$

Now we consider

$$f_2(x, y) = \ln \frac{a_1 e^x + a_2 e^y + a_3}{b_1 e^x + b_2 e^y + b_3}.$$

By inequality (6.52), we have

$$\left|\frac{\partial f_2(x, y)}{\partial x}\right| \le \frac{|a_1(b_2 e^y + b_3) - b_1(a_2 e^y + a_3)|}{(\sqrt{a_1(b_2 e^y + b_3)} + \sqrt{b_1(a_2 e^y + a_3)})^2}. \tag{6.53}$$

Let $t = e^y$ and

$$\varphi(t) = \frac{|a_1(b_2 t + b_3) - b_1(a_2 t + a_3)|}{(\sqrt{a_1(b_2 t + b_3)} + \sqrt{b_1(a_2 t + a_3)})^2}.$$

Then

$$\varphi'(t) = \sqrt{\frac{a_1 b_1}{(a_2 t + a_3)(b_2 t + b_3)}} \cdot \frac{a_3 b_2 - a_2 b_3}{(\sqrt{a_1(b_2 t + b_3)} + \sqrt{b_1(a_2 t + a_3)})^2}.$$

Consequently, $\varphi(t)$ increases monotonically if the inequality $a_3 b_2 - a_2 b_3 \ge 0$ holds and decreases monotonically in the case $a_3 b_2 - a_2 b_3 < 0$.

Since $t > 0$, inequality (6.53) implies that

$$\left|\frac{\partial f_2(x, y)}{\partial x}\right| \le \max\{|\varphi(0)|, |\varphi(\infty)|\} = \tag{6.54}$$

$$\max\left\{\frac{|a_1 b_3 - b_1 a_3|}{(\sqrt{a_1 b_3} + \sqrt{b_1 a_3})^2}, \frac{|a_1 b_2 - b_1 a_2|}{(\sqrt{a_1 b_2} + \sqrt{b_1 a_2})^2}\right\}.$$

Using (6.54), we similarly obtain the first inequality for $\partial f(x, y, z)/\partial x$ in assertion (a) in Lemma 6.8. The other two inequalities are proved similarly. The inequalities in assertion (b) are obtained from those in (a) if at least one of the parameters $a_1, a_2, a_3, a_4, b_1, b_2, b_3$, and b_4 is zero; in this case, inequalities (6.51) hold. $\qquad\square$

We note that every coordinate of the function F (see (6.44)) has form (6.50). Lemma 6.8 implies the next lemma.

Let $h = (h_0, h_1, h_2) \in R^3$. Denote

$$\|h\| = \max\{|h_0|, |h_1|, |h_2|\}.$$

Let function $h \mapsto F(h) = F(h, \theta)$ be defined by (6.44).

Lemma 6.9. *For any* $h = (h_0, h_1, h_2) \in R^3$ *the following inequalities hold:*

$$\left|\frac{\partial F_0}{\partial h_j}\right| \le \frac{|\theta^3 - 1|}{1 + \theta^3}, \quad \left|\frac{\partial F_1}{\partial h_j}\right| \le \frac{|\theta^2 - 1|}{1 + \theta^2}, \quad \left|\frac{\partial F_2}{\partial h_j}\right| \le \frac{|\theta - 1|}{1 + \theta}, \; j = 0, 1, 2. \quad (6.55)$$

$$\|F(h, \theta) - F(l, \theta)\| \le 3\frac{|\theta^3 - 1|}{1 + \theta^3}\|h - l\|, \quad h, l \in R^3. \qquad (6.56)$$

Proof. Inequalities (6.55) are consequences of those in Lemma 6.8. To prove (6.56), we write

$$\|F(h) - F(l)\| = \max_{i=0,1,2}\{|F_i(h) - F_i(l)|\} \le$$

$$\max_{i=0,1,2}\{|(F_i)'_{h_0}||h_0 - l_0| + |(F_i)'_{h_1}||h_1 - l_1| + |(F_i)'_{h_2}||h_2 - l_2|\}$$

$$\le 3\frac{|\theta^3 - 1|}{\theta^3 + 1}\|h - l\|. \qquad \square$$

Using Lemma 6.9, one can prove the following theorem.

Theorem 6.9. *For arbitrary infinite paths* π_1 *and* π_2, *there is a vector function* $h^{\pi_1 \pi_2}$ *which satisfies relations (6.35).*

We map (π_1, π_2) to the pair $(t, s) \in [0, 1]^2$. It can be shown that $h^{\pi_1(t)\pi_2(s)}$ are different for different $(t, s) \in D = \{(u, v) \in [0, 1]^2 : u \le v\}$.

Denoted by $\mu(t, s)$, the Gibbs measure associated with $h^{\pi_1(t)\pi_2(s)}$. Then the following theorem can be proved.

Theorem 6.10. *For each* $(t, s) \in D$, *there is a unique extreme SGM* $\mu(t, s)$, *for four-state SOS model on Cayley tree.*

Commentaries and references. In the case of a cubic lattice SOS models were analyzed in [156] where an analogue of the so-called Dinaburg–Mazel–Sinai theory was developed. Besides interesting phase transitions in these models, the attention to them is motivated by applications, in particular in the theory of communication networks; see, e.g., [127], [207].

Compared with the Potts model (see Chapter 5), the SOS model has 'less symmetry' and therefore more diverse structure of phases. For example, in the ferromagnetic case it is intuitively plausible that the ground states corresponding to 'middle-level' surfaces will be 'dominant'. This observation was made formal in [156] for the model on a cubic lattice.

We note that for $m = 2$, the SOS model coincides with the Blume-Capel model [72]. In [223], for arbitrary $m \geq 1$, Proposition 6.1 is proved and the case $m = 2$ is studied. In [221] the case $m = 3$ is studied. We note that the phase transition critical value β_{cr} of β depends on k, m, J. In this chapter we only found this value for $m = 2$ and $m = 3$, the case of arbitrary $m \geq 4$ have not been studied yet. Note that $\forall\, m, k \geq 2$ there exist symmetric solutions $h = (h_0, h_1, ..., h_{m-1})$ to (6.7), with $h_0 = 0$ and $h_i = h_{m-i}$, $i = 1, 2, ..., m - 1$. Using this mirror-symmetry one can reduce the finding problem of the critical value $\beta_{\mathrm{c}} = \beta_{\mathrm{cr}}(k, m, J)$ to solution of a polynomial equation $P_m(\theta, k) = 0$, (where m is the highest degree of θ). But solution of this equation for $m \geq 4$ is a difficult problem. To avoid such difficulties, for the SOS model, the contour method on a Cayley tree was used in [220] to prove that there are $m + 1$ distinct Gibbs measures at sufficiently low temperatures.

Chapter 7

Models with hard constraints

In this chapter we consider models which have "hard constraints". Following [35] such models are defined by the space $\mathrm{Hom}(\Gamma^k, H)$ of homomorphisms from a Cayley tree Γ^k to a fixed finite constraint graph H. For any assignment λ of positive real activities to the nodes of H, there is at least one Gibbs measure on $\mathrm{Hom}(\Gamma^k, H)$, there may be more than one (phase transition). We mainly consider the case where graph H contains two vertices or three vertices. In such simple cases, a hard core model with two spin values and several hard core models with three spin values will be discussed. In the last section of this chapter we give a model with two spin values (without hard constraints), but with interaction radius equal to two. We show that this model can be "transformed" to a nearest-neighbor interaction model with 8 spin values and with hard constraints on the Cayley tree. In each case we construct several kind of Gibbs measures of these models.

7.1 Definitions

In this chapter we consider models which exhibit what physicists sometimes call "hard constraints" -forbidden configurations, in which (for example) adjacent particles are not permitted to have certain pairs of spins. In the classical (ferromagnetic) Ising model, adjacent particles are discouraged from having opposing spins, since such opposition increases the energy of a configuration, making it a less likely state; this is a "soft" constraint.

Remark 7.1. Note that, the definitions which we are going to give here, in [35] given for general graphs, instead of Γ^k. But here we consider the graph as a Cayley tree.

145

Let $\Gamma^k = (V, L)$ be a Cayley tree of order $k \geq 1$. We model hard constraints by means of a fixed finite graph H, whose nodes may be thought of as different spins, or, as we prefer, different "colors." Adjacent (nearest-neighbor) sites of Γ^k may receive colors i and j only if i and j are adjacent nodes of H; in particular both sites may have color i only when in H there is a loop at node i. Thus a legal (admissible) configuration is a homomorphism from Γ^k to H, that is, a map φ from V to the nodes of H such that if u is adjacent to v in Γ^k (written $u \sim v$ or $\langle u, v \rangle$) then $\varphi(u) \sim \varphi(v)$ in H.

Among legal configurations, relative likelihood is determined by positive reals, called "activities," assigned to the colors. Thus, suppose two legal configurations differ only at site u; if the activity of the color of u in the first configuration is twice that of the color of u in the second, then the first configuration is twice as likely.

As mentioned above, the constraint graph H is finite, with nodes $1, 2, \ldots, n$ and H will often have loops at some or all of its nodes. To avoid degeneracies we always assume that H is connected and has at least two nodes.

We denote by $\mathrm{Hom}(\Gamma^k, H)$ the graph whose nodes are homomorphisms from Γ^k to H, with $\alpha \sim \beta$ when α and β differ on at most one site of Γ^k.

We denote the image under α of u by $\alpha(u)$, so that for $\alpha \in \mathrm{Hom}(\Gamma^k, H)$, $u \sim v$ in Γ^k implies $\alpha(u) \sim \alpha(v)$ in H.

Example 7.1. If H is the graph K_n^{loop} on $\{1, \ldots, n\}$ containing all possible loops and edges then there are no constraints, and $\mathrm{Hom}(\Gamma^k, H)$ contains all maps from Γ^k to $\{1, \ldots, n\}$.

Recall that a bipartite graph (or bigraph) is a graph whose vertices can be divided into two disjoint sets V_1 and V_2 such that every edge connects a vertex in V_1 to one in V_2; that is, V_1 and V_2 are independent sets. Equivalently, a bipartite graph is a graph that does not contain any odd-length cycles [29].

Example 7.2. If H is the loopless complete graph K_n, then $\mathrm{Hom}(\Gamma^k, H)$ is the set of all proper n-colorings of Γ^k, a coloring being proper if no two adjacent sites get the same color. Since trees are bipartite and connected, $\mathrm{Hom}(\Gamma^k, K_2)$ has just two elements.

Example 7.3. Let H be the graph on nodes $0, 1$ with edge $\{0, 1\}$ and loop at node 0. This constraint graph is precisely the one in effect for the two-state hard-core model (see below for details).

7.1.1 *Gibbs measures*

A set of activities for a constraint graph H is a function $\lambda : H \to R_+$ from the nodes of H to the positive reals, two such being regarded as equivalent if they differ by a constant factor. The value λ_i of λ at a node i is called its "activity," and will represent the relative probability of i as an image.

For finite subset $G \subset \Gamma^k$, when H and λ are given, we define the probability measure m_G on $\mathrm{Hom}(G, H)$, with respect to a random map φ, by

$$\mathrm{Pr}_{m_G}(\varphi = \alpha) \equiv m_G(\{\alpha\}) \equiv \frac{1}{Z} \prod_{u \in G} \lambda_{\alpha(u)},$$

where Z is the normalizing constant,

$$Z \equiv \sum_{\varphi \in \mathrm{Hom}(G,H)} \prod_{u \in G} \lambda_{\varphi(u)}.$$

Now we shall define Gibbs measure for models with hard constraints. If U is a patch (finite subset of Γ^k) and $\varphi \in \mathrm{Hom}(\Gamma^k, H)$, we denote by φ_U the restriction of φ to U; thus $\varphi_U \in \mathrm{Hom}(U, H)$. If A is an event of the form

$$A = \{\varphi \in \mathrm{Hom}(\Gamma^k, H) : \varphi_U \in F\}$$

for some patch U and some $F \subset \mathrm{Hom}(U, H)$, then we call A a "patch event." Equip $\mathrm{Hom}(\Gamma^k, H)$ with the σ-field, denoted by $\mathcal{F}$, generated by the patch events, and consider henceforth only probability measures μ on $(\mathrm{Hom}(\Gamma^k, H), \mathcal{F})$.

To define Gibbs measure we shall use DLR condition (see Chapter 2), which ensures that the conditional behavior of μ on patches is exactly what it should be. We fix and suppress reference to H and λ, and define $U^+ \equiv U \cup \partial U$ for any patch U, where ∂U is the set of sites in $\Gamma^k \setminus U$ which are adjacent to at least one site of U.

Definition 7.1. A measure μ on $\mathrm{Hom}(\Gamma^k, H)$ is a Gibbs measure if for any finite $U \subset \Gamma^k$, and almost every $\psi \in \mathrm{Hom}(\Gamma^k, H)$,

$$\mathrm{Pr}_\mu \left(\varphi_U = \psi_U \mid \varphi_{\Gamma^k \setminus U} = \psi_{\Gamma^k \setminus U} \right) = \mathrm{Pr}_{m_{U^+}} \left(\varphi_U = \psi_U \mid \varphi_{\partial U} = \psi_{\partial U} \right).$$

This is usual definition of a Gibbs measure: the probability distribution of a random φ inside a patch U depends only on its value on the boundary of U, and is the same as if U and its boundary comprised all of the lattice.

Note that, for some constraint graphs H (constituting an easily defined class known as the "cop-win," or "dismantlable," graphs) the Gibbs condition can be verified merely by checking patches U which consist only of a single site.

Note that Gibbs measure exists on the Cayley tree for any H and λ, and indeed in far more general circumstances. The following is a theorem of Dobrushin [58], specialized to our framework.

Theorem 7.1. *Let H be a constraint graph with the set of activities λ, and G any countable, locally finite graph for which $\mathrm{Hom}(G, H)$ is nonempty. Then there is a Gibbs measure on $\mathrm{Hom}(G, H)$.*

Proof. Let α be any fixed homomorphism from G to H. For each i, let μ_i be the discrete measure which is positive only on homomorphisms that agree with α on $G \setminus U_i$, and which satisfies

$$\mathrm{Pr}_{\mu_i}\left(\varphi_{U_i^+} = g\right) = \mathrm{Pr}_{m_{U_i^+}}\left(\varphi = g | \varphi_{\partial U_i} = g_{\partial U_i}\right),$$

for every $g \in \mathrm{Hom}(U_i^+, H)$ that agrees with α on ∂U_i. Since the space $\mathcal{M}$ of all probability measures on $\mathrm{Hom}(G, H)$ is compact, it remains only to choose a convergent subsequence of $\mu_1, \mu_2, \ldots$ and to observe that its limit is indeed a Gibbs measure. $\qquad\square$

Definition 7.2. Let $\mathcal{A}(\Gamma^k)$ be the automorphism group of Γ^k, and for any subset $S \subset \mathrm{Hom}(\Gamma^k, H)$ and $\kappa \in \mathcal{A}(\Gamma^k)$ let $S \circ \kappa \equiv \{\varphi \circ \kappa : \varphi \in S\}$. A measure μ on $\mathrm{Hom}(\Gamma^k, H)$ is called (translation)-invariant if, for any μ-measurable $S \subset \mathrm{Hom}(\Gamma^k, H)$ and any $k \in \mathcal{A}(\Gamma^k)$, we have $\mu(S \circ \kappa) = \mu(S)$.

For any site u in tree Γ^k let $\Gamma_1^k(u), \Gamma_2^k(u), \ldots, \Gamma_{k+1}^k(u)$ be the connected components of $\Gamma^k \setminus \{u\}$.

Definition 7.3. A Gibbs measure μ on $\mathrm{Hom}(\Gamma^k, H)$, is *simple* (or *splitting*) if, for any site $u \in \Gamma^k$ and any node $i \in H$, the μ-measures of

$$\varphi_{\Gamma_1^k(u)}, \quad \varphi_{\Gamma_2^k(u)}, \quad \ldots, \quad \varphi_{\Gamma_{k+1}^k(u)}$$

are mutually independent given $\varphi(u) = i$.

We will say that constraint graph H exhibit a phase transition if there are values of the activity vector λ to which correspond more than one (splitting), translation-invariant Gibbs measure on $\mathrm{Hom}(\Gamma^k, H)$. Since in fact there are always activities for which only one such measure exists, it is unimportant whether the phase transition is regarded as something which turns one Gibbs measure into another for the same set of activities, or something which changes activities from the region of unique Gibbs measures to a region where there is more than one.

In the following sections we shall consider several examples of models with hard constraints which have phase transitions.

7.2 Two-state hard core model

In this section we consider H as in Example 7.3, i.e., the constraint graph H consists of a looped node 0 connected by a edge to an unlooped node 1. We will take $\lambda_0 = 1$ so that the set of activities for H is specified by $\lambda \equiv \lambda_1$. In this case we have a nearest-neighbor hard-core model, with activity (fugacity) $\lambda > 0$, on a Cayley tree of order k. This model arises as a simple example of a loss network with a nearest-neighbor exclusion.

Our goal is to present results on the nature of a phase transition for the nearest-neighbor hard-core model. In this model one assigns, to the sites of the tree $x \in V$, values $\sigma(x) = 0, 1$; value $\sigma(x) = 1$ means that site x is "occupied" and $\sigma(x) = 0$ that x is "vacant". A configuration σ on the tree (i.e., in V) is a collection $\{\sigma(x), x \in V\}$ considered also as a function $V \to \{0, 1\}$.

We call σ an admissible (legal) configuration on the Cayley tree if the product $\sigma(x)\sigma(y) = 0$ for any nearest-neighbor pair x, y from V.

Denote the set of admissible configurations by Ω. The set of all Gibbs measures of the hard core model is denoted by $\mathcal{G}$.

7.2.1 *Construction of splitting (simple) Gibbs measures*

We shall consider splitting (simple) Gibbs measures, which in addition to the aforementioned Markov property, they satisfy the following condition: given values $\sigma(x)$, $x \in V_n$, of an admissible configuration $\sigma \in \Omega$ over set V_n, its values $\sigma(y)$ at sites $y \in W_{n+1}$ are conditionally independent. A formal definition follows.

For any configuration σ_n of Ω_{V_n} of admissible configurations we set

$$\sharp\sigma_n = \sum_{x \in V_n} \sigma_n(x),$$

the number of occupied sites in V_n by σ_n.

Let $z : x \to z_x = (z_{0,x}, z_{1,x}) \in R_+^2$ be a vector-valued function on V. Given $n = 1, 2, \ldots$, consider the probability distribution $\mu^{(n)}$ on the set

Ω_{V_n} defined by

$$\mu^{(n)}(\sigma_n) = \frac{1}{Z_n} \lambda^{\#\sigma_n} \prod_{x \in W_n} z_{\sigma_n(x), x}. \tag{7.1}$$

Here

$$Z_n = \sum_{\varphi_n \in \Omega_{V_n}} \lambda^{\#\varphi_n} \prod_{x \in W_n} z_{\varphi_n(x), x}. \tag{7.2}$$

We say that the probability measures $\mu^{(n)}$ are compatible if for all $n \geq 1$ and $\sigma_{n-1} \in \Omega_{V_{n-1}}$:

$$\sum_{\omega_n \in \Omega_{W_n}} \mu^{(n)}(\sigma_{n-1} \vee \omega_n)\mathbf{1}(\sigma_{n-1} \vee \omega_n \in \Omega_{V_n}) = \mu^{(n-1)}(\sigma_{n-1}), \tag{7.3}$$

where the symbol $\vee$ denotes concatenation of configurations.

This condition implies the existence of a unique measure μ defined on Ω such that, for all n and $\sigma_n \in \Omega_{V_n}$, $\mu(\{\sigma|_{V_n} = \sigma_n\}) = \mu^{(n)}(\sigma_n)$.

Definition 7.4. Measure μ defined by (7.1), (7.3) is called a splitting (hard core) Gibbs measure, corresponding to the function z.

The following statement describes conditions on the function z that ensure compatibility of measures $\mu^{(n)}$.

Proposition 7.1. *Probability measures* $\mu^{(n)}, n = 1, 2, \ldots,$ *in (7.1) are compatible iff for any* $x \in V$ *the following equation holds:*

$$z'_x = \prod_{y \in S(x)} \frac{1}{1 + \lambda z'_y}, \tag{7.4}$$

here $z'_x = \frac{z_{1,x}}{z_{0,x}}$.

Proof. Write

$$\text{LHS of (7.3)} = \frac{1}{Z_n} \lambda^{\#\sigma_{n-1}} \prod_{x \in W_{n-1}} \prod_{y \in S(x)} \left(z_{0,y} + \mathbf{1}(\sigma_{n-1}(x) = 0)\lambda z_{1,y} \right). \tag{7.5}$$

Sufficiency. Suppose that (7.4) holds. It is equivalent to the representations

$$\prod_{y \in S(x)} (z_{0,y} + \lambda z_{1,y}) = a(x)z_{0,x}, \qquad \prod_{y \in S(x)} z_{0,y} = a(x)z_{1,x},$$

for some $a(x) > 0$, $x \in V$. Setting $A_n = \prod_{x \in W_n} a(x)$ and substituting (7.1) into (7.5), we get

$$\text{RHS of (7.5)} = \frac{1}{Z_n} \lambda^{\#\sigma_{n-1}} \prod_{x \in W_{n-1}} z_{\sigma_{n-1}(x), x} a(x) =$$

$$\frac{A_{n-1}}{Z_n}\lambda^{\sharp\sigma_{n-1}}\prod_{x\in W_{n-1}}z_{\sigma_{n-1}(x),x}.$$

We should have

$$\sum_{\sigma_{n-1}\in\Omega_{V_{n-1}}}\sum_{\omega_n\in\Omega_{W_n}}\mathbf{1}(\sigma_{n-1}\vee\omega_n\in\Omega_{V_n})\mu^{(n)}(\sigma_{n-1}\vee\omega_n)=1,$$

hence $A_{n-1}/Z_n=1/Z_{n-1}$, and (7.3) holds.

Necessity. Suppose that (7.3) holds; we want to prove (7.4). Substituting (7.1) in (7.3) and using (7.5), we obtain that for any $\sigma_{n-1}\in\Omega_{V_{n-1}}$

$$\frac{Z_{n-1}}{Z_n}\prod_{x\in W_{n-1}}\prod_{y\in S(x)}(z_{0,y}+\mathbf{1}(\sigma_{n-1}(x)=0)\lambda z_{1,y})=\prod_{x\in W_{n-1}}z_{\sigma_{n-1},x}.$$

In particular, comparing a pair of configurations $\sigma_{n-1},\sigma'_{n-1}$ different at a single site $x\in W_{n-1}$ yields (7.4). $\qquad\square$

Proposition 7.2. *Any measure μ with local distributions $\mu^{(n)}$ satisfying (7.1), (7.4) is a splitting (hard core) Gibbs measure.*

Proof. Straightforward. $\qquad\square$

7.2.2 Uniqueness of a translation-invariant splitting Gibbs measure

Without loss of generality, we set hereafter $z_{0,x}\equiv 1$ and $z_x=z'_x=z_{1,x}>0$. Then condition (7.4) reads

$$z_x=\prod_{y\in S(x)}\frac{1}{1+\lambda z_y}. \tag{7.6}$$

Then for any function z_x satisfying functional equation (7.6) there exists a unique hard core splitting Gibbs measure μ and vice versa. However, the analysis of solutions to (7.6) is rather tricky. It is natural to begin with translation-invariant solutions where $z_x=z$ is constant >0, $x\neq x^0$.

In this case we obtain, from (7.6), the following equation:

$$z=f(z),\quad\text{where}\quad f(z)=(1+\lambda z)^{-k}. \tag{7.7}$$

Function f is analytic for $z>0$, with value 1 at $z=0$, and decreases monotonically to 0 as $z\to\infty$. Thus (7.7) has a unique positive solution z^*. So:

Theorem 7.2. *For any $\lambda>0$, the translation-invariant hard-core splitting Gibbs measures is unique.*

The translation-invariant hard-core splitting Gibbs measure is denoted by μ^*.

7.2.3 *Periodic hard core splitting Gibbs measures*

Using the similar definitions as Definitions 2.1 and 2.2 one can define periodic Gibbs measure for the hard core model. Moreover, one can also prove the analogies of Theorems 2.3 and 2.4. Then it remains to describe $G_k^{(2)}$-periodic splitting Gibbs measures, where $G_k^{(2)}$ is the subgroup which contains all words of even length.

They correspond to functions

$$z_x = \begin{cases} z_1, & \text{if } x \in G_k^{(2)}, \\[2mm] z_2, & \text{if } x \in G_k \setminus G_k^{(2)}. \end{cases}$$

In this case we have from (7.6):

$$z_1 = (1 + \lambda z_2)^{-k}, \quad z_2 = (1 + \lambda z_1)^{-k}. \tag{7.8}$$

That is, each of z_1, z_2, satisfy

$$z = f(f(z)), \quad f(z) = (1 + \lambda z)^{-k}. \tag{7.9}$$

Theorem 7.3. *For*

$$\lambda \leq \lambda_{\mathrm{cr}} = \frac{k^k}{(k-1)^{k+1}} \tag{7.10}$$

$G_k^{(2)}$*-periodic hard core splitting Gibbs measure coincides with* μ^*.

If $\lambda > \lambda_{\mathrm{cr}}$ *then there exist at least three* $G_k^{(2)}$*-periodic Gibbs measures* μ_-, μ^*, μ_+, *one of which* μ^* *is translation invariant.*

Proof. The key observation was made in [127]: If z^* is the unique positive solution to (7.6) then

$$\frac{d}{dz} f(z^*) = -\frac{k\lambda z^*}{1 + \lambda z^*} \tag{7.11}$$

which is ≥ -1 iff (7.10) holds. So, under (7.10), $z = z^*$ is a stable fixed point of the map $f(z)$, and $\lim_{n\to\infty} f^{(n)}(z) = z^*$, for any $z > 0$. Here and below, $f^{(n)}$ is the nth iterate of the above map $f(z)$. Therefore, z^* is the unique positive solution to (7.9). On the other hand, under $\lambda > \lambda_{\mathrm{cr}}$, the fixed point z^* is unstable. Iterates $f^{(2n)}(z)$ remain $> z^*$, monotonically decrease and hence converge to a limit, $z_+ \geq z^*$ which solves (7.9). However, $z_+ > z^*$ as z^* is unstable. Then $z_- = f(z_+)$ is $< z^*$ and also solves (7.9). This completes the proof. $\square$

Proposition 7.3. *If $k \geq 2$, $\lambda > \lambda_{\mathrm{cr}}$ and z_x is a solution of (7.6) then*

$$z_- \leq z_x \leq z_+,$$

for any $x \in V$, where $z_- < z_+$ are defined uniquely by $z_+ = f(f(z_+))$ and $z_- = f(z_+)$.

Proof. Is similar to the proof of Proposition 2.1. $\qquad\square$

Clearly, the two translation-periodic splitting measures corresponding to solutions (z_-, z_+) and (z_+, z_-) are μ_- (minimal) and μ_+ (maximal).

The extremality of these measures can be deduced without using the fact that they are maximal and minimal (but using the minimality and maximality of the corresponding values $z_\pm$). Assume that μ_+ are non-extreme, i.e., are decomposed:

$$\mu_+ = \int \mu(\omega)\nu(d\omega).$$

Then for any vertex $x \in V$ we have

$$z_+ = \int z_x(\omega)\nu(d\omega). \tag{7.12}$$

As z_+ is an extreme point in the set $\{z_x : z_- \leq z_x \leq z_+\}$, (7.12) holds if $z_x(\omega) = z_+$ for almost all ω. Hence, μ_+ is extreme.

A simple calculation shows that as $\lambda \to \infty$, $z^* \to 0$. $z_- \to 0$ and $z_+ \to 1$. We conjecture that measures $\mu_\pm$ approach the Dirac delta-measures concentrated on the two "chess-board" type configurations $\omega_\pm$ and measure μ^* the half-sum of these.

Corollary 7.1. *Under condition (7.10), the whole set $\mathcal{G}$ of hard core measures is reduced to a single measure μ^*. This measure is extreme.*

Proof. Under (7.10), maximal and minimal measures $\mu_\pm$ coincide: or equivalently $z_- = z_+$. Hence, $\mathcal{G}$ is reduced to a single point. $\qquad\square$

From general results of [266], [267] it follows that the number of translation-periodic splitting measures equals one or three. Here we give a direct proof of the related fact for the hard core model.

Theorem 7.4. *For $k \geq 2$ and $\lambda > \lambda_{\mathrm{cr}}$, the number of hard core periodic splitting Gibbs measures is precisely three. They are μ_-, μ^* and μ_+.*

Proof. For periodic splitting Gibbs measures, the values z_1 and z_2 are solutions to

$$z_1 = f(z_2), \quad z_2 = f(z_1).$$

Here

$$f^{(2)}(z) = \left(\frac{(1 + \lambda z)^k}{\lambda + (1 + \lambda z)^k} \right)^k .$$

The equation $f^{(2)}(z) - z = 0$ will have more than three positive solutions only when $(d^2/dz^2)f^{(2)}(z) = 0$ has more than one positive root. However,

$$\frac{d^2}{dz^2} f^{(2)}(z) = \lambda^3 k^2 \frac{(1 + \lambda z)^{k^2-2}}{(\lambda + (1 + \lambda z)^k)^{k+2}} [(k^2 - 1)\lambda - (k + 1)(1 + \lambda z)^k],$$

which has only one positive root.

Hence, we obtain three distinct periodic splitting Gibbs measures. It is straightforward that they must be μ_-, μ^* and μ_+. $\qquad\square$

In the case $k = 2$, the proof does not need such calculations. In this case, the fixed points for the second iteration map $f^{(2)}$ are among the roots of a fifth degree polynomial $P^{(2)}(z)$. However, we should discard the roots of a third degree polynomial $P(z)$ that divides $P^{(2)}(z)$ and arises when we consider the fixed points for f. The ratio $P^{(2)}(z)/P(z)$ is a quadratic polynomial and has at most two roots. These roots must be positive and distinct.

7.2.4 *Extremality of the translation-invariant splitting Gibbs measure*

Results presented in this subsection are based on methods (reconstruction solvability) developed in Chapter 4.

Theorem 7.5. *For $k \geq 2$ and*

$$\lambda > \frac{1}{\sqrt{k} - 1} \left(\frac{\sqrt{k}}{\sqrt{k} - 1} \right)^k \tag{7.13}$$

the translation-invariant splitting Gibbs measure μ^* is not extreme.

Proof. We apply the "second eigenvalue" calculation from Chapter 4. In this calculation, one considers a one-parameter family of Markov chains with states 0, 1 and transition probabilities

$$\mathbf{P}(1,0) = 1, \quad \mathbf{P}(0,0) = \frac{\mu^*(\{\sigma(x) = \sigma(x') = 0\})}{\mu^*(\{\sigma(x) = 0\})} = \frac{1}{1 + \lambda z^*},$$

$$\mathbf{P}(0,1) = \frac{\mu^*(\{\sigma(x) = 0, \sigma(x') = 1\})}{\mu^*(\{\sigma(x) = 0\})} = \frac{\lambda z^*}{1 + \lambda z^*}.$$

Here, $z^* = z^*(\lambda)$ is the unique positive solution to (7.7). Matrix $\mathbf{P} = (\mathbf{P}(i,j), i, j = 0, 1)$ has eigenvalues $\Lambda_1 = 1$ and $\Lambda_2 = \lambda z^*/(1 + \lambda z^*)$.

A sufficient condition for non-extremality of μ^* is that $k\Lambda_2^2 > 1$, i.e., $\lambda z^*/(1 + \lambda z^*) > 1/\sqrt{k}$. Going back to (7.11), we see that this bound holds iff $(d/dz)f(z^*) \leq -\sqrt{k}$, i.e., $\lambda z^* > 1/(\sqrt{k} - 1)$. This immediately leads to bound (7.13). $\qquad\square$

The question when precisely measure μ^* becomes non-extreme (and what its decomposition is) is of a great interest. One could envisage two possibilities:

(a) μ^* is non-extreme for all $\lambda > \lambda_{\mathrm{cr}}$.

(b) there exists λ'_{cr} lying strictly in between such that μ^* is extreme for $\lambda < \lambda'_{\mathrm{cr}}$ and non-extreme for $\lambda > \lambda'_{\mathrm{cr}}$. Theorem 7.6 below shows that λ'_{cr} for large k grows at most as $\ln k(\ln k + \ln \ln k)$.

Also, the decomposition of μ^* into extreme measures looks pretty "weird" (it cannot be a half-sum of measures $\mu_\pm$ as this would destroy the splitting character of μ^*).

Theorem 7.6. *For any $\epsilon > 0$, there exists k_0 such that for $k \geq k_0$, measure μ^* is non-extreme for*

$$\lambda > e^{1+\epsilon} \ln k(\ln k + \ln \ln k + 1 + \epsilon). \tag{7.14}$$

Proof. The proof involves the analysis of a reconstruction algorithm of the sort proposed and studied in Chapter 4. In this algorithm, one assigns, recursively, value 1 to site x if all $y \in S(x)$ have values 0, and value 0 otherwise. This allows us to assign values 0 or 1 to the origin x^0, given $\sigma_n \in \Omega_{W_n}$, an admissible configuration on W_n. If $\lambda > \lambda'_{\mathrm{cr}}$, the reconstruction, in the sense of Chapter 4, is possible. This is known to be equivalent to non-extremality of μ^*. The analysis of the reconstruction algorithm reveals that the reconstruction works under condition (7.14) (for k large enough).

More precisely, one considers a 0, 1-valued Markov chain generated by measure μ^* for sites along any path on the tree. The transition probability matrix for the chain is

$$\begin{pmatrix} 1 - c & c \\ 1 & 0 \end{pmatrix},$$

where c is determined from $\lambda = c/(1 - c)^{k+1}$. One then shows that the above algorithm succeeds if

$$c > \frac{1}{k}(\ln k + \ln \ln k + 1 + \epsilon)$$

which, for k large, is equivalent to (7.14). $\qquad\square$

The following statement shows that the definition of the "second" critical point λ'_{cr} as $\inf\{\lambda : \text{measure } \mu^* \text{ is non-extreme}\}$ is correct.

Proposition 7.4. *If, for given k and λ^0, measure μ^* is non-extreme then it remains non-extreme for the same k and all $\lambda > \lambda^0$.*

Proof. Again one uses reconstruction techniques. Given k and λ^0, if there exists an algorithm reconstructing value $\sigma(x^0)$ from σ_{W_n} then it can be modified for any $\lambda \geq \lambda_0$ so that the reconstruction remains possible. $\square$

To conclude this subsection, we shall give main results of work by Martin [153]. One of the results of [153] is the following Theorem 7.7 which develops Theorem 7.6.

Theorem 7.7. *For $\lambda = 1$, measure μ^* is extreme for any k.*

Let $\mathbf{P} = (p_{ij})_{i,j=0,1}$ be a 2×2 stochastic matrix, which we regard as a transition matrix on the set $\{0, 1\}$.

In Proposition 4.1 of [170], Mossel and Peres show that reconstruction is impossible whenever

$$\frac{(p_{00} - p_{10})^2}{\min\{p_{00} + p_{10}, \ \ p_{01} + p_{11}\}} \leq \frac{1}{k}.$$

In [153] the bound is improved by the following.

Theorem 7.8. *Reconstruction is impossible whenever*

$$\left(\sqrt{p_{00}p_{11}} - \sqrt{p_{01}p_{10}}\right)^2 \leq \frac{1}{k}. \tag{7.15}$$

In [39] Brightwell and Winkler show that, as $k \to \infty$,

$$\frac{1 + o(1)}{\ln k} \leq \lambda'_{\mathrm{cr}} \leq (\ln k)^2 (1 + o(1)). \tag{7.16}$$

Martin [153] improved the lower bound by the following:

Theorem 7.9. $\lambda'_{\mathrm{cr}} > e - 1$ *for all k.*

7.2.5   *Weakly periodic Gibbs measures*

For a subgroup $\widehat{G}_k \subset G_k$, similarly as in case of Ising model, one can define weakly periodic solutions of (7.4):

The function $z_x, x \in G_k$ is called $\widehat{G}_k$-weakly periodic, if $z_x = z_{ij}$ for $x \in H_i, x_\downarrow \in H_j, \forall x \in G_k$ (cf. with Definition 2.3).

A Gibbs measure μ of hard-core model is called $\widehat{G}_k$-weakly periodic, if it corresponds to a $\widehat{G}_k$-weakly periodic solution z of (7.4).

In this subsection we take $\widehat{G}_k$ as a subgroup of index two, i.e., for $\emptyset \neq A \subset N_k = \{1, 2, ..., k+1\}$ we consider

$$\widehat{G}_k \equiv H_A = \{x \in G_k : \sum_{i \in A} w_x(a_i) - \text{even}\},$$

where $w_x(a_i)-$ number of a_i in word $x \in G_k$.

Note that H_A-weakly periodic function z has the form:

$$z_x = \begin{cases} z_1, \ x \in H_A, \ x_\downarrow \in H_A \\[2mm] z_2, \ x \in H_A, \ x_\downarrow \in G_k \backslash H_A \\[2mm] z_3, \ x \in G_k \backslash H_A, \ x_\downarrow \in H_A \\[2mm] z_4, \ x \in G_k \backslash H_A, x_\downarrow \in G_k \backslash H_A. \end{cases}$$

Then by (7.4) $z_i (i = 1, 2, 3, 4)$ satisfy the following system

$$\begin{cases} z_1 = \left(\frac{1+\lambda z_1}{1+\lambda z_3}\right)^i \cdot \frac{1}{(1+\lambda z_1)^k} \\[3mm] z_2 = \left(\frac{1+\lambda z_1}{1+\lambda z_3}\right)^{i-1} \cdot \frac{1}{(1+\lambda z_1)^k} \\[3mm] z_3 = \left(\frac{1+\lambda z_4}{1+\lambda z_2}\right)^{i-1} \cdot \frac{1}{(1+\lambda z_4)^k} \\[3mm] z_4 = \left(\frac{1+\lambda z_4}{1+\lambda z_2}\right)^i \cdot \frac{1}{(1+\lambda z_4)^k}, \end{cases} \qquad (7.17)$$

where $|A| = i$. Dividing the first equation to the second and third equation to the fourth, we obtain

$$\begin{cases} \frac{z_1}{z_2} = \frac{1+\lambda z_1}{1+\lambda z_3} \\[3mm] \frac{z_3}{z_4} = \frac{1+\lambda z_2}{1+\lambda z_4}. \end{cases} \qquad (7.18)$$

Using this system of equations, from (7.17) we get

$$\begin{cases} z_1 = \left(\frac{z_1}{z_2}\right)^i \cdot \frac{1}{(1+\lambda z_1)^k} \\[3mm] z_2 = \left(\frac{z_1}{z_2}\right)^{i-1} \cdot \frac{1}{(1+\lambda z_1)^k} \\[3mm] z_3 = \left(\frac{z_4}{z_3}\right)^{i-1} \cdot \frac{1}{(1+\lambda z_4)^k} \\[3mm] z_4 = \left(\frac{z_4}{z_3}\right)^i \cdot \frac{1}{(1+\lambda z_4)^k}. \end{cases} \qquad (7.19)$$

From the first equation of (7.19) we find z_2, and from the fourth equation find z_3 then by (7.17), we obtain

$$\begin{cases} z_1 = \varphi(z_4) \cdot \psi(z_1) \\ z_4 = \varphi(z_1) \cdot \psi(z_4), \end{cases} \tag{7.20}$$

where

$$\varphi(z) = \frac{(1+\lambda z)^k}{((1+\lambda z)^{k/i} + \lambda z^{1-1/i})^i}, \quad \psi(z) = \frac{1}{(1+\lambda z)^{k-i}}.$$

Lemma 7.1. *1) If $z_1 = z_4, z_2 = z_3$ for a solution of (7.17) then $z_1 = z_2 = z_3 = z_4$.*

2) $z_1 = z_4$ iff $z_2 = z_3$.

Proof. 1) If $z_2 = z_3$, then from the first equation of (7.18) we have $z_1/z_2 = (1+\lambda z_1)/(1+\lambda z_2)$, consequently $z_1 = z_2$. Moreover, by condition we have $z_1 = z_4$, hence $z_1 = z_2 = z_3 = z_4$.

2) If $z_1 = z_4$, from (7.18) we get $z_3/z_2 = (1+\lambda z_2)/(1+\lambda z_3)$, i.e., $z_2 = z_3$. If $z_2 = z_3$, then $z_1/z_4 = (1+\lambda z_1)/(1+\lambda z_4)$, consequently $z_1 = z_4$. $\qquad\square$

The following theorem gives complete description of H_A-weakly periodic Gibbs measures of the two state hard-core model.

Theorem 7.10. *For any $k \geq 1$, $i \leq k$, (where $i = |A|$), and any $\lambda > 0$ the H_A-weakly periodic Gibbs measure of hard-core model is unique. Moreover, it coincides with the unique translation invariant Gibbs measure.*

Proof. It suffices to prove that the system of equations (7.20) has solutions (z_1, z_4) only with $z_1 = z_4$. From (7.20) we get

$$z_1 - z_4 = \psi(z_1) \cdot \varphi'(\xi) \cdot (z_4 - z_1) + \varphi(z_1) \cdot \psi'(\eta) \cdot (z_1 - z_4),$$

where $\xi \in (z_1, z_4), \eta \in (z_1, z_4)$. Consequently

$$(z_1 - z_4) \cdot [1 + \psi(z_1) \cdot \varphi'(\xi) - \varphi(z_1) \cdot \psi'(\eta)] = 0. \tag{7.21}$$

From this equation we get $z_1 = z_4$ or

$$1 + F(z_1, \xi, \eta) = 0, \tag{7.22}$$

where

$$F(z_1, \xi, \eta) = \psi(z_1) \cdot \varphi'(\xi) - \varphi(z_1) \cdot \psi'(\eta).$$

Here $\psi'(\eta) = -\frac{\lambda(k-i)}{(1+\lambda\eta)^{k-i+1}} \leq 0$, i.e., function $\psi(\eta)$ is a decreasing function, since $i \leq k, \lambda > 0, \eta > 0$. For $\xi \in (z_1, z_4)$ we have

$$\varphi'(\xi) = \frac{\lambda(1+\lambda\xi)^{k-1}[\lambda\xi(k-i+1) - i + 1]}{\xi^{1/i}[(1+\lambda\xi)^{k/i} + \lambda\xi^{1-1/i}]^{i+1}}.$$

Consider two cases for (i, k):

Case $i \leq \frac{k+1}{2}$: We shall show that equation (7.22) has no solution, i.e.,

$$F(z_1, \xi, \eta) > -1. \tag{7.23}$$

First by Proposition 7.3 for $x \in V$ one has

$$(1 + \lambda)^{-k} < z_x < 1. \tag{7.24}$$

Using this we get

$$F(z_1, \xi, \eta) > \frac{\lambda(1 + \frac{\lambda}{(1+\lambda)^k})^{k-1}[\frac{\lambda(k-i+1)}{(1+\lambda)^k} - i + 1]}{(1 + \lambda)^{k-i}[(1 + \lambda)^{\frac{k}{i}} + \lambda]^{i+1}}$$

$$+ \frac{\lambda(k - i)(1 + \frac{\lambda}{(1+\lambda)^k})^k}{[(1 + \lambda)^{\frac{k}{i}} + \lambda]^i(1 + \lambda)^{k-i+1}}$$

$$> \frac{\lambda((1 + \lambda)^k + \lambda)^{k-1}[\lambda(k - i + 1) + (1 - i)(1 + \lambda)^k]}{(1 + \lambda)^{k^2+k-i+1}[(1 + \lambda)^k + \lambda]^{i+1}}$$

$$+ \frac{\lambda(k - i)((1 + \lambda)^k + \lambda)^k}{(1 + \lambda)^{k^2+k-i+1}[(1 + \lambda)^k + \lambda]^i} > -1.$$

Consequently

$$\lambda((1 + \lambda)^k + \lambda)^{k-i-2}[\lambda(k - i + 1) + (1 - i)(1 + \lambda)^k]+$$

$$\lambda(k - i)((1 + \lambda)^k + \lambda)^{k-i} + (1 + \lambda)^{k^2+k-i+1} > 0.$$

Denoting $a = 1 + \lambda$, we get from the last inequality

$$\lambda(a^k + \lambda)^{k-i-2}[\lambda(k-i+1)+(1-i)a^k] + \lambda(k-i)(a^k + \lambda)^{k-i} + a^{k^2+k-i+1} > 0$$

or

$$\lambda(a^k + \lambda)^{k-i}[\lambda(k - i + 1) + (k - i)(a^k + \lambda)^2 - (i - 1)a^k]+$$

$$a^{k^2+k-i+1}(a^k + \lambda)^2 > 0.$$

Now we shall show that $(k - i)(a^k + \lambda)^2 - (i - 1)a^k$ is positive. Indeed,

$$(k-i)(a^k + \lambda)^2 - (i-1)a^k = a^k[(k-i)a^k - (i-1)] + 2(k-i)\lambda a^k + (k-i)\lambda^2.$$

It is easy to see that

$$(k-i)a^k - (i-1) = (k-i)(1+\lambda)^k - (i-1) = k-i+(k-i)k\lambda + ... + (k-i)\lambda^k - (i-1)$$

is positive if $i \leq \frac{k+1}{2}$. Consequently, $\varphi'(\xi) > -1$, this means that system (7.21) for $i \leq \frac{k+1}{2}$ has solutions only with $z_1 = z_4$. Hence by Lemma 7.1 we get $z_1 = z_2 = z_3 = z_4$, i.e., the solution is translation invariant.

Case $\frac{k+1}{2} < i \leq k$: Note that the case $i = k$ is obvious. Consider case $\frac{k+1}{2} < i \leq k - 1$. In this case, as previous case we get

$$\psi(z_1) \cdot \varphi'(\xi) - \varphi(z_1) \cdot \psi'(\eta)$$

$$> \frac{\lambda[(1+\lambda)^k + \lambda]^{k-1} \cdot [\lambda(k+1-i) - (i-1)(1+\lambda)^k]}{(1+\lambda)^{k^2+k-i}[(1+\lambda)^{\frac{k}{i}} + \lambda]^{i+1}} +$$

$$\frac{\lambda(k-i)[(1+\lambda)^k + \lambda]^k}{(1+\lambda)^{k^2+k-i+1}[(1+\lambda)^{\frac{k}{i}} + \lambda]^i}.$$

Using inequalities (7.24) one obtains

$$\psi(z_1) \cdot \varphi'(\xi) - \varphi(z_1) \cdot \psi'(\eta)$$

$$> \frac{\lambda[(1+\lambda)^k + \lambda]^{k-1} \cdot [\lambda(k+1-(k-1)) - (k-1-1)(1+\lambda)^k]}{(1+\lambda)^{k^2+k-\frac{k+1}{2}} \cdot [(1+\lambda)^{\frac{k}{\frac{k+1}{2}}} + \lambda]^{k-1+1}}$$

$$+ \frac{\lambda(k-(k-1)) \cdot [(1+\lambda)^k + \lambda]^k}{(1+\lambda)^{k^2+k-\frac{k+1}{2}+1} \cdot [(1+\lambda)^{\frac{k}{\frac{k+1}{2}}} + \lambda]^{k-1}}$$

$$> \frac{\lambda[(1+\lambda)^k + \lambda]^{k-1} \cdot [2\lambda - (k-2)(1+\lambda)^k]}{(1+\lambda)^{2k^2+k+1} \cdot [(1+\lambda)^k + \lambda]^k} +$$

$$\frac{\lambda \cdot [(1+\lambda)^k + \lambda]^k}{(1+\lambda)^{2k^2+k+1} \cdot [(1+\lambda)^k + \lambda]^k}$$

$$= \frac{\lambda \cdot [2\lambda - (k-2)(1+\lambda)^k]}{(1+\lambda)^{2k^2+k+1} \cdot [(1+\lambda)^k + \lambda]} + \frac{\lambda}{(1+\lambda)^{2k^2+k+1}} > -1.$$

Consequently

$$\lambda \cdot [2\lambda - (k-2)(1+\lambda)^k] + \lambda \cdot [(1+\lambda)^k + \lambda] + (1+\lambda)^{2k^2+k+1} \cdot [(1+\lambda)^k + \lambda]$$

$$= 3\lambda^2 + (1+\lambda)^k \cdot [\lambda + \lambda(1+\lambda)^{2k^2+1} + (1+\lambda)^{2k^2+k+1} - (k-2)\lambda] > 0.$$

Here we used $(1+\lambda)^{2k^2+k+1} - (k-2)\lambda > 0$. $\qquad\qquad\square$

7.2.6 *The model with two fugacities*

In definition of the activity (fugacity) λ defined above they were positive real functions $\lambda : H \to R_+$, but the values do not depend on sites of the Cayley tree.

A site is called odd (resp. even) if it is at odd (resp. even) distance from the origin x^0.

In this subsection we consider hard core model with two values of fugacity, λ_e and λ_o depending on sites of the Cayley tree: Value λ_e is assigned to the even and λ_o to the odd sites of the tree.

We are interested in periodic splitting Gibbs measures characterized by values z_e and z_o. They satisfy equations analogous to (7.8):

$$z_e = (1 + \lambda_o z_o)^{-k}, \quad z_o = (1 + \lambda_e z_e)^{-k}. \tag{7.25}$$

Denote

$$f_e(x) = (1 + \lambda_e x)^{-k}, \quad f_o(x) = (1 + \lambda_o x)^{-k}, \quad x \ge 0. \tag{7.26}$$

From (7.25) we get

$$z_e = f_o\left(f_e(z_e)\right). \tag{7.27}$$

Note that $f_o(x) = f_e((\lambda_o/\lambda_e)x)$. Therefore, if we let $z_e = (\lambda_o/\lambda_e)x$ then (7.27) implies

$$\frac{\lambda_o}{\lambda_e}x = F(x), \quad \text{where} \ \ F(x) = f_o(f_o(x)), \ x \ge 0. \tag{7.28}$$

The analysis of equation (7.28) is carried in:

Theorem 7.11.

(1) If $k = 1$ or $k \ge 2$ and $\lambda_o \le (1/(k-1))(k/(k-1))^k$ then a positive solution to (7.28) is unique.

(2) If $k > 1$ and $\lambda_o > (1/(k-1))(k/(k-1))^k$ then there exist values $\eta_1 = \eta_1(\lambda_o, k)$ and $\eta_2 = \eta_2(\lambda_o, k)$ such that $0 < \eta_1 < \eta_2$ and if $\lambda_o/\eta_2 < \lambda_e < \lambda_o/\eta_1$ then (7.28) has three distinct positive solutions.

(3) If $\lambda_e = \lambda_o/\eta_1$ or $\lambda_e = \lambda_o/\eta_2$ and $\lambda_o > (1/(k-1))(k/(k-1))^k$ then (7.28) has two distinct positive solutions.

The quantities η_1 and η_2 are determined by the formula

$$\eta_i = \frac{1}{x_i}F(x_i), \quad i = 1, 2, \tag{7.29}$$

where x_1 and x_2 are the positive solutions to the equation

$$(1 + \lambda_o x)^{k+1} - (k^2 - 1)\lambda_o^2 x + \lambda_o = 0. \tag{7.30}$$

Proof. We have

$$F'(x) = \frac{k^2\lambda_o^2(1 + \lambda_o(1 + \lambda_o x)^{-k})^{-k}}{(1 + \lambda_o x)^{k+1} + \lambda_o^2 x + \lambda_o};$$

$$F''(x) = \frac{k^2(k + 1)\lambda_o^3(1 + \lambda_o x)^{k^2 - 2}}{(\lambda_o + (1 + \lambda_o x)^k)^{k+2}}[(k - 1)\lambda_o - (1 + \lambda_o x)^k].$$

If $k = 1$ then $F''(x) < 0$ and F is concave increasing. Hence for $k = 1$, (7.28) has only one solution. For $k \geq 2$, F is convex for $x < \lambda_o[((k - 1)\lambda_o)^{1/k} - 1]$ and concave for $x > \lambda_o[((k - 1)\lambda_o)^{1/k} - 1]$; thus there are at most three solutions. On the other hand, it is easy to see that (7.28) has more than one solution if and only if there is more than one solution to the equation $xF'(x) = F(x)$ which is equivalent to equation (7.30).

Now consider (7.30). Denoting $u = 1 + \lambda_o x$, it follows from (7.30) that

$$u^{k+1} = (k^2 - 1)\lambda_o u - k^2\lambda_o, \; u > 1. \tag{7.31}$$

As function $u \to u^{k+1}$ is concave increasing, we conclude that (7.31) has a unique positive solution, say u_*, if u_* satisfies

$$\begin{cases} u_*^{k+1} = (k^2 - 1)\lambda_o u_* - k^2\lambda_o, \\ (k + 1)u_*^k = (k^2 - 1)\lambda_o. \end{cases} \tag{7.32}$$

In other words, if (7.32) is satisfied for $u_* > 0$ then u_* is a unique positive solution to (7.31).

From (7.32) we obtain $u_* = k/(k - 1)$ and $\lambda_o = (1/(k - 1))(k/(k - 1))^k$. We conclude that (7.31) has two solutions for $\lambda_o > (1/(k - 1))(k/(k - 1))^k$. This completes the proof of Theorem 7.11. $\square$

Denote

$$\mathbf{M} = \left\{(\lambda_o, \lambda_e) \in R_+^2 : \lambda_o > \frac{1}{k - 1}\left(\frac{k}{k - 1}\right)^k, \; \frac{\lambda_o}{\eta_2} < \lambda_e < \frac{\lambda_o}{\eta_1}\right\}.$$

We obtain the following assertion about solutions to system (7.25).

Corollary 7.2. *For $(\lambda_o, \lambda_e) \in \mathbf{M}$, (7.25) has three solutions, $(z_e^{(i)}, z_o^{(i)}) = ((\lambda_o/\lambda_e)x_i^*, f_o(x_i^*))$, $i = 1, 2, 3$, where x_i^* are the positive solutions to (7.28).*

We summarize the above findings in:

Theorem 7.12. *In the case of two fugacities, λ_o and λ_e, set $\mathcal{G}$ contains a unique measure when $(\lambda_o, \lambda_e) \in R_+^2 \setminus \mathbf{M}$, three splitting periodic hard core Gibbs measures when $(\lambda_o, \lambda_e) \in \mathbf{M}$. In the latter case, two of the three measures are always extreme: one is maximal and another minimal in $\mathcal{G}$.*

It is easy to see that domain $\mathbf{M}$ is symmetric relative to the diagonal $\lambda_o = \lambda_e$. The "upper" boundary function $\lambda \in (\lambda_{cr}, \infty) \to \frac{\lambda}{\eta_1}$ is monotone increasing and $O(\lambda^k)$, as $\lambda \to \infty$, while the "lower" function $\lambda \in (\lambda_{cr}, \infty) \to \frac{\lambda}{\eta_2}$ is monotone increasing and $O(\lambda^{1/k})$.

The maximal measure is denoted by $\mu_+(\lambda_o, \lambda_e)$ and the minimal by $\mu_-(\lambda_o, \lambda_e)$. It is easy to see that $\mu_+(\lambda_o, \lambda_e)$ is the space-shift of $\mu_-(\lambda_o, \lambda_e)$ and vice versa. The third measure is again denoted by $\mu^*(\lambda_o, \lambda_e)$, it has the property that $\mu^*(\lambda_e, \lambda_o)$ is the space-shift of $\mu^*(\lambda_o, \lambda_e)$.

By repeating constructions from the previous subsection, we can obtain additional information about measure μ^*.

Theorem 7.13. *Measure μ^* is non-extreme in an open domain $\mathbf{D} \subset \mathbf{M}$ containing the diagonal ray $\lambda_o = \lambda_e > \frac{1}{\sqrt{k}-1}\left(\frac{\sqrt{k}}{\sqrt{k}-1}\right)^k$.*

Finally, repeating the Bleher-Ganikhodjaev construction yields:

Theorem 7.14. *For $(\lambda_o, \lambda_e) \in \mathbf{M}$, there exists a continuum of non-periodic extreme Gibbs hard core measures.*

7.3 Node-weighted random walk as a tool

In this section consider random walks on graphs, which will be useful to study a simple, invariant Gibbs measures. We are interested in ergodic (irreducible, aperiodic) random walks. A graph is called *ergodic* if it is connected and non-bipartite (equivalently, if it is connected and contains a loop or odd cycle).

Let H be a fixed ergodic graph on nodes $\{1, 2, \ldots, n\}$ with adjacency matrix A, and let $P = \{p_{ij}\}$ be the transition matrix of a Markov chain M whose states are the nodes on H. We say that M is H-based if P is positive exactly where A is; in other words, if $i \to j$ is an allowable transition of M exactly when $\{i, j\}$ is an edge of H. In that case M will indeed be an ergodic Markov chain, with a unique stationary distribution π. A natural subclass of the H-based Markov chains is obtained by assigning positive real weights w_{ij} to the edges of H, then putting

$$p_{ij} = \frac{w_{ij}}{\sum_{k \sim i} w_{ik}}$$

for $i \sim j$, and $p_{ij} = 0$ otherwise. This produces a *reversible* Markov chain i.e., $\pi_i p_{ij} = \pi_j p_{ji}$ and every reversible H-based Markov chain arises in this manner. It turns out that of greatest interest here is an even more restrictive class, obtained by attaching real weights w_i to each node of H, then putting

$$p_{ij} = \frac{w_j}{\sum_{k \sim i} w_k}$$

for $i \sim j$, and $p_{ij} = 0$ otherwise. Such a Markov chain is called a *node-weighted random walk* on H with weights $w = (w_1, \ldots, w_n)$.

Setting $w_{ij} = w_i w_j$ shows that node-weighted random walks are also reversible, but it is easily verified that the converse fails; in fact the class of node-weighted random walks enjoys only $n - 1$ degrees of freedom while the edge-weights live in R^{m-1}, where m is the number of edges of H. In both cases, two sets of weights give the same Markov chain if and only if they differ by a constant factors; this relies on H being ergodic. Henceforth, we regard two weight vectors differing by a constant multiple as equivalent.

Given a set w of weights for the nodes of H, it is convenient to define

$$z_i \equiv \sum_{j \sim i} w_j$$

thus $p_{ij} = w_j / z_i$ for any $i \sim j$. Consequently, for any $j \in H$,

$$\sum_{i \sim j} p_{ij} z_i w_i = \sum_{i \sim j} \frac{w_j}{z_i} z_i w_i = w_j \cdot \sum_{i \sim j} w_i = z_j w_j,$$

hence $\pi_i = c z_i w_i$ for some constant c.

The stationary distributions for H-based Markov chains live in the hyperplane

$$\left\{ s = (s_1, \ldots, s_n) : \sum_{i=1}^{n} s_i = 1 \right\}$$

of dimension $n - 1$, as do the node-weighted random walks, so it is natural to ask how the stationary distributions of node-weighted random walks compare to those of the more general H-based Markov chains.

The answer is the following.

Theorem 7.15. *Let H be an ergodic graph and M an H-based Markov chain. Then there is exactly one node-weighted random walk on H with the same stationary distribution as M.*

Proof. Suppose that there are two different node-weighted random walks on H, given by weights w and w', with the same stationary distribution π; we wish to derive a contradiction.

Denote by $f : R^n \to R^n$ the map whose coordinates are given by $f_i(w) = z_i w_i$. Let A be the set of nodes j of H for which w_j/w_j' is minimal; A cannot be all of H else w and w' would be equivalent.

Since H is not bipartite there is an edge e contained either in A or $H \setminus A$; and since H is connected, one can include e in a path $\mathcal{P}$ which begins in A and ends in $H \setminus A$. In $\mathcal{P}$ we can find a node j_2 which is simultaneously adjacent to some $j_1 \in A$ and some $j_3 \notin A$. (Since e might be a loop, these nodes are not necessarily all distinct.)

Let $p_{i,j}$ denote the transition probability from i to j in H weighted by w, and similarly $p_{i,j}'$ for the weights w'. Then $p_{i,j} \leq p_{i,j}'$ for any $j \in A$, and $p_{j_2,j_1} < p_{j_2,j_1}'$ since j_2 has a neighbor outside of A. It follows that

$$\sum_{j \sim j_1} p_{j,j_1} \pi_j < \sum_{j \sim j_1} p_{j,j_1}' \pi_j$$

but this is impossible because both sides are equal to π_{j_1}.

To show that f is onto, we need to consider which stationary distributions are achievable by H-based Markov chains. Let P be the (unbounded) polyhedron in R^n consisting of all s such that $s_i \geq 0$ for each node i of H, and $s_I \leq s_{N(I)}$ for each independent set I of H, where $s_A \equiv \sum_{i \in A} s_i$ for any set of nodes A, and $N(A) \equiv \{j : j \sim i \text{ for some } i \in A\}$.

Let π be the stationary distribution π of an H-based Markov chain, and c any positive real. We claim $c\pi$ lies in the interior $\hat{P}$ of P. We have $\pi_i > 0$ for each node i. For any independent set $I \subset H$, $\pi_I \leq \pi_{N(I)}$ since in a random walk, every visit to I is followed immediately by a visit to $N(I)$. Equality is not possible because the ergodicity of H implies that $N(I)$ can also be visited from $H \setminus I$.

Choose now any $s \in \hat{P}$, say at ℓ^1-distance ϵ from the boundary $P \setminus \hat{P}$. It suffices to show that there is a set of weights w for H with $f(w) = s$. It is convenient at this time to introduce the notion of a "push." Suppose w is a set of weights for which

$$|f(w), s|_1 - \sum_{i=1}^{n} |z_i w_i - s_i| > 0$$

and j is a node for which $z_j w_j > s_j$. It turns out that w_j can be lowered without increasing $|f(w), s|_1$.

Indeed, let $\delta > 0$ be a small number and put $w_j' = w_j - \delta$, $w_i' = w_i$ for $i \neq j$. Then if j has no loop,

$$|f_j(w') - s_j| = |f_j(w) - s_j| - \delta z_j,$$

otherwise

$$|f_j(w') - s_j| = |f_j(w) - s_j| - \delta(z_j + w_j - \delta) < |f_j(w) - s_j| - \delta z_j.$$

Consequently,

$$\sum_{i \in N(j)} |f_i(w') - s_i| \leq \sum_{i \in N(j)} (|f_i(w) - s_i| + w_i \delta) =$$

$$\left(\sum_{i \in N(j)} |f_i(w) - s_i| \right) + \delta z_j$$

with equality only when $f_i(w) \leq s_i$ for every $i \in N(j)$. Thus

$$|f(w'), s|_1 \leq |f(w), s|_1$$

and the inequality is strict if j is looped or if any neighbor i of j has $z_i w_i > s_i$.

Let $\mathcal{R}$ be the range of f and let x be a nearest point in the closure $\mathcal{R}^{\mathrm{Cl}}$ of $\mathcal{R}$ to s again in the ℓ^1 metric.

Among the weight sets w for which $f(w) = x$, we choose one which maximizes the size of the set

$$\Delta \equiv \{i \in H : f_i(w) \neq s_i\}.$$

Then $\Delta \neq \emptyset$, since $f(w) \neq s$, and in fact we claim Δ is all of H. Were this not so, then on account of the connectedness of H, there would be an edge $\{j, k\}$ of H with $j \in \Delta$ but $k \notin \Delta$. Supposing that $f_j(w) > s_j$, we push w_j down by a small quantity δ to get a weight set w' with $\Delta' = \Delta \cup N(j)$ and $|f(w'), s|_1 \leq |f(w), s|_1$. Since $k \in N(j) \setminus \Delta$, this contradicts the choice of w. A similar argument applies if $f_j(w) < s_j$.

Now that we have $\Delta = H$ we may again use the ergodicity of H to conclude that there are two neighboring (but not necessarily distinct) nodes j and k with $f_j(w) > s_j$ and $f_k(w) > s_k$, or with $f_j(w) < s_j$ and $f_k(w) < s_k$; we may assume the former. Then a push down on w_j decreases $|f(w), s|_1$, this time contradicting the choice of x.

We are reduced now to the case where x, the nearest point in $\mathcal{R}^{\mathrm{Cl}}$ to s, is not in R itself but is instead the ℓ^1-limit of a sequence of points in $\mathcal{R}$. Since f is continuous we may, by taking subsequences n times, select a sequence $w(1), w(2), \ldots$ of weight vectors so that $f(w(t)) \to x$ and each $w(t)_i$ tends to a limit u_i in the compact interval $[0, \infty]$ as $t \to \infty$.

Let B be the set of nodes i for which $u_i = \infty$ and S be set of nodes i for which $u_i = 0$. Note that B and S cannot both be empty since then $w(t)$ has a legal weight vector w as its limit, and $f(w) = x \in \mathcal{R}$.

We claim that $x \notin \hat{P}$ (and therefore x lies on the boundary ∂P of P). This will certainly follow if $B = \emptyset$, because then $f_i(w(t)) \to 0$ for any $i \in S$.

Note that $N(B) \subset S$, because if there are $i \in B$, $j \notin S$ with $i \sim j$ then $f_i(w(t)) \to \infty$ (an impossibility since x is only finitely far from s). In particular, B is an independent set. Moreover, every node $i \in S$ does have a neighbor in B else $f_i(w(t)) \to 0$ as before; thus the neighborhood of B is exactly S.

We have

$$\sum_{i \in N(B)} f_i(w(t)) - \sum_{j \in B} f_j(w(t)) = \sum_{i \in S} z(t)_i w(t)_i - \sum_{j \in B} z(t)_j w(t)_j =$$

$$\sum_{\substack{i,j: \\ i \in S, j \sim i}} w(t)_i w(t)_j - \sum_{\substack{i,j: \\ i \in S, j \in B, j \sim i}} w(t)_i w(t)_j =$$

$$\sum_{\substack{i,j: \\ i \in S, j \notin B, j \sim i}} w(t)_i w(t)_j \to 0,$$

thus $\lim f(w(t))$ fails one of the constraints and must lie in ∂P as claimed.

Thus for any $s \in \hat{P}$, either $s \in \mathcal{R}$ or the nearest point of $\partial \mathcal{R}^{\mathrm{Cl}}$ to s is in ∂P. We claim the latter case cannot in fact arise.

Assume the contrary and let $s \in \hat{P} \setminus \mathcal{R}$, with nearest point $x \in \mathcal{R}^{\mathrm{Cl}}$, $x \in \partial P$. Let π be any point of $\mathcal{R}$, e.g. the stationary distribution for the simple (uniform weights) random walk on H. Let $a = |s, \partial P|_1 \leq |s, x|_1$ and $b = |\pi, \partial P|_1$, and consider the line segment

$$L \equiv \{\rho s + (1 - \rho)\pi : 0 \leq \rho \leq 1\}$$

from s to π. We claim that $|L, \partial P|_1 \geq \min\{a, b\}$. Indeed, suppose that u is a point in R^n with $|u, v|_1 = c < \min\{a, b\}$ for some $v = \rho s + (1 - \rho)\pi \in L$; then

$$|(s + u - v), s|_1 = |(\pi + u - v), \pi|_1 = c$$

so $s + u - v$ and $\pi + u - v$ are both points of $\hat{P}$. But

$$u = \rho(s + u - v) + (1 - \rho)(\pi + u - v)$$

hence $y \in \hat{P}$ by convexity of P, which proves the claim.

Finally, let y and y' be two points on L with $|y, y'|_1 < \min\{a, b\}$ and $y \in \mathcal{R}$, $y' \notin \mathcal{R}$. Then supposedly the nearest point of $\mathcal{R}^{\mathrm{Cl}}$ to y' is on ∂P at distance at least $\min\{a, b\}$ from y' but this is nonsense, since y is closer. This contradiction completes the proof of the theorem. $\qquad\square$

A constraint graph H may fail to be ergodic, by virtue of being bipartite (with, say, parts U and V); then a set of node weights for H has only $n-2$ degrees of freedom since multiplying the weights on the nodes in U by a constant has no effect on the random walk. The random walk still has a unique stationary distribution π satisfying $\pi_j = \sum_{i \sim j} \pi_i p_{ij}$ for all j. The stationary distributions of H-based Markov chains now live in the subspace

$$\left\{ \pi \in R^n : \sum_{i \in U} \pi_i = \sum_{i \in V} \pi_i = \frac{1}{2} \right\}$$

but the essentials of the proof of Theorem 7.15 go through, yielding the following addendum.

Theorem 7.16. *Let H be a connected, bipartite graph, and M an H-based Markov chain. Then there is exactly one node-weighted random walk on H with the same stationary distribution as M.*

7.4   A Gibbs measure associated to a k-branching node-weighted random walk

In this section we show that a node-weighted random walk on H yields a simple, invariant Gibbs measure on $\mathrm{Hom}(\Gamma^k, H)$; and conversely, that every simple, invariant Gibbs measure on $\mathrm{Hom}(\Gamma^k, H)$ can be obtained in this way.

Note that in statistical mechanics Γ^k is sometimes called the k-branching "Cayley tree" or "Bethe lattice." Baxter [18] makes a subtle distinction between the Cayley tree and the Bethe lattice, by declaring that statistical properties of the latter are to be determined by averaging only over those sites in a finite k-ary tree which are far from the leaves. This is necessary because for $k > 1$ the size of the boundary of any finite subset of Γ^k is at least a constant fraction of the size of the subset. Although the tree Γ^k is a planar graph it is for some purposes infinite in dimension, since (for $k > 1$) the number of sites at distance d from a given site, namely $(k+1)k^{d-1}$, exceeds any power of d.

Let S_k be the set of finite strings of integers in $\{1, \ldots, k\}$, except that 0 is permitted as the first entry in a string; the empty string $\emptyset$ is also allowed. A k-branching random walk on H is a function σ from S_k to H defined recursively as follows: $\sigma(\emptyset)$ is a node of H drawn from the distribution π (so that $\Pr(i)$ is proportional to $z_i w_i$), and if $\sigma(s) = i$ is defined and s' is an extension of s by one symbol, then $\sigma(s')$ is chosen randomly (and independently) from the neighbors of i with probabilities proportional to their weights. Note that a branching random walk is like an ordinary, stationary random walk on H except that the usual token is replaced by an amoeba which divides into k independent amoebas at each step. The amoeba starts at a node chosen from the stationary distribution, and divides $k+1$ ways at the first step only. In the case $k = 1$ the branching random walk begins (say) at the origin of the number line and walks both left and right from there; this produces an ordinary stationary random walk on account of reversibility.

Since S_k has the structure of the tree Γ^k it is evident that the k-branching random walk on H, together with a labeling of Γ^k by S_k, produces a random "coloring" φ of Γ^k by H. A moment's reflection will show that $\varphi \in \mathrm{Hom}(\Gamma^k, H)$, and thus the branching random walk produces a measure μ on $\mathrm{Hom}(\Gamma^k, H)$. Our interest in branching random walks is explained by the following.

Theorem 7.17. *Let H be a fixed connected constraint graph with node-weights w then the measure μ induced on $\mathrm{Hom}(\Gamma^k, H)$ by the k-branching node-weighted random walk on H is a simple, invariant Gibbs measure, for some set λ of activities on H. Conversely, if H, k and λ are given, the every simple, invariant Gibbs measure on $\mathrm{Hom}(\Gamma^k, H)$ is given by the k-branching random walk on H with nodes weighted by some w.*

Proof. Let H and w be given, and let us first check that μ does not depend on the labeling of Γ^k by S_k, and is therefore an invariant measure. On account of the symmetry of the random walk definition and the fact that Γ^k is connected, we need only check that for two neighboring sites u and v of Γ^k, μ is the same whether u is chosen as root or v is.

We may also choose $\varphi(u)$ and $\varphi(v)$ as the first two "colors" and the rest of the procedure is the same; so it suffices to check that for any nodes i and j of H, the probability that $\varphi(u) = i$ and $\varphi(v) = j$ is the same with either root choice. But these two probabilities are

$$\pi_i p_{ij} = z_i w_i \frac{w_j}{z_i} = w_i w_j = z_j w_j \frac{w_i}{z_j} = \pi_j p_{ji}$$

as desired. In fact, we are using only reversibility here; any edge-weighted random walk would also yield an invariant (and simple) measure. It is the Gibbs condition which fails when the branching random walk is not of the node-weighted variety.

Now we shall show that μ is simple: if we condition on $\varphi(u) = i$ then, using invariance to put the root at u, the independence of φ on the $k + 1$

components of $\Gamma^k \setminus \{u\}$ is evident from the definition of the branching random walk. In fact any branching Markov chain, even a non-reversible one, yields a simple measure.

Define a mapping $\Phi : R_+^{n-1} \to R_+^{n-1}$ by

$$\Phi(w)_i = \frac{w_i}{z_i^k}.$$

We shall show that μ is a Gibbs measure on $\mathrm{Hom}(\Gamma^k, H)$ with set of activities $\lambda = \Phi(w)$. Let U be any finite set of sites in Γ^k with exterior boundary ∂U. On account of invariance of labeling, we may assume that the root x (corresponding to the empty string in S_k) does not lie in $U^+ = U \cup \partial U$.

Let $g \in \mathrm{Hom}(U^+, H)$; we want to show that the probability that a branching random walk φ matches g on U, given that it matches on ∂U, is the same as the corresponding conditional probability for the measure m_{U^+} defined from λ.

Let Γ be the subtree of Γ^k induced by U^+ and the root x; for any $f \in \mathrm{Hom}(\Gamma, H)$,

$$\Pr(\varphi_\Gamma = f) = \pi_{f(x)} \prod_{u \to v} p_{f(u), f(v)} = z_{f(x)} w_{f(x)} \prod_{u \to v} \frac{w_{f(v)}}{z_{f(u)}},$$

where $u \to v$ means that v succeeds u in the tree construction. The factors $z_{f(u)}$ corresponding to sites u in U each occur as denominator k times in the above expression, since each site in U has all of its k successors in Γ; and each $w_{f(u)}$ occurs once as a numerator as well. It follows that if we compare $\Pr(\varphi_\Gamma = f)$ with $\Pr(\varphi_\Gamma = f')$, where f' differs from f only on U, then the value of the first is proportional to

$$\prod_{u \in U} \frac{w_{f(u)}}{z_{f(u)}^k} = \prod_{u \in U} \lambda_{f(u)}$$

which means that μ coincides with the finite measure, as desired.

Now assume that μ is a simple, invariant Gibbs measure on $\mathrm{Hom}(\Gamma^k, H)$ with set of activities λ, with the intent of showing that μ arises from a node-weighted branching random walk on H.

We shall construct a μ-random φ, site by site. Fix a labeling on Γ^k by S_k and choose $\varphi(x) = \varphi(\emptyset)$ from the *a priori* distribution σ of colors of the root x and therefore, by invariance, of any other site. We next color the neighbor $y = < 0 >$ of x according to the conditional distribution matrix $P = \{p_{ij}\}$ given by

$$p_{ij} = \Pr(\varphi(y) = j | \varphi(x) = i);$$

by invariance of μ, P is the same for any pair of neighboring site. It follows that $\sigma = \sigma \cdot P$, and moreover that P is the transition matrix of a reversible Markov chain, since the roles of x and y can be interchanged.

Now we proceed to the rest of the neighbors of x, then to the sites at distance 2 from x (i.e., strings of length 2 in S_k), etc., coloring each conditionally according to all sites so far colored.

We claim that the distribution of the colors $\varphi(s)$ of $s =< s_1, s_2, \ldots, s_r >$ depends only on $\varphi(< s_1, \ldots, s_{r-1} >)$; because μ is simple and all sites so far colored are in components of $\Gamma_k \setminus \{< s_1, \ldots, s_{r-1} >\}$ other than the component containing s. Thus the value of $\varphi(u)$ is given by P for every site $u \neq x$, and it follows that μ arises from a k-branching Markov chain on H, starting at distribution σ.

It is easy to see that for any (not necessarily distinct) nodes i, j of H, there will be pairs (u, v) of adjacent sites with $\varphi(u) = i$ and $\varphi(v) = j$ if and only if $i \sim j$ in H. Hence P is an H-based Markov chain, and there is a unique distribution π satisfying $\pi \cdot P = \pi$; thus $\sigma = \pi$.

It remains to show that P is a node-weighted random walk, and it turns out that a special case of the Gibbs (or DRL) condition for one-site patches suffices. Let j and j' be nodes of H which have a common neighbor i, and suppose that all of the neighbors of the root x have color i. Such a configuration will occur with positive probability and according to the Gibbs condition for $U = \{x\}$,

$$\frac{\Pr(\varphi(x) = j')}{\Pr(\varphi(x) = j)} = \frac{\lambda_{j'}}{\lambda_j}$$

but

$$\Pr(\varphi(x) = j) = \frac{\pi_j p_{ji}^{k+1}}{\sum_{r \sim i} \pi_r p_{ri}^{k+1}}$$

and similarly for j',

$$\frac{\Pr(\varphi(x) = j')}{\Pr(\varphi(x) = j)} = \frac{\pi_{j'} p_{j'i}^{k+1}}{\pi_j p_{ji}^{k+1}}.$$

Consequently, the ratio

$$\frac{p_{j'i}}{p_{ji}} = \left(\frac{\lambda_{j'} \pi_j}{\lambda_j \pi_{j'}} \right)^{1/k+1}$$

is independent of i.

Since P is reversible we have $p_{ij} = \pi_j p_{ji} / \pi_i$, and

$$\frac{p_{ij'}}{p_{ij}} = \frac{\pi_{j'} p_{j'i}}{\pi_j p_{ji}}$$

is also independent of i, and it follows that P is a node-weighted random walk on H. $\square$

If φ is a coloring in $\mathrm{Hom}(\Gamma^k, H)$ and there is a vector $\rho = <\rho_1, \ldots, \rho_n>$ such that for any root of Γ^k and any node i of H

$$\lim_{r \to \infty} \frac{|\{x \in \Gamma^k(r) : \varphi(x) = i\}|}{|\Gamma^k(r)|} = \rho_i$$

then one says that φ has density ρ. If a measure μ on $\mathrm{Hom}(\Gamma^k, H)$ has the property that with μ-probability 1 every coloring φ has the same density ρ, then one says μ itself has density ρ.

Since the stationary distribution of a random walk is unaffected by branching, the measure induced by one of our node-weighted branching random walks on an ergodic graph has density π; putting together Theorems 7.17 and 7.15 gives the following.

Corollary 7.3. *If H is ergodic then any simple, invariant Gibbs measure on $\mathrm{Hom}(\Gamma^k, H)$ has a density ρ and moreover is uniquely determined by ρ.*

Suppose now that H, k, and λ are given. Is there a simple, invariant Gibbs measure on $\mathrm{Hom}(\Gamma^k, H)$? Theorem 7.17 reduces the problem to the technical one of showing that for any $\lambda \in R_+^n$, there is a set of weights $w \in R_+^n$ such that

$$\lambda_i = \frac{w_i}{z_i^k}$$

for each node i of H.

Theorem 7.18. *For every $k \geq 2$, every constraint graph H and every set λ of activities for H, there is a node-weighted branching random walk on Γ^k which induces a simple, invariant Gibbs measure on $\mathrm{Hom}(\Gamma^k, H)$.*

Proof. Assume that both the weights and the activities sum to 1 on H. We may do this, since if we have a weight function w (summing to 1) with $\lambda_i = cw_i/z_i^k$ for all i, then multiplying all weights by $c^{1/(k-1)}$ gives a new weight function inducing λ.

Let S be the simplex $\{x \in R^n : x \geq 0, \sum_i x_i = 1\}$, and let $\hat{S}$ be its $(n-1)$-dimensional interior. For $w \in \hat{S}$, we define $(g(w))_i = w_i/Cz_i^k$, where $C = \sum_i w_i/z_i^k$. Thus $g(w) \in \hat{S}$ for every $w \in \hat{S}$; we need to prove that g is a surjection.

For $t > 0$, we define the function $g_t : \hat{S} \to \hat{S}$ by

$$g_t(w) = \frac{tw + (w_1 w_2 \ldots w_n)g(w)}{t + w_1 w_2 \ldots w_n}.$$

Note that, for $w \in \hat{S}$, $g_t(w)$ is in $\hat{S}$. Also, g_t is continuous on $\hat{S}$.

For w on the boundary of S (i.e., with at least one entry 0), we define $g_t(w) = w$. We shall show that this constitutes a continuous extension of g_t to the boundary of S. Indeed, if w is on the boundary of S with, say, $w_i = 0$, and v is in $\hat{S}$, then

$$g_t(w) - g_t(v) = \frac{w(t + v_1 v_2 \ldots v_n) - tv - (v_1 v_2 \ldots v_n) g(v)}{t + v_1 v_2 \ldots v_n}.$$

Consequently,

$$\|g_t(w) - g_t(v)\|_1 \leq \frac{t\|w - v\|_1 + 2 v_1 v_2 \ldots v_n}{t} \leq \|w - v\|_1 + \frac{2 v_i}{t},$$

since $\|w - g(v)\|_1 \leq 2$. If $v^{(r)}$ is a sequence of points in $\hat{S}$ converging to w, then $v_i^{(r)}$ converges to 0, and so $g_t(v^{(r)})$ converges to $g_t(w)$.

Hence each g_t ($t > 0$) is a continuous function from the simplex S to itself, equal to the identity on the boundary of S. If $x \in S$ were not in the range of g_t, then the map obtained by composing g_t with projection from x to ∂S would be a retraction of S to ∂S, contradicting Brouwer's Fixed-Point Theorem (see [62], Chapter XVI.2). Thus each g_t is a surjection.

For given any $\lambda \in \hat{S}$, and any sequence t_k tending to 0, choose points $w^{(r)} \in \hat{S}$ such that $g_{t_r}(w^{(r)}) = \lambda$. By taking a subsequence if necessary, we may assume that the sequence $w^{(r)}$ tends to a limit $w \in S$. We shall show that $w \in \hat{S}$, and that $g(w) = \lambda$. For each r, set $z_i^{(r)} = \sum_{j \sim i} w_j^{(r)}$, as normal, and $C^{(r)} = \sum_i w_i^{(r)} / (z_i^{(r)})^k$, so that $(g(w^{(r)}))_i = w_i^{(r)} / (z_i^{(r)})^k C^{(r)}$.

Assume first that some w_i is equal to 0. Of course, not all the w_j are 0, so we may assume that i is adjacent to a node j with $w_j \neq 0$. Now λ_i / λ_j is a fixed positive quantity, which is equal to $(g_{t_r}(w^{(r)}))_i / (g_{t_r}(w^{(r)}))_j$ for any r. This ratio is given by

$$\frac{t_r w_i^{(r)} + w_1^{(r)} w_2^{(r)} \ldots w_n^{(r)} (w_i^{(r)} / (z_i^{(r)})^k C^{(r)})}{t_r w_j^{(r)} + w_1^{(r)} w_2^{(r)} \ldots w_n^{(r)} (w_j^{(r)} / (z_j^{(r)})^k C^{(r)})} =$$

$$\frac{w_i^{(r)}}{w_j^{(r)}} \left[\frac{t_r C^{(r)} + w_1^{(r)} w_2^{(r)} \ldots w_n^{(r)} / (z_i^{(r)})^k}{t_r C^{(r)} + w_1^{(r)} w_2^{(r)} \ldots w_n^{(r)} / (z_j^{(r)})^k} \right].$$

The first fraction above tends to zero as r tends to infinity, so the second must tend to infinity. This means that $z_j^{(r)} / z_i^{(r)}$ tends to infinity as $r \to \infty$. But $z_i^{(r)}$ is at least $w_j^{(r)}$, which is bounded away from 0 as $r \to \infty$, while $z_j^{(r)}$ is at most 1. This is a contradiction, so we conclude that $w \in \hat{S}$.

Consider $g(w^{(r)}) - \lambda = g(w^{(r)}) - g_{t_r}(w^{(r)})$. This is equal to

$$\frac{t_r(g(w^{(r)}) - w^{(r)})}{t_r + w_1^{(r)} w_2^{(r)} \ldots w_n^{(r)}}.$$

As $r \to \infty$, the numerator tends to 0, while the denominator tends to $w_1 w_2 \ldots w_n > 0$, so we have that $g(w^{(r)}) \to \lambda$ as $r \to \infty$. But, by continuity of g, $g(w^{(r)}) \to g(w)$, so $g(w) = \lambda$, as claimed. $\qquad\square$

7.5   Cases of uniqueness of Gibbs measure

In this section we shall show that for any H there is a unique simple invariant Gibbs measure on $\mathrm{Hom}(\Gamma^k, H)$ when $k = 1$, or, for arbitrary k and sufficiently unbalanced sets of activities λ.

Theorem 7.19. *For any connected constraint graph H and any set of activities λ, there is a unique simple invariant Gibbs measure on $\mathrm{Hom}(\Gamma^1, H)$.*

Proof. We show that there is precisely one set of weights (up to equivalence) for H which yield the activity set λ. Let $A = \{a_{ij}\}$ be the adjacency matrix of H, and let $B = \{b_{ij}\}$ where $b_{ij} = \lambda_i a_{ij}$.

It follows from the Perron-Frobenius Theorem (see, e.g., [241]) that B has a positive left eigenvector w which is unique up to a constant multiple; say $w \cdot B = cw$. Consequently, for any node i of H,

$$\frac{w_i}{z_i} = \frac{w_i}{(w \cdot A)_i} = \frac{\lambda_i w_i}{(w \cdot B)_i} = \frac{\lambda_i}{c}$$

hence w is a valid set of weights. Reversing the argument shows that any w for which w_i / z_i is proportional to λ_i is a left eigenvector for B.

In the case where H is bipartite with parts U and V, and w and w' are two equivalent sets of weights which do not differ by a constant factor, we seem to have proved that different sets of activities emerge. This cannot be the case; in the bipartite case the activities also lose a degree of freedom. Multiplying the activities of the nodes in U by a constant factor does not change the specification. $\qquad\square$

The following theorem shows that for $k \geq 2$ there is always a region of activities where the simple invariant Gibbs measure is unique.

Theorem 7.20. *For any k and H there is a set of activities λ for which there is only one simple invariant Gibbs measure on $\mathrm{Hom}(\Gamma^k, H)$.*

Proof. Let n be the number of vertices of H, and let M be a large constant. Choose one node of H, a looped node if there is one, and denote this node as 1. For each node i of H, set $\lambda_i = M^{-d(i,1)}$, where d denotes the graph distance in H. We shall show that, for sufficiently large M

(depending on n and k), this set of activities corresponds to just one k-branching random walk, and hence just one simple invariant Gibbs measure on $\mathrm{Hom}(\Gamma^k, H)$.

Let w be a weight-vector for H such that $\lambda_i = w_i/z_i^k$ for every node i of H. First we show that every node other than 1 has a neighbor of much higher weight. For any constant $K > 1$ such that $M > K^{k+1}n^k$, suppose that node $i \neq 1$ has the property that $w_j \leq Kw_i$ for all $j \sim i$ in H. Then $z_i \leq nKw_i$, whereas $z_j \geq w_i$ for all $j \sim i$. There is some vertex $r \sim i$ with $d(r,1) = d(i,1) - 1$, so $\lambda_r/\lambda_i = M$. But

$$\frac{\lambda_r}{\lambda_i} = \frac{w_r}{w_i}\frac{z_i^k}{z_r^k} \leq K\frac{n^k K^k w_i^k}{w_i^k} = K^{k+1}n^k < M,$$

a contradiction.

Hence every node i other than 1 has a neighbor of weight greater than Kw_i. In particular, node 1 has the largest weight and $w_1 > Kw_i$ for every $i \neq 1$. Also $w_i/z_i < 1/K$ for every $i \neq 1$.

For a neighbor i of 1 we have that $\lambda_i = M^{-1}$, and that z_i lies between w_1 and $w_1\alpha$, where $\alpha = 1 + (n-1)/K$. Thus $w_i = \lambda_i z_i^k$ lies between w^k/M and $w_1^k\alpha^k/M$.

Turning back to node 1, we have to consider two cases depending on whether or not it is looped.

If node 1 is looped, then all we will need is that $w_1 < z_1 \leq w_1\alpha$.

If node 1 is unlooped, let ρ denote its degree in H, so that

$$\frac{\rho w_1^k}{M} \leq z_1 \leq \frac{\rho w_1^k \alpha^k}{M}.$$

We have $w_1 = z_1^k$, which yields

$$\left(\frac{M}{\rho\alpha^k}\right)^{k/(k^2-1)} \leq w_1 \leq \left(\frac{M}{\rho}\right)^{k/(k^2-1)}.$$

Using these with the estimates for w_i, $i \sim 1$, gives the inequalities

$$\left(\frac{M}{\rho}\right)^{1/(k+1)}\alpha^{-k-2} < \frac{w_1}{z_1} < \left(\frac{M}{\rho}\right)^{1/(k+1)}\alpha^{k+1},$$

$$\left(\frac{M}{\rho}\right)^{k/(k+1)}\frac{1}{M}\alpha^{-2} < \frac{w_i}{z_i} < \left(\frac{M}{\rho}\right)^{k/(k+1)}\frac{1}{M}\alpha^{k+2},$$

for every node $i \sim 1$.

We shall use the following lemma, which is a straightforward consequence of the multidimensional Mean Value Theorem (see, for instance, [205], Theorem 12.2).

Lemma 7.2. *Let E be an open convex set in R^n, and let $\Phi : E \to R^m$ be a function whose components $\Phi_1, \ldots, \Phi_m$ are differentiable on E. Suppose that, for every n-tuple of points $x^{(1)}, \ldots, x^{(n)}$ in E, the determinant of the matrix A given by*

$$a_{ij} = \left.\frac{\partial \Phi_i}{\partial x_j}\right|_{x^{(i)}}$$

is non-zero. Then Φ is 1-1 on E.

Note that the condition of the lemma is stronger than merely saying that the Jacobian is non-zero throughout the region E, which would not be strong enough to guarantee that Φ is 1-1 on E.

We shall apply this lemma to the function Φ given by $\Phi_i(w) = w_i/z_i^k$, with some suitable region E containing all positive solutions w with $\Phi(w) = \lambda$. We have

$$\frac{\partial \Phi_i}{\partial w_j} = \begin{cases} 0, & i \neq j, \quad i \nsim j \\ -kw_i z_i^{-k-1}, & i \neq j, \quad i \sim j \\ z_i^{-k}, & i = j, i \ \text{not looped} \\ z_i^{-k} - kw_i z_i^{-k-1}, & i = j, \quad i \ \text{looped.} \end{cases}$$

Consider any n-tuple of weight vectors $w^{(1)}, \ldots, w^{(n)}$, and let $z^{(i)}$ be the corresponding z-vectors. The matrix A of the lemma, evaluated at these vectors, is thus given by

$$A = \begin{pmatrix} (z_1^{(1)})^{-k} & & 0 \\ & \ddots & \\ 0 & & (z_n^{(n)})^{-k} \end{pmatrix} B$$

where

$$B = I - k \begin{pmatrix} w_1^{(1)}/z_1^{(1)} & & 0 \\ & \ddots & \\ 0 & & w_n^{(n)}/z_n^{(n)} \end{pmatrix} A_H,$$

where A_H is the adjacency matrix of the graph H.

The diagonal matrix in the expression for A is non-singular whenever all the $z_i^{(i)}$ are non-zero, so we need only consider the matrix B.

Consider first the easier case where the node 1 of H is looped. Set $\varepsilon = 1/(2k^2 n!)$, and let E be the region of R^n where: $w_i > 0$ for all i, $w_i < \varepsilon z_i$ for all $i \neq 1$, and $z_1 > w_1 > (2/k)z_1$. Thus it is clearly an open convex region. It was showed that all solutions w of $\Phi(w) = \lambda$ lie in this region, provided $K > 1/\varepsilon$ and $\alpha = [1 + (n-1)/K] < k/2$. For any set of points $w^{(i)}$ inside E, the entries of the matrix B satisfy

$$b_{11} < 1 - k \cdot \frac{2}{k} = -1$$

$$b_{ii} > 1 - k\varepsilon > \frac{1}{2}, \quad i \neq 1$$

$$|b_{1i}| < k, \quad i \neq 1$$

$$|b_{ji}| < k\varepsilon, \quad 1 \neq j \neq i.$$

The expansion of the determinant of B contains the diagonal term $\prod b_{ii}$, and $n! - 1$ terms of modulus at most $2k^2 \varepsilon \prod b_{ii}$. Choosing a term from the first row instead of b_{11} multiplies the product by a factor of at most k; all other terms involve at least one other off-diagonal entry, which is smaller than the diagonal entries by a factor of at least $2k\varepsilon$.

Thus, since $2k^2 \varepsilon (n! - 1) < 1$, the determinant of B is non-zero. By Lemma 7.2, this means that, for M sufficiently large, the function Φ is 1-1 on E, and hence there is at most one solution to $\Phi(w) = \lambda$ in E and none elsewhere. This completes the proof in the case where node 1 is looped (and indeed in the case where H has a looped node).

Now consider the case where node 1, along with all the other nodes, is unlooped. In this case, we need the more precise estimates for the ratios w_i/z_i derived earlier. In this case, we set $\varepsilon = 1/(2k^3 n!)$, and define E as the region on R^n such that $w_i > 0$ for all i, $w_i < \varepsilon z_i$ for all $i \neq 1$,

$$\left(\frac{M}{\rho}\right)^{1/(k+1)} (1+\varepsilon)^{-1} < \frac{w_1}{z_1} < \left(\frac{M}{\rho}\right)^{1/(k+1)} (1+\varepsilon),$$

$$\left(\frac{M}{\rho}\right)^{k/(k+1)} \frac{1}{M}(1+\varepsilon)^{-1} < \frac{w_i}{z_i} < \left(\frac{M}{\rho}\right)^{k/(k+1)} \frac{1}{M}(1+\varepsilon),$$

for every neighbor i of 1.

It is obvious that E is a convex open region, and all solutions of $\Phi(w) = \lambda$ lie inside E, provided $K > 1/\varepsilon$ and $\alpha^{k+2} < 1 + \varepsilon$, both of which can be guaranteed by choosing M sufficiently large. Again, we choose an n-tuple of weight-vectors $w^{(1)}, \ldots, w^{(n)}$ inside E, and evaluate the matrix B at this n-tuple.

The matrix B has 1s on the leading diagonal (since H has no loops). All other entries, except those on the first row, are at most $k\varepsilon$ modulus. For each $i \sim 1$, the entry b_{1i} is within a multiplicative factor $(1 + \varepsilon)$ of $-k(M/\rho)^{1/(k+1)}$, and the entry b_{i1} is within a multiplicative factor $(1 + \varepsilon)$ of $-k(M/\rho)^{k/(k+1)}/M$. Other entries in the first row and column are 0.

The expansion of the determinant of B has one term $\prod b_{ii} = 1$, and ρ terms

$$b_{1i}b_{i1} \prod_{j \neq 1, i} b_{jj}, \quad i \sim 1,$$

which are all within a multiplicative factor $(1 + \varepsilon)^2$ or k^2/ρ, and all occur with signature -1 in the expansion. All other terms have absolute value at most $k\varepsilon \max(1, (1 + \varepsilon)^2 k^2/\rho) < 2k^3\varepsilon$. The determinant of B is thus at most $1 - k^2/(1 + \varepsilon)^2 + 2k^3\varepsilon n! < 0$. By Lemma 7.2 we have that, for M sufficiently large, there is at most one solution to $\Phi(w) = \lambda$ inside E, which completes the proof. $\qquad\square$

7.6   Non-uniqueness of Gibbs measure: sterile and fertile graphs

This section answers the main question: For which constraint graphs can we get multiple simple invariant Gibbs measures?

We have seen that for a fixed constraint graph H and positive activity set λ, each simple, invariant Gibbs measure on $\mathrm{Hom}(\Gamma^k, H)$ corresponds precisely to a branching random walk on H with certain positive weights w on the nodes. We know also that for any $k \geq 1$ and any ergodic H, there are activity sets for which the branching random walk is unique; and that when $k = 1$, it is always unique.

It is thus natural to ask: When $k > 1$, for which H is there a set of activities yielding multiple branching random walks? The answer, which turns out not to depend on k, is given by the theorem below.

Theorem 7.21. *Fix $k > 1$ and let H be any constraint graph. Suppose that H satisfies the following two conditions:*

(a) Every looped node of H is adjacent to all other nodes of H;
(b) With its loops deleted, H is a complete multipartite graph.

Then for every positive activity set on H, there is a unique invariant Gibbs measure on the space $Hom(\Gamma^k, H)$.

If H fails either condition (a) or condition (b) then there is a set of activities λ on H for which $Hom(\Gamma^k, H)$ has at least two simple, invariant Gibbs measures, and therefore λ can be obtained by more than one branching random walk.

A graph H which satisfies (a) and (b) above will be called *sterile*, otherwise we will say that H is *fertile*.

Proof. We proceed in three stages. First we show that for sterile graphs H, the map Φ described in the proof of Theorem 7.17, taking a set of weights w to its corresponding activity set λ, is injective. Then we show that any fertile H must admit at least one of a certain list of seven small graphs as an induced subgraph. Finally, we show that for any $k > 1$, a graph H containing one of the seven small induced subgraphs can be provided with two sets of weights giving the same set of positive activities. In view of Theorems 7.15 and 7.16, these sets of weights give different stationary probabilities, hence very different Gibbs measures.

As before, the map Φ is defined by

$$\Phi(w)_i \equiv \lambda_i = \frac{w_i}{z_i^k}.$$

Let H be a graph for which Φ is injective; let H' be obtained from H by splitting an unlooped node j of H into two new unlooped nodes, j_1 and j_2, each with the same neighborhood previously enjoyed by j. Then $z'_{j_1} = z'_{j_2}$ for any set of weights w' on H', so if u' and v' are distinct sets of weights for H' with the same image λ' under Φ, then we must have $u'_r \neq v'_r$ for some node $r \neq j$ of H. Then if we define u on H by $u_i = u'_i$ for $i \neq j$ and $u_j = u_{j_1} + u_{j_2}$, and similarly for v and λ, we have $\Phi(u) = \Phi(v) = \lambda$, a contradiction. Hence Φ is also injective for H'.

Thus we reduced to showing injectivity of Φ when H is a complete graph on nodes $\{1, 2, \ldots, n\}$ where (say) nodes 1 through m are looped.

Putting $\lambda = \Phi(w)$ and

$$Z \equiv \sum_{i=1}^{n} w_i$$

we get

$$Z = \left(\frac{w_i}{\lambda_i}\right)^{1/k} = \left(\frac{w_j}{\lambda_j}\right)^{1/k} + w_j$$

for every i, j with $1 \leq i \leq m < j \leq n$. Since each of these expressions for Z is strictly increasing in its corresponding variable w_i or w_j, we must have $w < w'$ or $w > w'$ (say, the former) if w and w' are distinct vectors with $\Phi(w) = \Phi(w') = \lambda$.

But this cannot be, since if i minimizes w_i'/w_i then $z_i'/z_i \geq w_i'/w_i$ hence $\Phi(w')_i < \Phi(w)_i$. Thus the first part of the proof is complete.

For the second, we begin by providing the seven minimal fertile graphs, illustrated in Fig. 7.1. Each of them has three or four nodes, with node set, edges, and loops as follows: the *stick*: $\{a, b, c, d\}$; $\{a, b\}$, $\{b, c\}$, $\{c, d\}$; no loops

the *pipe*: $\{a, b, c\}$; $\{a, b\}$, $\{b, c\}$; loop at a
the *wrench*: $\{a, b, c\}$; $\{a, b\}$, $\{b, c\}$; loops at a and b
the *wand*: $\{a, b, c\}$; $\{a, b\}$, $\{b, c\}$; loops at a and c
the *hinge*: $\{a, b, c\}$; $\{a, b\}$, $\{b, c\}$; loops at a, b and c
the *key*: $\{a, b, c, d\}$; $\{a, b\}$, $\{a, c\}$, $\{b, c\}$, $\{c, d\}$; no loops
the *gun*: $\{a, b, c, d\}$; $\{a, b\}$, $\{a, c\}$, $\{b, c\}$, $\{c, d\}$; loop at c.

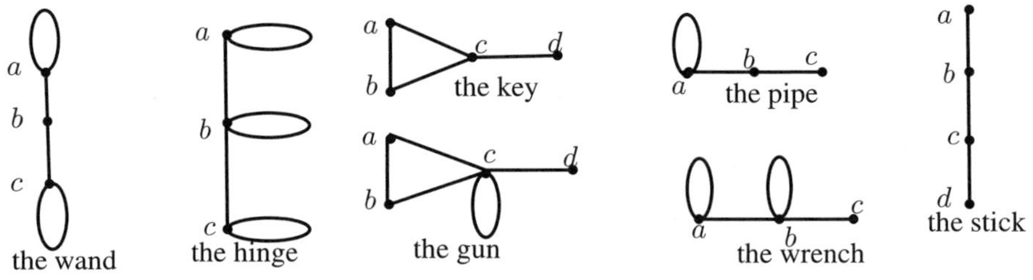

Fig. 7.1   The seven minimal fertile graphs.

The following lemma is useful.

Lemma 7.3. *Let J be the graph on three nodes with one edge and no loops. Then any finite unlooped graph H which does not induce a copy of J is complete multipartite.*

Proof.   Note that the complement of J is a path P_3 on three nodes, and every component of a graph with no induced P_3 must have diameter 1. Hence the complement of any H which does not induce J must be a disjoint union of complete graphs. $\square$

Assume that H is a connected graph which fails condition (a); then H has a looped node, which we may call a, at distance 2 from some node c.

Then a, c and any node b adjacent to both induce either a pipe, a wand, a wrench or a hinge depending upon which of b and c are looped.

Suppose next that H satisfies condition (a) but not (b), and has some looped nodes, say 1 through m, which must therefore constitute a clique. The remaining, unlooped nodes of H cannot constitute a multipartite graph else H would satisfy (b), hence by Lemma 7.3, they induce a copy of the graph J. But J plus any of the looped nodes of H constitute an induced gun.

Finally, assume H has no loops and is not multipartite. Since H induces a copy of J we can find an edge $\{a, b\}$ and a node d of distance precisely 2 from the set $\{a, b\}$; we may assume there is a node c adjacent to both b and d. Then $\{a, b, c, d\}$ induces either a stick or a key.

Now we must show that any graph H containing one of the seven small graphs admits more than one simple invariant Gibbs measure. We will show first that, for each of the seven graphs, there is a set of activities that is induced by more than one node-weighted random walk. Then we show that this property is preserved when moving to a larger graph. We break the seven graphs into three groups; we deal with all the "asymmetric" graphs–the wrench, pipe, key, and gun–together, then the wand and hinge together, and finally the stick separately.

7.6.1 *The Asymmetric Graphs*

For all the four asymmetric graphs, we define two different sets of node-weights, depending on two parameters s and γ, as well as on the branching factor k of the Cayley tree, and then show that certain values of γ give rise to node- weights that induce the same activity sets–and hence the same Gibbs measure–in both cases. This then gives a family of examples, parametrized by s, for each of the four graphs.

For given real $s > 1$ and $\gamma > 0$, and integer $k \geq 2$, set

$$w_1 = s^k, \quad w_2 = 1, \quad w_3 = \gamma,$$

$$w_1' = 1, \quad w_2' = \beta \equiv \frac{s^k + 1}{s} - 1, \quad w_3' = \gamma\beta^k.$$

Note that, if $s > 1$, then $s(\beta - 1) = (s^k - s^2) + (s^2 - 2s - 1) > 0$, so $\beta > 1$. For the wrench and the pipe, we attach weight w_1 (w_1') to a, w_2 (w_2') to b and w_3 (w_3') to c. For the key and the gun, we attach weight w_1 (w_1') to both a and b, w_2 (w_2') to c and w_3 (w_3') to d.

As before, we have $\lambda_i = w_i/z_i^k$ and $\lambda_i' = w_i'/(z_i')^k$, where z_i (z_i') is the sum of w_j (w_j') over all neighbors j of i. We wish to show that there are conditions ensuring that $\lambda_i = \lambda_i'$ for all i.

In all cases, all bar one of the conditions is automatically satisfied by specification of the w_i and w_i'. Precisely, at node a, we have

$$\lambda_a = \frac{w_1}{(w_1 + w_2)^k} = \frac{s^k}{(1 + s^k)^k} = \frac{w_1'}{(w_1' + w_2')^k} = \lambda_a'$$

this also means that the condition is satisfied at node b in the key and gun, and at c for the wrench and pipe we have

$$\lambda_c = \frac{w_3}{w_2^k} = \gamma = \frac{w_3'}{(w_3')^k} = \lambda_c',$$

and similarly for d in the key and gun.

For the remaining equation, we need to consider each of the four graphs separately.

For the wrench, we require that

$$\lambda_b = \frac{w_2}{(w_1 + w_2 + w_3)^k} = \frac{w_2'}{(w_1' + w_2' + w_3')^k},$$

which gives

$$1 + \beta + \gamma\beta^k = \beta^{1/k}(s^k + 1 + \gamma),$$

or

$$\gamma = \frac{\beta^{1/k}(s^k + 1) - (\beta + 1)}{\beta^k - \beta^{1/k}}.$$

We need to check that this specifies a positive value of λ; for this we recall that s and β are both greater than 1, and note in addition that

$$\beta^{1/k}(s^k + 1) = (\beta + 1)s\beta^{1/k} > \beta + 1.$$

Thus, for each $s > 1$, we have a value of $\gamma = \gamma(s)$ giving two different sets of weights, but the same set of activities. These two sets of weights give rise to two different Gibbs measures.

The three other asymmetric graphs are handled in exactly the same way; we obtain these values for γ.

$$\text{Pipe}: \ \gamma = \frac{\beta^{1/k}s^k - 1}{\beta^k - \beta^{1/k}},$$

$$\text{Key}: \quad \gamma = \frac{2(\beta^{1/k} s^k - 1)}{\beta^k - \beta^{1/k}},$$

$$\text{Gun}: \quad \gamma = \frac{\beta^{1/k}(2s^k + 1) - (\beta + 2)}{\beta^k - \beta^{1/k}}.$$

Note that in each case, the specified value of γ is positive, so we obtain two different node-weighted random walks producing the same set of activities.

7.6.2 *The Wand and the Hinge*

In both of these cases, we give weight 1 to node a, weight $s > 1$ to node c, and weight β to node b. If we can do this so that $\lambda_a = \lambda_c$, then exchanging the weights on a and c will give us a different set of node-weights and the same set of activities. It is evidently immaterial whether b has a loop, so the two graphs–wand and hinge–can be treated identically. The condition we have to satisfy is

$$\lambda_a = \frac{1}{(1 + \beta)^k} = \frac{s}{(s + \beta)^k} = \lambda_c.$$

From this we get $s + \beta = s^{1/k}(1 + \beta)$, or

$$\beta = \frac{s - s^{1/k}}{s^{1/k} - 1}.$$

Note that this specifies a positive value of β. Thus, for any $s > 1$, this asymmetric system of weights does indeed produce a symmetric set of activities, as required.

Incidentally, the choice $s = 2^k$ gives $\beta = 2^k - 2$; a particularly simple example of the phenomenon of multiple Gibbs measures.

7.6.3 *The Stick*

As mentioned above, it is enough to exhibit an asymmetric system of node-weights inducing a symmetric set of activities, and again we actually produce a one-parameter family of such node-weighting.

Set $w_b = 1$, $w_c = s > 1$, $w_a = \alpha$, and $w_d = \alpha s^k$. From this choice we get $\lambda_d = w_d / w_c^k = \alpha = w_a / w_b^k = \lambda_a$, so it remains to choose α to ensure that $\lambda_b = \lambda_c$. The condition is that

$$\frac{1}{(\alpha + s)^k} = \frac{s}{(1 + \alpha s^k)^k},$$

or $1 + \alpha^k = s^{1/k}(\alpha + s)$. This is true when

$$\alpha = \frac{s^{1+1/k} - 1}{s^k - s^{1/k}},$$

which is positive since $s > 1$.

Since (unlike the other six graphs) the stick is bipartite, we need to observe that allegedly asymmetrical system of weights is not accidentally equivalent to its reverse. It suffices to note that

$$\frac{w_c}{w_a} = \frac{s}{\alpha} \neq \frac{s^{-k}}{\alpha} = \frac{w_b}{w_d}.$$

The final part of the proof is to show that, if J is a graph with a set of activities $\lambda^{(J)}$ induced by more than one node-weighted random walk, and H contains J as an induced subgraph, then H also has a set of activities λ induced by more than one node-weighted random walk. The basic idea is: set $\lambda = \lambda^{(J)}$ on J, and λ very small elsewhere; then any system of weights on J inducing $\lambda^{(J)}$ should be close to a system inducing λ on H.

We are not allowed to take either weights or activities equal to 0, and it is necessary that we obtain exactly the same λ for two different systems of weights, so we need to show that, if the weights w give rise to a set of activities near the target λ, then a nearby system of weights w' hits λ exactly. We wish to show that the map Φ is locally a bijection. We are in the process of proving exactly that the map is not globally 1-1, so we will need some condition to ensure that we are not near a "fold" in Φ. To continue the proof we need several lemmas:

Lemma 7.4. *For real numbers a and b, and positive integer s,*
    (i) $|a^s - b^s| \leq |a - b|s(\max\{|a|, |b|\})^{s-1}$.
    (ii) $|a^s - b^s - sb^{s-1}(a - b)| \leq |a - b|^2(s(s-1)/2)(\max\{|a|, |b|\})^{s-2}$.

Proof. Part (i) follows from the following

$$a^s - b^s = (a - b)(a^{s-1} + a^{s-2}b + \cdots + b^{s-1}).$$

Similarly, part (ii) follows from

$$a^s - b^s - sb^{s-1}(a - b) = (a - b)^2(a^{s-2} + 2a^{s-3}b + \cdots + (s - 1)b^{s-2}). \qquad \square$$

Next give a technical lemma that is closely related to what we need. The general theme is that, if λ and μ are two sets of activities, close to each other, and λ is induced by a weight function w, then μ is also induced by a nearby weight function v. For the purposes of this lemma, we will allow the various activities and weights to take the value 0 (or even, in the case of the function v we seek, negative values).

Let $\| \cdot \|$ be the ℓ^1 norm, i.e., the sum of the absolute values of the coordinates; and the norm of a matrix B is defined as $\max \|Bx\|/\|x\|$, which turns out to be the maximum of the norms of the columns of B.

Lemma 7.5. *Let H be a graph with adjacency matrix A. Suppose that we are given an integer $k \geq 2$, two functions λ and μ from H to the non-negative reals, a positive real number ε, and a function w from H to the non-negative reals such that $w_i = \lambda_i(Aw)_i^k$. (Here we treat w as a vector, so that Aw is just the function we have been denoting z.) Let D be the diagonal matrix whose ith diagonal entry is $\mu_i k(Aw)_i^{k-1}$. Suppose that the matrix $I - DA$ is invertible, with inverse B, and set*

$$K = 3k(k-1)\|\mu\|\|B\|^2\|w\|^k(\|w\| + 2\varepsilon\|B\|\|w\|^k)^{k-2}.$$

Finally suppose that $2K\varepsilon < 1$, and that $\|\mu - \lambda\| < \varepsilon$. Then there is a function v from H to R with $v_i = \mu_i(Av)_i^k$, such that

$$\|v - w\| \leq 2\|w\|^k\|B\|\varepsilon.$$

Proof. For any vectors x and y, we apply the Lemma 7.4 to obtain

$$|(Ax)_i^k - (Ay)_i^k - k(Ay)_i^k(A(x - y))_i| \leq$$

$$|(A(x - y))_i|^2 \frac{k(k-1)}{2}(\max\{(Ax)_i, (Ay)_i\})^{k-2}.$$

Note that $|(Ax)_i| \leq \|x\|$, for any i, since A is an adjacency matrix; thus we have

$$|(Ax)_i^k - (Ay)_i^k - k(Ay)_i^k(A(x - y))_i| \leq$$

$$\|x - y\|^2 \frac{k(k-1)}{2}(\max\{\|x\|, \|y\|\})^{k-2}.$$

Similarly, from the first part of lemma we get, for any vector y,

$$\|(Ay)_i^{k-1} - (Aw)_i^{k-1}\| \leq \|y - w\|(k-1)(\max\{\|y\|, \|w\|\})^{k-2}.$$

Define a function $g : R^n \to R^n$ by

$$(g(x))_i = \mu_i(Ax)_i^k,$$

so we are looking for a non-negative vector v with $g(v) = v$.

For x and y arbitrary vectors, and w the specific vector we are given, we have

$$\|g(x) - g(y) - DA(x - y)\| =$$

$$\sum_{i \in H} \mu_i |(Ax)_i^k - (Ay)_i^k - k(Aw)_i^{k-1}(A(x-y))_i| \le$$

$$\sum_{i \in H} \mu_i \left[|(Ax)_i^k - (Ay)_i^k - k(Ay)_i^{k-1}(A(x-y))_i| + \right.$$

$$k|(A(x-y))_i||(Ay)_i^{k-1} - (Aw)_i^{k-1}|\Big] \le$$

$$\sum_{i \in H} \mu_i \left[\|x-y\|^2 (k(k-1)/2)(\max\{\|x\|, \|y\|\})^{k-2} + \right.$$

$$k\|x-y\|\|y-w\|(k-1)(\max\{\|y\|, \|w\|\})^{k-2} \Big] \le$$

$$\|\mu\| \frac{k(k-1)}{2} (\max\{\|x\|, \|y\|, \|w\|\})^{k-2} (\|x-y\|^2 + 2\|x-y\|\|y-w\|).$$

Now we shall describe an iterative procedure that gets us from w to the required v. Set $v^{(0)} = w$ and, for $j \ge 1$,

$$v^{(j)} = v^{(j-1)} + B(g(v^{(j-1)}) - v^{(j-1)}).$$

The claim that the following hold for any j:

(i) $\|g(v^{(j)}) - v^{(j)}\| \le \|w\|^k \varepsilon 2^{-j}$,

(ii) $\|v^{(j)} - v^{(j-1)}\| \le \|B\|\|w\|^k \varepsilon 2^{1-j}$,

(iii) $\|v^{(j)} - w\| \le 2\|B\|\|w\|^k \varepsilon$.

Once we have verified (i)-(iii) for every j, we will be done; condition (ii) implies that the $v^{(j)}$ tend to limit v; condition (i), plus the continuity of g, gives that $g(v) = v$; condition (iii) gives that v is close to w, as in the statement.

For $j = 0$, we have that

$$|(g(w) - w)_i| = |\mu_i - \lambda_i||(Aw)_i|^k \le |\mu_i - \lambda_i|\|w\|^k,$$

consequently

$$\|g(w) - w\| \le \|\mu - \lambda\|\|w\|^k < \varepsilon\|w\|^k.$$

This means that (i) is true for $j = 0$ as is (iii).

Now assume that (i) holds for $j - 1$, then we have

$$\|v^{(j)} - v^{(j-1)}\| = \|B(g(v^{(j-1)})) - v^{(j-1)}\| \le$$

$$\|B\|\|g(v^{(j-1)}) - v^{(j-1)}\| \le \|B\| \le \|w\|^{k-1}\varepsilon 2^{1-j},$$

hence (ii) holds for j.

Assume now (ii) holds up to j, then we have

$$\|v^{(j)} - w\| \le \sum_{l=1}^{j} \|v^{(l)} - v^{(l-1)}\| \le$$

$$\|B\|\|w\|^k \varepsilon (1 + \frac{1}{2} + \cdots + \frac{1}{2^{j-1}}) \le 2\|B\|\|w\|^k \varepsilon,$$

i.e., (iii) holds for j.

Finally, supposing that (ii) and (iii) hold for j, and (i) holds for $j - 1$, we have

$$g(v^{(j)}) - v^{(j)} = [g(v^{(j)}) - g(v^{(j-1)}) - DA(v^{(j)} - v^{(j-1)})] -$$

$$(I - DA)(v^{(j)} - v^{(j-1)}) + [g(v^{(j-1)}) - v^{(j-1)}].$$

But we have

$$(I - DA)(v^{(j)} - v^{(j-1)}) = B^{-1}B(g(v^{(j-1)}) - v^{(j-1)}),$$

by definition of $v^{(j)}$ and of B. So the final two terms above cancel, and we have

$$\|g(v^{(j)}) - v^{(j)}\| = \|g(v^{(j)}) - g(v^{(j-1)}) - DA(v^{(j)} - v^{(j-1)})\| \le$$

$$\|\mu\| \frac{k(k-1)}{2} (\max\{\|v^{(j)}\|, \|v^{(j-1)}\|, \|w\|\})^{k-2} \times$$

$$(\|v^{(j)} - v^{(j-1)}\|^2 + 2\|v^{(j)} - v^{(j-1)}\|\|v^{(j-1)} - w\| <$$

$$\|\mu\| \frac{k(k-1)}{2} (\|w\| + 2\varepsilon\|B\|\|w\|^k)^{k-2} \|B\|\|w\|^k \varepsilon 2^{1-j} \times 6\varepsilon\|B\|\|w\|^k =$$

$$\varepsilon^2 2^{1-j} \|w\|^k K < \varepsilon 2^{-j} \|w\|^k.$$

This verifies (i) for j, completing the inductive proof of (i)-(iii), and thus the proof of the lemma. $\qquad\square$

Lemma 7.6. *Suppose*

- *that J, with adjacency matrix A_J, is an induced subgraph of a graph H, with adjacency matrix A_H;*
- *that we are given an integer $k \ge 2$, a positive real number δ, a set of activities λ on J, and a weight vector w on J inducing λ, i.e., with $w_i = \lambda_i (A_J w)_i^k$;*

- *set D to be the diagonal matrix whose ith diagonal entry is $\lambda_i k (A_J w)_i^{k-1}$; and suppose that $I - D A_J$ is invertible;*
- *set λ^+ to be the function on the nodes of H which agrees with λ on J and is 0 elsewhere, and w^+ similarly.*

Then there is some positive real number ε such that, whenever μ is a set of activities on H such that $\|\mu - \lambda^+\| < \varepsilon$, there is a node-weighting v on H inducing μ, i.e., with $v_i = \mu_i (A_H v)_i^k$, such that $\|v - w^+\| < \delta$.

Proof. Suppose that the node set of H is $[n]$ and, without loss of generality, that the node set of the subgraph J is $[m]$. This means that A_J is the top $m \times m$ submatrix of the $n \times n$ matrix A_H. We shall use I_n and I_m to denote identity matrices of sizes n and m.

Note first that we have $w_i^+ = \lambda_i^+ (A_H w^+)_i^k$ for every $i \in H$. For the moment, we leave ϵ unspecified, and take any set of activities μ on H with $\|\mu - \lambda^+\| < \varepsilon$. We shall apply the previous lemma to w^+, λ^+ and μ. To do this, we need to consider the diagonal matrix D_μ of the lemma, whose ith entry is $\mu_i k (A_H w^+)_i^{k-1}$; we need a bound on $\|(I_n - D_\mu A_H)^{-1}\|$ that does not depend on the particular choice of μ.

Set D^+ to be the $n \times n$ diagonal matrix whose top $m \times m$ submatrix is D and which is 0 elsewhere; in other words the ith diagonal entry of D^+ is $\lambda_i^+ k (A_H w^+)_i^{k-1}$. The ith diagonal entry of $D_\mu - D^+$ is then equal to $(\mu_i - \lambda_i^+) k (A_H w^+)_i^{k-1}$, which has absolute value at most $\varepsilon k \|w\|^{k-1}$. This is also therefore an upper bound on $\|D_\mu - D^+\|$.

The hypothesis states that $I_m - D A_J$ is invertible, with inverse B_J, say, not depending on μ. Suppose A_H has the structure

$$\begin{pmatrix} A_J & C \\ C^t & E \end{pmatrix},$$

then one can verify that the inverse of $I_n - D^+ A_H$ is

$$B_H = \begin{pmatrix} B_J & B_J D C \\ 0 & I_{n-m} \end{pmatrix}.$$

Note that $(I_n - D_\mu A_H) B_H = I_n - (D_\mu - D^+) A_H B_H$. It is well-known and easy to check that, for $\|X\| < 1$, the inverse of $I - X$ is given by the convergent sum $I + X + X^2 + \ldots$, of norm at most $(1 - \|X\|)^{-1}$. If we choose ε so that $\varepsilon k \|w\|^{k-1} \|A_H\| \|B_H\| < 1/2$, then we have that $X = (D_\mu - D^+) A_H B_H$ has norm less than $1/2$, so $I_n - X$ has inverse Y

with norm at most 2. Then $I_n - D_\mu A_H$ has inverse $B_\mu = B_H X$ of norm at most $2\|B_H\|$.

Set

$$K_H = 3k(k-1)(2\|\lambda\|)(2\|B_H\|)^2\|w\|^k(\|w\| + 2\|B_H\|\|w\|^k)^{k-2},$$

and note that, provided $\varepsilon \le \|\lambda\|$ and $\varepsilon < 1/2$, this is at least the quantity K defined in the previous lemma.

If we also demand that $2K_H\varepsilon < 1$, then we may apply that lemma to obtain that there is a function v from the nodes of H to R such that $v_i = \mu_i(A_H v)_i^k$ and $\|v - w^+\| \le 4\|w\|^k\|B_H\|\varepsilon$.

If we also choose ε so that $4\|w\|^k\|B_H\|\varepsilon < \delta$, then we are done, except that we must be sure that all the v_i are positive. This is trivial if k is even; if k is odd it is true whenever $\varepsilon < n^{-k}/\delta^{k-1}$, and ε is less than any of the w_i for $i \in J$. Indeed, suppose i (necessarily not in J) has the largest negative value of v_i, then we have $(A_H v)_i = (v_i/\mu_i)^{1/k}$, $v_i > -\delta$ and $\mu_i < \varepsilon$, so there is some neighbor j of i with $v_j \le 1/n(v_i/\mu_i)^{1/k} < 1/n(v_i/\varepsilon)^{1/k} < v_i(1/n)\delta^{1/k-1}\varepsilon^{-1/k} < v_i$, contradicting the choice of i. $\qquad\square$

Now continue the proof of the main theorem. If H contains one of the seven small graphs J as a subgraph, take two different node-weightings w and w' on J inducing the same set of activities λ, producing invertible matrices $I - DA_J$ and $I - D'A_J$ respectively, and choose $\delta < \frac{1}{2}\|w - w'\|$. The previous lemma guarantees the existence of positive ε and ε' such that, if μ is any set of activities on H with $\|\mu - \lambda^+\| < \min\{\varepsilon, \varepsilon'\}$, then there are node-weightings v and v' on H, both inducing μ, with $\|v - w\|$ and $\|v' - w'\|$ both at least δ. The choice of δ then guarantees that v and v' are different.

It remains to check that, in each of the seven cases, and for each k, the node-weightings we gave earlier do indeed produce invertible matrices $I - DA_J$ and $I - D'A_J$, at least for some values of the parameters s. It turns out to be easiest to check that this is the case as s tends to ∞.

Note that, in the definition of the matrix D from the weight function w, the ith diagonal entry d_i is $\lambda_i k(Aw)_i^{k-1} = kw_i/z_i$, where $z_i = (Aw)_i$, as usual. In all cases, most of the d_i will tend to 0 as $s \to \infty$.

First consider the four asymmetric graphs, and the weighting w. As $s \to \infty$, it is easy to see that the parameter γ tends to 0, for any value of k. Thus $d_i = kw_i/z_i$ tends to 0 if i is a node with weight 1 (adjacent to a node of weight s^k) a weight γ (adjacent to a node of weight 1). If i is a node of weight s^k, then it is adjacent to one (other) such node, and so

$d_i \to k$ as $s \to \infty$. The matrices $I - DA_J$ thus tend to

$$\begin{pmatrix} 1 - k & -k & 0 \\ 0 & 1 & 0 \\ 0 & 0 & 1 \end{pmatrix}$$

for the pipe and wrench and

$$\begin{pmatrix} k - 1 & -k & -k & 0 \\ -k & 1 - k & -k & 0 \\ 0 & 0 & 1 & 0 \\ 0 & 0 & 0 & 1 \end{pmatrix}$$

for the key and gun, both of which are non-singular for $k \geq 2$.

The same argument goes through for the wand and hinge; the node-weightings we gave there tend to 0 except on the one weighted node c, where they tend to k, so the limit matrix will simply be a relabeling of the first matrix above.

The case of the weighting w' on the four asymmetric graphs is a little more complex. Take the pipe as an example; the others are similar. We have

$$|I - DA_J| = \begin{vmatrix} 1 - d_a & -d_a & 0 \\ -d_b & 1 & -d_b \\ 0 & -d_c & 1 \end{vmatrix} = \begin{vmatrix} 1 - d_a & -d_a & 0 \\ -d_b & 1 - d_b d_c & -d_b \\ 0 & 0 & 1 \end{vmatrix},$$

doing a column operation. The point is that, in this case, $d_a = k/(\beta+1)$ and $d_b = k\beta/(1 + \gamma\beta^k)$ both tend to 0 as $s \to \infty$, and although $d_c = k\gamma\beta^k/\beta$; tends to infinity, $d_b d_c$ tends to k^2. By the above calculation it is clear that the matrix $I - DA_J$ is non-singular for sufficiently large s.

Finally, the same technique works for the stick; again we have that all the d_i tend to 0 except for d_d, which tends to infinity, but with $d_c d_d$ tending to k^2, the same column operation then shows that the matrix is non-singular for s sufficiently large. This completes the proof of the main theorem. $\qquad\qquad\square$

7.6.4   *The hinge*

Here we look at a phase transition exhibited by $\mathrm{Hom}(\Gamma^2, H)$, where H is the hinge. We associate nodes a, b, c of the hinge with colors red, yellow and green, the only non-adjacency in H being between red and green. Thus $\mathrm{Hom}(\Gamma^2, H)$ is the space of all red-yellow-green colorings of the sites of Γ^2 in which no red site is adjacent to a green one.

The hinge constraint corresponds to a discrete version of the Widom-Rowlinson model, in which two gases (whose particles are represented by red and green) compete for space and are not permitted to occupy adjacent sites; see, e.g., [36], [262], [263].

The above developed methods can be used to obtain a result of [262], namely that when $\lambda_a = \lambda_c = \gamma$ and $\lambda_b = 1$, then there is a unique simple invariant Gibbs measure when

$$\gamma \leq \frac{1}{k-1} \left(\frac{k+1}{k} \right)^k.$$

For any lattice G, and H the hinge, putting $a < b < c$ imposes on $\mathrm{Hom}(G, H)$ the structure of a partially ordered set, in fact a distributive lattice (with "lattice" now used in the algebraic sense). This nice structure can be used to show that if there is only one simple invariant Gibbs measure, then there is only one Gibbs measure of any kind; see [36]. Here, the hinge is chosen for our example just because it is particularly easy to visualize.

Consider the weights $w_a = 1$, $w_b = 2$, and $w_c = 4$, giving

$$\lambda_a = \frac{1}{(1+2)^2} = \frac{1}{9}, \quad \lambda_b = \frac{2}{(1+2+4)^2} = \frac{2}{49}, \quad \lambda_c = \frac{4}{(2+4)^2} = \frac{1}{9}.$$

Let us compute two conditional probabilities, to see how it can be that asymmetric weights give rise to symmetric activities. Suppose the predecessor and both successors to some site $x \in \Gamma^2$ are all yellow; then from Bayes' formula the probability that x is red is

$$\frac{\frac{1}{1+2+4} \cdot \left(\frac{2}{1+2} \right)^2}{\frac{1}{1+2+4} \cdot \left(\frac{2}{1+2} \right)^2 + \frac{2}{1+2+4} \cdot \left(\frac{2}{1+2+4} \right)^2 + \frac{4}{1+2+4} \cdot \left(\frac{2}{2+4} \right)^2} = \frac{49}{116}.$$

While the probability that x is green is

$$\frac{\frac{4}{1+2+4} \cdot \left(\frac{2}{2+4} \right)^2}{\frac{1}{1+2+4} \cdot \left(\frac{2}{1+2} \right)^2 + \frac{2}{1+2+4} \cdot \left(\frac{2}{1+2+4} \right)^2 + \frac{4}{1+2+4} \cdot \left(\frac{2}{2+4} \right)^2} = \frac{49}{116}.$$

The stationary probabilities for this node-weighted random walk are, in contrast to the activities, even less balanced than the weights:

$$\pi_a = \frac{1 \cdot (1+2)}{1 \cdot (1+2) + 2 \cdot (1+2+4) + 4 \cdot (2+4)} = \frac{3}{41},$$

$$\pi_b = \frac{14}{41}, \quad \pi_c = \frac{24}{41}.$$

The measure induced by this random walk is called the "September" phase.

Note that the reverse weighting, $w_a = 4$, $w_b = 2$, and $w_c = 1$, yields the same set of activities and the reverse stationary probabilities; this is the "November" phase.

In addition to the above weighting and is reverse, there is a third, symmetric, weighting which generates the same set of activities. Its values are

$$w_a = w_c \approx 2.5866, \quad w_b \approx 2.2383$$

giving stationary distribution

$$\pi_a = \pi_c \approx \frac{3}{10}, \quad \pi_b \approx \frac{2}{5}.$$

The corresponding phase is called the "October" phase.

7.7   Fertile three-state hard core models

In this section we shall consider nearest-neighbor hard-core models, with three states, on a Cayley tree. In these models one assigns to each site x, values $\sigma(x) \in \{0, 1, 2\}$. Values $\sigma(x) = 1, 2$ mean that site x is 'occupied' and $\sigma(x) = 0$ that x is 'vacant'.

In this section we consider the fertile graphs (see previous section) with three vertices $0, 1, 2$ (on the set of values of $\sigma(x)$).

Denote $O = \{wrench, wand, hinge, pipe\}$.

First we recall the definitions of Section 7.1 in a convenience way: For $H \in O$ we call σ a H-admissible configuration (on the tree, in V_n or W_n) if $\{\sigma(x), \sigma(y)\}$ is an edge of H $\forall$ nearest-neighbor pair x, y (from V, V_n or W_n, respectively). Denote the set of H-admissible configurations by Ω^H ($\Omega^H_{V_n}$ and $\Omega^H_{W_n}$).

A set of activities (see Section 7.1) for a graph H is a function $\lambda : H \to R_+$ from the vertices of H to the positive reals. The value λ_i of λ at a vertex $i \in \{0, 1, 2\}$ is called its "activity".

For a given H and λ we define Hamiltonian of the (H-)hard core model as

$$\mathcal{H}^\lambda_H(\sigma) = \begin{cases} \sum_{x \in V} \ln \lambda_{\sigma(x)}, & \text{if } \sigma \in \Omega^H, \\ +\infty & \text{otherwise.} \end{cases} \tag{7.33}$$

In the previous section it was proven that (i) for every sterile graph H and any positive activity set on H there is a unique invariant Gibbs measure on Ω^H; (ii) for any fertile graph H there is a set of activities λ on H for which Ω^H has at least two simple, invariant Gibbs measures.

In this section we shall consider the case $\lambda_0 = 1$, $\lambda_1 = \lambda_2 = \lambda > 0$ and describe the corresponding translation-invariant, periodic and some non-periodic Gibbs measures.

7.7.1 *System of functional equations*

For $\sigma_n \in \Omega^H_{V_n}$ we define:

$$\#\sigma_n = \sum_{x \in V_n} \mathbf{1}(\sigma_n(x) \geq 1)$$

the number of occupied sites in σ_n.

Let $z : x \mapsto z_x = (z_{0,x}, z_{1,x}, z_{2,x}) \in R^3_+$ be a vector-valued function on V. Given $n = 1, 2, \ldots$, and $\lambda > 0$ consider the probability distribution $\mu^{(n)}$ on $\Omega^H_{V_n}$ defined by

$$\mu^{(n)}(\sigma_n) = \frac{1}{Z_n} \lambda^{\#\sigma_n} \prod_{x \in W_n} z_{\sigma(x),x}. \tag{7.34}$$

Here Z_n is the corresponding partition function:

$$Z_n = \sum_{\widetilde{\sigma}_n \in \Omega^H_{V_n}} \lambda^{\#\widetilde{\sigma}_n} \prod_{x \in W_n} z_{\widetilde{\sigma}(x),x}.$$

The probability distributions $\mu^{(n)}$ are compatible if $\forall\, n \geq 1$ and $\sigma_{n-1} \in \Omega^H_{V_{n-1}}$:

$$\sum_{\omega_n \in \Omega^H_{W_n}} \mu^{(n)}(\sigma_{n-1} \vee \omega_n)\mathbf{1}(\sigma_{n-1} \vee \omega_n \in \Omega^H_{V_n}) = \mu^{(n-1)}(\sigma_{n-1}). \tag{7.35}$$

Definition 7.5. Measure μ defined by (7.34), (7.35) is called a $(H\text{-})$hard core Gibbs measure with $\lambda > 0$, corresponding to function $z : x \in V \setminus \{x^0\} \mapsto z_x$. The set of such measures (for all possible choices of z) is denoted by $\mathcal{S}_H$.

For graph H denote by $L(H)$ the set of its edges and by $A \equiv A^H = \left(a_{ij}\right)_{i,j=0,1,2}$ the adjacency matrix of H, i.e.,

$$a_{ij} \equiv a_{ij}^H = \begin{cases} 1, & \text{if } \{i,j\} \in L(H), \\ 0 & \text{otherwise}. \end{cases}$$

The following statement describes conditions on z_x guaranteeing compatibility of distributions $\mu^{(n)}$.

Theorem 7.22. *Probability distributions $\mu^{(n)}$, $n = 1, 2, \ldots$, in (7.34) are compatible iff for any $x \in V$ the following system of equations holds:*

$$z'_{1,x} = \lambda \prod_{y \in S(x)} \frac{a_{10} + a_{11} z'_{1,y} + a_{12} z'_{2,y}}{a_{00} + a_{01} z'_{1,y} + a_{02} z'_{2,y}},$$

$$z'_{2,x} = \lambda \prod_{y \in S(x)} \frac{a_{20} + a_{21} z'_{1,y} + a_{22} z'_{2,y}}{a_{00} + a_{01} z'_{1,y} + a_{02} z'_{2,y}}, \tag{7.36}$$

where $z'_{i,x} = \lambda z_{i,x}/z_{0,x}$, $i = 1, 2$.

Proof. Left-hand side of (7.35) can be written as:

$$\frac{1}{Z_n} \lambda^{\#\sigma_{n-1}} \prod_{x \in W_{n-1}} \prod_{y \in S(x)} \left(a_{\sigma_{n-1}(x)0} z_{0,y} + a_{\sigma_{n-1}(x)1} \lambda z_{1,y} + a_{\sigma_{n-1}(x)2} \lambda z_{2,y} \right).$$

$$\tag{7.37}$$

Sufficiency. Suppose that (7.36) holds. It is equivalent to the representations

$$\prod_{y \in S(x)} \left(a_{i0} z_{0,y} + \lambda a_{i1} z_{1,y} + \lambda a_{i2} z_{2,y} \right) = a(x) z_{i,x}, \quad i = 0, 1, 2 \tag{7.38}$$

for some function $a(x) > 0$, $x \in V$. Setting $A_n = \prod_{x \in W_n} a(x)$ and substituting (7.34) into LHS of (7.35), we get (7.37) and by (7.38) we have

$$\frac{1}{Z_n} \lambda^{\#\sigma_{n-1}} \prod_{x \in W_{n-1}} z_{\sigma_{n-1}(x),x} a(x) = \frac{A_{n-1}}{Z_n} \lambda^{\#\sigma_{n-1}} \prod_{x \in W_{n-1}} z_{\sigma_{n-1}(x),x}.$$

We should have

$$\sum_{\sigma_{n-1} \in \Omega_{V_{n-1}}^H} \sum_{\omega_n \in \Omega_{W_n}^H} \mathbf{1}\left(\sigma_{n-1} \vee \omega_n \in \Omega_{V_n}^H \right) \mu^{(n)}(\sigma_{n-1} \vee \omega_n) = 1,$$

hence $A_{n-1}/Z_n = 1/Z_{n-1}$, and (7.35) holds.

Necessity. Suppose that (7.35) holds; we want to prove (7.36). Substituting (7.34) in (7.35) and using (7.37), we obtain that $\forall \, \sigma_{n-1} \in \Omega_{V_{n-1}}^H$:

$$\frac{1}{Z_n} \lambda^{\#\sigma_{n-1}} \prod_{x \in W_{n-1}} \prod_{y \in S(x)} \left(a_{\sigma_{n-1}(x)0} z_{0,y} + a_{\sigma_{n-1}(x)1} \lambda z_{1,y} + a_{\sigma_{n-1}(x)2} \lambda z_{2,y} \right) =$$

$$\prod_{x \in W_{n-1}} z_{\sigma_{n-1}(x),x}.$$

From this equality follows

$$\frac{Z_{n-1}}{Z_n} \prod_{x \in W_{n-1}} \prod_{y \in S(x)} \left(a_{i0} z_{0,y} + \lambda a_{i1} z_{1,y} + \lambda a_{i2} z_{2,y} \right) = \tag{7.39}$$

$$\prod_{x \in W_{n-1}} z_{i,x}, \quad i = 0, 1, 2.$$

Denoting $z'_{i,x} = \lambda z_{i,x}/z_{0,x}$, $i = 1, 2$ from (7.39) we get (7.36). $\qquad\square$

7.7.2 *Translation-invariant Gibbs measures*

We set in future $z_{0,x} \equiv 1$ and $z_{i,x} = z'_{i,x} > 0$, $i = 1, 2$. Then $\forall$ function $x \in V \mapsto z_x = (z_{1,x}, z_{2,x})$ satisfying

$$z_{i,x} = \lambda \prod_{y \in S(x)} \frac{a_{i0} + a_{i1} z_{1,y} + a_{i2} z_{2,y}}{a_{00} + a_{01} z_{1,y} + a_{02} z_{2,y}}, \quad i = 1, 2 \tag{7.40}$$

there exists a unique H-hard core Gibbs measure μ and vice versa. We start with the translation-invariant solutions such that $z_x = z \in R_+^2$ does not depend on x.

7.7.2.1 *The Case wrench*

In this case we obtain, from (7.40), the following system of equations:

$$z_1 = \lambda \left(\frac{1 + z_1}{1 + z_1 + z_2} \right)^k, \quad z_2 = \lambda \left(\frac{1}{1 + z_1 + z_2} \right)^k. \tag{7.41}$$

It is easy to see that

$$z_2 = z_1 \left(1 + z_1 \right)^{-k}. \tag{7.42}$$

Using this equality from (7.41) we get

$$\lambda^{-1} z = \left(\frac{(1 + z)^{k+1}}{z + (1 + z)^{k+1}} \right)^k, \tag{7.43}$$

with $z = z_1$.

Proposition 7.5. *For any $\lambda > 0$, $k \geq 1$ the system of equations (7.41) has a unique positive solution (with $z_1, z_2 > 0$).*

Proof. We shall prove that equation (7.43) has a unique positive solution. Denote

$$f(z) \equiv f(z, k) = \left(\frac{(1 + z)^{k+1}}{z + (1 + z)^{k+1}} \right)^k.$$

We have $f(0) = 1$ and

$$f'(z) = \frac{k(1 + z)^{k^2 + k - 1}}{(z + (1 + z)^{k+1})^{k+1}} (kz - 1).$$

For $0 < z < 1/k$, f decreases monotonically from 1 to

$$f(1/k) = ((k + 1)^{k+1}((k + 1)^{k+1} + k^k)^{-1})^k.$$

For $z > 1/k$, f is monotonically increasing to 1 as $z \to \infty$; thus there are at most three solutions of (7.43). On the other hand, it is easy to see that (7.43) has more than one solution if and only if there is more than one solution to the equation $zf'(z) = f(z)$ which is equivalent to

$$(1 + z)^{k+2} = (k^2 - 1)z^2 - (k + 1)z. \tag{7.44}$$

It is straightforward that equation (7.44) has no positive solutions. Thus (7.43) has a unique positive solution $z^* = z_1^*$. Using (7.42) we get unique z_2^*. $\qquad\square$

We then obtain

Theorem 7.23. *For any* $\lambda > 0$, *translation-invariant wrench-hard-core Gibbs measures* μ^* *is unique.*

The following proposition gives an estimation for an arbitrary solution of system (7.40):

Proposition 7.6. *If* $z_x = (z_{1,x}, z_{2,x})$ *is a solution of (7.40) in the case wrench then* $z_i^- \leq z_{i,x} \leq z_i^+$, *for any* $i = 1, 2$, $x \in V$, *where* $(z_1^-, z_1^+, z_2^-, z_2^+)$ *is a solution of*

$$
\begin{cases}
z_1^- = \lambda \left(\dfrac{1+z_1^-}{1+z_1^-+z_2^+} \right)^k, \\[2mm]
z_1^+ = \lambda \left(\dfrac{1+z_1^+}{1+z_1^++z_2^-} \right)^k, \\[2mm]
z_2^- = \lambda \left(\dfrac{1}{1+z_1^++z_2^+} \right)^k, \\[2mm]
z_2^+ = \lambda \left(\dfrac{1}{1+z_1^-+z_2^-} \right)^k.
\end{cases}
\tag{7.45}
$$

Proof. It is clear that $0 < z_{i,x} < \lambda$, $i = 1, 2$, for any $x \in V$. We rewrite (7.40) in case wrench as

$$z_{1,x} = \lambda \prod_{i=1}^{k} \frac{1 + z_{1,x_i}}{1 + z_{1,x_i} + z_{2,x_i}}, \quad z_{1,x} = \lambda \prod_{i=1}^{k} \frac{1}{1 + z_{1,x_i} + z_{2,x_i}},$$

where x_j, $j = 1, 2, \ldots, k$ are direct successors of x. Denote

$$f_1(u_1, \ldots, u_k, v_1, \ldots, v_k) = \lambda \prod_{i=1}^{k} \frac{1 + u_i}{1 + u_i + v_i},$$

$$f_2(u_1, \ldots, u_k, v_1, \ldots, v_k) = \lambda \prod_{i=1}^{k} \frac{1}{1 + u_i + v_i},$$

with $0 < u_j < \lambda, \;\; 0 < v_j < \lambda$. It is not difficult to see that

$$\frac{\lambda}{(1+\lambda)^k} < f_1(u_1,\ldots,u_k,v_1,\ldots,v_k) < \lambda,$$

$$\frac{\lambda}{(1+2\lambda)^k} < f_2(u_1,\ldots,u_k,v_1,\ldots,v_k) < \lambda.$$

Thus for $z_{i,x}$ we get

$$\frac{\lambda}{(1+\lambda)^k} < z_{1,x} < \lambda, \quad \frac{\lambda}{(1+2\lambda)^k} < z_{2,x} < \lambda.$$

Now consider $f_i(u_1,...,u_k,v_1,...,v_k)$ with

$$\frac{\lambda}{(1+\lambda)^k} < u_j < \lambda, \quad \frac{\lambda}{(1+2\lambda)^k} < v_j < \lambda, \;\; j=1,\ldots,k.$$

Iterating this procedure we can obtain the following

$$z_{i,n}^- < z_{i,x} < z_{i,n}^+, \quad i=1,2,$$

where $z_{i,n}^{\pm}$, $i=1,2$, $n=1,2,\ldots$ satisfy

$$\begin{cases} z_{1,n+1}^- = \lambda \left(\dfrac{1+z_{1,n}^-}{1+z_{1,n}^-+z_{2,n}^+} \right)^k, \\[2ex] z_{1,n+1}^+ = \lambda \left(\dfrac{1+z_{1,n}^+}{1+z_{1,n}^++z_{2,n}^-} \right)^k, \\[2ex] z_{2,n+1}^- = \lambda \left(\dfrac{1}{1+z_{1,n}^++z_{2,n}^+} \right)^k, \\[2ex] z_{2,n+1}^+ = \lambda \left(\dfrac{1}{1+z_{1,n}^-+z_{2,n}^-} \right)^k, \end{cases}$$

with $z_{1,1}^- = \frac{\lambda}{(1+\lambda)^k}$, $z_{1,1}^+ = z_{2,1}^+ = \lambda$ and $z_{2,1}^- = \frac{\lambda}{(1+2\lambda)^k}$. It is easy to see that $z_{i,n}^-$, (resp. $z_{i,n}^+$), $i=1,2$ are increasing (decreasing) and bounded sequences. Thus there exist

$$\lim_{n\to\infty} z_{i,n}^{\pm} = z_i^{\pm}, \quad i=1,2. \qquad \square$$

Proposition 7.7. *If* $z = (z_1^-, z_1^+, z_2^-, z_2^+)$ *a solution of (7.45) then* $z_1^- = z_1^+$ *iff* $z_2^- = z_2^+$.

Proof. From (7.45) we have

$$\begin{cases} z_1^- - z_1^+ = \lambda A \left((1+z_1^-)(z_2^- - z_2^+) + z_2^+(z_1^- - z_1^+) \right), \\[1ex] z_2^- - z_2^+ = \lambda B \left((z_2^- - z_2^+) + (z_1^- - z_1^+) \right), \end{cases} \tag{7.46}$$

where $A = A(z) > 0$, $B = B(z) > 0$. Rewrite (7.46) in the following form:

$$\begin{cases} \lambda A(1 + z_1^-)(z_2^- - z_2^+) + (\lambda A z_2^+ - 1)(z_1^- - z_1^+) = 0, \\ (\lambda B - 1)(z_2^- - z_2^+) + \lambda B(z_1^- - z_1^+) = 0. \end{cases} \tag{7.47}$$

If $z_1^- = z_1^+$ then from the first equation of (7.47) we get $z_2^- = z_2^+$. If $z_2^- = z_2^+$ then from the second equation of (7.47) we have $z_1^- = z_1^+$. $\qquad \square$

Corollary 7.4. *If system (7.45) has a unique solution, then system (7.40) in case wrench also has a unique solution. Moreover this solution is $z_x = (z_1^*, z_2^*)$, $x \in V$ where (z_1^*, z_2^*) is the unique solution of (7.41).*

7.7.2.2   The case hinge

In this case assuming $z_x = z$ we obtain from (7.40) the following system of equations:

$$\begin{cases} z_1 = \lambda \left(\frac{1+z_1}{1+z_1+z_2} \right)^k, \\ z_2 = \lambda \left(\frac{1+z_2}{1+z_1+z_2} \right)^k. \end{cases} \tag{7.48}$$

Subtracting from the first equation of system (7.48) the second one we get

$$(z_1 - z_2) \left[1 - \lambda \frac{(1 + z_1)^{k-1} + ... + (1 + z_2)^{k-1}}{(1 + z_1 + z_2)^k} \right] = 0.$$

Consequently, we have $z_1 = z_2$ or

$$(1 + z_1 + z_2)^k = \lambda \left((1 + z_1)^{k-1} + ... + (1 + z_2)^{k-1} \right), \tag{7.49}$$

for $z_1 \neq z_2$. For $z_1 = z_2 = z$ from system (7.48) we have

$$\lambda^{-1} z = f(z) = \left(\frac{1+z}{1+2z} \right)^k. \tag{7.50}$$

The function $f(z)$ is decreasing for $z > 0$ which implies that equation (7.50) has a unique solution $z^* = z^*(k, \lambda)$ for any $\lambda > 0$.

If (7.49) is satisfied then we assume $k = 2$ and from (7.49) we have

$$1 + z_1 + z_2 = \frac{\lambda + \sqrt{\lambda^2 + 4\lambda}}{2}. \tag{7.51}$$

Using this equality from the first equation of system (7.48) we have for $k = 2$

$$z_1^{(1)} = \left(\frac{1 + \sqrt{1 - 4a^2}}{2a} \right)^2, \quad z_1^{(2)} = \left(\frac{1 - \sqrt{1 - 4a^2}}{2a} \right)^2, \tag{7.52}$$

if $\lambda > 9/4$ where $a = 2(\sqrt{\lambda} + \sqrt{\lambda+4})^{-1}$.

Using the second equation we also have $z_2^{(1)}, z_2^{(2)} \in \{z_1^{(1)}, z_1^{(2)}\}$. Since $z_1 \neq z_2$ we conclude that $z_1 = z_1^{(1)}, z_2 = z_1^{(2)}$ and $z_1 = z_1^{(2)}, z_2 = z_1^{(1)}$. It is easy to check that these solutions satisfy the condition (7.51).

Thus if $k = 2$, $\lambda > \frac{9}{4}$ then the system (7.48) has three solutions (z^*, z^*), $(z_1^{(1)}, z_1^{(2)})$, $(z_1^{(2)}, z_1^{(1)})$, where z^* is the unique solution of (7.50) and $z_1^{(i)}$, $i = 1, 2$ is defined in (7.52). Note that $z_1^{(1)} = \frac{1}{z_2^{(2)}}$. Consequently by Theorem 7.22 we get the following

Theorem 7.24. *If $k = 2$ then for the hinge case*

1) *for $\lambda \leq \frac{9}{4}$ there exists unique hard-core translation-invariant Gibbs measure μ_0;*

2) *for $\lambda > \frac{9}{4}$ there are at least three hard-core translation-invariant Gibbs measures μ_i, $i = 0, 1, 2$.*

Remark 7.2. The value $\lambda = \lambda_{\mathrm{cr}} = \frac{9}{4}$ is exactly the critical value for $k = 2$. Clearly $\lambda_{\mathrm{cr}} < 4 = \lambda_{\mathrm{cr}}^{\mathrm{HC}}$ for $k = 2$. Here $\lambda_{\mathrm{cr}}^{\mathrm{HC}} = \frac{1}{k-1}(\frac{k}{k-1})^k$ is the critical value for two-state hard-core model.

Proposition 7.8. *If $z_x = (z_{1,x}, z_{2,x})$ is a solution of (7.40) in the case hinge then $z_i^- \leq z_{i,x} \leq z_i^+$, for any $i = 1, 2$, $x \in V$, where $(z_1^-, z_1^+, z_2^-, z_2^+)$ is a solution of*

$$\begin{cases} z_1^- = \lambda\left(\dfrac{1+z_1^-}{1+z_1^-+z_2^+}\right)^k, \\[2mm] z_1^+ = \lambda\left(\dfrac{1+z_1^+}{1+z_1^++z_2^-}\right)^k, \\[2mm] z_2^- = \lambda\left(\dfrac{1+z_2^-}{1+z_1^++z_2^-}\right)^k, \\[2mm] z_2^+ = \lambda\left(\dfrac{1+z_2^+}{1+z_1^-+z_2^+}\right)^k. \end{cases} \tag{7.53}$$

Proof. Is very similar to that of Proposition 7.6. $\qquad\square$

As in the case wrench we can prove the following statements:

Proposition 7.9. *If $z = (z_1^-, z_1^+, z_2^-, z_2^+)$ a solution of (7.53) then $z_1^- = z_1^+$ iff $z_2^- = z_2^+$.*

Corollary 7.5. *If the system (7.53) has unique solution then system (7.40) in case hinge also has unique solution. Moreover, this solution is the unique solution of (7.48).*

Now we shall find exact values of $z_i^-, z_i^+, i = 1, 2$ for $k = 2$.

Consider the system consisting of the first and last equations in system (7.53):

$$z_1^- = \lambda \left(\frac{1 + z_1^-}{1 + z_1^- + z_2^+} \right)^k, \quad z_2^+ = \lambda \left(\frac{1 + z_2^+}{1 + z_1^- + z_2^+} \right)^k. \tag{7.54}$$

If $z_1^- = z_2^+$ then this system has a unique solution. In case the $z_1^- \neq z_2^+$ we get

$$(1 + z_1^- + z_2^+)^k = \lambda \left((1 + z_1^-)^{k-1} + ... + (1 + z_2^+)^{k-1} \right). \tag{7.55}$$

If $k = 2$ then from (7.55) we have

$$1 + z_1^- + z_2^+ = \frac{\lambda + \sqrt{\lambda^2 + 4\lambda}}{2}. \tag{7.56}$$

Using this equality, from the first equation in system (7.54) for $k = 2$ and $\lambda > 9/4$ we obtain

$$(z_1^-)^{(1)} = \left(\frac{1 + \sqrt{1 - 4a^2}}{2a} \right)^2, \quad (z_1^-)^{(2)} = \left(\frac{1 - \sqrt{1 - 4a^2}}{2a} \right)^2, \tag{7.57}$$

where $a = 2(\sqrt{\lambda} + \sqrt{\lambda + 4})^{-1}$.

Using $(z_1^-)^{(i)}, i = 1, 2$ and (7.56) we get

$$(z_2^+)^{(1)} = \left(\frac{1 - \sqrt{1 - 4a^2}}{2a} \right)^2, \quad (z_2^+)^{(2)} = \left(\frac{1 + \sqrt{1 - 4a^2}}{2a} \right)^2.$$

Similarly, from the second and third equalities of (7.53) we get

$$z_1^+, z_2^- \in M = \left\{ \left(\frac{1 + \sqrt{1 - 4a^2}}{2a} \right)^2, \left(\frac{1 - \sqrt{1 - 4a^2}}{2a} \right)^2 \right\}.$$

Note that $(z_i^\pm)^{(1)} = \frac{1}{(z_i^\mp)^{(2)}}, \quad i = 1, 2.$

We have thus proved the following

Proposition 7.10. *If $k = 2$ then for the case hinge,*

1) for $\lambda \leq \frac{9}{4}$, system (7.53) has unique solution z^*;

2) for $\lambda > \frac{9}{4}$, system (7.53) has three solutions

$$z_1^* = (z^-, \frac{1}{z^-}, z^-, \frac{1}{z^-}), \quad z_2^* = (\frac{1}{z^-}, \frac{1}{z^-}, z^-, z^-), \quad z_3^* = (z^-, z^-, \frac{1}{z^-}, \frac{1}{z^-}),$$

where $z^- = \left(\frac{1 - \sqrt{1 - 4a^2}}{2a} \right)^2.$

Note that we have the inequalities $0 < a < \frac{1}{2}$ and $z^- < 1$ for $\lambda > \frac{9}{4}$.

Corollary 7.6. *If $k = 2$, $\lambda > \frac{9}{4}$ then for any solution of (7.40) (in the case hinge) we have $z^- \leq z_{i,x} \leq \frac{1}{z^-}$, $i = 1, 2$.*

Remark 7.3. In obtaining the exact solution of system (7.53) for $k = 2$, we used the independence of the first and last equations in system (7.53) and also the independence of the second and third equations. But this is inapplicable to the case pipe and wrench. An analogue of Corollary 7.6 in these cases is therefore unclear.

7.7.2.3 *The Case wand*

In this case from (7.40) for $z_x = z$ we have

$$z_1 = \lambda \left(\frac{1 + z_1}{z_1 + z_2} \right)^k, \quad z_2 = \lambda \left(\frac{1 + z_2}{z_1 + z_2} \right)^k. \tag{7.58}$$

This case is quite similar to the case hinge, and if $k = 2$, $\lambda > 1$, then it can be proved similarly that system (7.58) has three solutions, which are given by the formulas for hinge with a in (7.52) replaced with $a = 2(\sqrt{\lambda} + \sqrt{\lambda + 8})^{-1}$.

Hence, we can formulate a theorem similar to Theorem 7.24 with $\lambda_{cr} = 9/4$ replaced by $\lambda_{cr} = 1$. But we do not have analogues of Propositions 7.8-7.10 in the case wand.

7.7.2.4 *The case pipe*

In this case from (7.40) for $z_x = z$ we have

$$z_1 = \lambda \left(\frac{1 + z_2}{1 + z_1} \right)^k, \quad z_2 = \lambda \left(\frac{z_1}{1 + z_1} \right)^k. \tag{7.59}$$

From this, we obtain ($x = z_2$)

$$\lambda^{-1} x = f(x) = \left(\frac{\sqrt[k+1]{x(1+x)^k}}{1 + \sqrt[k+1]{x(1+x)^k}} \right)^k. \tag{7.60}$$

We have

$$f'(x) = \frac{k}{k+1} \cdot \frac{(k+1)x + 1}{x(x+1)} \cdot \frac{(\sqrt[k+1]{x(1+x)^k})^k}{(1 + \sqrt[k+1]{x(1+x)^k})^{k+1}} > 0.$$

Note that equation (7.60) has at least one positive solution because f is an increasing function and $f(0) = 0$, $f(+\infty) = 1$. It is easy to see that equation (7.60) has more than one positive solution if and only if there is more than positive solution to $xf'(x) = f(x)$, which is equivalent to the equation

$$(k^2 - 1)x = \varphi(x) = (k+1)(1+x)\sqrt[k+1]{x(1+x)^k} + 1. \tag{7.61}$$

Repeating this argument, we find that (7.61) has more than one solution if and only if $x\varphi'(x) = \varphi(x)$. This equation has the form

$$(k+1)x = \psi(x) = \frac{1}{\sqrt[k+1]{x(1+x)^k}} + k. \tag{7.62}$$

Since the function $\psi(x)$ is decreasing, equation (7.62) has a unique solution. Consequently system (7.59) has a unique solution.

We have thus proved the following theorem.

Theorem 7.25. *For the case pipe for any $\lambda > 0$, and $k \geq 1$, the translation-invariant pipe-hard-core Gibbs measure is unique.*

For the case pipe, one can prove the following propositions which are analogues of Propositions 7.8 and 7.9.

Proposition 7.11. *If $z_x = (z_{1,x}, z_{2,x})$ is a solution of (7.40) in the case pipe then $z_i^- \leq z_{i,x} \leq z_i^+$, for any $i = 1, 2$, $x \in V$, where $(z_1^-, z_1^+, z_2^-, z_2^+)$ is a solution to*

$$z_1^- = \lambda\left(\frac{1 + z_2^-}{1 + z_1^+}\right)^k, \quad z_1^+ = \lambda\left(\frac{1 + z_2^+}{1 + z_1^-}\right)^k, \tag{7.63}$$

$$z_2^- = \lambda\left(\frac{z_1^-}{1 + z_1^-}\right)^k, \quad z_2^+ = \lambda\left(\frac{z_1^+}{1 + z_1^+}\right)^k.$$

Proposition 7.12. *If $z = (z_1^-, z_1^+, z_2^-, z_2^+)$ is a solution of (7.63) then $z_1^- = z_1^+$ iff $z_2^- = z_2^+$.*

Remark 7.4. 1) For the case pipe, we have no analogue of Proposition 7.10 and Corollary 7.6 since in this case there is no independence (mentioned in Remark 7.3) between equations of (7.63).

2) The next two subsections are devoted to description of periodic and some non-periodic Gibbs measures for cases wrench and hinge. Results of these subsections can be similarly proved for case pipe. But for the case wand one needs to prove an analogue of Proposition 7.9.

7.7.3 *Periodic Gibbs measures*

7.7.3.1 *The case wrench*

First of all we note that the definition of periodicity is similar to the definitions which we used in previous chapters. Moreover that analogues of Theorems 2.3 and 2.4 are also true for each hard-core model which we are considering in this section. Thus it remains only to study the $G_k^{(2)}$-periodic Gibbs measures, (where $G_k^{(2)}$ is the subgroup which contains all word of even lengths). For the wrench case this problem is reduced to describing the solutions of the system:

$$\begin{cases} z_1 = \lambda \left(\frac{1+t_1}{1+t_1+t_2} \right)^k, \\[2mm] z_2 = \lambda \left(\frac{1}{1+t_1+t_2} \right)^k, \\[2mm] t_1 = \lambda \left(\frac{1+z_1}{1+z_1+z_2} \right)^k, \\[2mm] t_2 = \lambda \left(\frac{1}{1+z_1+z_2} \right)^k. \end{cases} \tag{7.64}$$

The analysis of solutions to system (7.64) is rather tricky.

Let $z^* = z^*(\lambda) = (z_1^*, z_2^*)$ be the unique solution to (7.41).

The instability condition for z^* is

$$k^2 \cdot \frac{z_1^* z_2^*}{(1 + z_1^*)(1 + z_1^* + z_2^*)} > 1. \tag{7.65}$$

The left-hand side is the product of the eigenvalues of the Jacobian at $z = z^*$:

$$\Lambda_{1,2} = \Lambda_{1,2}(\lambda) =$$

$$\frac{-k}{2(1 + z_1^*)(1 + z_1^* + z_2^*)} \left(z_2^* \pm \sqrt{(z_2^*(2z_1^* + 1))^2 + 4z_1^* z_2^*(1 + z_1^*)^2} \right).$$

Observe that, (7.65) is a necessary condition for the existence of more than one solution to (7.64).

Now we shall reduce the system (7.64) to equation $\gamma(\gamma(x)) = x$ for some function γ and will apply the following lemma (see [126], p.70).

Lemma 7.7. *Let $f : [0,1] \to [0,1]$ be a continuous function with a fixed point $\xi \in (0,1)$. Assume that f is differentiable at ξ and that $f'(\xi) < -1$. Then there exist x_0, x_1, $0 \le x_0 < \xi < x_1 \le 1$, such that $f(x_0) = x_1$ and $f(x_1) = x_0$.*

From (7.64) we have

$$z_2 = z_1(1 + t_1)^{-k}, \quad t_2 = t_1(1 + z_1)^{-k},$$

and

$$z_1 = \lambda \cdot \left(\frac{1 + t_1}{1 + t_1 + t_1(1 + z_1)^{-k}} \right)^k, \quad t_1 = \lambda \cdot \left(\frac{1 + z_1}{1 + z_1 + z_1(1 + t_1)^{-k}} \right)^k.$$

$$(7.66)$$

Denote

$$\gamma(x) = \gamma(x, \lambda, k) = \frac{(\lambda^{1/k} - x^{1/k})(1 + x)^k}{x^{1/k} - (\lambda^{1/k} - x^{1/k})(1 + x)^k}, \quad 0 < x < \lambda.$$

Then (7.66) can be rewritten in the following form

$$x = \gamma(y), \quad y = \gamma(x).$$
$$(7.67)$$

Hence now we need to solve the following

$$\gamma(\gamma(x)) = x.$$
$$(7.68)$$

It is easy to see the following properties of γ

1) There is a unique $a \in (0, \lambda)$ such that $\gamma(a \pm 0) = \pm\infty$ and $\gamma(x) > 0$ only for $x \in (a, \lambda)$.

2) γ is a decreasing function on (a, λ).

3) There are a_1, $a_2 \in (a, \lambda)$ such that $\gamma(a_1) = a$, $\gamma(a_2) = a_1$.

4) γ has a unique fixed point $x_* \in (a_2, a_1)$.

5) There is $a_3 \in (a_2, a_1)$ such that $\gamma(a_3) = \lambda$ and $\gamma(\gamma(x)) > 0$ iff $x \in (a_3, a_1)$.

Using Lemma 7.7 for $\gamma(x)$, $x \in [a_3, a_1]$ one can prove the following

Theorem 7.26. *For*

$$\lambda \in \{\lambda : \gamma'(x_*) < -1\} = \{\lambda : (1 + x_*)^{k+2} > (k-1)x_*((k+1)x_* + 1)\} \quad (7.69)$$

there are three $G_k^{(2)}$*-periodic measures* μ_0, μ_*, μ_1. *Which correspond to* $(x_0, x_1), (x_*, x_*), (x_1, x_0)$ *of (7.67).*

7.7.3.2   *The case hinge*

In this case also there are only periodic measures with period two, precisely, $G_k^{(2)}$-periodic Gibbs measures. Such measures correspond to solutions of

$$z_1 = \lambda \left(\frac{1 + t_1}{1 + t_1 + t_2} \right)^k, \quad z_2 = \lambda \left(\frac{1 + t_2}{1 + t_1 + t_2} \right)^k, \quad (7.70)$$

$$t_1 = \lambda\left(\frac{1+z_1}{1+z_1+z_2}\right)^k, \quad t_2 = \lambda\left(\frac{1+z_2}{1+z_1+z_2}\right)^k.$$

If $z_1 = z_2 = z$, $t_1 = t_2 = t$ then (7.70) reduces to the following system

$$z = \lambda\left(\frac{1+t}{1+2t}\right)^k, \quad t = \lambda\left(\frac{1+z}{1+2z}\right)^k. \tag{7.71}$$

Denote $\gamma(x) = \lambda\left(\frac{1+x}{1+2x}\right)^k$. Then from (7.71) we have

$$z = \gamma(t), \quad t = \gamma(z). \tag{7.72}$$

Note that the equation $x = \gamma(x)$ has unique solution $x^* = x^*(k, \lambda)$, for any $k \geq 1$ and $\lambda > 0$.

Theorem 7.27. *For $k \geq 6$ and*

$$\lambda \in \left\{\lambda : \frac{k-3-\sqrt{(k-3)^2-8}}{4} < x^* < \frac{k-3+\sqrt{(k-3)^2-8}}{4}\right\} \tag{7.73}$$

there are three $G_k^{(2)}$-periodic Gibbs measures μ_0, μ_*, μ_1. Which corresponds to three solutions (x_0, x_1), (x_*, x_*), (x_1, x_0) of (7.72).

Proof. Note that function $\gamma(x)$ is decreasing for any $x > 0$. By Lemma 7.7, if x^* satisfies

$$\gamma(x^*) = x^*, \quad \gamma'(x^*) < -1, \tag{7.74}$$

then (7.71) has two solutions. From (7.74) it follows that

$$2(x^*)^2 + (3-k)x^* + 1 < 0. \tag{7.75}$$

Solving this inequality we get $k \geq 6$ and (7.73). $\square$

7.7.4 *Non-Periodic Gibbs measures: the case hinge*

In this subsection we shall use a construction similar to Bleher-Ganikhodjaev construction.

For the case hinge we write (7.40) in the following form

$$h_{1,x} = \ln\lambda + \sum_{y \in S(x)} \ln \frac{1 + \exp(h_{1,y})}{1 + \exp(h_{1,y}) + \exp(h_{2,y})}, \tag{7.76}$$

$$h_{2,x} = \ln\lambda + \sum_{y \in S(x)} \ln \frac{1 + \exp(h_{2,y})}{1 + \exp(h_{1,y}) + \exp(h_{2,y})},$$

where $h_{i,x} = \ln z_{i,x}$, $i = 1, 2$. We show that system of equations (7.76) in the case hinge admit uncountably many non-translational-invariant solutions.

Take an arbitrary infinite path $\pi = \{x^0 = x_0 < x_1 < x_2 < ...\}$ on the Cayley tree starting at the origin $x_0 = x^0$. Establish a 1-1 correspondence between such paths and real numbers $t \in [0, 1]$. Write $\pi = \pi(t)$ when it is desirable to stress the dependence upon t. Map path π to a function $h^\pi : x \in V \mapsto h_x^\pi$ satisfying (7.76). Note that π splits Cayley tree Γ^k into two subgraphs Γ_1^k and Γ_2^k.

For $k = 2$, $\lambda > \frac{9}{4}$ the function h^π is defined by

$$h_x^\pi = \begin{cases} \ln(z^-), & \text{if } x \in \Gamma_1^k, \\[2mm] -\ln(z^-), & \text{if } x \in \Gamma_2^k, \end{cases} \tag{7.77}$$

where $z^- = \left(\frac{1 - \sqrt{1 - 4a^2}}{2a} \right)^2$, (see Proposition 7.10).

Define function $h = (h_1, h_2) \mapsto F(h) = (F_1(h), F_2(h))$ where

$$F_1(h) = \ln \frac{1 + \exp(h_1)}{1 + \exp(h_1) + \exp(h_2)}, \quad F_2(h) = \ln \frac{1 + \exp(h_2)}{1 + \exp(h_1) + \exp(h_2)}. \tag{7.78}$$

Proposition 7.13. *For $k = 2$, $\lambda > \frac{9}{4}$ and any $h = (h_1, h_2) \in [\ln z^-; -\ln z^-]^2$ (recall $z^- < 1$) the following inequalities hold:*

a)

$$\left| \frac{\partial F_1}{\partial h_1} \right| \le \frac{1}{(\sqrt{z^- + 1} + \sqrt{z^-})^2}; \quad \left| \frac{\partial F_2}{\partial h_2} \right| \le \frac{1}{(\sqrt{z^- + 1} + \sqrt{z^-})^2};$$

$$\left| \frac{\partial F_1}{\partial h_2} \right| \le \frac{1}{1 + z^- + (z^-)^2}; \quad \left| \frac{\partial F_2}{\partial h_1} \right| \le \frac{1}{1 + z^- + (z^-)^2};$$

b)

$$\|F(h) - F(l)\| \le \frac{2}{1 + z^- + (z^-)^2} \|h - l\|.$$

Proof. a) Using Lemma 6.8 and the inequality $\ln z^- \le h_i \le -\ln z^-$, $i = 1, 2$ we obtain

$$\left| \frac{\partial F_1}{\partial h_1} \right| \le \frac{\exp(h_2)}{(\sqrt{\exp(h_2) + 1} + 1)^2} = \psi(h_2).$$

The function $\psi(x)$ is increasing, therefore

$$\left| \frac{\partial F_1}{\partial h_1} \right| \le \psi(-\ln z^-) = \frac{1}{(\sqrt{z^- + 1} + \sqrt{z^-})^2}.$$

The proof for $\left|\frac{\partial F_2}{\partial h_2}\right|$ is similar.

We now consider

$$\left|\frac{\partial F_1}{\partial h_2}\right| = \frac{\exp(h_2)}{1 + \exp(h_1) + \exp(h_2)} = \varphi(h_1, h_2).$$

Since $\varphi'_{h_2} > 0$ and $\varphi'_{h_1} < 0$ we have

$$\left|\frac{\partial F_1}{\partial h_2}\right| \leq \max \varphi(h_1, h_2) = \varphi(\ln z^-, -\ln z^-) = \frac{1}{1 + z^- + (z^-)^2}.$$

Similarly one can show that

$$\left|\frac{\partial F_2}{\partial h_1}\right| \leq \frac{1}{1 + z^- + (z^-)^2}.$$

b) For $z^- < 1$ it is easy to see that $\frac{1}{(\sqrt{z^-+1}+\sqrt{z^-})^2} < \frac{1}{1+z^-+(z^-)^2}$. Using this inequality we obtain

$$\|F(h) - F(l)\| = \max_{i=1,2}\{|F_i(h) - F_i(l)|\} \leq$$

$$\max_{i=1,2}\{|(F_i)'_{h_1}||h_1 - l_1| + |(F_i)'_{h_2}||h_2 - l_2|\} \leq \frac{2}{1 + z^- + (z^-)^2}\|h - l\|.$$

This completes the proof. $\qquad\square$

If $\frac{2}{1+z^-+(z^-)^2} < 1$, i.e., $z^- > \frac{\sqrt{5}-1}{2}$ then one can use Proposition 7.13 to prove the following

Theorem 7.28. *If $k = 2$, and $9/5 < \lambda < 2(\sqrt{5} - 1)$ (i.e, $\frac{\sqrt{5}-1}{2} < z^- < 1$) then for any infinite path π there exists a unique function h^π satisfying (7.76) and (7.77).*

In the standard way one can prove that functions $h^{\pi(t)}$ are different for different $t \in [0; 1]$. Now let $\mu(t)$ denote the Gibbs measure corresponding to function $h^{\pi(t)}$, $t \in [0; 1]$. Similarly to Theorem 2.12, one can prove the following:

Theorem 7.29. *If conditions of Theorem 7.28 are satisfied then for any $t \in [0; 1]$, there exists a unique hinge-hard core Gibbs measure $\mu(t)$. Moreover, the Gibbs measures μ_1, μ_2 (see Theorem 7.24) are specified as $\mu(0) = \mu_1$ and $\mu(1) = \mu_2$.*

Because measures $\mu(t)$ are different for different $t \in [0, 1]$ we obtain a continuum of distinct Gibbs measures which are non-periodic.

7.8　Eight state hard-core model associated to a model with interaction radius two

Consider a finite set Φ. Let Ω be the set of configurations σ on V, i.e., the set of functions $x \in V \to \sigma(x) \in \Phi$. For a subset $A \subset V$ we denote by Ω_A the set of all configurations on A. By $|A|$ denote the number of elements of A.

Consider the generalized Kronecker symbol to be the function

$$U(\sigma_A) : \Omega_A \to \{|A| - 1, |A| - 2, \ldots, |A| - \min\{|A|, |\Phi|\}\},$$

defined as

$$U(\sigma_A) = |A| - |\sigma_A \cap \Phi|, \tag{7.79}$$

where $A \subset V$, and $|\sigma_A \cap \Phi|$ is the number of different values of $\sigma_A(x), x \in A$ (see [227]).

In this section we consider the case where $\Phi = \{-1, \ 1\}$ and $|A| = 4$.

We let M denote the set of all balls $b(x) = \{y \in V : d(x, y) \leqslant 1\}$ of unit radius.

Define the Hamiltonian as

$$\mathcal{H}(\sigma) = -J \sum_{b \in M} U(\sigma_b), \tag{7.80}$$

where $J > 0$.

7.8.1　*The system of functional equations*

Recall that for $x \in G_k$ we set $x_\downarrow = \{y \in G_k :\, < x, y >\} \setminus S(x)$, where $S(x)$ is the set of all direct successors of a point $x \in V$.

In this section, we consider the case $k = 2$.

Let

$$b(x) = \{x, xa_1, xa_2, xa_3\}, \quad \sigma_{b(x)} = \{\sigma(x), \sigma(xa_1), \sigma(xa_2), \sigma(xa_3)\}.$$

Consider a probability distribution $\mu^{(n)}$ on Ω_{V_n} :

$$\mu^{(n)}(\sigma_n) = Z_n^{-1} \exp\left\{ -\beta \mathcal{H}(\sigma_n) + \sum_{x \in W_n} h_{b(x),\, \sigma_{b(x)}}^{\sigma_{b(x_\downarrow)}} \right\}, \tag{7.81}$$

where $\sigma_n \in \Omega_{V_n}$

$$Z_n = \sum_{\overline{\sigma}_n \in \Omega_{V_n}} \exp\left\{ -\beta \mathcal{H}(\overline{\sigma}_n) + \sum_{x \in W_n} h_{b(x),\, \overline{\sigma}_{b(x)}}^{\overline{\sigma}_{b(x_\downarrow)}} \right\}$$

and $h_{b,\sigma}^{\bar{\sigma}} \in R.$

We consider the consistency condition for $\mu^{(n)}$:

$$\sum_{\sigma^{(n)}} \mu^{(n)}(\sigma_{n-1}, \sigma^{(n)}) = \mu^{(n-1)}(\sigma_{n-1}) \tag{7.82}$$

for all $n \geqslant 1$ and $\sigma_{n-1} \in \Omega_{V_{n-1}}$.

Consider the configurations

$$\sigma_0 = \{+,+,+,+\}, \quad \sigma_1 = \{+,-,+,+\}$$

$$\sigma_2 = \{+,+,-,-\}, \quad \sigma_3 = \{+,-,-,-\},$$

$$-\sigma_0 = \{-,-,-,-\}, \quad -\sigma_1 = \{-,-,-,+\},$$

$$-\sigma_2 = \{-,-,+,+\}, \quad -\sigma_3 = \{-,+,+,+\}$$

on a unit ball.

We set

$$h_{b,\sigma_0}^{\sigma_0} = h_{b,0}, \quad h_{b,\sigma_1}^{\sigma_0} = h_{b,1}, \quad h_{b,\sigma_2}^{\sigma_0} = h_{b,2}, \quad h_{b,-\sigma_1}^{\sigma_1} = h_{b,3},$$

$$h_{b,-\sigma_2}^{\sigma_1} = h_{b,4}, \quad h_{b,-\sigma_3}^{\sigma_1} = h_{b,5} \quad h_{b,\sigma_1}^{-\sigma_1} = h_{b,6}, \quad h_{b,\sigma_2}^{-\sigma_1} = h_{b,7}, \tag{7.83}$$

$$h_{b,\sigma_3}^{-\sigma_1} = h_{b,8} \quad h_{b,-\sigma_0}^{-\sigma_0} = h_{b,9}, \quad h_{b,-\sigma_1}^{-\sigma_0} = h_{b,10}, \quad h_{b,-\sigma_2}^{-\sigma_0} = h_{b,11}.$$

Given $a \in M$, we let b and c be "direct successors" of the ball a, i.e., the balls b and c have there centers in the set of direct successors of the center of a.

Theorem 7.30. *Let $k = 2$. A probability distribution $\mu^{(n)}(\sigma_n)$, $n = 1, 2, \ldots$ in (7.81) is consistent if and only if we have the following equalities for any $a \in M$:*

$$y_{a,0} = y_{a,6} = \frac{\lambda y_{b,0} + y_{b,1} + y_{b,2}}{y_{b,0} + y_{b,1} + y_{b,2}} \cdot \frac{\lambda y_{c,0} + y_{c,1} + y_{c,2}}{y_{c,0} + y_{c,1} + y_{c,2}}$$

$$y_{a,1} = y_{a,7} = \frac{\lambda y_{b,0} + y_{b,1} + y_{b,2}}{y_{b,0} + y_{b,1} + y_{b,2}} \cdot \frac{y_{c,3} + y_{c,4} + 1}{y_{c,0} + y_{c,1} + y_{c,2}},$$

$$y_{a,2} = y_{a,8} = \frac{y_{b,3} + y_{b,4} + 1}{y_{b,0} + y_{b,1} + y_{b,2}} \cdot \frac{y_{c,3} + y_{c,4} + 1}{y_{c,0} + y_{c,1} + y_{c,2}}, \tag{7.84}$$

$$y_{a,3} = y_{a,9} = \frac{\lambda y_{b,3} + y_{b,4} + 1}{y_{b,0} + y_{b,1} + y_{b,2}} \cdot \frac{\lambda y_{c,3} + y_{c,4} + 1}{y_{c,0} + y_{c,1} + y_{c,2}},$$

$$y_{a,4} = y_{a,10} = \frac{\lambda y_{c,3} + y_{c,4} + 1}{y_{c,0} + y_{c,1} + y_{c,2}}, \quad y_{a,5} = 1,$$

where $\lambda = e^{J\beta}$, $\beta = \frac{1}{T}$ and $y_{a,i} = \exp(h_{a,i} - h_{a,11})$, $i = 0, 1, \ldots, 10$.

Proof. Let x and y be neighbor vertices, and let a configuration $\sigma_{b(x)} \in \{\pm\sigma_0, \pm\sigma_1, \pm\sigma_2, \pm\sigma_3\}$ be defined on the ball $b(x)$. The configuration $\sigma_{b(y)}$ then depends on $\sigma_{b(x)}$. We collect all possible values of $\sigma_{b(y)}$ for a fixed $\sigma_{b(x)}$ in the following table:

$\sigma_{b(x)}$	$\sigma_{b(y)}, \sigma(y) = +1$	$\sigma_{b(y)}, \sigma(y) = -1$
σ_0	$\sigma_0, \sigma_1, \sigma_2$	$\emptyset$
σ_1	$\sigma_0, \sigma_1, \sigma_2$	$-\sigma_1, -\sigma_2, -\sigma_3$
σ_2	$\sigma_0, \sigma_1, \sigma_2$	$-\sigma_1, -\sigma_2, -\sigma_3$
σ_3	$\emptyset$	$-\sigma_1, -\sigma_2, -\sigma_3$
$-\sigma_0$	$\emptyset$	$-\sigma_0, -\sigma_1, -\sigma_2$
$-\sigma_1$	$\sigma_1, \sigma_2, \sigma_3$	$-\sigma_0, -\sigma_1, -\sigma_2$
$-\sigma_2$	$\sigma_1, \sigma_2, \sigma_3$	$-\sigma_0, -\sigma_1, -\sigma_2$
$-\sigma_3$	$\sigma_1, \sigma_2, \sigma_3$	$\emptyset$

Remark 7.5. Note that by this table we can define a constraint graph H with eight nodes. The set of nodes is $\{\sigma_0, \sigma_1, \sigma_2, \sigma_3, -\sigma_0, -\sigma_1, -\sigma_2, -\sigma_3\}$ and the adjacency matrix of the constraint graph is

$$\begin{pmatrix} 1 & 1 & 1 & 0 & 0 & 0 & 0 & 0 \\ 1 & 1 & 1 & 0 & 0 & 1 & 1 & 1 \\ 1 & 1 & 1 & 0 & 0 & 1 & 1 & 1 \\ 0 & 0 & 0 & 0 & 0 & 1 & 1 & 1 \\ 0 & 0 & 0 & 0 & 1 & 1 & 1 & 0 \\ 0 & 1 & 1 & 1 & 1 & 1 & 1 & 0 \\ 0 & 1 & 1 & 1 & 1 & 1 & 1 & 0 \\ 0 & 1 & 1 & 1 & 0 & 0 & 0 & 0 \end{pmatrix}.$$

Thus instead of the model (7.80) one can consider a model with hard constraints graph H, and configurations φ with values from the set $\{\pm\sigma_0, \pm\sigma_1, \pm\sigma_2, \pm\sigma_3\}$.

Denote $\overline{M}_n = \{b \in M : d(b, \bar{e}) \le n\}$, and $\overline{W}_n = \{b \in M : d(b, \bar{e}) = n\}$, where $\bar{e}$ is the unit ball centered at e and e is the unit element of the group G_k.

We then have $d(b_1, b_2) = d(c_{b_1}, c_{b_2})$, where c_b is the center of ball b.

Let $\sigma_n = \{\sigma_b, b \in \overline{M}_n\}$, $\sigma^{(n)} = \{\sigma_b, b \in \overline{W}_n\}$. Then the Hamiltonian can be written as

$$\mathcal{H}(\sigma_n) = -J \sum_{b \in \overline{V}_n} U(\sigma_b) = \mathcal{H}(\sigma_{n-1}) - J \sum_{b \in \overline{W}_n} U(\sigma_b).$$

We first prove the necessity of the theorem conditions. From formula (7.82), we obtain

$$\sum_{\sigma^{(n)}} \mu_n(\sigma_{n-1},\, \sigma^{(n)})$$

$$= Z_n^{-1} \sum_{\sigma^{(n)}} \exp\left\{ \beta J \mathcal{H}(\sigma_{n-1}) - J\beta \sum_{b\in \overline{W}_n} U(\sigma_b) + \sum_{b\in \overline{W}_n} h_{b,\,\sigma_b}^{\sigma_{b\downarrow}} \right\}$$

$$= Z_{n-1}^{-1} \exp\left\{ -\beta\mathcal{H}(\sigma_{n-1}) + \sum_{b\in \overline{W}_{n-1}} h_{a,\,\sigma_a}^{\sigma_{a\downarrow}} \right\}. \tag{7.85}$$

Therefore
$$L_n \sum_{\sigma^{(n)}} \prod_{b\in \overline{W}_{n-1}} \prod_{a\in S(b)} \exp\left\{ -J\beta U(\sigma_a) + h_{a,\,\sigma_a}^{\sigma_{a\downarrow}} \right\} = \prod_{b\in \overline{W}_{n-1}} \exp\left\{ h_{b,\,\sigma_b}^{\sigma_{b\downarrow}} \right\},$$

where $L_n = \frac{Z_{n-1}}{Z_n}$. Hence,

$$L_n \prod_{b\in \overline{W}_{n-1}} \prod_{a\in S(b)} \sum_{\sigma_a^{(n)}} \exp\left\{ -J\beta U(\sigma_a) + h_{a,\,\sigma_a}^{\sigma_{a\downarrow}} \right\} \mathbf{1}_{\sigma_n \in \Omega}$$

$$= \prod_{b\in \overline{W}_{n-1}} \exp\left\{ h_{b,\,\sigma_b}^{\sigma_{b\downarrow}} \right\}. \tag{7.86}$$

Using (7.86) by the data in table we then obtain

$$\exp(h_{a,\sigma_0}^{\sigma_i}) = L_n \cdot (e^{-\beta U(\sigma_0)+h_{b,\sigma_0}^{\sigma_0}} + e^{-\beta U(\sigma_1)+h_{b,\sigma_1}^{\sigma_0}} + e^{-\beta U(\sigma_2)+h_{b,\sigma_2}^{\sigma_0}}) \times$$
$$(e^{-\beta U(\sigma_0)+h_{c,\sigma_0}^{\sigma_0}} + e^{-\beta U(\sigma_1)+h_{c,\sigma_1}^{\sigma_0}} + e^{-\beta U(\sigma_2)+h_{c,\sigma_2}^{\sigma_0}}), i = 0, 1, 2;$$

$$\exp(h_{a,\sigma_1}^{\sigma_i}) = L_n \cdot (e^{-\beta U(\sigma_0)+h_{b,\sigma_0}^{\sigma_1}} + e^{-\beta U(\sigma_1)+h_{b,\sigma_1}^{\sigma_1}} + e^{-\beta U(\sigma_2)+h_{b,\sigma_2}^{\sigma_1}}) \times$$
$$(e^{-\beta U(-\sigma_1)+h_{c,-\sigma_1}^{\sigma_1}} + e^{-\beta U(-\sigma_2)+h_{c,-\sigma_2}^{\sigma_1}} + e^{-\beta U(-\sigma_3)+h_{c,-\sigma_3}^{\sigma_1}}), i = 0, 1, 2;$$

$$\exp(h_{a,\sigma_1}^{-\sigma_j}) = L_n \cdot (e^{-\beta U(\sigma_0)+h_{b,\sigma_0}^{\sigma_1}} + e^{-\beta U(\sigma_1)+h_{b,\sigma_1}^{\sigma_1}} + e^{-\beta U(\sigma_2)+h_{b,\sigma_2}^{\sigma_1}}) \times$$
$$(e^{-\beta U(\sigma_0)+h_{c,\sigma_0}^{\sigma_1}} + e^{-\beta U(\sigma_1)+h_{c,\sigma_1}^{\sigma_1}} + e^{-\beta U(\sigma_2)+h_{c,\sigma_2}^{\sigma_1}}), j = 1, 2, 3;$$

$$\exp(h_{a,\sigma_2}^{\sigma_i}) = L_n \cdot (e^{-\beta U(-\sigma_1)+h_{b,-\sigma_1}^{\sigma_2}} + e^{-\beta U(-\sigma_2)+h_{b,-\sigma_2}^{\sigma_2}} + e^{-\beta U(-\sigma_3)+h_{b,-\sigma_3}^{\sigma_2}}) \times$$
$$(e^{-\beta U(-\sigma_1)+h_{c,-\sigma_1}^{\sigma_2}} + e^{-\beta U(-\sigma_2)+h_{c,-\sigma_2}^{\sigma_2}} + e^{-\beta U(-\sigma_3)+h_{c,-\sigma_3}^{\sigma_2}}), i = 0, 1, 2;$$

$$\exp(h_{a,\sigma_2}^{-\sigma_j}) = L_n \cdot (e^{-\beta U(\sigma_0)+h_{b,\sigma_0}^{\sigma_2}} + e^{-\beta U(\sigma_1)+h_{b,\sigma_1}^{\sigma_2}} + e^{-\beta U(\sigma_2)+h_{b,\sigma_2}^{\sigma_2}}) \times$$
$$(e^{-\beta U(-\sigma_1)+h_{c,-\sigma_1}^{\sigma_2}} + e^{-\beta U(-\sigma_2)+h_{c,-\sigma_2}^{\sigma_2}} + e^{-\beta U(-\sigma_3)+h_{c,-\sigma_3}^{\sigma_2}}), j = 1, 2, 3;$$

$$\exp(h_{a,\sigma_3}^{-\sigma_j}) = L_n \cdot (e^{-\beta U(-\sigma_1)+h_{b,-\sigma_1}^{\sigma_3}} + e^{-\beta U(-\sigma_2)+h_{b,-\sigma_2}^{\sigma_3}} + e^{-\beta U(-\sigma_3)+h_{b,-\sigma_3}^{\sigma_3}}) \times$$
$$(e^{-\beta U(-\sigma_1)+h_{c,-\sigma_1}^{\sigma_3}} + e^{-\beta U(-\sigma_2)+h_{c,-\sigma_2}^{\sigma_3}} + e^{-\beta U(-\sigma_3)+h_{c,-\sigma_3}^{\sigma_3}}), j = 1,2,3;$$

$$\exp(h_{a,-\sigma_0}^{-\sigma_j}) = L_n \cdot (e^{-\beta U(-\sigma_0)+h_{b,-\sigma_0}^{-\sigma_0}} + e^{-\beta U(-\sigma_1)+h_{b,-\sigma_1}^{-\sigma_0}} + e^{-\beta U(-\sigma_2)+h_{b,-\sigma_2}^{-\sigma_0}}) \times$$
$$(e^{-\beta U(-\sigma_0)+h_{c,-\sigma_0}^{-\sigma_0}} + e^{-\beta U(-\sigma_1)+h_{c,-\sigma_1}^{-\sigma_0}} + e^{-\beta U(-\sigma_2)+h_{c,-\sigma_2}^{-\sigma_0}}), j = 0,1,2;$$

$$\exp(h_{a,-\sigma_1}^{\sigma_i}) = L_n \cdot (e^{-\beta U(-\sigma_0)+h_{b,-\sigma_0}^{-\sigma_1}} + e^{-\beta U(-\sigma_1)+h_{b,-\sigma_1}^{-\sigma_1}} + e^{-\beta U(-\sigma_2)+h_{b,-\sigma_2}^{-\sigma_1}}) \times$$
$$(e^{-\beta U(-\sigma_0)+h_{c,-\sigma_0}^{-\sigma_1}} + e^{-\beta U(-\sigma_1)+h_{c,-\sigma_1}^{-\sigma_1}} + e^{-\beta U(-\sigma_2)+h_{c,-\sigma_2}^{-\sigma_1}}), i = 1,2,3;$$

$$\exp(h_{a,-\sigma_1}^{-\sigma_j}) = L_n \cdot (e^{-\beta U(\sigma_1)+h_{b,\sigma_1}^{-\sigma_1}} + e^{-\beta U(\sigma_2)+h_{b,\sigma_2}^{-\sigma_1}} + e^{-\beta U(\sigma_3)+h_{b,\sigma_3}^{-\sigma_1}}) \times$$
$$(e^{-\beta U(-\sigma_0)+h_{c,-\sigma_0}^{-\sigma_1}} + e^{-\beta U(-\sigma_1)+h_{c,-\sigma_1}^{-\sigma_1}} + e^{-\beta U(-\sigma_2)+h_{c,-\sigma_2}^{-\sigma_1}}), j = 0,1,2;$$

$$\exp(h_{a,-\sigma_2}^{\sigma_i}) = L_n \cdot (e^{-\beta U(\sigma_1)+h_{b,\sigma_1}^{-\sigma_2}} + e^{-\beta U(\sigma_2)+h_{b,\sigma_2}^{-\sigma_2}} + e^{-\beta U(\sigma_3)+h_{b,\sigma_3}^{-\sigma_2}}) \times$$
$$(e^{-\beta U(-\sigma_0)+h_{c,-\sigma_0}^{-\sigma_2}} + e^{-\beta U(-\sigma_1)+h_{c,-\sigma_1}^{-\sigma_2}} + e^{-\beta U(-\sigma_2)+h_{c,-\sigma_2}^{-\sigma_2}}), i = 1,2,3;$$

$$\exp(h_{a,-\sigma_2}^{-\sigma_j}) = L_n \cdot (e^{-\beta U(\sigma_1)+h_{b,\sigma_1}^{-\sigma_2}} + e^{-\beta U(\sigma_2)+h_{b,\sigma_2}^{-\sigma_2}} + e^{-\beta U(\sigma_3)+h_{b,\sigma_3}^{-\sigma_2}}) \times$$
$$(e^{-\beta U(\sigma_1)+h_{c,\sigma_1}^{-\sigma_2}} + e^{-\beta U(\sigma_2)+h_{c,\sigma_2}^{-\sigma_2}} + e^{-\beta U(\sigma_3)+h_{c,\sigma_3}^{-\sigma_2}}), j = 0,1,2;$$

$$\exp(h_{a,-\sigma_3}^{\sigma_i}) = L_n \cdot (e^{-\beta U(\sigma_1)+h_{b,\sigma_1}^{-\sigma_3}} + e^{-\beta U(\sigma_2)+h_{b,\sigma_2}^{-\sigma_3}} + e^{-\beta U(\sigma_3)+h_{b,\sigma_3}^{-\sigma_3}}) \times$$
$$(e^{-\beta U(\sigma_1)+h_{c,\sigma_1}^{-\sigma_3}} + e^{-\beta U(\sigma_2)+h_{c,\sigma_2}^{-\sigma_3}} + e^{-\beta U(\sigma_3)+h_{c,\sigma_3}^{-\sigma_3}}), i = 1,2,3.$$

Consequently, for any ball a we have

$$h_{a,\sigma_0}^{\sigma_0} = h_{a,\sigma_0}^{\sigma_1} = h_{a,\sigma_0}^{\sigma_2} = h_{a,0}, \quad h_{a,\sigma_1}^{\sigma_0} = h_{a,\sigma_1}^{\sigma_1} = h_{a,\sigma_1}^{\sigma_2} = h_{a,1},$$

$$h_{a,\sigma_2}^{\sigma_0} = h_{a,\sigma_2}^{\sigma_1} = h_{a,\sigma_2}^{\sigma_2} = h_{a,2}, \quad h_{a,-\sigma_1}^{\sigma_1} = h_{a,-\sigma_1}^{\sigma_2} = h_{a,-\sigma_1}^{\sigma_3} = h_{a,3},$$

$$h_{a,-\sigma_2}^{\sigma_1} = h_{a,-\sigma_2}^{\sigma_2} = h_{a,-\sigma_2}^{\sigma_3} = h_{a,4}, \quad h_{a,-\sigma_3}^{\sigma_1} = h_{a,-\sigma_3}^{\sigma_2} = h_{a,-\sigma_3}^{\sigma_3} = h_{a,5},$$

$$h_{a,\sigma_1}^{-\sigma_1} = h_{a,\sigma_1}^{-\sigma_2} = h_{a,\sigma_1}^{-\sigma_3} = h_{a,6}, \quad h_{a,\sigma_2}^{-\sigma_1} = h_{a,\sigma_2}^{-\sigma_2} = h_{a,\sigma_2}^{-\sigma_3} = h_{a,7},$$

$$h_{a,\sigma_3}^{-\sigma_1} = h_{a,\sigma_3}^{-\sigma_2} = h_{a,\sigma_3}^{-\sigma_3} = h_{a,8}, \quad h_{a,-\sigma_0}^{-\sigma_0} = h_{a,-\sigma_0}^{-\sigma_1} = h_{a,-\sigma_0}^{-\sigma_2} = h_{a,9},$$

$$h_{a,-\sigma_1}^{-\sigma_0} = h_{a,-\sigma_1}^{-\sigma_1} = h_{a,-\sigma_1}^{-\sigma_2} = h_{a,10}, \quad h_{a,-\sigma_2}^{-\sigma_0} = h_{a,-\sigma_2}^{-\sigma_1} = h_{a,-\sigma_2}^{-\sigma_2} = h_{a,11}.$$

We set $z_{a,i} = \exp(h_{a,i})$. Then

$$z_{a,0} = \lambda^4 L_n \cdot (\lambda z_{b,0} + z_{b,1} + z_{b,2}) \times (\lambda z_{c,0} + z_{c,1} + z_{c,2});$$
$$z_{a,1} = \lambda^4 L_n \cdot (\lambda z_{b,0} + z_{b,1} + z_{b,2}) \times (z_{c,3} + z_{c,4} + z_{c,5});$$
$$z_{a,2} = \lambda^4 L_n \cdot (z_{b,3} + z_{b,4} + z_{b,5}) \times (z_{c,3} + z_{c,4} + z_{c,5});$$
$$z_{a,3} = \lambda^4 L_n \cdot (\lambda z_{b,9} + z_{b,10} + z_{b,11}) \times (\lambda z_{c,9} + z_{c,10} + z_{c,11});$$
$$z_{a,4} = \lambda^4 L_n \cdot (z_{b,6} + z_{b,7} + z_{b,8}) \times (\lambda z_{c,9} + z_{c,10} + z_{c,11});$$
$$z_{a,5} = \lambda^4 L_n \cdot (z_{b,6} + z_{b,7} + z_{b,8}) \times (z_{c,6} + z_{c,7} + z_{c,8});$$
$$z_{a,6} = \lambda^4 L_n \cdot (\lambda z_{b,0} + z_{b,1} + z_{b,2}) \times (\lambda z_{c,0} + z_{c,1} + z_{c,2});$$
$$z_{a,7} = \lambda^4 L_n \cdot (\lambda z_{b,0} + z_{b,1} + z_{b,2}) \times (z_{c,3} + z_{c,4} + z_{c,5});$$
$$z_{a,8} = \lambda^4 L_n \cdot (z_{b,3} + z_{b,4} + z_{b,5}) \times (z_{c,3} + z_{c,4} + z_{c,5});$$
$$z_{a,9} = \lambda^4 L_n \cdot (\lambda z_{b,9} + z_{b,10} + z_{b,11}) \times (\lambda z_{c,9} + z_{c,10} + z_{c,11});$$
$$z_{a,10} = \lambda^4 L_n \cdot (z_{b,6} + z_{b,7} + z_{b,8}) \times (\lambda z_{c,9} + z_{c,10} + z_{c,11});$$
$$z_{a,11} = \lambda^4 L_n \cdot (z_{b,6} + z_{b,7} + z_{b,8}) \times (z_{c,6} + z_{c,7} + z_{c,8}).$$

$$(7.87)$$

We set $y_{a,i} = \dfrac{z_{a,i}}{z_{a,11}}$, $i = 0, 1, ..., 10$ and obtain

$$
\begin{cases}
y_{a,0} = \dfrac{\lambda y_{b,0} + y_{b,1} + y_{b,2}}{y_{b,6} + y_{b,7} + y_{b,8}} \cdot \dfrac{\lambda y_{c,0} + y_{c,1} + y_{c,2}}{y_{c,6} + y_{c,7} + y_{c,8}} \\[2ex]
y_{a,1} = \dfrac{\lambda y_{b,0} + y_{b,1} + y_{b,2}}{y_{b,6} + y_{b,7} + y_{b,8}} \cdot \dfrac{y_{c,3} + y_{c,4} + y_{c,5}}{y_{c,6} + y_{c,7} + y_{c,8}} \\[2ex]
y_{a,2} = \dfrac{y_{b,3} + y_{b,4} + y_{b,5}}{y_{b,6} + y_{b,7} + y_{b,8}} \cdot \dfrac{y_{c,3} + y_{c,4} + y_{c,5}}{y_{c,6} + y_{c,7} + y_{c,8}} \\[2ex]
y_{a,3} = \dfrac{\lambda y_{b,9} + y_{b,10} + 1}{y_{b,6} + y_{b,7} + y_{b,8}} \cdot \dfrac{\lambda y_{c,9} + y_{c,10} + 1}{y_{c,6} + y_{c,7} + y_{c,8}} \\[2ex]
y_{a,4} = \dfrac{\lambda y_{c,9} + y_{c,10} + 1}{y_{c,6} + y_{c,7} + y_{c,8}} \\[2ex]
y_{a,5} = 1 \\[2ex]
y_{a,6} = \dfrac{\lambda y_{b,0} + y_{b,1} + y_{b,2}}{y_{b,6} + y_{b,7} + y_{b,8}} \cdot \dfrac{\lambda y_{c,0} + y_{c,1} + y_{c,2}}{y_{c,6} + y_{c,7} + y_{c,8}} \\[2ex]
y_{a,7} = \dfrac{\lambda y_{b,0} + y_{b,1} + y_{b,2}}{y_{b,6} + y_{b,7} + y_{b,8}} \cdot \dfrac{y_{c,3} + y_{c,4} + y_{c,5}}{y_{c,6} + y_{c,7} + y_{c,8}} \\[2ex]
y_{a,8} = \dfrac{y_{b,3} + y_{b,4} + y_{b,5}}{y_{b,6} + y_{b,7} + y_{b,8}} \cdot \dfrac{y_{c,3} + y_{c,4} + y_{c,5}}{y_{c,6} + y_{c,7} + y_{c,8}} \\[2ex]
y_{a,9} = \dfrac{\lambda y_{b,9} + y_{b,10} + 1}{y_{b,6} + y_{b,7} + y_{b,8}} \cdot \dfrac{\lambda y_{c,9} + y_{c,10} + 1}{y_{c,6} + y_{c,7} + y_{c,8}} \\[2ex]
y_{a,10} = \dfrac{\lambda y_{c,9} + y_{c,10} + 1}{y_{c,6} + y_{c,7} + y_{c,8}}.
\end{cases}
$$

Note that

$$y_{a,0} = y_{a,6}\,,\ \ y_{a,1} = y_{a,7}\,,\ \ y_{a,2} = y_{a,8}\,,\ \ y_{a,3} = y_{a,9}\,,\ \ y_{a,4} = y_{a,10}\,,\ \ y_{a,5} = 1. \tag{7.88}$$

Consequently,

$$\begin{cases} y_{a,0} = \dfrac{\lambda y_{b,0} + y_{b,1} + y_{b,2}}{y_{b,0} + y_{b,1} + y_{b,2}} \cdot \dfrac{\lambda y_{c,0} + y_{c,1} + y_{c,2}}{y_{c,0} + y_{c,1} + y_{c,2}} \\[3mm] y_{a,1} = \dfrac{\lambda y_{b,0} + y_{b,1} + y_{b,2}}{y_{b,0} + y_{b,1} + y_{b,2}} \cdot \dfrac{y_{c,3} + y_{c,4} + 1}{y_{c,0} + y_{c,1} + y_{c,2}} \\[3mm] y_{a,2} = \dfrac{y_{b,3} + y_{b,4} + 1}{y_{b,0} + y_{b,1} + y_{b,2}} \cdot \dfrac{y_{c,3} + y_{c,4} + 1}{y_{c,0} + y_{c,1} + y_{c,2}} \\[3mm] y_{a,3} = \dfrac{\lambda y_{b,3} + y_{b,4} + 1}{y_{b,0} + y_{b,1} + y_{b,2}} \cdot \dfrac{\lambda y_{c,3} + y_{c,4} + 1}{y_{c,0} + y_{c,1} + y_{c,2}} \\[3mm] y_{a,4} = \dfrac{\lambda y_{c,3} + y_{c,4} + 1}{y_{c,0} + y_{c,1} + y_{c,2}}. \end{cases} \tag{7.89}$$

We prove the sufficiency by taking (7.88) into account and obtaining formulas (7.85)–(7.87), i.e., formula (7.82), from system (7.89). $\qquad\square$

7.8.2   *Translation-invariant solutions*

In this subsection, we consider translation-invariant measures for model (7.80) on the Cayley tree. Such Gibbs measures correspond to solutions of system (7.84) with $y_{a,i} = y_i$, for any $a \in M$ and $i = 0, 1, 2, 3, 4$.

Setting $y_{a,i} = y_i \in R_+, \forall a \in M$ in (7.89), we obtain

$$\begin{cases} y_0 = \left(\dfrac{\lambda y_0 + y_1 + y_2}{y_0 + y_1 + y_2}\right)^2 \\[3mm] y_1 = \left(\dfrac{\lambda y_0 + y_1 + y_2}{y_0 + y_1 + y_2}\right) \cdot \left(\dfrac{y_3 + y_4 + 1}{y_0 + y_1 + y_2}\right) \\[3mm] y_2 = \left(\dfrac{y_3 + y_4 + 1}{y_0 + y_1 + y_2}\right)^2 \\[3mm] y_3 = \left(\dfrac{\lambda y_3 + y_4 + 1}{y_0 + y_1 + y_2}\right)^2 \\[3mm] y_4 = \dfrac{\lambda y_3 + y_4 + 1}{y_0 + y_1 + y_2}, \end{cases} \tag{7.90}$$

where $\lambda > 0$.

Theorem 7.31. *For model (7.80), there exists λ_{cr} (≈ 1.8055) such that we have at least one translation-invariant Gibbs measure for $\frac{11-\sqrt{21}}{8} < \lambda < \lambda_{cr}$,* at least two such measures for $\lambda = \lambda_{cr}$, at least three translation-invariant Gibbs measures for $\lambda_{cr} < \lambda < \frac{11+\sqrt{21}}{8}$.

Proof. From system (7.90) after simple algebra, we obtain

$$y_4 = y = \{(\lambda y^2 + y + 1)^2[\lambda y^2 + y + 1 \pm \sqrt{D}]^2$$
$$+2(\lambda - 1)y^2(y^2 + y + 1)(\lambda y^2 + y + 1) \times [\lambda y^2 + y + 1 \pm \sqrt{D}]$$
$$+4(\lambda - 1)^2 y^4(y^2 + y + 1)^2\} / \{4(\lambda - 1)^2(\lambda y^2 + y + 1)^3\} = \varphi_{\pm}(y, \lambda),$$

where $D = (\lambda y^2 + y + 1)(\lambda y^2 + (5 - 4\lambda)y + 1)$.

We note that the condition $D \geq 0$ implies $\frac{11-\sqrt{21}}{8} \leq \lambda \leq \frac{11+\sqrt{21}}{8}$. It can be easily seen that $\varphi_-(1, \lambda) - 1 < 0$ and $\varphi_-(+\infty, \lambda) = +\infty$ for any $\lambda > 0$. The equation $\varphi_-(y, \lambda) = y$ therefore admits not less than one solution for any $\lambda \geq \frac{11-\sqrt{21}}{8} \geq 0$. At $\lambda = 1.86$, we have $\varphi_+(0, \lambda) > 0$, $\varphi_+(1, \lambda) - 1 < 0$, and $\varphi_+(+\infty, \lambda) = +\infty$. The equation $y = \varphi_+(y)$ has therefore not less than two solutions. One can obtain the approximate value $\lambda_{cr} \approx 1.8055$ using computer simulations. $\square$

Now we assume that

$$y_{a,0} = y_{a,3}, \ y_{a,1} = y_{a,4}, \ y_{a,2} = 1, \forall a \in M. \tag{7.91}$$

From (7.87) we then obtain

$$\begin{cases} y_{a,0} = \dfrac{\lambda y_{b,0} + y_{b,1} + 1}{y_{b,0} + y_{b,1} + 1} \cdot \dfrac{\lambda y_{c,0} + y_{c,1} + 1}{y_{c,0} + y_{c,1} + 1} \\[3mm] y_{a,1} = \dfrac{\lambda y_{b,0} + y_{b,1} + 1}{y_{b,0} + y_{b,1} + 1}. \end{cases} \tag{7.92}$$

The following lemmas can be proved similarly as proof of Propositions 7.8 and 7.9.

Lemma 7.8.

1) *Let* $\lambda > 1$. *If* $y_a = (y_{a,0}, y_{a,1})$, $a \in M$ *is a solution of system (7.92), then* $y_i^- \leqslant y_{a,i} \leqslant y_i^+$ *for any* $i = 0, 1$, $a \in M$, *where* $(y_0^-, y_0^+, y_1^-, y_1^+)$ *is a solution of the system*

$$\begin{cases} y_0^- = \left(\dfrac{\lambda y_0^- + y_1^+ + 1}{y_0^- + y_1^+ + 1} \right)^2 \\[4mm] y_0^+ = \left(\dfrac{\lambda y_0^+ + y_1^- + 1}{y_0^+ + y_1^- + 1} \right)^2 \\[4mm] y_1^- = \dfrac{\lambda y_0^- + y_1^+ + 1}{y_0^- + y_1^+ + 1} \\[4mm] y_1^+ = \dfrac{\lambda y_0^+ + y_1^- + 1}{y_0^+ + y_1^- + 1}. \end{cases} \tag{7.93}$$

2) *Let* $\lambda < 1$. *If* $y_a = (y_{a,0}, y_{a,1})$ $a \in M$ *is a solution of (7.92), then* $y_i^- \leqslant y_{a,i} \leqslant y_i^+$ *for any* $i = 0, 1$, $a \in M$, *where* $(y_0^-, y_0^+, y_1^-, y_1^+)$ *is a solution of the following system*

$$
\begin{cases}
y_0^- = \left(\dfrac{\lambda y_0^+ + y_1^- + 1}{y_0^+ + y_1^- + 1} \right)^2 \\[4mm]
y_0^+ = \left(\dfrac{\lambda y_0^- + y_1^+ + 1}{y_0^- + y_1^+ + 1} \right)^2 \\[4mm]
y_1^- = \dfrac{\lambda y_0^+ + y_1^- + 1}{y_0^+ + y_1^- + 1} \\[4mm]
y_1^+ = \dfrac{\lambda y_0^- + y_1^+ + 1}{y_0^- + y_1^+ + 1}.
\end{cases}
$$

Lemma 7.9. *If* $y = (y_0^-, y_0^+, y_1^-, y_1^+)$ *is a solution of (7.93), then* $y_0^- = y_0^+$ *if and only if* $y_1^- = y_1^+$.

7.8.3 *Periodic solutions*

In this subsection, we consider periodic Gibbs measures in the case where condition (7.91) is satisfied. We write system (7.92) in the form

$$
\begin{cases}
h_{a,0} = \ln \left(\dfrac{\lambda e^{h_{b,0}} + e^{h_{b,1}} + 1}{e^{h_{b,0}} + e^{h_{b,1}} + 1} \cdot \dfrac{\lambda e^{h_{c,0}} + e^{h_{c,1}} + 1}{e^{h_{c,0}} + e^{h_{c,1}} + 1} \right) \\[4mm]
h_{a,1} = \ln \dfrac{\lambda e^{h_{b,0}} + e^{h_{b,1}} + 1}{e^{h_{b,0}} + e^{h_{b,1}} + 1},
\end{cases}
\tag{7.94}
$$

where $h_{a,i} = \ln y_{a,i}$, $i = 0, 1$ and investigate its periodic solutions.

We define $F : h = (h_0, h_1) \to F(h) \in R$ as

$$
F(h) = \frac{\lambda e^{h_0} + e^{h_1} + 1}{e^{h_0} + e^{h_1} + 1}.
$$

Let $G_k^{(2)}$ be a subgroup of the group G_k, consisting of words of even length.

Let $h_{a,i} = \begin{cases} h_i, & \text{if } c_a \in G_2^{(2)} \\ h_i', & \text{if } c_a \in G_2 \backslash G_2^{(2)} \end{cases} \qquad i = 0, 1.$

From (7.94) we obtain then

$$
\begin{cases}
h_0 = 2\ln \dfrac{\lambda e^{h_0'} + e^{h_1'} + 1}{e^{h_0'} + e^{h_1'} + 1} \\[2mm]
h_0' = 2\ln \dfrac{\lambda e^{h_0} + e^{h_1} + 1}{e^{h_0} + e^{h_1} + 1} \\[2mm]
h_1 = \ln \dfrac{\lambda e^{h_0'} + e^{h_1'} + 1}{e^{h_0'} + e^{h_1'} + 1} \\[2mm]
h_1' = \ln \dfrac{\lambda e^{h_0} + e^{h_1} + 1}{e^{h_0} + e^{h_1} + 1}.
\end{cases}
\tag{7.95}
$$

Theorem 7.32. *For all $\lambda > 0$ with condition (7.91) satisfied, there exists a unique $G_2^{(2)}$-periodic Gibbs measure in model (7.80). Moreover, this measure coincides with the translation invariant Gibbs measure.*

Proof. We consider the case where $h_0 = h_0'$ and $h_1 = h_1'$. From system (7.95), we obtain

$$
\begin{cases}
h_0 = 2\ln \dfrac{\lambda e^{h_0} + e^{h_1} + 1}{e^{h_0} + e^{h_1} + 1} \\[2mm]
h_1 = \ln \dfrac{\lambda e^{h_0} + e^{h_1} + 1}{e^{h_0} + e^{h_1} + 1} .
\end{cases}
\tag{7.96}
$$

This system of equations has a unique solution (h_0^*, h_1^*). Indeed, from (7.96), we have $e^{3h_1} + (1 - \lambda)e^{2h_1} - 1 = 0$. Let $e^{h_1} = z$. Then $Q(z) = z^3 + (1 - \lambda)z^2 - 1 = 0$.

It is known that the number of positive roots of the equation $z^3 + (1 - \lambda)z^2 - 1 = 0$ does not exceed the number of sign changes of its coefficients, i.e., $1, 1 - \lambda, -1$ (see [203]). Hence, the equation $Q(z) = 0$ has a unique solution.

We now consider the case where $h_0 \neq h_0'$ and $h_1 \neq h_1'$. From system (7.95), we obtain

$$
\begin{cases}
h_1 = \ln \dfrac{\lambda e^{2h_1'} + e^{h_1'} + 1}{e^{2h_1'} + e^{h_1'} + 1} \\[2mm]
h_1' = \ln \dfrac{\lambda e^{2h_1} + e^{h_1} + 1}{e^{2h_1} + e^{h_1} + 1},
\end{cases}
\tag{7.97}
$$

whence for $z = e^{h_1}$, we obtain the polynomial equation:

$$
P(z) = (\lambda^2 + \lambda + 1)z^5 - (\lambda^3 - 2\lambda - 2)z^4 - (2\lambda^2 - 2\lambda - 3)z^3
$$

$$
-(2\lambda^2 + 2\lambda - 1)z^2 - (2\lambda + 1)z - \lambda - 2 = 0.
$$

It is easy to see that

$$P(z) = Q(z) \cdot K(z), \quad \text{where} \quad K(z) = (\lambda^2 + \lambda + 1)z^2 + (2\lambda + 1)z + \lambda + 2.$$

Because the equation $Q(z) = 0$ has a unique solution and the equation $K(z) = 0$ has no solutions, the equation $P(z) = 0$ has a unique solution. $\square$

We let $\omega_x(a_i)$ denote the number of letters a_i, $i = 1, \ldots, k+1$, in the irreducible form of writing the word x.

Let $\varnothing \neq A \subset N_2 = \{1, 2, 3\}$ and $H_A = \{x \in G_2 : \sum_{i \in A} \omega_x(a_i) \text{ is even}\}$.

Obviously, H_A is a subgroup of index two (see Chapter 1).

Let

$$h_{a,j} = \begin{cases} h_j, & c_a \in H_A \\ h'_j, & c_a \in G_2 \backslash H_A \end{cases}, \quad j = 0, 1.$$

We note that

$$h_0 = \begin{cases} F(h) + F(h') \ \text{ or } \ 2F(h) \ \text{ if } \ |A| = 1, \\ 2F(h') \ \text{ or } \ F(h) + F(h') \ \text{ if } \ |A| = 2. \end{cases}$$

$$h'_0 = \begin{cases} F(h) + F(h') \ \text{ or } \ 2F(h') \ \text{ if } \ |A| = 1, \\ 2F(h) \ \text{ or } \ F(h) + F(h') \text{ if } \ |A| = 2. \end{cases} \tag{7.98}$$

$$h_1 = F(h) \ \text{ or } \ F(h'),$$

$$h'_1 = F(h') \ \text{ or } \ F(h).$$

The case $|A| = 3$ corresponds to the case $H_A = G_2^{(2)}$, considered above.

Theorem 7.33. *For model (7.80) under condition (7.91), we have uncountably many H_A-periodic Gibbs measures. Moreover, these measures are neither translation-invariant nor $G_2^{(2)}$-periodic measures.*

Proof. From equation (7.98), we find that a system has solutions if and only if $F(h) = F(h')$.

Hence $t_0(s_1 - t_1) + (t_0 - s_0)(t_1 + 1) = 0$, where $t_i = e^{h_i}, s_i = e^{h'_i}, i = 0, 1$. Setting $s_1 - t_1 = \alpha$, we obtain $s_1 = t_1 + \alpha$, $s_0 = t_0 + \frac{\alpha t_0}{t_1 + 1}$. From (7.98) we get the condition $t_0 = \frac{1 - t_1^2}{t_1 - \lambda}$ and $t_1^3 + (1 - \lambda)t_1^2 - 1 = 0$. The last equation has the only positive root t_1^*. We therefore have $t_1 = t_1^*$, $t_0 = t_0^* = \frac{1 - (t_1^*)^2}{t_1^* - \lambda}$, $s_1 = s_1^*(\alpha) = t_1^* + \alpha$, $s_0 = s_0^*(\alpha) = t_0^* + \frac{\alpha t_0^*}{t_1^* + 1}$, $\forall \alpha \in R$.

For any $\alpha \in R$ we therefore obtain the solution

$$(\ln t_0^*, \ \ln t_1^*, \ \ln s_0^*(\alpha), \ \ln s_1^*(\alpha))$$

of system (7.98). Moreover, these solutions differ for different α. $\square$

Remark 7.6. The result of Theorem 7.33 is new for model (7.80) because the previously studied models (see, Chapters 2-6) did not admit the existence of H_A-periodic Gibbs measures for $H_A \neq G_k^{(2)}$.

Commentaries and references. Hard constraints arise in fields as diverse as combinatorics, statistical mechanics and telecommunications. In particular, the hard core model arises in the study of random independent sets of a graph [38], [79], the study of gas molecules on a lattice [18], and in the analysis of multi-casting in telecommunication networks [128], [149], [163].

A hard core model on d-dimensional lattice $\mathbb{Z}^d$, was introduced and studied by Mazel and Suhov in [156], motivated by applications in statistical physics.

On the tree, hard core model was studied in [163] as an idealized example of multi-casting on a regular tree network, each of whose edges has the same capacity C. In [81], for a generalization of hard core model on a Cayley tree is considered. This model is a special case of a loss network (see [128] for a general survey of loss networks and [163], [150] for connections with this particular model).

This chapter is devoted to the study of Gibbs measures of models with hard constraints on trees, which is based on much recent papers (see [35]-[39], [153], [154], [225], [231], [251]).

The subsection concerning weakly periodic Gibbs measures of this chapter is a result of papers [129], [234]. We note that the Ising model on Cayley tree of order $k \geq 4$ have many (non-periodic) H_A-weakly periodic Gibbs measures (see Theorem 2.6). But Theorem 7.10 says that for hard core model on Cayley tree there is no any non-trivial H_A-weakly periodic Gibbs measure. This is one of crucial differences between Ising and hard core models.

The main part of the chapter is due to the paper [35]. Two-state hard core model was studied in many papers mentioned above, but in this chapter we presented the results of [251]. In [154] and [231] several three-state fertile hard core models were studied. The results of the last section is due to [231], where a model with interaction radius two was studied, it was reduced to the eight-state hard core model with interaction radius one.

In the language of [35], the model considered in [81] corresponds to a constraint graph which is fertile. However, the emphasis of work [81] is quite different, which devoted to identify regions where multiple Gibbs measures (not necessarily simple and invariant) exist for the particular choice of activity vector.

In [36] the hard-core and Widom-Rowlinson models on Cayley tree were studied. These models considered as examples of non-monotonic behavior in symmetric systems exhibiting more than one critical point at which spontaneous symmetry breaking appears or disappears. In that paper the critical behavior of the Widom-Rowlinson model on Cayley tree is studied.

For other results related to hard core models on trees see [47], [80], [114], [120].

Chapter 8

Potts model with countable set of spin values

This chapter is devoted to a nearest-neighbor Potts model, with countable spin values $0, 1, \ldots$, and non-zero external field, on a Cayley tree of order k. We study translation-invariant 'splitting' Gibbs measures, which depend on k and a probability measure ν (with $\nu(i) > 0$ on the set of all non-negative integer numbers $\Phi = \{0, 1, ...\}$). This problem is reduced to the description of the solutions of some infinite system of equations. For any $k \geq 1$ and any fixed probability measure ν we show that the set of translation-invariant splitting Gibbs measures contains at most one point, independently on parameters of the Potts model with countable set of spin values on a Cayley tree. Also we give a description of the class of measures ν on Φ such that with respect to each element of this class the infinite system of equations has unique solution $\{a^i, i = 1, 2, ...\}$, where $a \in (0, 1)$.

8.1   An infinite system of functional equations

Let $\Gamma^k = (V, L)$ be a Cayley tree of order $k \geq 1$. In this chapter we consider model where the spin takes values in the set of all non-negative integer numbers $\Phi := \{0, 1, \ldots\}$, and is assigned to the vertices of the tree. A configuration σ on V is then defined as a function $x \in V \mapsto \sigma(x) \in \Phi$; the set of all configurations is Φ^V.

The (formal) Hamiltonian of the Potts model is :

$$H(\sigma) = -J \sum_{\langle x,y \rangle \in L} \delta_{\sigma(x)\sigma(y)} - \alpha \sum_{x \in V} \delta_{0\sigma(x)}, \tag{8.1}$$

where $J, \alpha \in R$ are constants. As usually, $\langle x, y \rangle$ stands for nearest neighbor vertices and δ is the Kroneker's symbol.

For $A \subset V$ denote by Φ^A the configuration space on A. Let $h : x \mapsto$

221

$h_x = (h_{0,x}, h_{1,x}, ...) \in R^\infty$ be a real sequence-valued function of $x \in V \setminus \{x^0\}$.

Fix a probability measure $\nu = \{\nu(i) > 0, i \in \Phi\}$.

Given $n = 1, 2, \ldots$, consider the probability distribution μ_n on Φ^{V_n} defined by

$$\mu^{(n)}(\sigma_n) = Z_n^{-1} \exp\left(-\beta H(\sigma_n) + \sum_{x \in W_n} h_{\sigma(x),x}\right) \prod_{x \in V_n} \nu(\sigma(x)). \qquad (8.2)$$

Here, $\sigma_n : x \in V_n \mapsto \sigma(x)$ and Z_n is the corresponding partition function:

$$Z_n = \sum_{\widetilde{\sigma}_n \in \Phi^{V_n}} \exp\left(-\beta H(\widetilde{\sigma}_n) + \sum_{x \in W_n} h_{\widetilde{\sigma}(x),x}\right) \prod_{x \in V_n} \nu(\widetilde{\sigma}(x)). \qquad (8.3)$$

Remark 8.1. Note that Z_n is finite, since ν is a probability measure and $\exp(-\beta H(\widetilde{\sigma}_n) + \sum_{x \in W_n} h_{\widetilde{\sigma}(x),x})$ is bounded on Φ^{V_n}.

As usual, the probability distributions $\mu^{(n)}$ are compatible if for any $n \geq 1$ and $\sigma_{n-1} \in \Phi^{V_{n-1}}$:

$$\sum_{\omega_n \in \Phi^{W_n}} \mu^{(n)}(\sigma_{n-1} \vee \omega_n) = \mu^{(n-1)}(\sigma_{n-1}). \qquad (8.4)$$

Here $\sigma_{n-1} \vee \omega_n \in \Phi^{V_n}$ is the concatenation of σ_{n-1} and ω_n.

The measure defined by (8.2) and (8.4) is called a *splitting Gibbs measure* corresponding to Hamiltonian (8.1) and function $x \mapsto h_x$, $x \neq x^0$.

Proposition 8.1. *Probability distributions* $\mu^{(n)}(\sigma_n)$, $n = 1, 2, \ldots$, *in (8.2) are compatible iff for any* $x \in V \setminus \{x^0\}$ *the following equation holds:*

$$h_x^* = \sum_{y \in S(x)} F(h_y^*, \theta). \qquad (8.5)$$

Here, $\theta = \exp(J\beta)$, $h_x^* = (h_{1,x} - h_{0,x} - \alpha\beta + \ln \frac{\nu(1)}{\nu(0)}, h_{2,x} - h_{0,x} - \alpha\beta + \ln \frac{\nu(2)}{\nu(0)}, ...)$ *and the function* $F(\cdot, \theta) : R^\infty \to R^\infty$ *is* $F(h, \theta) = (F_1(h, \theta), F_2(h, \theta), ...)$, *with*

$$F_i(h, \theta) = -\alpha\beta + \ln \frac{\nu(i)}{\nu(0)} + \ln \frac{(\theta - 1)\exp(h_i) + \sum_{j=1}^\infty \exp(h_j) + 1}{\theta + \sum_{j=1}^\infty \exp(h_j)},$$

$h = (h_1, h_2, ...)$, $i = 1, 2, \ldots$.

Proof. *Necessity.* Suppose that (8.4) holds; we shall prove (8.5). Substituting (8.2) into (8.4), obtain that for any configurations $\sigma_{n-1}\colon x \in V_{n-1} \mapsto \sigma_{n-1}(x) \in \Phi$:

$$\frac{Z_{n-1}}{Z_n} \sum_{\omega_n \in \Phi^{W_n}} \exp\left(\sum_{x \in W_{n-1}} \sum_{y \in S(x)} (J\beta\delta_{\sigma_{n-1}(x)\omega_n(y)} + \alpha\delta_{0\omega_n(y)} + h_{\omega_n(y),y})\right) \times$$

$$\prod_{y \in W_n} \nu(\omega_n(y)) = \exp\left(\sum_{x \in W_{n-1}} h_{\sigma_{n-1}(x),x}\right), \qquad (8.6)$$

where $\omega_n\colon x \in W_n \mapsto \omega_n(x)$.

From (8.6) we get:

$$\frac{Z_{n-1}}{Z_n} \sum_{\omega_n \in \Phi^{W_n}} \prod_{x \in W_{n-1}} \prod_{y \in S(x)} \exp\left(J\beta\delta_{\sigma_{n-1}(x)\omega_n(y)} + \alpha\beta\delta_{0\omega_n(y)}+\right.$$

$$h_{\omega_n(y),y} + \ln\nu(\omega_n(y))\big) = \prod_{x \in W_{n-1}} \exp\left(h_{\sigma_{n-1}(x),x}\right).$$

Consequently, for any $i \in \Phi$,

$$\prod_{y \in S(x)} \frac{\sum_{j \in \Phi} \exp\left(J\beta\delta_{ij} + \alpha\beta\delta_{0j} + h_{j,y} + \ln\nu(j)\right)}{\sum_{j \in \Phi} \exp\left(J\beta\delta_{0j} + \alpha\beta\delta_{0j} + h_{j,y} + \ln\nu(j)\right)} = \exp\left(h_{i,x} - h_{0,x}\right),$$

so that

$$\prod_{y \in S(x)} \frac{1 + \sum_{j=1}^{\infty} \exp\left(h_{j,y}^*\right) + (\theta - 1)\exp\left(h_{i,y}^*\right)}{\theta + \sum_{j=1}^{\infty} \exp\left(h_{j,y}^*\right)} =$$

$$\exp\left(h_{i,x}^* + \alpha\beta - \ln\frac{\nu(i)}{\nu(0)}\right),$$

where $h_{i,x}^* = h_{i,x} - h_{0,x} + \ln\frac{\nu(i)}{\nu(0)} - \alpha\beta$, which implies (8.5).

Sufficiency. Suppose that (8.5) holds. Then

$$\prod_{y \in S(x)} \sum_{j \in \Phi} \exp\left(J\beta\delta_{ij} + \alpha\beta\delta_{0j} + h_{j,y} + \ln\nu(j)\right) = a(x)\exp\left(h_{i,x}\right), i = 0, 1, \ldots$$

$$(8.7)$$

for some function $a(x) > 0, x \in V$.

We have

$$\text{LHS of } (8.4) = \frac{1}{Z_n} \exp(-\beta H(\sigma_{n-1})) \prod_{x \in V_{n-1}} \nu(\sigma(x))\times$$

$$\prod_{x \in W_{n-1}} \prod_{y \in S(x)} \sum_{j \in \Phi} \exp\left(J\beta\delta_{\sigma_{n-1}(x)j} + \alpha\beta\delta_{0j} + h_{j,y} + \ln\nu(j)\right). \qquad (8.8)$$

Substituting (8.7) into (8.8) and denoting $A_n(x) = \prod_{x \in W_{n-1}} a(x)$, we get

$$\text{RHS of } (8.8) = \frac{A_{n-1}}{Z_n} \exp(-\beta H(\sigma_{n-1})) \prod_{x \in V_{n-1}} \nu(\sigma(x)) \prod_{x \in W_{n-1}} h_{\sigma_{n-1}(x),x}. \qquad (8.9)$$

Since $\mu^{(n)}$, $n \geq 1$ is a probability, we should have

$$\sum_{\sigma_{n-1}} \sum_{\sigma^{(n)}} \mu^{(n)}(\sigma_{n-1}, \sigma^{(n)}) = 1.$$

Hence from (8.9) we get $Z_{n-1}A_{n-1} = Z_n$, and (8.4) holds. $\qquad \square$

8.2 Translation-invariant solutions

Assume $h_x = h = (h_1, h_2, ...)$ for any $x \in V$. In this case we obtain from (8.5):

$$h_i = -\alpha\beta + \ln\frac{\nu(i)}{\nu(0)} + k\ln\frac{(\theta-1)\exp(h_i) + \sum_{j=1}^{\infty}\exp(h_j) + 1}{\theta + \sum_{j=1}^{\infty}\exp(h_j)}, i = 1, 2, ... \qquad (8.10)$$

Set $u_i = \exp(h_i), i = 1, 2, ...$ From (8.10) we have

$$u_i = \frac{\nu(i)}{\nu(0)}\exp(-\alpha\beta)\left(\frac{(\theta-1)u_i + \sum_{j=1}^{\infty}u_j + 1}{\theta + \sum_{j=1}^{\infty}u_j}\right)^k, i = 1, 2, ... \qquad (8.11)$$

In this section we give full analysis of the system of equations (8.11).

8.2.1 *The set of solutions $\{u_i\}$ with $\sum_{j=1}^{\infty} u_j = \infty$*

In this subsection we shall describe solutions of (8.11) with property $\sum_{j=1}^{\infty} u_j = \infty$. In this case from (8.11) we get

$$u_i = \frac{\nu(i)}{\nu(0)}\exp(-\alpha\beta), i = 1, 2, ... \qquad (8.12)$$

Since $\sum_{j=0}^{\infty} \nu(j) = 1$ by (8.12) we get

$$\sum_{j=1}^{\infty} u_j = \frac{1 - \nu(0)}{\nu(0)\exp(\alpha\beta)} < +\infty.$$

Thus there is no solution of (8.11) with $\sum_{j=1}^{\infty} u_j = \infty$.

8.2.2 *The set of solutions with $\sum_{j=1}^{\infty} u_j < +\infty$*

Now we want to describe solutions of (8.11) with $\sum_{j=1}^{\infty} u_j = A < +\infty$, where A is some fixed positive number.

In this case from (8.11) we obtain

$$\eta_i u_i = \left(\frac{(\theta - 1)u_i + A + 1}{\theta + A} \right)^k, \tag{8.13}$$

where $\eta_i = \frac{\nu(0)}{\nu(i)} \exp(\alpha\beta)$. Denote $B_i = \eta_i(\theta + A)^k$. Note that

$$u_i \to 0 \quad \text{and} \quad B_i \to \infty \quad \text{as} \quad i \to \infty. \tag{8.14}$$

From (8.13) we obtain

$$B_i u_i = ((\theta - 1)u_i + A + 1)^k, i = 1, 2, ... \tag{8.15}$$

8.2.2.1 *Case $\theta > 1$*

As function $u \to ((\theta - 1)u + A + 1)^k$ is concave increasing, we conclude that (8.15) has a unique positive solution, say u_i^* if u_i^* satisfies the following equations

$$\begin{cases} B_i u_i^* = ((\theta - 1)u_i^* + A + 1)^k, \\ B_i = k(\theta - 1)((\theta - 1)u_i^* + A + 1)^{k-1}. \end{cases} \tag{8.16}$$

In other words, if (8.16) is satisfied for $u_i^* > 0$ then u_i^* is a unique positive solution to (8.15).

Assume $k \geq 2$. From (8.16) we have $u_i^* = u^*$ where $u^* = \frac{A+1}{(k-1)(\theta-1)}$ and $B_i = B^*$ where

$$B^* = (\theta - 1)\left(\frac{A+1}{k-1} \right)^{k-1} k^k.$$

We conclude that (8.15) has two solutions $0 < u_{i,1}^* < u_{i,2}^*$ for $B_i > B^*$.

Note that B^* and u^* do not depend on i and for $B_i > B^*$ we have

$$0 < u_{i,1}^* < u^* < u_{i,2}^*, \quad \text{for any} \quad i = 1, 2, ... \tag{8.17}$$

By (8.17) we have $\sum_{i=1}^{\infty} u_{i,2}^* = \infty$ thus $u_{i,2}^*$ does not satisfy the convergence condition.

Note that $u_{i,1}^*$ depends on $k, J, \alpha, \beta, \nu(i), \nu(0)$. Assume (condition on $\nu(i), i = 0, 1, ...$)

$$\sum_{i=1}^{\infty} u_{i,1}^* = A. \tag{8.18}$$

Thus we have proved

Theorem 8.1. *If for any* $i = 1, 2, ...,$ $B_i > B^*$ *and* ν *the condition (8.18) is satisfied then the system of equations (8.11) has unique solution* $\{u^*_{i,1}\}$ *with* $\sum_{i=1}^{\infty} u^*_{i,1} = A.$

In order to demonstrate the conditions of Theorem 8.1, we will consider cases $k = 1$ and $k = 2$ separately.

Assume $k = 1$. In this case from (8.15) we get

$$u_i = \frac{(A+1)\nu(i)}{\nu(0)e^{\alpha\beta}(\theta + A) - (\theta - 1)\nu(i)},$$

which is positive if $\frac{\nu(0)}{\nu(i)}e^{\alpha\beta}(\theta + A) > \theta - 1$. This condition corresponds to $B_i > B^*$ for $k = 1$. In this case the condition (8.18) can be written as

$$\sum_{i=1}^{\infty} \frac{(A+1)\nu(i)}{\nu(0)e^{\alpha\beta}(\theta + A) - (\theta - 1)\nu(i)} = A. \tag{8.19}$$

Assume $k = 2$. In this case condition $B_i > B^*$ has the form

$$B_i = \frac{\nu(0)}{\nu(i)}e^{\alpha\beta}(\theta + A)^2 > 4(\theta - 1)(A + 1). \tag{8.20}$$

The solutions $u^*_{i,m}$, $m = 1, 2$ are

$$u^*_{i,1} = \frac{B_i - 2(\theta - 1)(A + 1) - \sqrt{B_i[B_i - 4(\theta - 1)(A + 1)]}}{2(\theta - 1)^2},$$

$$u^*_{i,2} = \frac{B_i - 2(\theta - 1)(A + 1) + \sqrt{B_i[B_i - 4(\theta - 1)(A + 1)]}}{2(\theta - 1)^2}.$$

By (8.14) we have $u^*_{i,2} \to \infty$ if $i \to \infty$. The condition (8.18) (for $k = 2$) on ν can be rewritten as

$$\sum_{i=1}^{\infty} \Bigg(\nu(0)(\theta + A)^2 e^{\alpha\beta} - 2(\theta - 1)(A + 1)\nu(i) -$$

$$(\theta + A)\sqrt{\nu(0)e^{\alpha\beta}[\nu(0)e^{\alpha\beta}(\theta + A) - 4(\theta - 1)(A + 1)\nu(i)]} \Bigg) (2(\theta - 1)^2\nu(i))^{-1} = A. \tag{8.21}$$

8.2.2.2 *Case $\theta \leq 1$*

If $\theta = 1$ then from (8.11) we obtain $u_i = \frac{\nu(i)}{\nu(0)e^{\alpha\beta}}$. This is a unique solution of (8.11). Note that this case corresponds to zero interaction case and is not interesting.

If $\theta < 1$ then function $\varphi(u) = ((\theta-1)u+A+1)^k$ is convex and decreasing for odd k and equation (8.15) has unique solution. For even k the function φ is decreasing for $u < \frac{A+1}{1-\theta}$ and increasing for $u > \frac{A+1}{1-\theta}$.

Thus for $\theta < 1$ the equation (8.11) has unique solution u_i' which may satisfy condition $\sum_{i=1}^{\infty} u_i' = A$. Note that for $\theta < 1$ we need not have a condition like $B_i > B^*$. Here we just need to have condition (8.18).

Denote by $\mathcal{G}(H)$ the set of all splitting translation-invariant Gibbs measures for Hamiltonian (8.1).

Summarizing, we obtain the following

Theorem 8.2. *For any parameters $\alpha, J \in R$, $k \geq 1$, $\beta > 0$ and any fixed probability measure ν on Φ the set $\mathcal{G}(H)$ contains at most one point.*

Remark 8.2.

1. It is known (see, for example [139]) that the Potts model with $q \geq 2$ spin values on $\mathbb{Z}^d, d \geq 2$ undergoes a first-order phase transition at a certain transition temperature $T_{\mathrm{cr}} = T_{\mathrm{cr}}(q)$, provided q is large enough. Namely, the model (on $\mathbb{Z}^d$) has q different Gibbs measures for temperatures $T < T_{\mathrm{cr}}$, $q+1$ measures at $T = T_{\mathrm{cr}}$ and one measure for $T > T_{\mathrm{cr}}$.

2. Note that (see Chapter 5) for the ferromagnetic Potts model with q spin values on Cayley tree *for any $q \geq 2$* (even for $q = 2$, i.e., for the Ising model, see Chapter 2) there are $q + 1$ distinct translation-invariant Gibbs measures. Namely, there are two critical temperatures $0 < T_{\mathrm{cr}}' < T_{\mathrm{cr}}$ such that (i) for $T \in (T_{\mathrm{cr}}', T_{\mathrm{cr}}]$ there are $q + 1$ extreme Gibbs measures. One of them, say μ_0, (with $\mu_0(\sigma(x) = i) = \frac{1}{q}, i = 1, ..., q$) is called unordered Gibbs measure; (ii) $T \in (0, T_{\mathrm{cr}}']$ the $q + 1$ Gibbs measures still exist but the measure μ_0 is not extreme. (iii) for $T > T_{\mathrm{cr}}$ there is one Gibbs measure.

3. Theorem 8.2 shows that the result is not true if $q \to \infty$.

8.3   Exponential solutions

Thus by results of previous section equation (8.11) has at most one solution for any fixed measure ν.

It is interesting to describe exact value of such solution and corresponding measure ν.

In this section we shall describe solutions of (8.11) such that $u_i = a^i$ for some $a \in (0, 1)$ and the corresponding measure ν.

In this case $\sum_{i=1}^{\infty} u_i = A = \frac{a}{1-a}$.

8.3.1   *Case $\theta > 1$*

From (8.11) we have

$$\nu(i) \equiv \nu(i, a) = \nu(0)a^i e^{\alpha\beta}\left(\frac{a + (1-a)\theta}{(\theta - 1)(1-a)a^i + 1}\right)^k. \tag{8.22}$$

Now we shall choose a such that $\sum_{i=1}^{\infty} \nu(i) < +\infty$. Consider

$$\frac{\nu(i+1)}{\nu(i)} = a \cdot \left(\frac{(\theta - 1)(1-a)a^i + 1}{(\theta - 1)(1-a)a^{i+1} + 1}\right)^k. \tag{8.23}$$

Using d'Alembert's convergence condition we should get

$$\frac{\nu(i+1)}{\nu(i)} \leq q < 1. \tag{8.24}$$

If $k = 1$ from (8.23) we have $a \in (0, 1)$.

If $k \geq 2$ using AM-GM inequality from (8.23) we have

$$\text{RHS of (8.23)} \leq \left(\frac{a[(\theta - 1)(1-a)a^{i+1} + 1] + k[(\theta - 1)(1-a)a^i + 1]}{(k+1)[(\theta - 1)(1-a)a^{i+1} + 1]}\right)^{k+1}$$

$$\leq \left(\frac{a+k}{k+1} \cdot T_i\right)^{k+1},$$

where

$$T_i = \frac{(\theta - 1)(1-a)a^i + 1}{(\theta - 1)(1-a)a^{i+1} + 1}, \quad i = 1, 2, ...$$

It is easy to see that $T_{i+1} < T_i$, $i = 1, 2, ...$ Hence in order to obtain (8.24) it is sufficient to solve $\frac{a+k}{k+1} \cdot T_1 < 1$ which is equivalent to

$$(a - 1)\left(a^2 - a + \frac{1}{k(\theta - 1)}\right) < 0. \tag{8.25}$$

From (8.25) we get $a \in (0, 1)$ if $1 < \theta \leq 1 + \frac{4}{k}$ and $a \in (0, a_-^*) \cup (a_+^*, 1)$ if $\theta > 1 + \frac{4}{k}$, where $a_\pm^* = \frac{1 \pm \sqrt{1 - 4((\theta-1)k)^{-1}}}{2}$.

8.3.2 *Case $\theta \leq 1$*

Denote $b_i = (\theta - 1)(1 - a)a^i + 1$. It is easy to see that $b_{i+1} \geq b_i > 0$ for any $i = 1, 2, ...$, $\theta \leq 1$ and $a \in (0, 1)$. Using $b_{i+1} > b_i$ from (8.23) we get (8.24) for any $a \in (0, 1)$.

Thus we have

Theorem 8.3.

(i) If $k = 1$ or $k \geq 2$ and $1 < \theta \leq 1 + \frac{4}{k}$ (resp. $\theta > 1 + \frac{4}{k}$) then for any $a \in (0, 1)$ (resp. $a \in (0, a_-^*) \cup (a_+^*, 1)$) and $\nu(i) = \nu(i, a)$ (see (8.22)) there exists unique translation-invariant Gibbs measure μ_a which corresponds to solution $\{u_i = a^i\}$ of (8.11).

(ii) If $\theta \leq 1$ then for any $a \in (0, 1)$ and $\nu(i) = \nu(i, a)$ there exists unique translation-invariant Gibbs measure μ_a.

Commentaries and references In [266], a countable state space Markov random fields and Markov chains on trees were constructed, and using of entrance laws for specifications Zachary extended and generalized results of [204], [250]. In [97] the Potts model with a countable set Φ of spin values on $\mathbb{Z}^d$ was studied. It was proved that with respect to Poisson distribution on Φ the set of limiting Gibbs measures is not empty.

This chapter is based on papers [98] and [100].

Note that other models with countable set of spin values on Cayley trees are not studied yet.

Chapter 9

Models with uncountable set of spin values

In this chapter we consider models with nearest-neighbor interactions and with the set $[0,1]$ of spin values, on a Cayley tree of order $k \geq 1$. We reduce the problem of describing the "splitting Gibbs measures" of the model to the description of the solutions of some non-linear integral equation. For $k = 1$ we show that the integral equation has a unique solution. In case $k \geq 2$ some models (with the set $[0,1]$ of spin values) which have a unique splitting Gibbs measure are constructed. Also for the Potts model with uncountable set of spin values it is proven that there is unique splitting Gibbs measure. For arbitrary $k \geq 2$ we find a sufficient condition under which the integral equation has unique solution; hence under this condition the corresponding model has unique splitting Gibbs measure. Finally, we construct several models with the set $[0,1]$ of spin values and show that each of the constructed model has at least two translational-invariant Gibbs measures.

9.1 Definitions

We consider models where the spin takes values in the set $[0,1]$, and is assigned to the vertexes of the Cayley tree.

For $A \subseteq V$ a configuration σ_A on A is an arbitrary function $\sigma_A : A \to [0,1]$. Denote $\Omega_A = [0,1]^A$ the set of all configurations on A. A configuration σ on V is then defined as a function $x \in V \mapsto \sigma(x) \in [0,1]$; the set of all configurations is $[0,1]^V$. The (formal) Hamiltonian of the model is:

$$H(\sigma) = -J \sum_{\langle x,y \rangle \in L} \xi_{\sigma(x)\sigma(y)}, \tag{9.1}$$

where $J \in R \setminus \{0\}$ and $\xi : (u,v) \in [0,1]^2 \to \xi_{uv} \in R$ is a given bounded,

231

measurable function. As usual, $\langle x, y \rangle$ stands for nearest neighbor vertices.

Let λ be the Lebesgue measure on $[0,1]$. On the set of all configurations on A the a priori measure λ_A is introduced as the $|A|$ fold product of the measure λ. Here and further on $|A|$ denotes the cardinality of A. We consider a standard sigma-algebra $\mathcal{B}$ of subsets of $\Omega = [0,1]^V$ generated by the measurable cylinder subsets. A probability measure μ on $(\Omega, \mathcal{B})$ is called a Gibbs measure (with Hamiltonian H) if it satisfies the DLR equation, namely for any $n = 1, 2, \ldots$ and $\sigma_n \in \Omega_{V_n}$:

$$\mu\left(\left\{\sigma \in \Omega : \left.\sigma\right|_{V_n} = \sigma_n\right\}\right) = \int_\Omega \mu(\mathrm{d}\omega)\nu^{V_n}_{\omega|_{W_{n+1}}}(\sigma_n),$$

where $\nu^{V_n}_{\omega|_{W_{n+1}}}$ is the conditional Gibbs density

$$\nu^{V_n}_{\omega|_{W_{n+1}}}(\sigma_n) = \frac{1}{Z_n\left(\omega|_{W_{n+1}}\right)} \exp\left(-\beta H\left(\sigma_n \,\|\, \omega|_{W_{n+1}}\right)\right),$$

and $\beta = \frac{1}{T}$, $T > 0$ is temperature. Here and below, W_l stands for a 'sphere' and V_l for a 'ball' on the tree, of radius $l = 1, 2, \ldots$, centered at a fixed vertex x^0 (an origin):

$$W_l = \{x \in V : d(x, x^0) = l\}, \quad V_l = \{x \in V : d(x, x^0) \leq l\};$$

and

$$L_n = \{\langle x, y \rangle \in L : x, y \in V_n\};$$

distance $d(x, y)$, $x, y \in V$, is the length of (i.e., the number of edges in) the shortest path connecting x with y. Ω_{V_n} is the set of configurations in V_n (and Ω_{W_n} that in W_n; see below). Furthermore, $\left.\sigma\right|_{V_n}$ and $\left.\omega\right|_{W_{n+1}}$ denote the restrictions of configurations $\sigma, \omega \in \Omega$ to V_n and W_{n+1}, respectively. Next, $\sigma_n : x \in V_n \mapsto \sigma_n(x)$ is a configuration in V_n and $H\left(\sigma_n \,\|\, \omega|_{W_{n+1}}\right)$ is defined as the sum $H(\sigma_n) + U\left(\sigma_n, \omega|_{W_{n+1}}\right)$ where

$$H(\sigma_n) = -J \sum_{\langle x, y \rangle \in L_n} \xi_{\sigma_n(x)\sigma_n(y)},$$

$$U\left(\sigma_n, \omega|_{W_{n+1}}\right) = -J \sum_{\langle x, y \rangle : \, x \in V_n, y \in W_{n+1}} \xi_{\sigma_n(x)\omega(y)}.$$

Finally, $Z_n\left(\omega|_{W_{n+1}}\right)$ stands for the partition function in V_n, with the boundary condition $\omega|_{W_{n+1}}$:

$$Z_n\left(\omega|_{W_{n+1}}\right) = \int_{\Omega_{V_n}} \exp\left(-\beta H\left(\tilde{\sigma}_n \,\|\, \omega|_{W_{n+1}}\right)\right) \lambda_{V_n}(\mathrm{d}\tilde{\sigma}_n).$$

We use a standard definition of a translation-invariant measure. The main object of study in this chapter is translation-invariant Gibbs measure for the model (9.1) on Cayley tree.

9.2 An integral equation

Recall that we write $x < y$ if the path from x^0 to y goes through x. Call vertex y a direct successor of x if $y > x$ and x, y are nearest neighbors. Denote by $S(x)$ the set of direct successors of x. Observe that any vertex $x \neq x^0$ has k direct successors and x^0 has $k + 1$.

Let $h : \ x \in V \mapsto h_x = (h_{t,x}, t \in [0,1]) \in R^{[0,1]}$ be mapping of $x \in V \setminus \{x^0\}$ with $|h_{t,x}| < C$ where C is a constant which does not depend on t. Given $n = 1, 2, \ldots$, consider the probability distribution $\mu^{(n)}$ on Ω_{V_n} defined by

$$\mu^{(n)}(\sigma_n) = Z_n^{-1} \exp \left(-\beta H(\sigma_n) + \sum_{x \in W_n} h_{\sigma(x),x} \right). \qquad (9.2)$$

Here, as before, $\sigma_n : x \in V_n \mapsto \sigma(x)$ and Z_n is the corresponding partition function:

$$Z_n = \int_{\Omega_{V_n}} \exp \left(-\beta H(\widetilde{\sigma}_n) + \sum_{x \in W_n} h_{\widetilde{\sigma}(x),x} \right) \lambda_{V_n}(d\widetilde{\sigma}_n). \qquad (9.3)$$

Remark 9.1. Note that Z_n is finite, since λ is a probability measure and $\exp(-\beta H(\widetilde{\sigma}_n) + \sum_{x \in W_n} h_{\widetilde{\sigma}(x),x})$ is bounded on Ω_{V_n}.

The probability distributions $\mu^{(n)}$ are compatible if for any $n \geq 1$ and $\sigma_{n-1} \in \Omega_{V_{n-1}}$:

$$\int_{\Omega_{W_n}} \mu^{(n)}(\sigma_{n-1} \vee \omega_n) \lambda_{W_n}(d(\omega_n)) = \mu^{(n-1)}(\sigma_{n-1}). \qquad (9.4)$$

Here $\sigma_{n-1} \vee \omega_n \in \Omega_{V_n}$ is the concatenation of σ_{n-1} and ω_n.

In this case there exists a unique measure μ on Ω_V such that, for any n and $\sigma_n \in \Omega_{V_n}$, $\mu \left(\left\{ \sigma \big|_{V_n} = \sigma_n \right\} \right) = \mu^{(n)}(\sigma_n)$.

Definition 9.1. The measure μ is called *splitting Gibbs measure* corresponding to Hamiltonian (9.1) and function $x \mapsto h_x$, $x \neq x^0$.

Proposition 9.1. *The probability distributions* $\mu^{(n)}(\sigma_n)$, $n = 1, 2, \ldots$, *in (9.2) are compatible iff for any* $x \in V \setminus \{x^0\}$ *the following equation holds:*

$$f(t,x) = \prod_{y \in S(x)} \frac{\int_0^1 \exp(J\beta \xi_{tu}) f(u,y) du}{\int_0^1 \exp(J\beta \xi_{0u}) f(u,y) du}. \qquad (9.5)$$

Here, and below $f(t,x) = \exp(h_{t,x} - h_{0,x})$, $t \in [0,1]$ *and* $du = \lambda(du)$ *is the Lebesgue measure.*

Proof. *Necessity.* Suppose that (9.4) holds; we shall prove (9.5). Substituting (9.2) into (9.4), obtain that for any configurations $\sigma_{n-1}\colon x \in V_{n-1} \mapsto \sigma_{n-1}(x) \in [0,1]$:

$$\frac{Z_{n-1}}{Z_n} \int_{\Omega_{W_n}} \exp\left(\sum_{x\in W_{n-1}} \sum_{y\in S(x)} (J\beta\xi_{\sigma_{n-1}(x)\omega_n(y)} + h_{\omega_n(y),y})\right) \lambda_{W_n}(d\omega_n) = \tag{9.6}$$

$$\exp\left(\sum_{x\in W_{n-1}} h_{\sigma_{n-1}(x),x}\right),$$

where $\omega_n\colon x \in W_n \mapsto \omega_n(x)$.

From (9.6) we get:

$$\frac{Z_{n-1}}{Z_n} \int_{\Omega_{W_n}} \prod_{x\in W_{n-1}} \prod_{y\in S(x)} \exp\left(J\beta\xi_{\sigma_{n-1}(x)\omega_n(y)} + h_{\omega_n(y),y}\right) d(\omega_n(y)) =$$

$$\prod_{x\in W_{n-1}} \exp\left(h_{\sigma_{n-1}(x),x}\right).$$

Consequently, for any $t \in [0,1]$,

$$\prod_{y\in S(x)} \frac{\int_0^1 \exp\left(J\beta\xi_{tu} + h_{u,y}\right)du}{\int_0^1 \exp\left(J\beta\xi_{0u} + h_{u,y}\right)du} = \exp\left(h_{t,x} - h_{0,x}\right),$$

which implies (9.5).

Sufficiency. Suppose that (9.5) holds. It is equivalent to the representations

$$\prod_{y\in S(x)} \int_0^1 \exp\left(J\beta\xi_{tu} + h_{u,y}\right)du = a(x) \exp\left(h_{t,x}\right), t \in [0,1] \tag{9.7}$$

for some function $a(x) > 0, x \in V$. We have

$$\text{LHS of } (9.4) = \frac{1}{Z_n} \exp(-\beta H(\sigma_{n-1}))\lambda_{V_{n-1}}(d(\sigma_n)) \times$$

$$\prod_{x\in W_{n-1}} \prod_{y\in S(x)} \int_0^1 \exp\left(J\beta\xi_{\sigma_{n-1}(x)u} + h_{u,y}\right)du. \tag{9.8}$$

Substituting (9.7) into (9.8) and denoting $A_n(x) = \prod_{x\in W_{n-1}} a(x)$, we get

$$\text{RHS of } (9.8) = \frac{A_{n-1}}{Z_n} \exp(-\beta H(\sigma_{n-1}))\lambda_{V_{n-1}}(d\sigma) \prod_{x\in W_{n-1}} h_{\sigma_{n-1}(x),x}. \tag{9.9}$$

Since $\mu^{(n)}$, $n \geq 1$ is a probability, we should have

$$\int_{\Omega_{V_{n-1}}} \lambda_{V_{n-1}}(d\sigma_{n-1}) \int_{\Omega_{W_n}} \lambda_{W_n}(d\omega_n)\mu^{(n)}(\sigma_{n-1},\omega_n) = 1.$$

Hence from (9.9) we get $Z_{n-1}A_{n-1} = Z_n$, and (9.4) holds. $\qquad\square$

From Proposition 9.1 it follows that for any $h = \{h_x \in R^{[0,1]},\ x \in V\}$ satisfying (9.5) there exists a unique Gibbs measure μ and vice versa. However, the analysis of solutions to (9.5) is not easy. This difficulty depends on the given function ξ. In the next sections we will find several conditions of such functions and give some solutions of the corresponding integral equations.

The following is an example:

9.2.1 *The Potts model with uncountable spin values*

Note that if $\xi_{tu} = \delta_{tu}$ where δ is the Kronecker's symbol then model (9.1) becomes the Potts model with uncountable set of spin values. It is easy to see that

$$\int_0^1 \exp\left(J\beta\delta_{tu}\right) f(u,y)du = \int_0^1 \exp\left(J\beta\delta_{0u}\right) f(u,y)du$$

for any $t \in [0,1], y \in V$. Consequently equation (9.5) has the unique solution $f(t,x) = 1, t \in [0,1], x \in V$ for any $k \geq 1$, $J \in R$, and any $\beta > 0$. Thus we have

Theorem 9.1. *The Potts model with uncountable set of spin values on the Cayley tree of order $k \geq 1$ has a unique splitting Gibbs measure for any $J \in R$ and $\beta > 0$.*

See Remark 8.2 for a comparison of the result with known results about ordinary Potts model.

9.3 Translational-invariant solutions

In this section we consider ξ_{tu} as a continuous function and we are going to solve equation (9.5) in the class of translational-invariant functions $f(t,x)$, i.e., $f(t,x) = f(t)$, for any $x \in V$. For such functions equation (9.5) can be written as

$$f(t) = \left(\frac{\int_0^1 K(t,u)f(u)du}{\int_0^1 K(0,u)f(u)du}\right)^k, \tag{9.10}$$

where $K(t, u) = \exp(J\beta \xi_{tu}) > 0, f(t) > 0, t, u \in [0, 1]$.

We shall find positive continuous solutions to (9.10), i.e., such that

$$f \in C^+[0, 1] = \{f \in C[0, 1] : f(x) > 0\}.$$

Note that equation (9.10) is not linear for any $k \geq 1$.

Define the linear operator $W : C[0, 1] \to C[0, 1]$ by

$$(Wf)(t) = \int_0^1 K(t, u)f(u)du \qquad (9.11)$$

and defined the linear functional $\omega : C[0, 1] \to R$ by

$$\omega(f) \equiv (Wf)(0) = \int_0^1 K(0, u)f(u)du. \qquad (9.12)$$

Then equation (9.10) can be written as

$$f(t) = (A_k f)(t) = \left(\frac{(Wf)(t)}{(Wf)(0)} \right)^k, \quad f \in C^+[0, 1], \ k \geq 1. \qquad (9.13)$$

9.3.1 *Case $k = 1$*

In this subsection we consider $k = 1$ and assume $K(\cdot, \cdot) \in C^+[0, 1]^2$ and $f(\cdot) \in C^+[0, 1]$.

Proposition 9.2. *If $f \in C^+[0, 1]$ is a solution to (9.10) then*

$$f(t) \geq \frac{\kappa^{\min}}{\kappa_0^{\max}}, \quad \text{for any} \ \ t \in [0, 1],$$

where $\kappa^{\min} = \inf_{t, u \in [0, 1]} K(t, u), \quad \kappa_0^{\max} = \sup_{u \in [0, 1]} K(0, u)$.

Proof. Straightforward. $\qquad\qquad\qquad\qquad\qquad\qquad\qquad\qquad\qquad\Box$

Denote

$$C_0^+ = \left\{ h \in C^+[0, 1] : h(t) \geq \frac{\kappa^{\min}}{\kappa_0^{\max}} \right\}.$$

The following lemma is also obvious

Lemma 9.1.

1) The set C_0^+ is a closed and convex subset of the space $C[0, 1]$.

2) The set C_0^+ is invariant w.r.t. operator A_1, i.e., $A_1(C_0^+) \subset C_0^+$.

Lemma 9.2. *Operator A_1 is continuous on C_0^+.*

Proof. Let $f \in C_0^+$ is an arbitrary element and $\{f_n\} \subset C_0^+$ such that $\lim_{n \to \infty} f_n = f$. We shall prove that $\|A_1 f_n - A_1 f\| \to 0$ as $n \to \infty$. We have

$$|A_1 f_n - A_1 f| \leq \frac{W f_n |\omega(f_n) - \omega(f)| + \omega(f_n)|W f_n - W f|}{\omega(f)\omega(f_n)}. \tag{9.14}$$

Since the functional $\omega(\cdot)$ and the operator $W(\cdot)$ are continuous on $C[0,1]$, for any small $\varepsilon > 0$ there exists $n_0 = n_0(\varepsilon) \in N$ such that

$$|\omega(f_n) - \omega(f)| < \varepsilon, \quad \|W f_n - W f\| < \varepsilon, \quad \forall n > n_0.$$

Consequently

$$\|A_1 f_n - A_1 f\| < \frac{\|W f_n\| + \omega(f_n)}{(\omega(f) - \varepsilon)\omega(f)} \cdot \varepsilon. \tag{9.15}$$

There are $M_i, i = 0,1,2$ such that $\omega(f) \geq M_0$, for all $f \in C_0^+$ and

$$\|W(f_n)\| \leq M_1, \quad \omega(f_n) \leq M_2, \quad n \in N.$$

Thus from (9.15) we get

$$\|A_1 f_n - A_1 f\| < \frac{M_1 + M_2}{(M_0 - \varepsilon)M_0} \cdot \varepsilon, \quad n > n_0.$$

This completes the proof. $\qquad\square$

Lemma 9.3. *The set $A_1(C_0^+)$ is relatively compact in $C[0,1]$.*

Proof. By Arzelá-Askoli's theorem (see [265], Chapter III, §3) it suffices to prove that all functions of $A_1(C_0^+)$ are uniformly continuous and there exists $M > 0$ such that

$$|h(t)| \leq M, \quad \forall t \in [0,1] \quad \text{and} \quad \forall h \in A_1(C_0^+).$$

Let $h \in A_1(C_0^+)$ be an arbitrary function, then for a function $f \in C_0^+$ we have $h = A_1 f$. Consequently

$$|h(t)| \leq \frac{\kappa^{\max}}{\kappa_0^{\min}}, \quad \forall t \in [0,1].$$

Now we shall prove that any $h \in A_1(C_0^+)$ is uniformly continuous. For arbitrary $t, t' \in [0,1]$ we have $(h = A_1 f)$

$$|h(t) - h(t')| \leq \frac{1}{\omega(f)} \int_0^1 |K(t,u) - K(t',u)| f(u) du. \tag{9.16}$$

Since the kernel $K(t,u)$ is uniformly continuous on $[0,1]^2$ we conclude that h is also a uniformly continuous function. This completes the proof. $\qquad\square$

By Lemmas 9.1-9.3 and Schauder's theorem (see [187], p.20) one obtains the following

Proposition 9.3. *The equation $A_1 f = f$ has at least one solution in $C^+[0,1]$.*

Now we shall prove that $A_1 f = f$ has a unique solution in $C^+[0,1]$. Since the equation $A_1 f = f$ is equivalent to

$$(Wf)(t) = \omega(f) \cdot f(t), \quad f \in C^+[0,1], \tag{9.17}$$

we shall study eigenvalues of the operator Wf.

Lemma 9.4. *If $\varphi_0 \in C^+[0,1]$ is an eigenfunction of the operator W, i.e., $W\varphi_0 = \lambda_0\varphi_0$, $\lambda_0 > 0$ then there are $a_1 > 0$ and $b_1 > 0$ such that*

$$a_1\omega_1(f)\varphi_0(t) \leq (Wf)(t) \leq b_1\omega_1(f)\varphi_0(t), \quad \forall t \in [0,1], \quad \forall f \in C^+[0,1], \tag{9.18}$$

where $\omega_1(f) = \int_0^1 f(u)du$.

Proof. Note that

$$a\omega_1(f) \leq Wf \leq b\omega_1(f), \quad f \in C[0,1], \tag{9.19}$$

where $a = \min_{t,u \in [0,1]} K(t,u)$ and $b = \max_{t,u \in [0,1]} K(t,u)$. We have

$$a\omega_1(\varphi_0) \leq W\varphi_0 = \lambda_0\varphi_0 \leq b\omega_1(\varphi_0).$$

Hence

$$\frac{\lambda_0\varphi_0(t)}{b\omega_1(\varphi_0)} \leq 1 \leq \frac{\lambda_0\varphi_0(t)}{a\omega_1(\varphi_0)}, \quad \forall t \in [0,1]. \tag{9.20}$$

Using (9.19) and (9.20) we get (9.18) with

$$a_1 = \frac{a\lambda_0}{b\omega_1(\varphi_0)} > 0, \quad b_1 = \frac{b\lambda_0}{a\omega_1(\varphi_0)} > 0. \qquad \square$$

Theorem 9.2. *If $\lambda_0 > 0$ is an eigenvalue of W then $Wf = \lambda_0 f$ has a unique solution $f \in C^+[0,1]$.*

Proof. Assume that there are two solutions $f_0 \in C^+[0,1]$ and $f_1 \in C^+[0,1]$, i.e., $Wf_i = \lambda_0 f_i$, $i = 0,1$. Denote

$$\delta_0 = \sup\{\delta \in [0,\infty) : f_0(t) - \delta f_1(t) \in C^+[0,1]\}.$$

We have

$$W(f_0 - \delta_0 f_1) = W(f_0) - \delta_0 W(f_1) = \lambda_0(f_0 - \delta_0 f_1) > 0.$$

By Lemma 9.4 we get

$$W(f_0 - \delta_0 f_1) \geq a_2 f_0(t) > a_2 \delta_0 f_1(t) \ \text{ with some } \ a_2 > 0,$$

where we used $a_2(f_0(t) - \delta_0 f_1(t)) > 0$.

Consequently

$$\lambda_0(f_0 - \delta_0 f_1) > a_2 \delta_0 f_1(t),$$

i.e.,

$$f_0(t) - \delta_0\left(1 + \frac{a_2}{\lambda_0}\right)f_1(t) > 0 \ \text{ for any } \ t \in [0,1].$$

This contradicts the maximality of δ_0. $\qquad\square$

Theorem 9.3. *The equation $A_1 f = f$ has a unique solution $f \in C^+[0,1]$.*

Proof. By Proposition 9.3 the equation has at least one solution. We shall prove its uniqueness. Assume that $A_1 f = f$ has two solutions f_0 and f_1, then there are $\lambda_0 = \lambda_0(f_0)$ and $\lambda_1 = \lambda_1(f_1)$ such that $W f_i = \lambda_i f_i$, $i = 0, 1$.

By Theorem 9.2 we have $\lambda_0 \neq \lambda_1$. Assume $\lambda_0 < \lambda_1$ (the case $\lambda_0 > \lambda_1$ is similar). Consider

$$h_\delta(t) = f_0(t) - \delta f_1(t), \ \delta \in [0, \infty)$$

and

$$\delta_0 = \sup\{\delta \in [0, \infty) : h_\delta(t) \in C^+[0,1]\}.$$

We have

$$W(h_{\delta_0})(t) = \lambda_0\left(f_0(t) - \delta_0 \frac{\lambda_1}{\lambda_0} f_1(t)\right) > 0, \ \forall t \in [0,1].$$

Since δ_0 is maximal we get $\frac{\lambda_1}{\lambda_0} \leq 1$, i.e., $\lambda_0 \geq \lambda_1$, this contradicts our assumption $\lambda_0 < \lambda_1$. $\qquad\square$

Example 9.1. If $K(t, u) = \alpha(t) + \alpha(u)$ where α is a given function, then one can easily check that

$$f(t) = \frac{\alpha(t) + \sqrt{\int_0^1 \alpha^2(u)du}}{\alpha(0) + \sqrt{\int_0^1 \alpha^2(u)du}}$$

is the unique solution of the equation $A_1 f = f$.

As a corollary of Theorem 9.3 and Proposition 9.1 we get

Theorem 9.4. *For model (9.1) with an arbitrary continuous function ξ_{tu} on $[0,1]^2$, $\forall J \in R$ and for any $\beta > 0$ on the Cayley tree of order 1 there exists a unique splitting Gibbs measure.*

9.3.2 Case $k \geq 2$

The analysis of solutions to (9.10) for $k \geq 2$ is not easy. In this subsection for $k \geq 2$ we shall consider several examples of $K(t, u) > 0$ which are easily solvable.

Proposition 9.4. *The function $f(t) \equiv 1$ is a solution to equation (9.10) iff*

$$\int_0^1 \left(\exp(J\beta\xi_{tu}) - \exp(J\beta\xi_{0u}) \right) du = 0, \ \ t \in [0, 1].$$

Proof. Denoting $A_t = \int_0^1 \exp(J\beta\xi_{tu})f(u)du$, $t \in [0, 1]$ one can see that equation (9.10) is equivalent to

$$f(t) - 1 = (A_t - A_0)A_0^{-k} \left(A_t^{k-1} + ... + A_0^{k-1} \right),$$

this completes the proof. $\square$

Example 9.2. For any $k \geq 1$ we shall consider one simple case: let $\xi_{tu} = a(t) + b(u)$, where $a(t)$ and $b(u)$ are arbitrary given functions. Very simple calculations show that equation (9.10) (even equation (9.5)) has unique solution $f(t, x) = f(t) = \exp(kJ\beta(a(t) - a(0)))$. Thus for the model (9.1) with $\xi_{tu} = a(t) + b(u)$ there is unique splitting Gibbs measure.

Example 9.3. Consider $K(t, u) = \alpha(t) + \alpha(u)$, i.e., $\xi_{tu} = \frac{1}{J\beta} \ln(\alpha(t) + \alpha(u))$, where α is a given positive function on $[0, 1]$. Then the unknown function f can be written as

$$f(t) = \left(\frac{\alpha(t)X + Y}{\alpha(0)X + Y} \right)^k = \left(\frac{\alpha(t)x + 1}{\alpha(0)x + 1} \right)^k,$$

where $X = \int_0^1 f(u)du$ and $Y = \int_0^1 \alpha(u)f(u)du$, $x = \frac{X}{Y}$. It is easy to see that x satisfies the equation

$$x = \frac{\sum_{j=0}^k \frac{a_j}{j!(k-j)!}x^j}{\sum_{j=0}^k \frac{a_{j+1}}{j!(k-j)!}x^j}, \ \ x > 0 \tag{9.21}$$

where $a_i = \int_0^1 \alpha^i(t)dt$, $i = 0, 1, ..., k+1$.

From (9.21) we get

$$\gamma(x) = a_{k+1}x^{k+1} + b_k x^k + b_{k-1}x^{k-1} + ... + b_1 x - 1 = 0, \tag{9.22}$$

where

$$b_j = \frac{k!(2j - k - 1)}{j!(k - j + 1)!} a_j, \ \ j = 1, ..., k.$$

It is well known (see [203], p.28) that the number of positive roots of the polynomial (9.22) does not exceed the number of sign changes of the sequence:

$$a_{k+1}, b_k, b_{k-1}, ..., b_1, -1. \tag{9.23}$$

It is obvious that $a_{k+1} > 0, b_j > 0$ if $j > \frac{k+1}{2}$ and $b_j < 0$ if $j < \frac{k+1}{2}$. Thus the number of positive roots of the polynomial (9.22) is at most one. Since $\gamma(0) = -1$ and $\gamma(+\infty) = +\infty$ we get that (9.22) has a unique positive root.

Consider the Hamiltonian

$$H(\sigma) = -\frac{1}{\beta} \sum_{\langle x,y \rangle \in L} \ln\left(\alpha(\sigma(x)) + \alpha(\sigma(y))\right), \tag{9.24}$$

where α is a given positive, integrable function.

Thus we have proved the following

Theorem 9.5. *For any $k \geq 1$ the model (9.24) has unique splitting Gibbs measure.*

Is there a kernel $K(t, u) > 0$ of equation (9.10) when the equation has more than one solutions? The answer is in the following sections.

9.4 A sufficient condition of uniqueness

In this section we find a sufficient condition under which equation (9.10) has a unique solution.

Then equation (9.10) can be written as

$$f(t) = (A_k f)(t) = ((Bf)(t))^k, \tag{9.25}$$

where

$$(Bf)(t) = \frac{(Wf)(t)}{(Wf)(0)}, \ f \in C_0^+[0,1], \ k \geq 1. \tag{9.26}$$

Denote

$$\mathcal{F}_k = \left\{ f \in C^+[0,1] : f(t) \geq \left(\frac{m}{M_0}\right)^k \right\}, k \in \mathbb{N},$$

where

$$m = \kappa^{\min} = \min_{t,u \in [0,1]} K(t, u), \quad M_0 = \kappa_0^{\max} = \max_{u \in [0,1]} K(0, u).$$

It is easy to see that $\mathcal{F}_k$ is a closed and convex subset of $C[0,1]$. Moreover this set is invariant with respect to operator A_k, i.e., $A_k(\mathcal{F}_k) \subset \mathcal{F}_k$.

Proposition 9.5. *The operator A_k is continuous on $\mathcal{F}_k$ for any $k \geq 2$.*

Proof.   For arbitrary $C > 0$ we denote

$$\mathcal{F}_0 = \left\{ f \in C^+[0,1] : f(t) \geq C,\ \forall t \in [0,1] \right\}.$$

It is obvious that the operator A_1 is continuous on the set $\mathcal{F}_0$ (see Lemma 9.2).

Let $f \in \mathcal{F}_k$ be an arbitrary element and $\{f_n\} \subset \mathcal{F}_k$ such that $\lim_{n \to \infty} f_n = f$. Since the operator A_1 is continuous we have $\lim_{n \to \infty} A_1 f_n = A_1 f$. Consequently, there exists $C_1 > 0$ such that $\|A_1 f_n\| \leq C_1$ for $n \in \mathbb{N}$. Moreover we have

$$(A_1 f)(t) \leq C_2 = \frac{M}{m_0},\ t \in [0,1],$$

where

$$M = \max_{t,u \in [0,1]} K(t,u),\ m_0 = \min_{u \in [0,1]} K(0,u).$$

We have

$$A_k f_n - A_k f = (Bf_n)^k - (Bf)^k = q_{k,n}(t)(A_1 f_n - A_1 f), \qquad (9.27)$$

where

$$q_{k,n}(t) = \sum_{j=0}^{k-1} (A_1 f_n)^{k-j-1}(t)(A_1 f)^j(t) > 0,\ t \in [0,1].$$

Consequently,

$$q_{k,n}(t) \leq C = \sum_{j=0}^{k-1} (C_1)^{k-j-1}(C_2)^j,\ t \in [0,1].$$

Hence

$$\|A_k f_n - A_k f\| \leq C\|A_1 f_n - A_1 f\|,\ n \in \mathbb{N}.$$

Since A_1 is continuous from the last inequality it follows that A_k is continuous on $\mathcal{F}_k$. $\qquad\square$

Denote

$$\mathcal{F}_k^0 = \left\{ f \in C^+[0,1] : \left(\frac{m}{M_0}\right)^k \leq f(t) \leq \left(\frac{M}{m_0}\right)^k \right\}.$$

Proposition 9.6. *Let $k \geq 2$. If $f \in C_0^+[0,1]$ is a solution of the equation $A_k f = f$, then $f \in \mathcal{F}_k^0$.*

Proof.   Straightforward. $\qquad\square$

Proposition 9.7. *Let $k \geq 2$. The set $A_k(\mathcal{F}_k^0)$ is relatively compact in* $C[0,1]$.

Proof. By Arzelá-Askoli's theorem it suffices to prove that the set of functions $A_k(\mathcal{F}_k^0)$ is equi-continuous and there exists $\gamma > 0$ such that

$$h(t) \leq \gamma, \quad \forall t \in [0,1] \ \text{ and } \ \forall h \in A_k(\mathcal{F}_k^0).$$

Let $h \in A_k(\mathcal{F}_k^0)$ be an arbitrary function, we have

$$0 < h(t) \leq \left(\frac{M}{m_0}\right)^k$$

and there exists a function $f \in \mathcal{F}_k^0$ such that $h = A_k f$.

Now we shall prove that $A_k(\mathcal{F}_k^0)$ is equi-continuous. For arbitrary $t, t' \in [0,1]$ we have $(h = A_k f)$

$$|h(t) - h(t')| = |(A_1 f)^k(t) - (A_1 f)^k(t')| =$$

$$\sum_{j=0}^{k-1} (A_1 f)^{k-j-1}(t)(A_1 f)^j(t')|(A_1 f)(t) - (A_1 f)(t')| \leq$$

$$k \left(\frac{M}{m_0}\right)^{k-1} \frac{1}{\omega(f)} \int_0^1 |K(t,u) - K(t',u)|f(u)du \leq$$

$$k \left(\frac{M}{m_0}\right)^{2k-1} \frac{1}{\omega(f)} \int_0^1 |K(t,u) - K(t',u)|du,$$

where $\omega(f)$ is defined in (9.12).

We have

$$\omega(f) \geq m_0 \cdot \left(\frac{m}{M_0}\right)^k, \ f \in \mathcal{F}_k^0.$$

Consequently,

$$|h(t) - h(t')| \leq \frac{k}{m_0} \left(\frac{M_0}{m}\right)^k \left(\frac{M}{m_0}\right)^{2k-1} \int_0^1 |K(t,u) - K(t',u)|du.$$

Since the kernel $K(t,u)$ is uniformly continuous on $[0,1]^2$, we conclude that $A_k(\mathcal{F}_k^0)$ is also equi-continuous. $\square$

By Propositions 9.5-9.7 and Schauder's theorem one gets the following

Theorem 9.6. *The equation $A_k f = f$ has at least one solution in $C_0^+[0,1]$ and the set of all solutions of the equation is a subset in $\mathcal{F}_k^0$.*

9.4.1   *The Hammerstein's non-linear equation*

For every $k \in \mathbb{N}$ we consider an integral operator H_k acting in $C^+[0,1]$ as follows:

$$(H_k f)(t) = \int_0^1 K(t,u) f^k(u) du.$$

If $k \geq 2$ then the operator H_k is a non-linear operator which is called Hammerstein's operator of order k. Moreover the linear operator equation $H_1 f = f$ has a unique positive solution f in $C[0,1]$ (see [140], p.80).

For a non-linear homogeneous operator A it is known that if there is one positive eigenfunction of the operator A then the number of positive eigenfunctions is continuum (see [140], p.186).

Denote

$$\mathcal{M}_0 = \left\{ f \in C^+[0,1] : f(0) = 1 \right\}.$$

Lemma 9.5. *The equation*

$$A_k f = f, \ k \geq 2 \tag{9.28}$$

has a strongly positive solution iff the equation

$$H_k f = \lambda f, \ k \geq 2 \tag{9.29}$$

has a strongly positive solution in $\mathcal{M}_0$.

Proof.   Necessity. Let $f_0 \in C_0^+[0,1]$ be a solution of equation (9.28). We have

$$(W f_0)(t) = \omega(f_0) \sqrt[k]{f_0(t)}.$$

From this equality we get

$$(H_k h)(t) = \lambda_0 h(t),$$

where $h(t) = \sqrt[k]{f_0(t)}$ and $\lambda_0 = \omega(f_0) > 0$.

It is easy to see that $h \in \mathcal{M}_0$ and $h(t)$ is an eigenfunction of the Hammerstein's operator H_k, corresponding to the positive eigenvalue λ_0.

Sufficiency. Let $k \geq 2$ and $h \in \mathcal{M}_0$ be an eigenfunction of the Hammerstein's operator. Then there is a number $\lambda_0 > 0$ such that $H_k h = \lambda_0 h$. From $h(0) = 1$ we get $\lambda_0 = (H_k h)(0) = \omega(h^k)$. Then

$$h(t) = \frac{H_k h}{\omega(h^k)}.$$

From this equality we get $A_k f_0 = f_0$ with $f_0 = h^k \in C_0^+[0,1]$. This completes the proof. $\qquad\square$

Theorem 9.7. *If $k \geq 2$ then every number $\lambda > 0$ is an eigenvalue of the Hammerstein's operator H_k.*

Proof. By Theorem 9.6 and Lemma 9.5 there exist $\lambda_0 > 0$ and $f_0 \in \mathcal{M}_0$ such that

$$H_k f_0 = \lambda_0 f_0.$$

Take $\lambda \in (0, +\infty)$, $\lambda \neq \lambda_0$. Define function $h_0(t) \in C_0^+[0,1]$ by

$$h_0(t) = \sqrt[k-1]{\frac{\lambda}{\lambda_0}} f_0(t), \quad t \in [0,1].$$

Then

$$H_k h_0 = H_k \left(\sqrt[k-1]{\frac{\lambda}{\lambda_0}} f_0 \right) = \lambda h_0.$$

This completes the proof. $\qquad\square$

Denote

$$\mathcal{K} = \left\{ f \in C^+[0,1] : M \cdot \min_{t \in [0,1]} f(t) \geq m \cdot \max_{t \in [0,1]} f(t) \right\},$$

$$\mathcal{P}_k = \left\{ f \in C[0,1] : \frac{m}{M} \cdot \left(\frac{1}{M}\right)^{\frac{1}{k-1}} \leq f(t) \leq \frac{M}{m} \cdot \left(\frac{1}{m}\right)^{\frac{1}{k-1}} \right\}, \ k \geq 2.$$

Proposition 9.8. *Let $k \geq 2$.*
a) The following holds

$$H_k(C^+[0,1]) \subset \mathcal{K}.$$

b) If a function $f_0 \in C_0^+[0,1]$ is a solution of the equation

$$H_k f = f \tag{9.30}$$

then $f_0 \in \mathcal{P}_k$.

Proof. a) Let $h \in H_k(C^+[0,1])$ be an arbitrary function. Then there exists a function $f \in C^+[0,1]$ such that $h = H_k f$. Since h is continuous on $[0,1]$, there are $t_1, t_2 \in [0,1]$ such that

$$h_{\min} = \min_{t \in [0,1]} h(t) = h(t_1) = (H_k f)(t_1),$$

$$h_{\max} = \max_{t \in [0,1]} h(t) = h(t_2) = (H_k f)(t_2).$$

Hence

$$h_{\min} \geq m \int_0^1 f^k(u)du \geq m \int_0^1 \frac{K(t_2, u)}{M} f^k(u)du = \frac{m}{M} h_{max},$$

i.e., $h \in \mathcal{K}$.

b) Let $f \in C_0^+[0,1]$ be a solution of equation (9.30). Then we have $\|f\| \leq M\|f\|^k$. Consequently,

$$\|f\| \geq \left(\frac{1}{M}\right)^{\frac{1}{k-1}}.$$

By property a) we have

$$f(t) \geq f_{\min} = \min_{t \in [0,1]} f(t) \geq \frac{m}{M}\|f\|.$$

Then we obtain

$$f(t) \geq \frac{m}{M} \left(\frac{1}{M}\right)^{\frac{1}{k-1}}.$$

Also we have

$$f(t) = (H_k f)(t) \geq m \int_0^1 f^k(u)du \geq m f_{\min}^k.$$

Then $f_{\min} \geq m f_{\min}^k$, i.e.,

$$f_{\min} \leq \left(\frac{1}{m}\right)^{\frac{1}{k-1}}.$$

Hence by the property a) we get

$$f(t) \leq f_{\max} \leq \frac{M}{m} f_{\min} \leq \frac{M}{m} \left(\frac{1}{m}\right)^{\frac{1}{k-1}}.$$

Thus we have $f \in \mathcal{P}_k$. $\qquad\qquad\qquad\qquad\qquad\qquad\qquad\qquad\square$

9.4.2   *The uniqueness of fixed point of the operators A_k and H_k*

Now we shall prove that $A_k f = f$ and $H_k f = f$ have a unique solution in $C_0^+[0,1]$.

Lemma 9.6. *Assume function $f \in C[0,1]$ changes its sign on $[0,1]$. Then for every $a \in \mathbb{R}$ the following inequality holds*

$$\|f_a\| \geq \frac{1}{n+1}\|f\|, \quad n \in \mathbb{N},$$

where $f_a = f_a(t) = f(t) - a$, $t \in [0,1]$.

Proof. By conditions of lemma there are $t_1, t_2 \in [0,1]$ such that

$$f_{\min} = f(t_1) < 0, \quad f_{\max} = f(t_2) > 0.$$

In case $a = 0$ the proof is obvious. We assume $a > 0$

a) Let $|f_{\min}| \geq f_{\max}$. Then $\|f\| = |f_{\min}| = |f(t_1)|$. Hence

$$\|f_a\| = \max\{|f(t_1) - a|, |f(t_2) - a|\} = |f(t_1) - a| >$$

$$|f(t_1)| = \|f\| \geq \frac{1}{n+1}\|f\|, \ n \in \mathbb{N}.$$

b) Let $|f_{\min}| < f_{\max}$ and $\frac{1}{2}\|f\| \geq a$. Then $\|f\| = f_{\max} = f(t_2)$ and $\|f\| - a \geq a > 0$. Consequently,

$$\|f_a\| = \max\{|f(t_1) - a|, |f(t_2) - a|\} \geq |f(t_2) - a| =$$

$$\|f\| - a \geq \frac{1}{2}\|f\| \geq \frac{1}{n+1}\|f\|, \ n \in \mathbb{N}.$$

c) Let $|f_{\min}| < f_{\max}$ and $\frac{1}{2}\|f\| < a$. Then $\|f\| = f(t_2)$ and

$$\|f_a\| = \max\{|f(t_1) - a|, |f(t_2) - a|\} \geq |f(t_1) - a| > a >$$

$$\frac{1}{2}\|f\| \geq \frac{1}{n+1}\|f\|, \ n \in \mathbb{N}.$$

Thus for $a > 0$ the proof is complete. For $a < 0$ we put $g_a(t) = g(t) - a'$ with $g(t) = -f(t)$ and $a' = -a > 0$. Then

$$\|f_a\| = \|g_a\| \geq \frac{1}{n+1}\|g\| = \frac{1}{n+1}\|f\|, \ n \in \mathbb{N}.$$

This completes the proof. $\qquad\square$

Theorem 9.8. *Let $k \geq 2$. If the kernel $K(t, u)$ satisfies the condition*

$$\left(\frac{M}{m}\right)^k - \left(\frac{m}{M}\right)^k < \frac{1}{k}, \tag{9.31}$$

then the operator H_k has a unique fixed point in $C_0^+[0,1]$.

Proof. By Theorem 9.7 the Hammerstein's equation $H_k f = f$ has at least one solution. Assume that there are two solutions $f_1 \in C_0^+[0,1]$ and $f_2 \in C_0^+[0,1]$, i.e., $H_k f_i = f_i$, $i = 1, 2$. Denote $f(t) = f_1(t) - f_2(t)$. Then by Theorem 46.6 of [141] the function $f(t)$ changes its sign on $[0,1]$.

From Lemma 9.6 we get

$$\max_{t \in [0,1]} \left| f(t) - \frac{k}{2}(\gamma_1 + \gamma_2) \int_0^1 f(s)ds \right| \geq \frac{1}{2}\|f\|,$$

where

$$\gamma_1 = \left(\frac{m}{M}\right)^k, \quad \gamma_2 = \left(\frac{M}{m}\right)^k.$$

By a mean value theorem we have

$$f(t) = \int_0^1 K(t, u) k \xi^{k-1}(u) f(u) du,$$

here $\xi \in C^+[0, 1]$ and

$$\min\{f_1(t), f_2(t)\} \le \xi(t) \le \max\{f_1(t), f_2(t)\}, \ t \in [0, 1].$$

By Proposition 9.8 we have $\xi \in \mathcal{P}_k$, i.e.,

$$\frac{m}{M}\left(\frac{1}{M}\right)^{\frac{1}{k-1}} \le \xi(t) \le \frac{M}{m}\left(\frac{1}{m}\right)^{\frac{1}{k-1}}, \ t \in [0, 1].$$

Hence

$$\gamma_1 \le K(t, u) \xi^{k-1}(u) \le \gamma_2, \ t, u \in [0, 1].$$

Therefore

$$\left| k \cdot K(t, u) \xi^{k-1}(u) - \frac{\gamma_1 + \gamma_2}{2} \right| \le \frac{\gamma_2 - \gamma_1}{2}.$$

Then

$$\left| f(t) - \frac{k}{2}(\gamma_1 + \gamma_2) \int_0^1 f(u) du \right| \le \frac{k}{2}(\gamma_2 - \gamma_1) \|f\|. \tag{9.32}$$

Assume the kernel $K(t, u)$ satisfies condition (9.31). Then $k(\gamma_2 - \gamma_1) < 1$ and the inequality (9.32) contradicts to Lemma 9.6. This completes the proof. $\qquad\square$

Theorem 9.9. *Let $k \ge 2$. If the kernel $K(t, u)$ satisfies the condition (9.31), then for every $\lambda > 0$ the Hammerstein's equation $H_k f = \lambda f$ has unique solution in $C_0^+[0, 1]$.*

Proof. Clearly the equation $H_k f = \lambda f$ is equivalent to the following equation

$$\int_0^1 K_\lambda(t, u) f^k(u) du = f(t), \tag{9.33}$$

where $K_\lambda(t, u) = \frac{1}{\lambda} K(t, u)$. The kernel $K_\lambda(t, u)$ satisfies the condition (9.31) with $\tilde{m} = \frac{m}{\lambda}$ and $\tilde{M} = \frac{M}{\lambda}$. Consequently, by Theorem 9.8 it follows that equation (9.33) has unique solution in $C_0^+[0, 1]$. $\qquad\square$

Theorem 9.10. *Let $k \geq 2$. If the kernel $K(t,u)$ satisfies the condition (9.31), then the equation $A_k f = f$ has unique solution in $C_0^+[0,1]$.*

Proof. Assume there are two solutions $f_1, f_2 \in C^+[0,1]$, $f_1 \neq f_2$, i.e., $A_k f_i = f_i$, $i = 1, 2$. By Lemma 9.5 the functions $h_i(t) = \sqrt[k]{f_i(t)}$, $t \in [0,1]$ are solutions of the Hammerstein's equation, i.e.,

$$H_k h_i = \lambda_i h_i, \quad i = 1, 2,$$

where $\lambda_i = \omega(f_i) > 0$ and $h_i \in \mathcal{M}_0$. On the other hand, Theorem 9.9 implies that $\lambda_1 \neq \lambda_2$. Let $h_0(t) \in C^+[0,1]$ be a fixed point of the Hammerstein's operator H_k. Then by Theorems 9.7 and 9.9 we get

$$h_i = \sqrt[k-1]{\lambda_i} h_0(t), \ i = 1, 2.$$

Consequently,

$$\frac{f_1(t)}{f_2(t)} = \gamma^k, \quad \text{with} \ \ \gamma = \sqrt[k-1]{\frac{\lambda_1}{\lambda_2}}.$$

Using this equality we obtain

$$f_1(t) = (A_k f_1)(t) = A_k(\gamma^k f_2) = A_k f_2(t) = f_2(t).$$

This completes the proof. $\qquad\square$

Consider the following Hamiltonian

$$H(\sigma) = -J \sum_{\langle x, y \rangle \in L} \xi_{\sigma(x)\sigma(y)} = - \sum_{\langle x, y \rangle \in L} \ln K(\sigma(x), \sigma(y)), \qquad (9.34)$$

where $J \in R \setminus \{0\}$ and $K(t,u)$ satisfies the condition (9.31). Then as a corollary of Proposition 9.5 and Theorem 9.10 we get the following

Theorem 9.11. *Let $k \geq 2$. If the function $K(t,u)$ of the Hamiltonian (9.34) satisfies the condition (9.31), then the model (9.34) has unique translational invariant Gibbs measure.*

Example 9.4. It is easy to see that the condition (9.31) is satisfied iff

$$\frac{M}{m} \leq \eta_k = \sqrt[k]{\frac{1 + \sqrt{4k^2 + 1}}{2k}}, \ k \geq 2.$$

Consider the following function

$$K(t,u) = \sum_{i=1}^{m} \sum_{j=1}^{n} c_{ij} t^i u^j + a, \ c_{ij} \geq 0, \ a > 0. \qquad (9.35)$$

For this function we have $m = a$, $M = \sum_{i=1}^{m} \sum_{j=1}^{n} c_{ij} + a$. The following is obvious

a) If $\frac{1}{a} \sum_{i=1}^{m} \sum_{j=1}^{n} c_{ij} \leq \eta_k - 1$ then for function (9.35) the condition (9.31) is satisfied.

b) If $\frac{1}{a} \sum_{i=1}^{m} \sum_{j=1}^{n} c_{ij} > \eta_k - 1$ then for function (9.35) the condition (9.31) is not satisfied.

9.4.3   *Physical interpretation*

Here we shall give some physical interpretation of the condition (9.31) and the Hamiltonian. Note that our interaction functions are defined as logarithms of certain kernels on which we put appropriate conditions.

It is known that the spin systems on trees have produced the first and most tractable examples of certain qualitative phenomena. The function ξ_{xy} is interpreted as *energy* function, which is as usual nonconstant, symmetric and continuous. In [194] the authors studied several models on general infinite trees, including the classical Heisenberg and Potts models. The model (9.1) (in particular (9.34)) naturally generalizes known models (for example rotor [194], spherical [270] and many other models) with nearest-neighbor interactions which have uncountable set of spin values.

The condition (9.31) can be reformulated with respect to temperature, or with respect to interaction parameter J, or with respect to energy function ξ_{xy}. Indeed

Condition on the temperature. We have

$$\frac{M}{m} = \frac{\max_{t,u \in [0,1]} K(t,u)}{\min_{t,u \in [0,1]} K(t,u)} = \frac{\max_{t,u \in [0,1]} e^{J\beta\xi_{tu}}}{\min_{t,u \in [0,1]} e^{J\beta\xi_{tu}}}$$

$$= \begin{cases} \exp\left(J\beta E\right), & \text{if } J > 0, \\ \exp\left(-J\beta E\right), & \text{if } J < 0 \end{cases} = \exp(|J|\beta E) < \eta_k, \qquad (9.36)$$

where $E = \max_{t,u \in [0,1]} \xi_{tu} - \min_{t,u \in [0,1]} \xi_{tu}$.

This gives

$$T = \beta^{-1} > \frac{|J|E}{\ln \eta_k},$$

i.e., if the temperature is greater than $\frac{|J|E}{\ln \eta_k}$ then there exists unique translation invariant Gibbs measure.

Condition on the interaction parameter J. Fix $T > 0$ and find J from our condition:

$$|J| < \frac{T}{E} \ln \eta_k, \quad \text{i.e.,} \quad -\frac{T}{E} \ln \eta_k < J < \frac{T}{E} \ln \eta_k.$$

9.5 Examples of Hamiltonians with non-unique Gibbs measure

9.5.1 *Case $k = 2$*

Consider the case $k = 2$ in the model (9.1) and

$$\xi_{t,u} = \frac{1}{\beta J} \ln \left(1 + \frac{14}{15} \cdot \sqrt[5]{4 \left(t - \frac{1}{2} \right) \left(u - \frac{1}{2} \right)} \right), \quad t, u \in [0, 1].$$

Then, for the kernel $K(t, u)$ of the Hammerstein's integral operator H_2 we have

$$K(t, u) = 1 + \frac{14}{15} \cdot \sqrt[5]{4 \left(t - \frac{1}{2} \right) \left(u - \frac{1}{2} \right)}.$$

Proposition 9.9. *The Hammerstein's operator H_2 :*

$$(H_2 f)(t) = \int_0^1 K(t, u) f^2(u) du$$

in the space $C[0, 1]$ has at least two strictly positive fixed points.

Proof. a) Let $f_1(t) \equiv 1$. Then we have

$$(H_2 f_1)(t) = 1 + \frac{14}{15} \cdot \sqrt[5]{4 \left(t - \frac{1}{2} \right)} \cdot \int_0^1 \left(u - \frac{1}{2} \right)^{\frac{1}{5}} du = 1 = f_1(t), \quad t \in [0, 1].$$

b) Denote

$$f_2(t) = \frac{3}{4} + \sqrt{\frac{21}{5}} \cdot \frac{\sqrt[5]{2}}{4} \cdot \left(t - \frac{1}{2} \right)^{\frac{1}{5}}, \quad t \in [0, 1].$$

Then $f_2 \in C[0, 1]$ and the function $f_2(t)$ is strictly positive. Put

$$a = \frac{14}{15} \cdot \sqrt[5]{4}, \quad b = \sqrt{\frac{21}{5}} \cdot \frac{\sqrt[5]{2}}{4}.$$

We have

$$H_2 f_2 = h_1(t) + h_2(t) + h_3(t) + \gamma,$$

where

$$h_1(t) = ab^2 \cdot \sqrt[5]{t - \frac{1}{2}} \cdot \int_0^1 \sqrt[5]{\left(u - \frac{1}{2} \right)^3} du,$$

$$h_2(t) = \frac{3ab}{2} \cdot \sqrt[5]{t - \frac{1}{2}} \cdot \int_0^1 \sqrt[5]{\left(u - \frac{1}{2}\right)^2}\, du,$$

$$h_3(t) = \frac{9a}{16} \cdot \sqrt[5]{t - \frac{1}{2}} \cdot \int_0^1 \sqrt[5]{u - \frac{1}{2}}\, du,$$

$$\gamma = \int_0^1 f_2^2(u)\, du.$$

It is clear that

$$h_1(t) = h_3(t) \equiv 0.$$

For the function $h_2(t)$ we obtain

$$h_2(t) = \frac{3ab}{2} \cdot \sqrt[5]{t - \frac{1}{2}} \cdot \int_{-1/2}^{1/2} u^{\frac{2}{5}}\, du = \frac{15ab}{14\sqrt[5]{4}} \cdot \sqrt[5]{t - \frac{1}{2}}\ .$$

Observe that

$$\gamma = \frac{5b^2}{7\sqrt[5]{4}} + \frac{9}{16}.$$

Consequently we have

$$(H_2 f_2)(t) = h_2(t) + \gamma = \frac{15ab}{14\sqrt[5]{4}} \cdot \sqrt[5]{t - \frac{1}{2}} + \frac{5b^2}{7\sqrt[5]{4}} + \frac{9}{16} =$$

$$\sqrt{\frac{21}{5}} \cdot \frac{\sqrt[5]{2}}{4} \cdot \sqrt[5]{t - \frac{1}{2}} + \frac{3}{4} = f_2(t). \qquad \square$$

Denote by μ_1 and μ_2 the translation-invariant Gibbs measures which by Proposition 9.1 correspond to solutions $f_1(t) = 1$ and $f_2(t) = \frac{3}{4} + \sqrt{\frac{21}{5}} \cdot \frac{\sqrt[5]{2}}{4} \cdot \left(t - \frac{1}{2}\right)^{\frac{1}{5}}$.

Thus we have proved the following

Theorem 9.12. *The model*

$$H(\sigma) = -\frac{1}{\beta} \sum_{\substack{<x,y> \\ x,y \in V}} \ln\left(1 + \frac{14}{15} \cdot \sqrt[5]{4\left(\sigma(x) - \frac{1}{2}\right)\left(\sigma(y) - \frac{1}{2}\right)}\right), \quad \sigma \in \Omega_V$$

on the Cayley tree Γ^2 has at least two translation-invariant Gibbs measures μ_1, μ_2.

9.5.2 *Case $k = 3$*

Now we shall consider the case $k = 3$ and

$$\xi_{t,u} = \frac{1}{\beta J}\ln\left(1 + \frac{1}{2}\cdot\sqrt[7]{4\left(t - \frac{1}{2}\right)\left(u - \frac{1}{2}\right)}\right), \quad t, u \in [0,1].$$

Then, for the kernel $K(t,u)$ of the operator H_3 we have

$$K(t,u) = 1 + \frac{1}{2}\sqrt[7]{4\left(t - \frac{1}{2}\right)\left(u - \frac{1}{2}\right)}.$$

Proposition 9.10. *The operator H_3 :*

$$(H_3 f)(t) = \int_0^1 \left(1 + \frac{1}{2}\cdot\sqrt[7]{4\left(t - \frac{1}{2}\right)\left(u - \frac{1}{2}\right)}\right) f^3(u)du$$

in the space $C[0,1]$ has at least two strictly positive fixed points.

Proof. a) Let $f_1(t) \equiv 1$. Then

$$(H_3 f_1)(t) = 1 + \frac{1}{2}\cdot\sqrt[7]{4\left(t - \frac{1}{2}\right)}\cdot\int_{-\frac{1}{2}}^{\frac{1}{2}} u^{\frac{1}{7}}du = 1 = f_1(t), \quad t \in [0,1].$$

b) We define the function f_2:

$$f_2(t) = \frac{1}{2}\left(\sqrt{\frac{57}{17}} + \sqrt{\frac{33}{119}}\cdot\sqrt[7]{2\left(t - \frac{1}{2}\right)}\right), \quad t \in [0,1].$$

Then $f_2 \in C[0,1]$ and the function $f_2(t)$ is strictly positive. Put

$$a = \frac{1}{2}\sqrt{\frac{57}{17}}, \quad b = \frac{1}{2}\sqrt{\frac{33}{119}}.$$

We have

$$(H_3 f_2)(t) = h_1(t) + h_2(t) + h_3(t) + h_4(t) + \gamma,$$

where

$$h_1(t) = \frac{a^3}{2}\varphi(t)\cdot\int_0^1 \sqrt[7]{u - \frac{1}{2}}\,du,$$

$$h_2(t) = \frac{3a^2 b}{2}\cdot\sqrt[7]{2}\varphi(t)\cdot\int_0^1 \sqrt[7]{\left(u - \frac{1}{2}\right)^2}\,du,$$

$$h_3(t) = \frac{3ab^2}{2} \cdot \sqrt[7]{4}\varphi(t) \cdot \int_0^1 \sqrt[7]{\left(u - \frac{1}{2}\right)^3}\, du,$$

$$h_4(t) = \frac{b^3}{2} \cdot \sqrt[7]{8}\varphi(t) \cdot \int_0^1 \sqrt[7]{\left(u - \frac{1}{2}\right)^4}\, du,$$

$$\gamma = \int_0^1 f_2^3(u)\, du.$$

Here $\varphi(t) = \sqrt[7]{4\left(t - \frac{1}{2}\right)}$, $t \in [0, 1]$.

It is clear that

$$h_1(t) = h_3(t) \equiv 0.$$

For the functions $h_2(t)$ and $h_4(t)$ we obtain that

$$h_2(t) = \frac{3a^2 b \sqrt[7]{2}}{2} \cdot \varphi(t) \int_{-\frac{1}{2}}^{\frac{1}{2}} u^{\frac{2}{7}}\, du = \frac{7a^2 b}{6\sqrt[7]{2}} \cdot \varphi(t),$$

$$h_4(t) = \frac{b^3 \sqrt[7]{8}}{2} \cdot \varphi(t) \int_{-\frac{1}{2}}^{\frac{1}{2}} u^{\frac{4}{7}}\, du = \frac{7b^3}{22\sqrt[7]{2}} \cdot \varphi(t).$$

Observe that

$$\gamma = a^3 + 3ab^2 \sqrt[7]{4} \cdot \int_{-\frac{1}{2}}^{\frac{1}{2}} u^{\frac{2}{7}}\, du = a^3 + \frac{7ab^2}{3} = a.$$

Consequently, we have

$$H_3 f_2 = h_2 + h_4 + a = a + \frac{7b}{2\sqrt[7]{2}}\left(\frac{a^2}{3} + \frac{b^2}{11}\right)\varphi(t) = a + \frac{b}{\sqrt[7]{2}}\varphi(t) = f_2(t). \qquad \square$$

From Proposition 9.10 and Proposition 9.1 we get

Theorem 9.13. *The model*

$$H(\sigma) = -\frac{1}{\beta} \sum_{\substack{<x,y> \\ x,y \in V}} \ln\left(1 + \frac{1}{2}\sqrt[7]{4\left(\sigma(x) - \frac{1}{2}\right)\left(\sigma(y) - \frac{1}{2}\right)}\right) \ , \quad \sigma \in \Omega_V$$

on the Cayley tree Γ^3 *has at least two translation-invariant Gibbs measures.*

9.5.3 *Case $k \geq 4$*

Let $k \in \mathbb{N}$ and $k \geq 2$. We consider sequences of continuous functions $P_n(x)$ $(n \in \mathbb{N})$ and $Q_m(x)$ $(m \in \mathbb{N},\ m > k)$ defined by

$$P_n(x) \equiv P_{n,k}(x) = \left(1 + \frac{x^{n-1}}{2}\right)^{k+1} - \left(1 - \frac{x^{n-1}}{2}\right)^{k+1}, \quad x \in R,$$

$$Q_m(x) \equiv Q_{m,k}(x) = (k+1)x^{m-k}, \quad m > k,\ x \in R.$$

Proposition 9.11. *Let $k \geq 2$. Then*

$$P_n(1) > Q_n(1), \tag{9.37}$$

for any $n \in \mathbb{N},\ n > k$.

Proof. Let $k \geq 2$ and $n > k$. We have

$$P_n(1) = \mu_k = \frac{3^{k+1} - 1}{2^{k+1}}, \quad Q_n = \eta_k = k+1.$$

In the case $k = 2$ we obtain, that

$$P_n(1) = \frac{13}{4} > Q_n(1) = 3.$$

We now suppose that the inequality (9.37) holds for $k = m > 2$. Then we show that the inequality (9.37) is also true for $k = m + 1$.

Obviously, that

$$\mu_{m+1} = \frac{3^{(m+1)+1} - 1}{2^{(m+1)+1}} > \frac{3^{(m+1)+1} - 3}{2^{m+1} \cdot 2} = \frac{3^{m+1} - 1}{2^{m+1}} \cdot \frac{3}{2}$$

$$= \mu_m \cdot \frac{3}{2} > (m+1) \cdot \frac{3}{2} > m + 2 = \eta_{m+1},$$

i.e., $\mu_{m+1} > \eta_{m+1}$. Thus we get

$$P_n(1) > Q_n(1)$$

for any $k \geq 2$ and $n > k$. $\square$

Proposition 9.12. *Let $k \geq 2$. The equation*

$$\left(1 + \frac{x}{2}\right)^{k+1} - \left(1 - \frac{x}{2}\right)^{k+1} - (k+1)x = 0, \quad x \geq 0 \tag{9.38}$$

has a unique solution $x = 0$.

Proof.   Let $k \geq 2$. Define the continuous function $\varphi(x)$:

$$\varphi(x) = \left(1 + \frac{x}{2}\right)^{k+1} - \left(1 - \frac{x}{2}\right)^{k+1} - (k+1)x, \quad x \in [0, \infty).$$

We have

$$\varphi'(x) = (k+1)\left(\frac{1}{2}\left(1 + \frac{x}{2}\right)^k + \frac{1}{2}\left(1 - \frac{x}{2}\right)^k - 1\right).$$

However,

$$\left(1 + \frac{x}{2}\right)^k + \left(1 - \frac{x}{2}\right)^k > 2, \quad \text{for all } x \in (0, \infty).$$

Consequently, we have $\varphi'(x) > 0$ for all $x \in (0, \infty)$, i.e., the function $\varphi(x)$ is increasing on $[0, \infty)$. So, the zero is a unique solution of equation (9.38).
$\square$

Proposition 9.13. *Let $k \geq 2$. Then for each $n \in \mathbb{N}$, $n > k$ the equation*

$$P_n(x) - Q_n(x) = 0 \tag{9.39}$$

has at least one solution $\xi = \xi(k; n)$ in (0,1).

Proof.   Let $k \geq 2$ and $n > k$. We have

$$\lim_{x \to 0+} \frac{P_n(x)}{Q_n(x)} = \frac{1}{k+1} \lim_{x \to 0+} \frac{\left(1 + \frac{x^{n-1}}{2}\right)^{k+1} - \left(1 - \frac{x^{n-1}}{2}\right)^{k+1}}{x^{n-k}} =$$

$$\frac{1}{k+1} \lim_{x \to 0+} \frac{\left(\left(1 + \frac{x^{n-1}}{2}\right) - \left(1 - \frac{x^{n-1}}{2}\right)\right) \sum_{j=0}^{k} \left(1 + \frac{x^{n-1}}{2}\right)^{k-j} \left(1 - \frac{x^{n-1}}{2}\right)^{j}}{x^{n-k}}$$

$$= \frac{1}{k+1} \lim_{x \to 0+} x^{k-1} \cdot \sum_{j=0}^{k} \left(1 + \frac{x^{n-1}}{2}\right)^{k-j} \left(1 - \frac{x^{n-1}}{2}\right)^{j} = 0.$$

Since the functions $P_n(x)$ and $Q_n(x)$ are continuous, there exists a number $\delta > 0$ such that

$$P_n(x) < Q_n(x) \quad \text{for all } x \in (0, \delta).$$

However $P_n(0) = Q_n(0) = 0$ and by Proposition 9.11 we have $P_n(1) > Q_n(1)$. Consequently, there exists a number $\xi = \xi(k; n) \in (0, 1)$ such that $P_n(\xi(k; n)) = Q_n(\xi(k; n)) = 0$.

Let $k \geq 2$ be a fixed number and suppose that $\{\xi(k;n)\}_{n>k} \subset (0,1)$ – some set of solutions of the following system of equations:

$$P_n(x) - Q_n(x) = 0, \ n \in \mathbb{N}, \ n > k.$$

We have $0 < \xi(k;n) < 1$ for all $n \in \mathbb{N}, \ n > k$. Consequently $0 < \xi(k;n)^{n-1} < 1$ for all $n > k$. Then there exists upper limit of the sequence $\xi(k;n)^{n-1}, \ n > k$, i.e., there exists a subsequence $\alpha_p = \xi(k;n_p)^{n_p-1}, \ p \in \mathbb{N}$ of the sequence $\xi(k;n)^{n-1}, \ n > k$ such that

$$\alpha = \lim_{n \to \infty} \sup \xi(k;n)^{n-1} = \lim_{p \to \infty} \xi(k;n_p)^{n_p-1} = \lim_{p \to \infty} \alpha_p.$$

Obviously, $0 \leq \alpha \leq 1$. Define the sequence $\beta_p, \ p \in \mathbb{N}$ by

$$\beta_p = \xi(k;n_p), \ p \in \mathbb{N}.$$

Then

$$\alpha_p = \beta_p^{n_p-1}, \ p \in \mathbb{N}. \qquad \square$$

Lemma 9.7. $\alpha = \lim_{p \to \infty} \alpha_p = 0.$

Proof. a) Assume $\alpha = 1$. Put

$$\beta = \lim_{p \to \infty} \sup \xi(k;n_p) = \lim_{p \to \infty} \sup \beta_p.$$

Then, there exists a subsequence $\{\beta_{p_q}\}_{q \in \mathbb{N}} \subset \{\beta_p\}_{p \in \mathbb{N}}$ such that

$$\lim_{q \to \infty} \beta_{p_q} = \beta.$$

We have $0 \leq \beta \leq 1$. If $0 \leq \beta < 1$, there exists $q_0 \in \mathbb{N}$ such that $\beta_{p_q} < \frac{1+\beta}{2}$ for all $q > q_0$. From that

$$0 \leq \alpha_{p_q} \leq \left(\frac{1+\beta}{2}\right)^{n_{p_q}-1}, \ q \in \mathbb{N}, \ q > q_0.$$

Therefore $\alpha = \lim_{q \to \infty} \alpha_{p_q} = 0$. The last equality is a contradiction to the assumption $\alpha = 1$. However, we obtain that $\beta = 1$. Then from the equality

$$P_{n_{p_q}}\left(\xi(k;n_{p_q})\right) = Q_{n_{p_q}}\left(\xi(k;n_{p_q})\right), \ q \in \mathbb{N} \tag{9.40}$$

as $q \to \infty$ we observe that

$$\left(1 + \frac{1}{2}\right)^{k+1} - \left(1 - \frac{1}{2}\right)^{k+1} = k+1,$$

i.e.,

$$P_m(1) = Q_m(1), \quad m > k.$$

The last equality is a contradiction to the assertion of Proposition 9.11. Thus, we have proved that $\alpha \neq 1$.

b) Assume that $0 < \alpha < 1$. In the case $0 \leq \beta < 1$ we get $\alpha = 0$. So $\beta = 1$. Then from (9.40) as $q \to \infty$ we get

$$\left(1 + \frac{\alpha}{2}\right)^{k+1} - \left(1 - \frac{\alpha}{2}\right)^{k+1} = (k+1)\alpha.$$

The last equality contradicts to the assertion of Proposition 9.12. Thus, we have proved that $\alpha \notin (0,1)$. Consequently, $\alpha = 0$. $\qquad\square$

Corollary 9.1. $\lim\limits_{p \to \infty} \beta_p = 1$.

Proof. From the equality (9.40) we get

$$\beta_p = \xi(k; n_p) = \sqrt[k-1]{\dfrac{k+1}{\sum\limits_{j=0}^{k} \left(1 + \frac{\alpha_p}{2}\right)^{k-j} \left(1 - \frac{\alpha_p}{2}\right)^{j}}} \,, \quad p \in \mathbb{N}.$$

Hence by Lemma 9.7 it follows that

$$\lim_{p \to \infty} \beta_p = 1.$$

$\qquad\square$

Define the sequence $C_n = C_n(k)$, $n > k \geq 2$:

$$C_n = \frac{\xi(k; n)^{3n-k-2}}{\frac{1}{2+k} \cdot \left[\left(1 + \frac{\xi(k;n)^{n-1}}{2}\right)^{k+2} - \left(1 - \frac{\xi(k;n)^{n-1}}{2}\right)^{k+2}\right] - \xi(k; n)^{n-k}}, \tag{9.41}$$

where $\xi(k; n) \in (0,1)$ is an arbitrary solution to equation (9.39).

Set

$$\gamma_p = \gamma_p(k) = C_{n_p}(k), \quad p \in \mathbb{N}.$$

Lemma 9.8. *For every* $k \in \mathbb{N}$, $k \geq 2$ *the following equality holds*

$$\lim_{p \to \infty} \gamma_p(k) = \frac{12}{k}.$$

Proof. We have

$$\gamma_p = \frac{\alpha_p^3 \cdot \beta_p^{1-k}}{\frac{1}{k+2}\cdot\left(\left(1+\frac{\alpha_p}{2}\right)^{k+2} - \left(1-\frac{\alpha_p}{2}\right)^{k+2}\right) - \xi(k;n_p)^{n_p-k}} =$$

$$\frac{\alpha_p^3 \cdot \beta_p^{1-k}}{\frac{1}{k+2}\cdot\left(\left(1+\frac{\alpha_p}{2}\right)^{k+2} - \left(1-\frac{\alpha_p}{2}\right)^{k+2}\right) - \frac{1}{k+1}\cdot\left(\left(1+\frac{\alpha_p}{2}\right)^{k+1} - \left(1-\frac{\alpha_p}{2}\right)^{k+1}\right)} \cdot$$

However

$$\left(1+\frac{\alpha_p}{2}\right)^{k+2} - \left(1-\frac{\alpha_p}{2}\right)^{k+2} = \sum_{j=0}^{k+2} C_{k+2}^j \cdot \left(\frac{\alpha_p}{2}\right)^j - \sum_{j=0}^{k+2} C_{k+2}^j \cdot \left(-\frac{\alpha_p}{2}\right)^j$$

$$= 2C_{k+2}^1 \cdot \frac{\alpha_p}{2} + 2C_{k+2}^3 \cdot \frac{\alpha_p^3}{2^3} + ... + 2C_{k+2}^{m_1} \cdot \frac{\alpha_p^{m_1}}{2^{m_1}},$$

where

$$m_1 \equiv m_1(k) = \begin{cases} k+2, & \text{if } k \text{ is odd} \\ k+1, & \text{if } k \text{ is even.} \end{cases}$$

Analogously we have

$$\left(1+\frac{\alpha_p}{2}\right)^{k+1} - \left(1-\frac{\alpha_p}{2}\right)^{k+1} = 2C_{k+1}^1 \cdot \frac{\alpha_p}{2} + 2C_{k+1}^3 \cdot \frac{\alpha_p^3}{2^3} + ... + 2C_{k+1}^{m_2} \cdot \frac{\alpha_p^{m_2}}{2^{m_2}},$$

where

$$m_2 \equiv m_2(k) = \begin{cases} k+1, & \text{if } k \text{ is even} \\ k, & \text{if } k \text{ is odd,} \end{cases}$$

i.e., $m_2 = 2m_0 - 1, \ m_0 \in \mathbb{N}$.

Therefore

$$\frac{1}{k+2}\cdot\left(\left(1+\frac{\alpha_p}{2}\right)^{k+2} - \left(1-\frac{\alpha_p}{2}\right)^{k+2}\right) -$$

$$\frac{1}{k+1}\cdot\left(\left(1+\frac{\alpha_p}{2}\right)^{k+1} - \left(1-\frac{\alpha_p}{2}\right)^{k+1}\right)$$

$$= \sum_{j=2}^{m_0} a_j \alpha_p^{2j-1} + a_{m_0+1}\alpha_p^{2m_0+1} = \alpha_p^3(a_2 + a_3\alpha_p^2 + a_4\alpha_p^4 + ... + a_{m_0+1}\alpha_p^{2(m_0-1)}),$$

where

$$a_j = \frac{2}{2^{2j-1}}\cdot\left(\frac{C_{k+2}^{2j-1}}{k+2} - \frac{C_{k+1}^{2j-1}}{k+1}\right), \ j = 2,3,... \ ,$$

$$a_{m_0+1} = \begin{cases} 0 & \text{if } m_1 = m_2, \\ \dfrac{1}{2^{2m_0}} \cdot \dfrac{C_{k+2}^{2m_0+1}}{k+2} & \text{if } m_2 < m_1. \end{cases}$$

Obviously

$$a_2 = \frac{k}{12}.$$

Thus we get

$$\gamma_p = \frac{\beta_p^{1-k}}{\frac{k}{12} + a_3\alpha_p^2 + a_4\alpha_p^4 + \ldots + a_{m_0+1}\alpha_p^{2(m_0-1)}}, \quad p \in \mathbb{N}.$$

Hence by Corollary 9.1 it follows that

$$\lim_{p\to\infty} \gamma_p = \frac{12}{k}.$$

$\square$

Corollary 9.2. *If $k \geq 4$ then $0 < \lim\limits_{p\to\infty} \gamma_p \leq 3$.*

For each $k \geq 4$ we define the set $\mathbb{N}_0(k)$:

$$\mathbb{N}_0(k) = \{p \in \mathbb{N} : |\gamma_p(k)| < 4\}.$$

Note that, the set $\mathbb{N}_0(k)$ is a countable subset in the set of all natural numbers. For each $p \in \mathbb{N}_0(k)$, $(k \geq 4)$ we define the continuous function $K_p(t, u; k)$ on $[0,1]^2$ by

$$K_p(t, u; k) = 1 + \gamma_p(k)\left(t - \frac{1}{2}\right)\left(u - \frac{1}{2}\right), \quad t, u \in [0,1].$$

By the inequality $|\gamma_p(k)| < 4$ it follows that, the function $K_p(t, u; k)$ is strictly positive.

Theorem 9.14. *Let $k \geq 4$. For each $p \in \mathbb{N}_0(k)$ the Hammerstein's equation*

$$\int_0^1 K_p(t, u; k)f^k(u)du = f(t) \tag{9.42}$$

in the $C[0,1]$ has at least two positive solutions.

Proof. Obviously the function $f_0(t) \equiv 1$ is a solution of equation (9.42). Define the strictly positive continuous function $f_1(t)$ on $[0,1]$ by

$$f_1(t) = \xi(k; n_p) + \xi(k; n_p)^{n_p}\left(t - \frac{1}{2}\right), \quad t \in [0,1].$$

We shall prove that the function $f_1(t)$ is also a solution of the Hammerstein's equation (9.42):

$$\int_0^1 K_p(t, u; k) f_1^k(u)\, du =$$

$$\int_0^1 \left(1 + \gamma_p(k)\left(t - \frac{1}{2}\right)\left(u - \frac{1}{2}\right)\right)\left(\xi(k; n_p) + \xi(k; n_p)^{n_p}\left(u - \frac{1}{2}\right)\right)^k du =$$

$$\int_{-1/2}^{1/2}\left(1 + \gamma_p(k)\left(t - \frac{1}{2}\right)u\right)\left(\beta_p + \beta_p^{n_p}u\right)^k du =$$

$$\int_{-1/2}^{1/2}\left(\beta_p + \beta_p^{n_p}u\right)^k du + \gamma_p(k)\left(t - \frac{1}{2}\right)\int_{-1/2}^{1/2} u\left(\beta_p + \beta_p^{n_p}u\right)^k du =$$

$$\frac{\beta_p^k}{\beta_p^{n_p-1}}\int_{-1/2}^{1/2}\left(1 + \beta_p^{n_p-1}u\right)^k d\left(1 + \beta_p^{n_p-1}u\right) + \gamma_p(k)\left(t - \frac{1}{2}\right) \times$$

$$\frac{\beta_p^k}{\beta_p^{n_p-1}}\int_{-1/2}^{1/2} u\left(1 + \beta_p^{n_p-1}u\right) d\left(1 + \beta_p^{n_p-1}u\right) = \frac{\beta_p^k}{\alpha_p}\cdot\frac{1}{k+1}\left(1 + \alpha_p u\right)^{k+1}\Big|_{-1/2}^{1/2}$$

$$+ \frac{\gamma_p(k)\beta_p^k}{\alpha_p^2}\left(t - \frac{1}{2}\right)\int_{-1/2}^{1/2}\left((1 + \alpha_p u)^{k+1} - (1 + \alpha_p u)^k\right) d(1 + \alpha_p u) =$$

$$\frac{\beta_p^k}{\alpha_p}\cdot\frac{1}{k+1}\left(\left(1 + \frac{\alpha_p}{2}\right)^{k+1} - \left(1 - \frac{\alpha_p}{2}\right)^{k+1}\right) +$$

$$\frac{\gamma_p(k)\beta_p^k}{\alpha_p^2}\left(t - \frac{1}{2}\right)\cdot\left(\frac{1}{k+2}(1 + \alpha_p u)^{k+2}\Big|_{-1/2}^{1/2} - \frac{1}{k+1}(1 + \alpha_p u)^{k+1}\Big|_{-1/2}^{1/2}\right) =$$

$$\frac{\beta_p^k}{\alpha_p}\cdot\frac{1}{k+1}\cdot(k+1)\beta_p^{n_p-k} + \frac{\gamma_p(k)\beta_p^k}{\alpha_p^2}\cdot\left(t - \frac{1}{2}\right) \times$$

$$\left[\frac{1}{k+2}\left(\left(1 + \frac{\alpha_p}{2}\right)^{k+2} - \left(1 - \frac{\alpha_p}{2}\right)^{k+2}\right) - \right.$$

$$\left.\frac{1}{k+1}\left(\left(1 + \frac{\alpha_p}{2}\right)^{k+1} - \left(1 - \frac{\alpha_p}{2}\right)^{k+1}\right)\right] = \frac{\beta_p^{n_p}}{\alpha_p} + \frac{\gamma_p(k)\beta_p^k}{\alpha_p^2} \times$$

$$\left(t-\frac{1}{2}\right)\cdot\left[\frac{1}{k+2}\left(\left(1+\frac{\alpha_p}{2}\right)^{k+2}-\left(1-\frac{\alpha_p}{2}\right)^{k+2}\right)-\frac{1}{k+1}(k+1)\beta_p^{n_p-k}\right]=$$

$$\beta_p+\frac{\gamma_p(k)\beta_p^k}{\alpha_p^2}\cdot\left(t-\frac{1}{2}\right)\left(\frac{1}{k+2}\left(\left(1+\frac{\alpha_p}{2}\right)^{k+2}-\left(1-\frac{\alpha_p}{2}\right)^{k+2}\right)-\beta_p^{n_p-k}\right)=$$

$$\beta_p+\frac{\beta_p^k}{\alpha_p^2}\cdot\alpha_p^3\beta_p^{1-k}\left(t-\frac{1}{2}\right)=\beta_p+\beta_p^{n_p}\cdot\left(t-\frac{1}{2}\right)=$$

$$\xi(k;n_p)+\xi(k;n_p)^{n_p}\left(t-\frac{1}{2}\right)=f_1(t).\qquad\square$$

From Theorem 9.14 and Proposition 9.1 we get the following theorem.

Theorem 9.15. *Let $k \geq 4$ and $p \in \mathbb{N}_0(k)$. The model*

$$H(\sigma) = -\frac{1}{\beta}\sum_{\substack{<x,y>\\x,y\in V}}\ln K_p(\sigma(x),\sigma(y);k), \quad \sigma\in\Omega_V$$

on the Cayley tree Γ^k has at least two translations-invariant Gibbs measures.

Commentaries and references. The chapter is based on results of very recently written papers [229], [68], [67].

Since the theory of non-linear integral equations is not well developed, we were only able to obtain some results about translation-invariant Gibbs measures of the models with uncountable set of spin values on Cayley trees. Therefore investigations of such models at starting point can be continued in many distinct directions. For example, one can define periodic Gibbs measures for such models and study periodic solutions of equation (9.5). I hope the integral equations considered in this chapter will open a new direction in the theory of non-linear integral equations. The results obtained in this direction will, of course, give new and interesting properties of Gibbs measures (phases) of physical systems with uncountable set of spin values.

Chapter 10

Contour arguments on Cayley trees

In the first section of this chapter we consider a one-dimensional model with nearest-neighbor interactions $I_n, n \in \mathbb{Z}$, and spin values ± 1. We show that under some conditions on parameters I_n the phase transition occurs for the model. We define a notion of "phase separation point" between two phases. We prove that the expectation value of the point is zero and its mean square fluctuation is bounded by a constant $C(\beta)$ which tends to $\frac{1}{4}$ if $\beta \to \infty$. Here $\beta = \frac{1}{T}$, $T > 0$-temperature.

In other sections of this chapter we consider a q-component model and the Ising model with competing two-step interactions on a Cayley tree of order $k \geq 1$. We constructively describe (periodic and weakly periodic) ground states and verify the Peierls condition for these models. We define the notion of a contour for the models on the Cayley tree. Using a contour argument we show the existence of several different Gibbs measures.

This chapter also contains a general contour argument for a finite range lattice models on Cayley tree with two basic properties: the existence of only a finite number of ground states and with Peierls type condition. We define a general contour for such models on the Cayley tree. By a contour argument we show the existence of s different (where s is the number of ground states) Gibbs measures.

10.1 One-dimensional models

In the section we consider the Hamiltonian

$$H(\sigma) = \sum_{l=(x-1,x):x\in\mathbb{Z}} I_x \mathbf{1}_{\sigma(x-1)\neq\sigma(x)}, \tag{10.1}$$

263

where $\mathbb{Z} = \{..., -1, 0, 1, ...\}$, $\sigma = \{\sigma(x) \in \{-1, 1\} : x \in \mathbb{Z}\} \in \Omega = \{-1, 1\}^{\mathbb{Z}}$, and $I_x \in R$ for any $x \in \mathbb{Z}$.

10.1.1   *Phase transition*

The goal of this subsection is to describe a condition on parameters of the model (10.1) under which the phase transition occurs.

Let us consider the sequence $\Lambda_n = [-n, n], n = 0, 1, ...$ and denote $\Lambda_n^c = \mathbb{Z} \setminus \Lambda_n$. Consider boundary condition $\sigma_n^{(+)} = \sigma_{\Lambda_n^c} = \{\sigma(x) = +1 : x \in \Lambda_n^c\}$. The energy $H_n^+(\sigma)$ of the configuration σ in the presence of boundary condition $\sigma_n^{(+)}$ is expressed by the formula

$$H_n^+(\sigma) = \sum_{l=(x-1,x):x\in\Lambda_n} I_x \mathbf{1}_{\sigma(x-1)\neq\sigma(x)} + I_{-n}\mathbf{1}_{\sigma(-n)\neq1} + I_{n+1}\mathbf{1}_{\sigma(n)\neq1}.$$

$$(10.2)$$

The Gibbs measure on $\Omega_n = \{-1, 1\}^{\Lambda_n}$ with boundary condition $\sigma_n^{(+)}$ is defined in the usual way

$$\mu_{n,\beta}^+(\sigma) = Z^{-1}(n, \beta, +) \exp(-\beta H_n^+(\sigma)), \qquad (10.3)$$

where $\beta = T^{-1}$, $T > 0-$ temperature and $Z(n, \beta, +)$ is the normalizing factor (statistical sum).

Denote by σ_n^+ the configuration on $\mathbb{Z}$ such that $\sigma_n^+(x) \equiv +1$ for any $x \in \Lambda_n^c$.

Put

$$A(\sigma_n^+) = \{x \in \mathbb{Z} : \sigma_n^+(x) = -1\},$$

$$\partial(\sigma_n^+) = \{l = (n-1, n) \in \mathbb{Z} \times \mathbb{Z} : \sigma(n-1) \neq \sigma(n)\}.$$

Note that there is a one-to-one correspondence between the set of all configurations σ_n^+ and the set of all subsets A of Λ_n.

Let $A'(\sigma_n^+)$ be the set of all maximal connected subsets of $A(\sigma_n^+)$.

Lemma 10.1. *Let $B \subset \mathbb{Z}$ be a fixed connected set and $p_\beta^+(B) = \mu_{n,\beta}^+\{\sigma_n^+ : B \in A'(\sigma_n^+)\}$. Then*

$$p_\beta^+(B) \leq \exp\left\{ -\beta\left[I_{n_B} + I_{N_B+1}\right] \right\},$$

where n_B (resp. N_B) is the left (resp. right) endpoint of B.

Proof. Denote $F_B = \{\sigma_n^+ : B \in A'(\sigma_n^+)\}$- the set of all configurations σ_n^+ on $\mathbb{Z}$ with "+"-boundary condition (i.e., $\sigma_n^+(x) \equiv 1$ for any $x \in \Lambda_n^c$) such that B is maximal connected set.

Denote also $F_B^- = \{\sigma_n^+ : B \cap A'(\sigma_n^+) = \emptyset\}$.

Define the map $\chi_B : F_B \to F_B^-$ as follows: for $\sigma_n \in F_B$ we destroy the set B changing the values $\sigma_n(x)$ inside of B to $+1$. The constructed configuration is $\chi_B(\sigma_n) \in F_B^-$.

For a given B the map χ_B is one-to-one.

It is clear that

$$A'(\sigma_n) = A'(\chi_B(\sigma_n)) \cup B, \quad \partial(\sigma_n) = \partial(\chi_B(\sigma_n)) \cup \{(n_B-1, n_B), (N_B, N_B+1)\}.$$

Thus we have

$$H_n^+(\sigma_n) - H_n^+(\chi_B(\sigma_n)) = I_{n_B} + I_{N_B+1}. \tag{10.4}$$

By definition we have

$$p_\beta^+(B) = \frac{\sum\limits_{\sigma_n \in F_B} \exp\{-\beta H_n^+(\sigma_n)\}}{\sum\limits_{\sigma_n} \exp\{-\beta H_n^+(\sigma_n)\}} \le \tag{10.5}$$

$$\frac{\sum\limits_{\sigma_n \in F_B} \exp\{-\beta H_n^+(\sigma_n)\}}{\sum\limits_{\sigma_n \in F_B^-} \exp\{-\beta H_n^+(\sigma_n)\}} = \frac{\sum\limits_{\sigma_n \in F_B} \exp\{-\beta H_n^+(\sigma_n)\}}{\sum\limits_{\sigma_n \in F_B} \exp\{-\beta H_n^+(\chi_B(\sigma_n))\}}.$$

Using (10.4) from (10.5) we get

$$p_\beta^+(B) = \frac{\sum\limits_{\sigma_n \in F_B} \exp\left\{-\beta H_n^+(\chi_B(\sigma_n)) - \beta\left[I_{n_B} + I_{N_B+1}\right]\right\}}{\sum\limits_{\sigma_n \in F_B} \exp\{-\beta H_n^+(\chi_B(\sigma_n))\}} =$$

$$\exp\left\{-\beta\left[I_{n_B} + I_{N_B+1}\right]\right\}. \qquad \square$$

Assume that for any $r \in \{1, 2, \ldots\}$ and $n \in \mathbb{Z}$ the Hamiltonian (10.1) satisfies the following condition

$$I_n + I_{n+r} \ge r. \tag{10.6}$$

Lemma 10.2. *Assume condition (10.6) is satisfied. Then for all sufficiently large β, there is a constant $C = C(\beta) > 0$, such that*

$$\mu_\beta^+\{\sigma_n : \; |B| > C \ln|\Lambda_n| \; \text{for some} \; B \in A'(\sigma_n)\} \to 0, \quad \text{as} \; |\Lambda_n| \to \infty,$$

where $|\cdot|$ denotes the number of elements.

Proof. Suppose $\beta > 1$, then by Lemma 10.1 and condition (10.6) we have

$$\mu_\beta^+\{\sigma_n : B \in A'(\sigma_n),\ t \in B, |B| = r\} = \sum_{B:\ t \in B,\ |B| = r} p_\beta^+(B) \le r \exp\{-\beta r\}.$$

Hence

$$\mu_\beta^+\{\sigma_n : B \in A'(\sigma_n),\ t \in B, |B| > C_1 \ln |\Lambda_n|\} \le \sum_{r \ge C_1 \ln |\Lambda_n|} r \exp\{-\beta r\} \le \tag{10.7}$$

$$\sum_{r \ge C_1 \ln |\Lambda_n|} \exp\{(1 - \beta)r\} = \frac{|\Lambda_n|^{C_1(1-\beta)}}{1 - e^{1-\beta}},$$

where C_1 will be defined later. Thus we have

$$\mu_\beta^+\{\sigma_n : \exists B \in A'(\sigma_n),\ |B| > C_1 \ln |\Lambda_n|\} \le \frac{|\Lambda_n|^{C_1(1-\beta)+1}}{1 - e^{1-\beta}}.$$

The last expression tends to zero if $|\Lambda_n| \to \infty$ and $C_1 > \frac{1}{\beta-1}$. $\quad\square$

Lemma 10.3. *Assume condition (10.6) is satisfied. Then*

$$\mu_\beta^+\{\sigma_n : \sigma_n(0) = -1\} \to 0, \quad as \ \beta \to \infty. \tag{10.8}$$

Proof. If $\sigma_n(0) = -1$, then 0 is a point for some $B \in A'(\sigma_n)$. Consequently,

$$\mu_\beta^+\{\sigma_n : 0 \in B,\ |B| < C_1 \ln |\Lambda_n|\} \le \sum_{r=1}^{C_1 \ln |\Lambda_n|} (e^{1-\beta})^r \le \frac{e^{1-\beta}}{1 - e^{1-\beta}}$$

and

$$\mu_\beta^+\{\sigma_n(0) = -1\} \le \mu_\beta^+\{\sigma_n : 0 \in B,\ B \in A'(\sigma_n)\} \le \tag{10.9}$$

$$\frac{e^{1-\beta}}{1 - e^{1-\beta}} + \frac{|\Lambda_n|^{C_1(1-\beta)+1}}{1 - e^{1-\beta}}.$$

For $|\Lambda_n| \to \infty$ and $\beta \to \infty$ from (10.9) we get (10.8). $\quad\square$

Theorem 10.1. *Assume condition (10.6) is satisfied. For all sufficiently large β there are at least two Gibbs measures for the model (10.1).*

Proof. Using a similar argument one can prove

$$\mu_\beta^-\{\sigma_n : \sigma_n(0) = 1\} \to 0, \quad as \ \beta \to \infty.$$

Consequently, for sufficiently large β we have

$$\mu_\beta^+\{\sigma_n : \sigma_n(0) = -1\} \ne \mu_\beta^-\{\sigma_n : \sigma_n(0) = -1\}.$$

This completes the proof. $\quad\square$

Denote

$$\mathcal{H} = \{H : H \text{ (see (10.1)) satisfies the condition (10.6)}\}$$

The following example shows that the set $\mathcal{H}$ is not empty.

Example 10.1. Consider Hamiltonian (10.1) with $I_m \geq |m|,\quad m \in \mathbb{Z}$. Then

$$I_m + I_{m+k} \geq |m| + |m + k| \geq k$$

for all $m \in \mathbb{Z}$ and $k \geq 1$. Thus the condition (10.6) is satisfied.

10.1.2 *Partition functions*

Note that the probability (with respect to measure $\mu_{n,\beta}^+$) of a subset Ω_n' of Ω_n is defined by

$$\mu_{n,\beta}^+(\Omega_n') = Z^{-1}(n, \beta, +) \sum_{\psi \in \Omega_n'} \exp(-\beta H_n^+(\psi)) = \frac{Z'(n, \beta, +)}{Z(n, \beta, +)}, \qquad (10.10)$$

where $Z'(n, \beta, +)$ is called a "crystal" partition function:

$$Z'(n, \beta, +) = \sum_{\psi \in \Omega_n'} \exp(-\beta H_n^+(\psi)). \qquad (10.11)$$

So to define the Gibbs measure and probability of an event of the system one has to compute the partition functions.

In this subsection we consider some (crystal) partition functions of the model and give the system of recursive equations for the functions. Under some conditions on parameters of the model we describe their solutions.

10.1.2.1 *Partition function of "+" and "±"-boundary conditions*

Consider two types of partition functions:

$$Z_n^+ = \sum_{\sigma_n \in \Omega_n} \exp\{-\beta H_n^+(\sigma_n)\}, \qquad (10.12)$$

$$Z_n^\pm = \sum_{\sigma_n \in \Omega_n} \exp\{-\beta H_n^\pm(\sigma_n)\}, \qquad (10.13)$$

where H_n^+ is defined by (10.2) and

$$H_n^\pm(\sigma_n) = H_n^+(\sigma_n) + I_{-n}\sigma(-n). \qquad (10.14)$$

Here for simplicity assume

$$I_n = I_{-n+1}, \quad \text{for any } n \in \mathbb{Z}. \qquad (10.15)$$

Proposition 10.1. *If condition (10.15) is satisfied then the partition functions (10.12) and (10.13) have the form*

$$Z_n^+ = \tfrac{1}{2}\left(\prod_{i=0}^{n}(1 + e^{-\beta I_{i+1}})^2 + \prod_{i=0}^{n}(1 - e^{-\beta I_{i+1}})^2\right),$$
$$Z_n^{\pm} = \tfrac{1}{2}\left(\prod_{i=0}^{n}(1 + e^{-\beta I_{i+1}})^2 - \prod_{i=0}^{n}(1 - e^{-\beta I_{i+1}})^2\right). \tag{10.16}$$

Proof. Under the condition (10.15) we get $Z_n^- = Z_n^+$ and $Z_n^{\pm} = Z_n^{\mp}$. Now from (10.12), (10.13) we obtain the following system of recursive equations

$$Z_n^+ = (1 + e^{-2\beta I_{n+1}})Z_{n-1}^+ + 2e^{-\beta I_{n+1}}Z_{n-1}^{\pm},$$
$$Z_n^{\pm} = (1 + e^{-2\beta I_{n+1}})Z_{n-1}^{\pm} + 2e^{-\beta I_{n+1}}Z_{n-1}^+. \tag{10.17}$$

Putting $X_n = Z_n^+ - Z_n^{\pm}$ and $Y_n = Z_n^+ + Z_n^{\pm}$ from (10.17) we get

$$X_n = (1 - e^{-\beta I_{n+1}})^2 X_{n-1},$$
$$Y_n = (1 + e^{-\beta I_{n+1}})^2 Y_{n-1}. \tag{10.18}$$

The equalities $X_0 = Z_0^+ - Z_0^{\pm} = (1 - e^{-\beta I_1})^2$, $Y_0 = (1 + e^{-\beta I_1})^2$ with (10.18) imply

$$X_n = \prod_{i=0}^{n}(1 - e^{-\beta I_{i+1}})^2, \quad Y_n = \prod_{i=0}^{n}(1 + e^{-\beta I_{i+1}})^2.$$

Hence we get (10.16). $\qquad\square$

For example, in a case of the usual Ising model, i.e., $I_m = I$, $\forall m \in \mathbb{Z}$ from (10.16) denoting $\tau = \exp(-\beta I)$ we get

$$Z_n^+ = \tfrac{1}{2}\left((1 + \tau)^{2(n+1)} + (1 - \tau)^{2(n+1)}\right),$$
$$Z_n^{\pm} = \tfrac{1}{2}\left((1 + \tau)^{2(n+1)} - (1 - \tau)^{2(n+1)}\right).$$

Using these equalities (for usual Ising model) it is easy to see that

$$\frac{Z_n^+}{Z_n^{\pm}} \to 1, \quad \text{if} \ \ n \to \infty.$$

This means that for the Ising model the partition functions Z_n^+ and $Z_n^{\pm}$ are asymptotically equal. This gives in fact uniqueness of limit Gibbs measure for the 1D Ising model. Such an asymptotical equality is true if I_m is a periodic function of m, i.e., $I_{m+p} = I_m$ for some $p \geq 1$ and all $m \in N$.

10.1.2.2 *Crystal partition functions*

In this subsection we are going to describe the crystal partition functions.

Denote $\Omega_{m,n} = \{-1,1\}^{[m,n]}$, where $[m,n] = \{m, m+1, ..., n\}$, $m, n \in \mathbb{Z}$, $n \geq m$. Put

$$N_\varepsilon(\sigma) = |\{x \in [m,n] : \sigma(x) = \varepsilon\}|, \quad \varepsilon = \pm 1,$$

where $|S|$ is the cardinal of the set S. For $r = 0, 1, ..., n - m + 1$ consider the following crystal partition functions:

$$Z_{m,n}^{\varepsilon,r} = \sum_{\sigma \in \Omega_{m,n} : N_{-\varepsilon}(\sigma) = r} e^{-\beta H^\varepsilon(\sigma)}, \quad \varepsilon = -, + \tag{10.19}$$

$$Z_{m,n}^{\pm,r} = \sum_{\sigma \in \Omega_{m,n} : N_+(\sigma) = r} e^{-\beta H^\pm(\sigma)}. \tag{10.20}$$

Note that $Z_{m,n}^{-,r} = Z_{m,n}^{+,r}$.

Denoting $X_{m,n}^r = Z_{m,n}^{-,r}$ and $Y_{m,n}^r = Z_{m,n}^{\pm,r}$ from (10.19) and (10.20) one easily gets the following system of (multi-variable) recursive equations

$$X_{m,n}^r = X_{m,n-1}^{r-1} + e^{-\beta I_{n+1}} Y_{m,n-1}^r,$$

$$Y_{m,n}^r = Y_{m,n-1}^r + e^{-\beta I_{n+1}} X_{m,n-1}^{r-1}, \tag{10.21}$$

where $r = 0, 1, ..., n - m + 1$, $m, n \in \mathbb{Z}$, $n \geq m$.

The system (10.21) can be reduced to a recursive equation with respect to $X_{m,n}^r$. Indeed, from the first equation of (10.21) we get

$$Y_{m,n-1}^r = e^{\beta I_{n+1}} \left(X_{m,n}^r - X_{m,n-1}^{r-1} \right). \tag{10.22}$$

Now from the second equation of (10.21) using (10.22) we get

$$X_{m,n}^r = X_{m,n-1}^{r-1} + e^{-\beta(I_{n+1} - I_n)}(X_{m,n-1}^r - X_{m,n-2}^{r-1}) + e^{-\beta(I_n + I_{n+1})} X_{m,n-2}^{r-1}, \tag{10.23}$$

where $r = 1, 2, ..., n - m + 1$, $n \geq m$ and

$$X_{m,m}^0 = 1, \quad X_{m,m}^1 = e^{-\beta(I_m + I_{m+1})}, \quad X_{m,m+1}^0 = 1,$$

$$X_{m,m+1}^1 = e^{-\beta(I_m + I_{m+1})} + e^{-\beta(I_{m+1} + I_{m+2})}, \quad X_{m,m+1}^2 = e^{-\beta(I_m + I_{m+2})}.$$

Iterating (10.23) one can obtain an expression for $X_{m,n}^r$. Then using (10.22) one can find $Y_{m,n}^r$. But these expressions would be in a very bulky form.

Now we shall illustrate such an expression for the Ising model, i.e., $I_m \equiv I$, $m \in \mathbb{Z}$. In this case the recurrence equation (10.23) becomes more simple

$$X_n^r = X_{n-1}^{r-1} + X_{n-1}^r + (\chi - 1)X_{n-2}^{r-1}, \tag{10.24}$$

where $X_n^r = X_{m,n_1}^r$ with $n_1 - m = n$, $\chi = e^{-2\beta I}$.

It is easy to see that for $r = 0, 1, 2, 3, 4$ the solutions are

$$X_n^0 = 1, \quad X_n^1 = n\chi, \quad X_n^2 = (n-1)\chi + \frac{(n-2)(n-1)}{2!}\chi^2,$$

$$X_n^3 = (n-2)\chi + (n-2)(n-3)\chi^2 + \frac{(n-2)(n-3)(n-4)}{3!}\chi^3,$$

$$X_n^4 = (n-3)\chi + \frac{3(n-3)(n-4)}{2}\chi^2 + \frac{(n-3)(n-4)(n-5)}{2}\chi^3 +$$

$$\frac{(n-3)(n-4)(n-5)(n-6)}{4!}\chi^4,$$

here we used the following formulas

$$\sum_{j=1}^{n} j^2 = \frac{1}{6}(2n^3 + 3n^2 + n), \quad \sum_{j=1}^{n} j^3 = \frac{1}{4}(n^4 + 2n^3 + n^2).$$

Note that X_n^r has a form

$$X_n^r = a_{1,n}^r \chi + a_{2,n}^r \chi^2 + \ldots + a_{r,n}^r \chi^r.$$

For the coefficients $a_{k,n}^r$, $0 \le k \le r \le n$, using (10.24) we obtain the following system of recursive equations

$$a_{1,n}^r = a_{1,n-1}^r + a_{1,n-1}^{r-1} - a_{1,n-2}^{r-1},$$

$$a_{k,n}^r = a_{k,n-1}^r + a_{k,n-1}^{r-1} + a_{k-1,n-2}^{r-1} - a_{k,n-2}^{r-1}, \quad k = 2, 3, \ldots, r-1, \quad (10.25)$$

$$a_{r,n}^r = a_{r,n-1}^r + a_{r-1,n-2}^{r-1}.$$

The following lemma gives solution to (10.25)

Lemma 10.4. *Solution of the system of recursive equations (10.25) is*

$$a_{k,n}^r = \binom{n-r+1}{k}\binom{r-1}{k-1}, \quad 0 \le k \le r \le n. \qquad (10.26)$$

Proof. We shall use mathematical induction (cf. with [188] pages 148-150). Let A_m denote all cases of (10.26) with $n + k + r = m$. Formulas given above for X_n^r, $r = 0, 1, 2, 3, 4$ show that the formula (10.26) is true for small values of m. Assuming that A_m holds, we are to prove A_{m+1} that is, equation (10.25), for any integers n, r and k whose sum is $m + 1$. Since RHS of (10.25) contains terms with $n + r + k \le m$ using the assumption of the induction for each term of RHS of (10.25) we get (10.26). $\qquad \square$

Thus the solution of (10.24) is given by

$$X_n^r = \sum_{k=1}^{r} \binom{n-r+1}{k} \binom{r-1}{k-1} \chi^k. \tag{10.27}$$

Remark 10.1. Note that for $\chi = 1$ (i.e., there is no interaction) the solution of (10.24) is $X_n^r = \binom{n}{r}$. Using (10.27) for $\chi = 1$ we obtain the following property of binomial coefficients

$$\binom{n}{r} = \sum_{k=1}^{r} \binom{n-r+1}{k} \binom{r-1}{k-1}. \tag{10.28}$$

This identity is known as the convolution identity of Vandermonde.

Since interaction (parameter I) of the 1D Ising model is translation-invariant (does not depend on the points of $\mathbb{Z}$), the unknown functions $X_{m,n}^r$, $Y_{m,n}^r$ of the system (10.21) depend on $n-m$ and r only (see (10.24)), consequently, instead of $n-m$ we can write n. Summarizing the results for the Ising model we have

Theorem 10.2. *For the Ising model the solution of the system of recursive equations (10.21) is*

$$X_n^r = \sum_{k=1}^{r} \binom{n-r+1}{k} \binom{r-1}{k-1} \chi^k,$$

$$Y_n^r = \frac{1}{\sqrt{\chi}} \left(X_{n+1}^r - X_n^{r-1} \right),$$

where n stands for $n-m$.

10.1.3 *Phase-separation point*

Fix $n \in \{0, 1, 2, ...\}$. Denote by Ω_n the set of all configurations on $\Lambda_n = \{-n, ..., n\}$, i.e., $\Omega_n = \{-1, 1\}^{\Lambda_n}$. For every $\sigma_n \in \Omega_n$ define $\sigma_n^{\pm} \in \{-1, 1\}^{\mathbb{Z}}$ as follows

$$\sigma_n^{\pm}(x) = \begin{cases} -1 & \text{if } x < -n \\ \sigma_n(x) & \text{if } x \in \Lambda_n \\ 1 & \text{if } x > n \end{cases} \qquad x \in \mathbb{Z}. \tag{10.29}$$

Let $\Omega_n^{\pm}$ be the set of all configurations defined by (10.29). Denote

$$\Omega_n^{(+)} = \{\sigma_n \in \Omega_n^{\pm} : |\{x \in \Lambda_n : \sigma_n(x) = 1\}| \geq n+1\};$$

$$\Omega_n^{(-)} = \{-\sigma_n : \sigma_n \in \Omega_n^{(+)}\}.$$

Clearly $\Omega_n^{(+)} \cap \Omega_n^{(-)} = \emptyset$ and $\Omega_n^{\pm} = \Omega_n^{(+)} \cup \Omega_n^{(-)}$.

Let $S : \Omega_n^{\pm} \to \Omega_n^{\pm}$ be operator such that

$$S(\sigma_n)(x) = -\sigma_n(-x), \quad x \in \mathbb{Z}. \tag{10.30}$$

It is easy to see that

$$S(\Omega_n^{(\pm)}) = \Omega_n^{(\mp)}, \tag{10.31}$$

i.e., the operator S is one-to-one map from $\Omega_n^{(+)}$ (resp. $\Omega_n^{(-)}$) to $\Omega_n^{(-)}$ (resp. $\Omega_n^{(+)}$).

Lemma 10.5. *The Hamiltonian (10.1) (under condition (10.15)) is invariant with respect to operator S, i.e., $H(S(\sigma)) = H(\sigma)$ for any $\sigma \in \Omega_n^{\pm}, n = 0, 1, \dots$.*

Proof. Note that operator S is the combination of the following two symmetry maps $U : \Omega_n^{\pm} \to \Omega_n^{\pm}$ such that $U(\sigma_n)(x) = -\sigma_n(x)$, and $V : \Omega_n^{\pm} \to \Omega_n^{\pm}$ such that $V(\sigma_n)(x) = \sigma_n(-x)$. Clearly, H is invariant with respect to U and V this completes the proof. $\square$

Denote $T_n = \{-n - \frac{1}{2}, -n + \frac{1}{2}, \dots, n - \frac{1}{2}, n + \frac{1}{2}\}$. Fix $\sigma_n \in \Omega_n^{\pm}$ and we say that $t \in T_n$ is an interface point for the configuration σ_n if $\sigma_n(t - \frac{1}{2}) \neq \sigma_n(t + \frac{1}{2})$. For any interface point $t \in T_n$ denote

$$l_t^- \equiv l_t^-(\sigma_n) = |\{x \in \Lambda_n : \sigma_n(x) = -1, x < t\}|,$$

$$r_t^+ \equiv r_t^+(\sigma_n) = |\{x \in \Lambda_n : \sigma_n(x) = 1, x > t\}|,$$

$$l_t^+ = n + t + \frac{1}{2} - l_t^-, \quad r_t^- = n - t + \frac{1}{2} - r_t^+.$$

$$\Delta_t = (l_t^-, r_t^+), \quad \|\Delta_t\| = l_t^- + r_t^+.$$

Definition 10.1. We define PSP $\gamma_n(\sigma_n) \in T_n$ as the following interface point

$$\gamma_n(\sigma_n) = \begin{cases} \max\{t_0 \in T_n : \|\Delta_{t_0}\| = \max_t \|\Delta_t\|\}, & \text{if } \sigma_n \in \Omega_n^-, \\ \min\{t_0 \in T_n : \|\Delta_{t_0}\| = \max_t \|\Delta_t\|\}, & \text{if } \sigma_n \in \Omega_n^+. \end{cases} \tag{10.32}$$

Lemma 10.6. *For any $\sigma_n \in \Omega_n^{\pm}$ we have*

$$\gamma_n(\sigma_n) = -\gamma_n(S(\sigma_n)). \tag{10.33}$$

Proof. Straightforward. $\square$

For $\theta \in T_n$ denote

$$P_n(\theta) = \mu_n^{\pm}\{\gamma_n : \gamma_n(\sigma_n) = \theta\},$$

where $\mu_n^{\pm}$ is the Gibbs measure with respect to $\pm$-boundary condition.

Lemma 10.7. *For any θ and $n \in N$ we have*

$$P_n(\theta) = P_n(-\theta).$$

Proof. The proof follows from Lemma 10.5, and equality (10.33). $\qquad\square$

As a consequence of Lemmas 10.5 and 10.7 we have

Lemma 10.8. *For any $n \in N$*

$$\mathbf{E}_{\mu_n^{\pm}}(\gamma_n) = 0,$$

where $\mathbf{E}_{\mu_n^{\pm}}$ is the expectation value of the random variable γ_n with respect to the Gibbs measure $\mu_n^{\pm}$.

For a given configuration σ_n denote by $\theta_1 < \theta_2 < ... < \theta_k$ the interface points generated by σ_n.

Theorem 10.3. *The following assertions are valid:*

1. If an interface point $t = \theta_1$, (resp. $t = \theta_k$) is PSP then

$$l_t^- \geq l_t^+ = 0, \quad r_t^+ > r_t^-, \quad (\text{resp. } l_t^- > l_t^+, \quad r_t^+ \geq r_t^- = 0). \quad (10.34)$$

2. If an interface point $t \in T_n, t \neq \theta_1, \theta_k$ is PSP then

$$l_t^- > l_t^+, \quad r_t^+ > r_t^-. \quad (10.35)$$

Proof. Consider case $\sigma_n \in \Omega_n^{(+)}$ and $t \neq \theta_1, \theta_k$ (all other cases can be proved similarly). Assume $l_t^- \leq l_t^+$ then

$$\|\Delta_t\| = l_t^- + r_t^+ < l_t^+ + r_t^+ = r_{\theta_1}^+ \leq \|\Delta_{\theta_1}\|.$$

Thus by definition we get $\gamma_n(\sigma_n) = \theta_1$, which contradicts to $t \neq \theta_1$. This completes the proof. $\qquad\square$

Remark 10.2. In general, for a given configuration σ_n a point t satisfying the conditions (10.34) and (10.35), is not unique. For example, take $\sigma_2 = \{\sigma_2(-2) = -1, \sigma_2(-1) = -1, \sigma_2(0) = 1, \sigma_2(1) = -1, \sigma_2(2) = 1\}$, the interface points $t = -0.5$ and $t = 1.5$ satisfy the condition (10.35). Thus the conditions (10.34) and (10.35) are necessary for t to be PSP but are not sufficient.

Summing over all configurations with a given θ we obtain the probability $P_n(\theta)$ of θ which can be written by

$$P_n(\theta) = \frac{e^{-\beta I_{\theta+1/2}} Y_{-n,\theta-1/2} Y_{\theta+1/2,n}}{Z_n^{\pm}}, \tag{10.36}$$

where $Z_n^{\pm}$ is defined by (10.16) and $Y_{-n,\theta-1/2}$ (resp. $Y_{\theta+1/2,n}$) is the "crystal" partition function which contains *only* sum of terms $\exp(-\beta H^+(\varphi))$ with $\varphi = \sigma' \in \{-1,1\}^{[-n,\theta-3/2]}$ (resp. $\varphi = \sigma'' \in \{-1,1\}^{[\theta+3/2,n]}$) such that the PSP of the total configuration $\sigma = \sigma' \cup \{\sigma(\theta-1/2) = -1, \ \sigma(\theta+1/2) = 1\} \cup \sigma''$ on $[-n,n]$ is θ.

Remark 10.3. In two-dimensional Ising model case an analog of the formula (10.36) is given in ([34], formula (3.2)). Comparing our formula (10.36) with the formula (3.2) we notice that the numerator of the formula (3.2) contains a product of "full" (all possible terms) partition functions with pure "+" boundary conditions (or "−" boundary conditions which is equivalent by symmetry) in the different connected components of $\mathbb{Z}^2$ which are separated by the phase separation curve. But in our setting the numerator of the formula (10.36) contains product of *crystal* partition functions which we have defined above. This is a remarkable difference between the notions of phase separation of one and two dimensional Ising models. In the sequel of this section we are going to estimate the crystal partition functions by "rarefied" partition functions.

By Lemma 10.7 it is enough to consider the case $\theta \geq \frac{1}{2}$. For $A \subset \mathbb{Z}$ we denote $\Omega_A = \{-1,1\}^A$ -the set of all configurations defined on A. Denote

$$H_{n,\theta}^-(\sigma) = \sum_{x=-n}^{\theta-\frac{3}{2}} I_x \mathbf{1}_{\sigma(x) \neq \sigma(x+1)} + I_{-n} \mathbf{1}_{\sigma(-n) \neq -1} + I_{\theta-\frac{1}{2}} \mathbf{1}_{\sigma(\theta-\frac{3}{2}) \neq -1},$$

where $\sigma \in \Omega_{\{-n,\dots,\theta-\frac{3}{2}\}}$;

$$H_{n,\theta}^+(\sigma) = \sum_{x=\theta+\frac{3}{2}}^{n} I_x \mathbf{1}_{\sigma(x) \neq \sigma(x+1)} + I_{n+1} \mathbf{1}_{\sigma(n) \neq 1} + I_{\theta+\frac{3}{2}} \mathbf{1}_{\sigma(\theta+\frac{3}{2}) \neq 1},$$

where $\sigma \in \Omega_{\{\theta+\frac{3}{2},\dots,n\}}$;

$$H_{n,\theta}^{\pm}(\sigma) = H_{n,\theta}^-(\sigma) - I_{-n}\sigma(-n);$$

$$H_{n,\theta}^{\mp}(\sigma) = H_{n,\theta}^+(\sigma) + I_{n+1}\sigma(n).$$

Now we are ready to define the "rarefied" partition functions, i.e.,

$$\overrightarrow{Z}_{n,\theta} = \sum_{\sigma \in \Omega_{\{\theta+\frac{3}{2},\dots,n\}}} \exp(-\beta H_{n,\theta}^+(\sigma));$$

$$\overrightarrow{Z}^{-}_{n,\theta} = \sum_{\sigma\in\Omega_{\{\theta+\frac{3}{2},\dots,n\}}} \exp(-\beta H^{\mp}_{n,\theta}(\sigma));$$

$$\overleftarrow{Z}_{n,\theta} = \sum_{\sigma\in\Omega_{\{-n,\dots,\theta-\frac{3}{2}\}}} \exp(-\beta H^{-}_{n,\theta}(\sigma));$$

$$\overleftarrow{Z}^{+}_{n,\theta} = \sum_{\sigma\in\Omega_{\{-n,\dots,\theta-\frac{3}{2}\}}} \exp(-\beta H^{\pm}_{n,\theta}(\sigma)).$$

Note that (see (10.36))

$$Y_{-n,\theta-1/2} \le \overleftarrow{Z}_{n,\theta}; \quad Y_{\theta+1/2,n} \le \overrightarrow{Z}_{n,\theta}. \tag{10.37}$$

It is easy to check that

$$\overrightarrow{Z}_{n,\theta} = \overrightarrow{Z}_{n-1,\theta} + e^{-\beta I_{n+1}} \overrightarrow{Z}^{-}_{n-1,\theta}, \quad n \ge \theta + \tfrac{3}{2}$$

$$\overrightarrow{Z}^{-}_{n,\theta} = \overrightarrow{Z}^{-}_{n-1,\theta} + e^{-\beta I_{n+1}} \overrightarrow{Z}_{n-1,\theta}, \tag{10.38}$$

$$\overrightarrow{Z}_{\theta+\frac{1}{2},\theta} = 1; \quad \overrightarrow{Z}^{-}_{\theta+\frac{1}{2},\theta} = e^{-\beta I_{\theta+\frac{3}{2}}}.$$

Denote $u_{n,\theta} = \overrightarrow{Z}_{n,\theta} - \overrightarrow{Z}^{-}_{n,\theta}$, $v_{n,\theta} = \overrightarrow{Z}_{n,\theta} + \overrightarrow{Z}^{-}_{n,\theta}$. Then from (10.38) we get

$$u_{n,\theta} = \left(1 - e^{-\beta I_{n+1}}\right) u_{n-1,\theta}, \quad n \ge \theta + \tfrac{3}{2},$$

$$u_{\theta+\frac{1}{2},\theta} = 1 - e^{-\beta I_{\theta+\frac{3}{2}}},$$

i.e.,

$$u_{n,\theta} = \prod_{i=\theta+\frac{1}{2}}^{n} \left(1 - e^{-\beta I_{i+1}}\right).$$

Similarly

$$v_{n,\theta} = \prod_{i=\theta+\frac{1}{2}}^{n} \left(1 + e^{-\beta I_{i+1}}\right).$$

Hence

$$\overrightarrow{Z}_{n,\theta} = \tfrac{1}{2}\left(\prod_{i=\theta+\frac{1}{2}}^{n}(1 + e^{-\beta I_{i+1}}) + \prod_{i=\theta+\frac{1}{2}}^{n}(1 - e^{-\beta I_{i+1}})\right);$$
$$\overrightarrow{Z}^{-}_{n,\theta} = \tfrac{1}{2}\left(\prod_{i=\theta+\frac{1}{2}}^{n}(1 + e^{-\beta I_{i+1}}) - \prod_{i=\theta+\frac{1}{2}}^{n}(1 - e^{-\beta I_{i+1}})\right). \tag{10.39}$$

Analogically, using condition (10.15) we get

$$\overleftarrow{Z}_{n,\theta} = \frac{1}{2}\left(\prod_{i=\theta+\frac{3}{2}}^{n} (1 + e^{-\beta I_{i+1}}) \prod_{i=1}^{\theta-\frac{3}{2}} (1 + e^{-\beta I_{i+1}})^2 + \right.$$

$$\left. \prod_{i=\theta+\frac{3}{2}}^{n} (1 - e^{-\beta I_{i+1}}) \prod_{i=1}^{\theta-\frac{3}{2}} (1 - e^{-\beta I_{i+1}})^2 \right) ; \qquad (10.40)$$

$$\overleftarrow{Z}^{+}_{n,\theta} = \frac{1}{2}\left(\prod_{i=\theta+\frac{3}{2}}^{n} (1 + e^{-\beta I_{i+1}}) \prod_{i=1}^{\theta-\frac{3}{2}} (1 + e^{-\beta I_{i+1}})^2 - \right.$$

$$\left. \prod_{i=\theta+\frac{3}{2}}^{n} (1 - e^{-\beta I_{i+1}}) \prod_{i=1}^{\theta-\frac{3}{2}} (1 - e^{-\beta I_{i+1}})^2 \right) .$$

Using formulae (10.16), (10.39), (10.40) and inequalities (10.37) from (10.36) one gets an upper bound of $P_n(\theta)$.

Variance of the PSP. In this subsection, for simplicity, we consider the following case

$$I_m = \begin{cases} m, & \text{if } m > 0, \\ -m + 1, & \text{if } m \le 0. \end{cases} \qquad (10.41)$$

By Lemmas 10.7 and 10.8 the variance of γ_n can be written as

$$\text{Var}(\gamma_n) = 2 \sum_{\theta=\frac{1}{2}}^{n+\frac{1}{2}} \theta^2 P_n(\theta). \qquad (10.42)$$

Theorem 10.4. *If interactions I_m satisfy (10.41) and β large enough then for any $n \ge 1$*

$$\frac{1}{4} \le \text{Var}(\gamma_n) \le\sim \frac{\tau A(\tau) \cosh\left(\frac{\tau^2}{1-\tau}\right)}{2 \sinh(2\tau)} \left(1 + \frac{3\tau(\tau+3)}{(1-\tau)^2} \right),$$

where $\tau = e^{-\beta}$,

$$A(\tau) = \cosh\left(\frac{\tau^2}{1-\tau}\right) \cosh\left(\tau(1+\tau)\right) - \sinh\left(\tau^2\right) \sinh\left(\tau(1+\tau)\right).$$

Proof. The lower bound easily follows from (10.42). We shall prove upper bound. It follows from (10.42), (10.36) and (10.37) that

$$\mathrm{Var}(\gamma_n) = 2 \sum_{\theta=\frac{1}{2}}^{n+\frac{1}{2}} \theta^2 \frac{e^{-\beta I_{\theta+1/2}} Y_{-n,\theta-1/2} Y_{\theta+1/2,n}}{Z_n^{\pm}} \leq$$

$$2 \sum_{\theta=\frac{1}{2}}^{n+\frac{1}{2}} \theta^2 \frac{e^{-\beta I_{\theta+1/2}} \overleftarrow{Z}_{n,\theta} \overrightarrow{Z}_{n,\theta}}{Z_n^{\pm}}.$$

By (10.41), from (10.16) we get

$$Z_n^{\pm} = \frac{1}{2}\left(\exp\left(2\sum_{i=0}^{n} \ln(1+\tau^{i+1}) \right) - \exp\left(2\sum_{i=0}^{n} \ln(1-\tau^{i+1}) \right) \right) \sim$$

$$\frac{1}{2}\left(\exp\left(2\sum_{i=0}^{n} \tau^{i+1} \right) - \exp\left(-2\sum_{i=0}^{n} \tau^{i+1} \right) \right) = \tag{10.43}$$

$$\sinh\left(\frac{2\tau(1-\tau^{n+1})}{1-\tau} \right) \geq \sinh(2\tau).$$

Here we used $\ln(1+\tau^i) \sim \tau^i$ for small τ (i.e., large β).

Similarly from (10.39) and (10.40), for $\theta \geq \frac{1}{2}$, we get

$$\overrightarrow{Z}_{n,\theta} \sim \cosh\left(\frac{\tau^{\theta+\frac{3}{2}}\left(1-\tau^{n-\theta+\frac{1}{2}}\right)}{1-\tau} \right) \leq \cosh\left(\frac{\tau^{\theta+\frac{3}{2}}}{1-\tau} \right) \leq \cosh\left(\frac{\tau^2}{1-\tau} \right);$$

$$\overleftarrow{Z}_{n,\theta} \sim \cosh\left(\frac{\tau^2\left(1-\tau^n\right)}{1-\tau} - \tau^{\theta+\frac{1}{2}}(1+\tau) \right) \leq A(\tau).$$

Hence

$$\mathrm{Var}(\gamma_n) \leq 2\tau \left(\frac{1}{4} + \sum_{m=1}^{\infty} (m+\frac{1}{2})^2 \tau^m \right) \frac{\cosh\left(\frac{\tau^2}{1-\tau}\right) A(\tau)}{\sinh(2\tau)}. \tag{10.44}$$

One can check that

$$\sum_{m=1}^{\infty} (m+\frac{1}{2})^2 \tau^m = \frac{3\tau(\tau+3)}{4(1-\tau)^2}. \tag{10.45}$$

Thus from (10.44) and (10.45) one gets the assertion of the Theorem. $\square$

Remark 10.4. The estimation $\frac{1}{4} \leq \mathrm{Var}(\gamma_n)$ is true for any interactions I_n, i.e., the condition (10.41) is not necessary.

Corollary 10.1. *For any n the following holds*

$$\lim_{\beta\to\infty} \mathrm{Var}(\gamma_n) = \frac{1}{4}.$$

10.2 q-component models

In this section we consider models where the spin takes values in the set $\Phi = \{v_1, v_2, ..., v_q\}$, $q \geq 2$. A *configuration* σ on V is then defined as a function $x \in V \to \sigma(x) \in \Phi$; the set of all configurations coincides with $\Omega = \Phi^V$.

The Hamiltonian of the q-component model has the form

$$H(\sigma) = \sum_{<x,y>\in L} \lambda(\sigma(x), \sigma(y)) + \sum_{x\in V} h(\sigma(x)) \qquad (10.46)$$

where $\lambda(v_i, v_j) = \lambda_{ij}$, $i, j = 1, ..., q$ is given by a symmetric matrix of order $q \times q$, $h(v_j) \equiv h_j \in R$, $j = 1, ..., q$ and $\sigma \in \Omega$.

10.2.1   *Contours for the q-component models on the Cayley tree*

Let $\Lambda \subset V$ be a finite set, $\Lambda' = V \setminus \Lambda$ and $\omega_\Lambda = \{\omega(x), x \in \Lambda'\}$, $\sigma_\Lambda = \{\sigma(x), x \in \Lambda\}$ a given configuration. The energy of the configuration σ_Λ has the form

$$H_\Lambda(\sigma_\Lambda|\omega_{\Lambda'}) = \sum_{<x,y>:\, x,y\in\Lambda} \lambda(\sigma(x), \sigma(y))+ \qquad (10.47)$$

$$\sum_{<x,y>:\, x\in\Lambda, y\in\Lambda'} \lambda(\sigma(x), \omega(y)) + \sum_{x\in\Lambda} h(\sigma(x)).$$

Let $\omega_{\Lambda'}^{(i)} \equiv v_i$, $i = 1, ..., q$ be a constant configuration outside Λ. For each i we extend the configuration σ_Λ inside Λ to the entire tree by the ith constant configuration and denote this configuration by $\sigma_\Lambda^{(i)}$ and $\Omega_\Lambda^{(i)} = \{\sigma_\Lambda^{(i)}\}$. Now we describe a boundary of the configuration $\sigma_\Lambda^{(i)}$.

Consider V_n and for a given configuration $\sigma_\Lambda^{(i)} \in \Omega_\Lambda^{(i)}$ denote $V_n^{(j)} \equiv V_n^{(j)}(\sigma_\Lambda^{(i)}) = \{t \in V_n : \sigma_\Lambda^{(i)}(t) = v_j\}, j = 1, ..., q$. Let $G^{n,j} = (V_n^{(j)}, L_n^{(j)})$ be the graph such that

$$L_n^{(j)} = \{l =< x, y >\in L : x, y \in V_n^{(j)}\}, \quad j = 1, ..., q.$$

It is clear, that for a fixed n the graph $G^{n,j}$ contains a finite number $(= m)$ of maximal connected subgraphs $G_r^{n,j}$, i.e.,

$$G^{n,j} = \{G_1^{n,j}, ..., G_m^{n,j}\}, \quad G_r^{n,j} = (V_{n,r}^{(j)}, L_{n,r}^{(j)}), \quad r = 1, ..., m.$$

Here $V_{n,r}^{(j)}$ is the set of vertices and $L_{n,r}^{(j)}$ the set of edges of $G_r^{n,j}$.

For a set A denote by $|A|$ the number of elements in A.

Two edges $l_1, l_2 \in L$, $(l_1 \neq l_2)$ are called *nearest neighboring edges* if $|i(l_1) \cap i(l_2)| = 1$, and we write $< l_1, l_2 >_1$.

For any connected component $K \subset \Gamma^k$ denote by $E(K)$ the set of edges of K and

$$b(K) = \{l \in L \setminus E(K) : \exists l_1 \in E(K) \text{ such that } < l, l_1 >_1\}.$$

Definition 10.2. An edge $l = < x, y > \in L_{n+1}$ is called a *boundary edge* of the configuration $\sigma_{V_n}^{(i)}$ if $\sigma_{V_n}^{(i)}(x) \neq \sigma_{V_n}^{(i)}(y)$. The set of boundary edges of the configuration is called *boundary* $\partial(\sigma_{V_n}^{(i)}) \equiv \Gamma$ of this configuration.

The boundary Γ consists of $\frac{q(q-1)}{2}$ parts

$$\partial_\varepsilon(\sigma_{V_n}^{(i)}) \equiv \Gamma_\varepsilon, \quad \varepsilon \in \{ij : i < j; i, j = 1, ..., q\} \equiv Q_q,$$

where, for instance Γ_{12} is the set of edges $l = < x, y >$ with $\sigma(x) = v_1$ and $\sigma(y) = v_2$.

The (finite) sets $b(G_r^{n,j}), j = 1, ..., q, r = 1, ..., m$ (together with indication for each edge of this set which part $\Gamma_\varepsilon, \varepsilon \in Q_q$ of the boundary contains this edge) are called *subcontours* of the boundary Γ.

The set $V_{n,r}^{(j)}, \quad j = 1, ..., q, r = 1, ..., m$ is called the *interior* $\text{Int} b(G_r^{n,j})$ of $b(G_r^{n,j})$.

The set of edges from a subcontour γ is denoted by $\text{supp}\gamma$. The configuration $\sigma_{V_n}^{(i)}$ takes the same value $v_j, j = 1, ..., q$ at all points of the connected component $G_r^{n,j}$. This value $v = v(G_r^{n,j})$ is called the *mark* of the subcontour and denoted by $v(\gamma)$, where $\gamma = b(G_r^{n,j})$.

The collection of subcontours $\tau = \tau(\sigma_{V_n}^{(i)}) = \{\gamma_r\}$ generated by the boundary $\Gamma = \Gamma(\sigma_{V_n}^{(i)})$ of the configuration $\sigma_{V_n}^{(i)}$ has the following properties

(a) Every subcontour $\gamma \in \tau$ lies inside the set V_{n+1}.

(b) For every two subcontours $\gamma_1, \gamma_2 \in \tau$ their supports $\text{supp}\gamma_1$ and $\text{supp}\gamma_2$ satisfy $|\text{supp}\gamma_1 \cap \text{supp}\gamma_2| \in \{0, 1\}$.

The subcontours $\gamma_1, \gamma_2 \in \tau$ are called *adjacent* if $|\text{supp}\gamma_1 \cap \text{supp}\gamma_2| = 1$.

(c) For any two adjacent subcontours $\gamma_1, \gamma_2 \in \tau$ we have $v(\gamma_1) \neq v(\gamma_2)$.

A set of subcontours $A \subset \tau$ is called *connected* if for any two subcontours $\gamma_1, \gamma_2 \in A$ there is a sequence of subcontours $\gamma_1 = \tilde{\gamma}_1, \tilde{\gamma}_2, ..., \tilde{\gamma}_l = \gamma_2$ in the set A such that for each $i = 1, ..., l-1$ the subcontours $\tilde{\gamma}_i$ and $\tilde{\gamma}_{i+1}$ are adjacent.

Definition 10.3. Any maximal connected set (component) of subcontours is called a *contour* of boundary Γ (see Fig. 10.1).

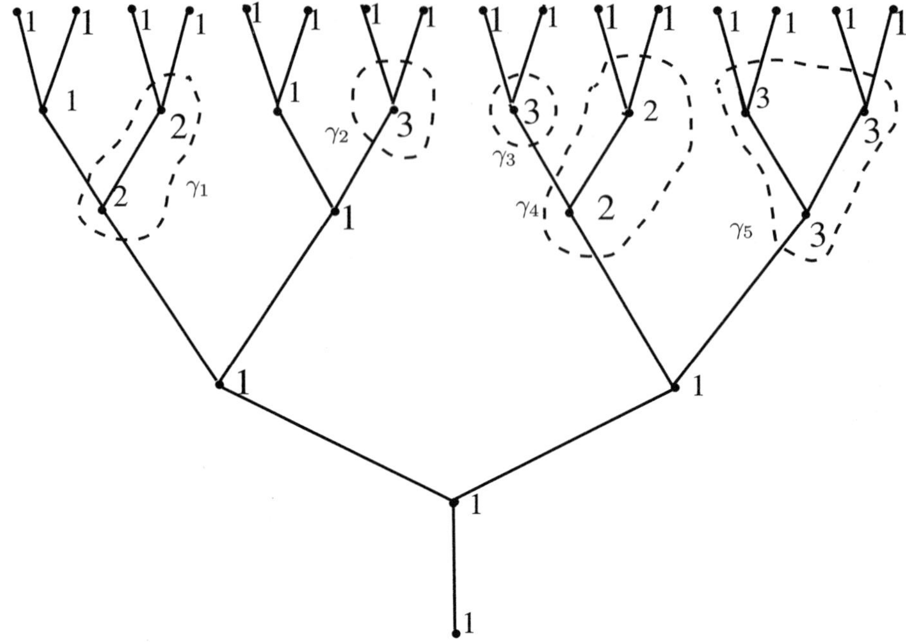

Fig. 10.1　Subcontours: γ_i, $i = 1, 2, 3, 4, 5$; Contours: $\Upsilon_1 = \{\gamma_1\}$, $\Upsilon_2 = \{\gamma_2\}$, $\Upsilon_3 = \{\gamma_3, \gamma_4\}$, $\Upsilon_4 = \{\gamma_5\}$.

Let $\Upsilon = \{\gamma_r, r = 1, 2, ...\}$ (where γ_r is subcontour) be a contour of Γ denote

$$\operatorname{Int}\Upsilon = \cup_j \operatorname{Int}\gamma_j; \quad \operatorname{supp}\Upsilon = \cup_j \operatorname{supp}\gamma_j;$$

$$\operatorname{ext}\Upsilon = V \setminus (\operatorname{Int}\Upsilon \cup \operatorname{supp}\Upsilon); \quad |\Upsilon| = |\operatorname{supp}\Upsilon|.$$

Let Υ be a contour, a subgraph of $\Gamma^k \setminus \operatorname{supp}\Upsilon$ is called *with v_j mark* if the neighboring spins of Υ have the value v_j. If the exterior $\operatorname{ext}\Upsilon$ has a v_j mark, then Υ is called *j-contour* and denoted by Υ^j. Denote by $\operatorname{Int}_j\Upsilon$ the union of all interior components with v_j mark of $\Gamma^k \setminus \operatorname{supp}\Upsilon$.

Using properties of subcontours it is easy to see that the collection of contours $\alpha = \{\Upsilon_r\}$ generated by the boundary Γ has the following properties

(I) Every contour $\Upsilon \in \alpha$ lies inside the set V_{n+1}.

(II) For every two contours $\Upsilon_1, \Upsilon_2 \in \alpha$ their supports $\operatorname{supp}\Upsilon_1$ and $\operatorname{supp}\Upsilon_2$ are disjoint.

(III) For every two contours $\Upsilon_1, \Upsilon_2 \in \alpha$ if $\mathrm{ext}\,\Upsilon_1$ is with v_j mark then $\mathrm{ext}\,\Upsilon_2$ is also with v_j mark $j = 1, ..., q$, i.e., $v(\mathrm{ext}\,\Upsilon_1) = v(\mathrm{ext}\,\Upsilon_2)$.

A collection of contours α with given marks that has the properties (I)-(III) is called a *configuration of contours*.

The set of all configurations α of contours inside V_{n+1} having the exterior mark v_i is denoted by $\mathcal{U}_n^{(i)}$.

As we have seen, the boundary of a configuration $\sigma_{V_n}^{(i)}$ of spin generates the configuration of contours $\alpha = \alpha(\sigma_{V_n}^{(i)}) \in \mathcal{U}_n^{(i)}$. The converse assertion is also true.

Lemma 10.9. *Every configuration of contours $\alpha = \{\Upsilon_r\}$ inside V_{n+1} with given marks and the exterior mark v_i is a boundary of some configuration of spins $\sigma_{V_n}^{(i)} \in \Omega_{V_n}^{(i)}$.*

Proof. Let $\alpha \in \mathcal{U}_n^{(i)}$. The configuration of spins $\sigma_{V_n}^{(i)} = \sigma_{V_n}^{(i)}(\alpha)$ is constructed as follows. Outside of all contours we put $\sigma_{V_n}^{(i)}(x) = v_i$. Inside of subcontour γ with mark v_j we put $\sigma_{V_n}^{(i)}(x) = v_j$. Obviously, the resulting configuration $\sigma_{V_n}^{(i)} \in \Omega_{V_n}^{(i)}$ and with outside of V_n equal v_i, also its boundary $\Gamma(\sigma_{V_n}^{(i)})$ coincides with the configuration α. $\qquad\qquad\square$

10.2.2 *Additional properties of the contours*

Let G be a graph, denote the vertex and edge set of the graph G by $V(G)$ and $E(G)$, respectively. For $A \subset V$ denote

$$\partial(A) = \{x \in V \setminus A : \exists y \in A, \quad \text{such that} \quad <x, y>\}.$$

Using the induction over n one can prove

Lemma 10.10. *Let K be a connected subgraph of the Cayley tree Γ^2 of order two, such that $|V(K)| = n$, then $|\partial V(K)| = n + 2$.*

Lemma 10.11. [30] *Let G be a countable graph of maximal degree $k + 1$ (i.e. each $x \in V(G)$ has at most $k + 1$ neighbors) and let $\tilde{N}_{n,G}(x)$ be the number of connected subgraphs $G' \subset G$ with $x \in V(G')$ and $|E(G')| = n$. Then*

$$\tilde{N}_{n,G}(x) \le (e \cdot k)^n.$$

For $x \in V$ we will write $x \in \Upsilon$ if there is $l \in \Upsilon$ such that $x \in i(l)$.
Denote $N_r(x) = |\{\Upsilon : x \in \Upsilon, |\Upsilon| = r\}|$.

Lemma 10.12. *If $k = 2$ (i.e., the Cayley tree of order two). Then*

$$N_r(x) \le \theta \cdot (4e)^{2r-1}, \tag{10.48}$$

where $\theta = \frac{1}{2(16e^2-1)}$.

Proof. Denote by K_Υ the minimal connected subgraph of Γ^2, which contains a contour Υ. It is easy to see that if $\Upsilon = \{\gamma_1, ..., \gamma_m\}$, $m \ge 1$, then

$$E(K_\Upsilon) = \operatorname{supp}\Upsilon \cup \left(\cup_{i=1}^m \{< x, y >: x, y \in \operatorname{Int}\gamma_i\} \right). \tag{10.49}$$

By Lemma 10.11 we have

$$|E(K_\Upsilon)| = |\Upsilon| + \sum_{i=1}^m (|\operatorname{Int}\gamma_i| - 1) = |\Upsilon| + \sum_{i=1}^m (|\gamma_i| - 3) = 2|\Upsilon| - 2m - 1. \tag{10.50}$$

Here we used $\sum_{i=1}^m |\gamma_i| = |\Upsilon| + m - 1$.

Since $\Upsilon \subseteq K_\Upsilon$ we get $|\Upsilon| \le |E(K_\Upsilon)| = 2|\Upsilon| - 2m - 1$. Consequently, $|\Upsilon| \ge 2m + 1$ which implies $1 \le m \le \frac{|\Upsilon|-1}{2}$. A combinatorial calculations show that

$$N_r(x) \le \sum_{m=1}^{[\frac{r-1}{2}]} \binom{2r - 2m - 1}{r} \tilde{N}_{2r-2m-1,\Gamma^2}(x), \tag{10.51}$$

where $[a]$ is the integer part of a. Using inequality $\binom{n}{r} \le 2^{n-1}$, $r \le n$ and Lemma 10.11 from (10.51) we get (10.48). $\square$

10.2.3 *The contour Hamiltonian*

The following lemma plays the key role in our analysis.

Lemma 10.13. *The energy $H_\Lambda^{(i)}(\sigma_\Lambda^{(i)}) \equiv H_\Lambda(\sigma_\Lambda^{(i)}|\sigma_{\Lambda'}^{(i)} \equiv v_i)$ (see (10.47)) has the form*

$$H_\Lambda^{(i)}(\sigma_\Lambda^{(i)}) = \sum_{\varepsilon \in Q_q} \rho_\varepsilon |\Gamma_\varepsilon| + \sum_{j=1}^q \eta_j |\tilde{V}_j| + (|W_{n+1}| + m - 1)\lambda_{ii} - \sum_{j=1}^q \lambda_{jj} m_j. \tag{10.52}$$

Here $\Lambda = V_n$, $m_j \equiv m_j(\sigma_\Lambda^{(i)}) = \left|\{\gamma : \gamma \subset \Gamma(\sigma_\Lambda^{(i)}), v(\gamma) = v_j\}\right|$; $Q_q = \{ij : i, j = 1, ..., q, i < j\}$, $\rho_\varepsilon = \lambda_\varepsilon - \lambda_{ii}, \varepsilon \in Q_q$; $\eta_j = h_j + \lambda_{jj}$; $m = \sum_{j=1}^q m_j$; $\tilde{V}_j \equiv \tilde{V}_j(\sigma_\Lambda^{(i)}) \subseteq V_n$ is the set of all points $x \in V_n$ with $\sigma_\Lambda^{(i)}(x) = v_j$.

Proof. Using the equalities $|\Gamma| = \sum_{\varepsilon \in Q_q} |\Gamma_\varepsilon|$, $|V_n| = \sum_{j=1}^{q} |\tilde{V}_j|$, $|V_{n+1}| = |V_n| + |W_{n+1}|$, $|\tilde{V}_j| = \sum_{r=1}^{m} |\mathrm{Int}_j \gamma_r|, j \neq i$ and the fact that if K is a connected subgraph of Γ^k then the number of edges of K is equal to $|K| - 1$ we get

$$H_\Lambda^{(i)}(\sigma_\Lambda^{(i)}) = \sum_{\varepsilon \in Q_q} \lambda_\varepsilon |\Gamma_\varepsilon| + \lambda_{ii}\left(|V_{n+1}| - 1 - |\Gamma| - \sum_{r=1}^{m}\left(|\mathrm{Int}\gamma_r| - 1\right)\right) +$$

$$\sum_{r=1}^{m}\left[\sum_{s=1}^{q} \lambda_{ss}(|\mathrm{Int}\gamma_r| - 1)\mathbf{1}_{(v(\gamma_r)=v_s)}\right] + \sum_{j=1}^{q} h_j|\tilde{V}_j| =$$

$$\sum_{\varepsilon \in Q_q}(\lambda_\varepsilon - \lambda_{ii})|\Gamma_\varepsilon| + \lambda_{ii}(|V_{n+1}| - 1) + \sum_{j=1}^{q}(\lambda_{jj} - \lambda_{ii})(|\tilde{V}_j| - m_j) + \sum_{j=1}^{q} h_j|\tilde{V}_j| =$$

$$\sum_{\varepsilon \in Q_q}\rho_\varepsilon|\Gamma_\varepsilon| + \sum_{j=1}^{q}\eta_j|\tilde{V}_j| + (|W_{n+1}| + m - 1)\lambda_{ii} - \sum_{j=1}^{q}\lambda_{jj}m_j. \qquad \square$$

Denote by $P_{\Lambda,\beta,\eta}^{(i)}(\sigma_\Lambda^{(i)}) = \mathbf{Z}^{-1}\exp\{-\beta H_\Lambda^{(i)}(\sigma_\Lambda^{(i)})\}$, $(i = 1,...,q, \eta = (\eta_1,...,\eta_q))$, the Gibbs measure given by $H_\Lambda^{(i)}$ on $\Omega_\Lambda^{(i)} = \{\sigma_\Lambda^{(i)}\}$, where $\mathbf{Z}$ is the normalizing factor. Assume that the constants $\rho_\varepsilon > 0, \varepsilon \in Q_q$ and the parameters $\eta_1,...,\eta_q$ can vary arbitrary.

If we add the same constant C to all $\eta_j, j = 1,...,q$, the Hamiltonian (10.52) increases by $C|\Lambda|$ and the measure $P_{\Lambda,\beta,\eta}^{(i)}$ does not change. Therefore we can assume that

$$\eta_1 + ... + \eta_q = 0. \tag{10.53}$$

Denote $(\sigma_\Lambda^{(i)})_{\min}$, the minimum point of $H_\Lambda^{(i)}$. If the constant configuration $\omega^{(i)}$ coincides, for each Λ, with the minimum configuration $(\sigma_\Lambda^{(i)})_{\min}$ of $H_\Lambda^{(i)}$, it is called the *ground state*. For instance, it is easy to see that $\omega^{(i)} \equiv v_i$ is the ground state if

$$\eta_i \leq \min_{j:j\neq i, j=1,...,q}\eta_j, \quad \lambda_{ii} \leq \min_{j:j\neq i,j=1,...,q}\lambda_{jj}. \tag{10.54}$$

Assume $\lambda_{ii} = \lambda$ for any $i = 1,...,q$ and denote

$$S_{i_1,...,i_r}^0 = \left\{(\eta_1,...,\eta_q): \sum_{j=1}^{q}\eta_j = 0, \eta_{i_1} = ... = \eta_{i_r} < \min_{j\notin\{i_1,...,i_r\}}\eta_j\right\}.$$

In this set each configuration $\omega^{(i_s)}, s = 1, ..., r$ is a ground state. For instance in $S^0_{1,2,...,q} = \{\eta_1 = ... = \eta_q = 0\}$ all configurations $\omega^{(i)}, i = 1, ..., q$ are ground states.

Remark 10.5. i) The phase diagram of the (limit) Gibbs measure is close to the phase diagram of ground states for sufficiently large β in a sufficiently small neighborhood of the point $\eta = (0, ..., 0)$. Namely, the phase diagram looks as follows: (1) There is a point $(\eta_1^0, ..., \eta_q^0)$ close to the point $(0, ..., 0)$, for which q Gibbs measures exist. (2) There are q lines emerging from this point and $q - 1$ different Gibbs measures "live" on each of them. (3) The lines confine q areas $S_1, ..., S_q$ in which $q - 2$ Gibbs measures "live" ... (q) There are areas in which 1 Gibbs measure "lives".

ii) The rigorous proof of the described phase diagram for large β will be obtained using Pirogov-Sinai theory on a Cayley tree. In the sequel of the section we shall study more simple cases of (10.46), i.e., Potts and SOS (solid on solid) models on Cayley tree.

10.2.4   *The Potts model*

In this section we consider the (ferromagnetic) Potts model on Cayley tree Γ^2 of order two, which is defined by (10.46) with $\lambda_{ii} = -1$, for any $i = 1, ..., q$, $\lambda_{ij} = 0$, if $i \neq j$ and $h_i = 0, i = 1, ..., q$. We shall prove that there are at least q Gibbs measures for the model. Note that the result was proved in [85] (see Chapter 5) using theory of Markov random fields and recurrent equations of this theory. Here we shall use our "contour method" on Cayley tree.

The energy $H_\Lambda(\sigma|\varphi)$ of the configuration σ in the presence of boundary configuration $\varphi = \{\varphi(x), x \in V \setminus \Lambda\}$ is expressed by the formula

$$H_\Lambda(\sigma|\varphi) = - \sum_{\substack{<x,y>: \ x,y \in \Lambda}} \delta_{\sigma(x)\sigma(y)} - \sum_{\substack{<x,y>: \ x \in \Lambda, \ y \in V \setminus \Lambda}} \delta_{\sigma(x)\varphi(y)}, \quad (10.55)$$

where δ is the Kronecker symbol.

The Gibbs measure on the space $\Omega_\Lambda = \{v_1, ..., v_q\}^\Lambda$ with boundary condition φ is defined in the usual way.

$$\mu_{\Lambda,\beta}(\sigma/\varphi) \equiv \mu_{\Lambda,\beta}^\varphi(\sigma) = \mathbf{Z}^{-1}(\Lambda, \beta, \varphi) \exp(-\beta H_\Lambda(\sigma|\varphi)), \quad (10.56)$$

where $\mathbf{Z}(\Lambda, \beta, \varphi)$ is the normalizing factor (statistical sum).

Let us consider a sequence of balls on Γ^2

$$V_1 \subset V_2 \subset ... \subset V_n \subset ..., \quad \cup V_n = V,$$

and q sequences of boundary conditions outside these balls:

$$\sigma_n^{(i)} \equiv v_i, n = 1, 2, ..., i = 1, ..., q.$$

By very similar argument of proof of Lemma 9.2 in [161] one can prove that each of q sequences of measures $\{\mu_{V_n,\beta}^{\sigma_n^{(i)}}, n = 1, 2, ...\}, i = 1, ..., q$ contains a convergent subsequence.

We denote the corresponding limits by $\mu_\beta^i, i = 1, ..., q$. Our purpose is to show for a sufficiently large β these measures are different.

Denote by $\sigma_\Lambda^{(i)}$ the configuration σ_Λ extended by $\sigma_{V\setminus\Lambda}^{(i)} \equiv v_i$ and by $\Gamma = \Gamma(\sigma_\Lambda^{(i)})$ its boundary as was explained above.

Lemma 10.14. *Let γ be a fixed contour and $p_i(\gamma) = \mu_\beta^i\{\sigma : \gamma \subset \Gamma(\sigma(V_n))\}$. Then*

$$p_i(\gamma) \le \exp\{-\beta|\gamma|\}.$$

Proof. For any σ, which coincides with $\sigma^{(i)} \equiv v_i$ outside of V_n by Lemma 10.11 we have

$$H_{V_n}(\sigma) = 1 - |V_{n+1}| + |\Gamma|,$$

here we have used $\rho_\varepsilon = 1, \varepsilon \in Q_q, \eta_j = -1, j = 1, ..., q$.

By definition we have (where $\Lambda = V_n$, $\sigma(\Lambda) = \sigma_\Lambda$)

$$p_i(\gamma) = \frac{\sum\limits_{\sigma(\Lambda):\gamma\subset\Gamma(\sigma(\Lambda))} \exp\{-\beta H_\Lambda(\sigma)\}}{\sum\limits_{\sigma(\Lambda)} \exp\{-\beta H_\Lambda(\sigma)\}} = \frac{\sum\limits_{\sigma(\Lambda):\gamma\subset\Gamma} \exp\{-\beta|\Gamma|\}}{\sum\limits_{\sigma(\Lambda)} \exp\{-\beta|\Gamma|\}}.$$

Denote $F_\gamma = \{\sigma(\Lambda) : \gamma \subset \Gamma(\sigma(\Lambda))\}$, $F_\gamma^- = \{\sigma(\Lambda) : \gamma \cap \Gamma(\sigma(\Lambda)) = \emptyset\}$. Define the map $\chi_\gamma; F_\gamma \to F_\gamma^-$ as follows: for $\sigma(\Lambda) \in F_\gamma$ we destroy the contour γ changing the values $\sigma(x)$ inside of γ to v_i. The constructed configuration is $\chi_\gamma(\sigma(\Lambda)) \in F_\gamma^-$.

It is clear that

$$\Gamma(\sigma(\Lambda)) = \Gamma(\chi_\gamma(\sigma(\Lambda))) \cup \gamma, \quad |\Gamma(\sigma(\Lambda))| = |\Gamma(\chi_\gamma(\sigma(\Lambda)))| + |\gamma|.$$

For a given γ (with corresponding marks) the map χ_γ is one-to-one map.

Further, we can write

$$p_i(\gamma) \le \frac{\sum\limits_{\sigma(\Lambda)\in F_\gamma} \exp\{-\beta|\Gamma(\sigma(\Lambda))|\}}{\sum\limits_{\sigma(\Lambda)\in F_\gamma} \exp\{-\beta|\Gamma(\chi_\gamma(\sigma(\Lambda)))|\}} = \exp\{-\beta|\gamma|\}.$$

$\square$

Lemma 10.15. *For $k = 2$ and all sufficiently large β, there is a constant $C = C(\beta) > 0$, such that*

$$\mu^i_\beta\{\sigma(\Lambda): \quad |\gamma| > C \ln|\Lambda| \ \text{for some} \ \gamma \subset \Gamma(\sigma(\Lambda))\} \to 0, \quad \text{as} \ |\Lambda| \to \infty.$$

Proof. Suppose $\beta > 2 + 4\ln 2$, then by Lemma 10.12

$$\mu^i_\beta\{\sigma(\Lambda): \gamma \subset \Gamma(\sigma(\Lambda)), \ t \in \gamma, |\gamma| = r\} \leq N_r(t) \cdot e^{-\beta r} \leq \theta \cdot (4e)^{2r-1} \cdot e^{-\beta r}.$$

$$\mu^i_\beta\{\sigma(\Lambda): \gamma \subset \Gamma(\sigma(\Lambda)), \ t \in \gamma, |\gamma| > C_1 \ln|\Lambda|\} \leq \frac{\theta}{4e} \sum_{r \geq C_1 \ln|\Lambda|} (16e^{2-\beta})^r =$$

$$\frac{\theta}{4e} \cdot \frac{(16 \cdot e^{2-\beta})^{C_1 \ln|\Lambda|}}{1 - 16e^{2-\beta}} = \frac{\theta}{4e} \cdot \frac{|\Lambda|^{C_1(\ln 16 + 2 - \beta)}}{1 - 16e^{2-\beta}},$$

where C_1 will be defined later. Thus, we have

$$\mu^i_\beta\{\sigma(\Lambda): \exists \gamma \subset \Gamma(\sigma(\Lambda)), \ |\gamma| > C_1 \ln|\Lambda|\} \leq \frac{\theta|\Lambda|^{C_1(\ln 16 + 2 - \beta) + 1}}{4e(1 - 16e^{2-\beta})}.$$

The last expression tends to zero if $|\Lambda| \to \infty$ and $C_1 > \frac{1}{\beta - 2 - \ln 16}$. $\Box$

Lemma 10.16. *If $k = 2$, $x \in \Lambda$. Then uniformly by Λ*

$$\mu^i_\beta\{\sigma(\Lambda): \sigma(x) = v_j\} \to 0, \quad j \neq i \ \text{as} \ \beta \to \infty.$$

Proof. If $\sigma(x) = v_j$, then x is a point for interior of some contour, we shall write this as $x \in \text{Int}_j\gamma$. Assume $t \in \gamma$ and $x \in \text{Int}_j\gamma$ then for any such contour we have $|\gamma| \geq |t| + 2$.

Consequently,

$$\mu^i_\beta\{\sigma(\Lambda): x \in \text{Int}_j\gamma, \ t \in \gamma, |\gamma| < C_1 \ln|\Lambda|\} \leq \frac{\theta(16e^{2-\beta})^{|t|+2}}{4e(1 - 16e^{2-\beta})}.$$

$$\mu^i_\beta\{\sigma(x) = v_j\} \leq \mu^i_\beta\{\sigma(\Lambda): x \in \text{Int}_j\gamma, \ \gamma \subset \Gamma(\sigma(\Lambda))\} \leq$$

$$\frac{\theta}{4e} \sum_{|t|=1}^{C_1 \ln|\Lambda|} \frac{(16 \cdot e^{2-\beta})^{|t|+2}}{1 - 16e^{2-\beta}} +$$

$$\mu^i_\beta\{\sigma(\Lambda): \exists \gamma \subset \Gamma(\sigma(\Lambda)), \text{such that}, |\gamma| \geq C_1 \ln|\Lambda|\} \leq$$

$$\frac{1024\theta e^{5-3\beta}}{(1 - 16e^{2-\beta})^2} + \frac{\theta|\Lambda|^{C_1(\ln 16 + 2 - \beta) + 1}}{4e(1 - 16e^{2-\beta})}. \tag{10.57}$$

For $|\Lambda| \to \infty$ and $\beta \to \infty$ from (10.57) we get $\mu^i_\beta\{\sigma(x) = v_j\} \to 0, j \neq i.\Box$

Theorem 10.5. *For all sufficiently large β there are at least q Gibbs measures for the ferromagnetic Potts model on Cayley tree of order two.*

Proof. By Lemma 10.16 we have

$$\lim_{\beta \to \infty} \mu_\beta^i \{\sigma(x) = v_j\} = \begin{cases} 0 \text{ if } i \neq j \\ 1 \text{ if } i = j. \end{cases} \tag{10.58}$$

Thus, from (10.58) for any $i, j \in \{1, ..., q\}, i \neq j$ and sufficiently large β we have

$$\mu_\beta^i(\sigma(x) = v_i) \neq \mu_\beta^j(\sigma(x) = v_i).$$

This completes the proof. $\qquad\square$

10.2.5 *The SOS model*

In this subsection we consider the solid-on-solid (SOS) model. This is a spin model, the spin takes values in the set $\Phi_1 = \{1, 2, ..., q\}$, $q \geq 2$.

The Hamiltonian of SOS model is a particular case of (10.46) with $\lambda_{ij} = -J|i - j|, i, j \in \Phi_1$ and $h_j = 0, j \in \Phi_1$.

In $J < 0$ case the ground states are 'flat' configurations, with $\sigma(x) \equiv j \in \Phi_1$, in $J > 0$ two 'contrasting' checker-board configurations where $|\sigma(x) - \sigma(y)| = m \; \forall \; \langle x, y \rangle$.

For simplicity (without loss of generality) we set $J = -1$. Now we shall prove an analogue of Lemma 10.14.

Denote by $H_\Lambda^{(i)}(\omega_\Lambda)$ the energy $H_\Lambda(\omega_\Lambda | \bar{\omega}_{V \backslash \Lambda}^{(i)})$ corresponding to the configuration $\bar{\omega}_{V \backslash \Lambda}^{(i)} \equiv i$ and by $P_{\Lambda, \beta}^{(i)}, \quad i = 1, ..., q$ the corresponding Gibbs measure.

Denote by $\omega_\Lambda^{(i)}$ the configuration ω_Λ extended by $\bar{\omega}_{V \backslash \Lambda}^{(i)}$ and by $\Gamma = \Gamma(\omega_\Lambda^{(i)})$ its boundary.

Lemma 10.17. *Let γ be a fixed contour and $p_\beta^{(i)}(\gamma) = P_\beta^{(i)} \{\omega : \gamma \subset \Gamma(\omega_\Lambda)\}$. Then*

$$p_\beta^{(i)}(\gamma) \leq \exp\{-\beta|\gamma|\}. \tag{10.59}$$

Proof. In the case of the SOS model by Lemma 10.13 we have

$$H_{V_n}(\omega) = \sum_{i,j \in \Phi_1 : i < j} (j - i) \cdot |\Gamma_{ij}|. \tag{10.60}$$

Note that γ consists of parts γ_ε, $\varepsilon \in Q_q$, where, for instance γ_{12} is the set of edges $l =\, < x, y >$ with $\omega(x) = 1$ and $\omega(y) = 2$.

It is clear that $|\gamma| = \sum_{\varepsilon \in Q_q} |\gamma_\varepsilon|$. For any fixed $\gamma \subset \Gamma$ define χ_γ as in the proof of Lemma 10.14.

Then it is easy to see

$$|\Gamma_{ij}| = |\Gamma_{ij}(\chi_\gamma(\omega_\Lambda))| + |\gamma_{ij}|, \qquad (10.61)$$

$$\sum_{i,j \in \Phi_1 : i < j} (j - i)|\gamma_{ij}| > \sum_{i,j \in \Phi_1 : i < j} |\gamma_{ij}| = |\gamma|. \qquad (10.62)$$

Using (10.60)-(10.62) by a similar argument of the proof of Lemma 10.14 we get (10.59). $\qquad\qquad \square$

Using Lemma 10.17 one can prove analogues of Lemmas 10.15 and 10.16 and also

Theorem 10.6. *For all sufficiently large β there are at least q Gibbs measures for the SOS model on the Cayley tree of order 2.*

Remark 10.6. In [31] the results of this section were generalized for q-component models on a Cayley tree of order $k \geq 2$. The generalizations were in two directions: (1) from $k = 2$ to any $k \geq 2$; (2) from concrete examples (Potts and SOS models) of q-component models to any q-component models (with nearest neighbor interactions). Since in the next sections of this chapter we shall give very general form of a contour argument, here we do not present results of [31].

10.3 An Ising model with competing two-step interactions

In this section we consider models where the spin takes values in the set $\Phi = \{-1, 1\}$. Let $A \subset V$ be a finite subset. Set $-\sigma_A = \{-\sigma_A(x), x \in A\}$.

We define a *periodic configuration* as a configuration $\sigma \in \Omega = \Phi^V$ which is invariant under a subgroup of shifts $G_k^* \subset G_k$ of finite index.

More precisely, a configuration $\sigma \in \Omega$ is called G_k^*-periodic if $\sigma(yx) = \sigma(x)$ for any $x \in G_k$ and $y \in G_k^*$.

For a given periodic configuration the index of the subgroup is called the *period of the configuration*. A configuration that is invariant with respect to all shifts is called *translational-invariant*.

The Hamiltonian of the Ising model with competing interactions has the form

$$H(\sigma) = J_1 \sum_{<x,y>} \sigma(x)\sigma(y) + J_2 \sum_{x,y \in V:\ d(x,y)=2} \sigma(x)\sigma(y), \qquad (10.63)$$

where $J_1, J_2 \in R$ are coupling constants and $\sigma \in \Omega$.

10.3.1 *Ground states*

For a pair of configurations σ and φ that coincide almost everywhere, i.e., everywhere except for a finite number of positions, we consider a relative Hamiltonian $H(\sigma, \varphi)$, the difference between the energies of the configurations σ, φ of the form

$$H(\sigma, \varphi) = J_1 \sum_{<x,y>} (\sigma(x)\sigma(y) - \varphi(x)\varphi(y)) + \tag{10.64}$$

$$J_2 \sum_{x,y \in V : d(x,y)=2} (\sigma(x)\sigma(y) - \varphi(x)\varphi(y)),$$

where $J = (J_1, J_2) \in R^2$ is an arbitrary fixed parameter.

Let M be the set of unit balls with vertices in V. We call the restriction of a configuration σ to the ball $b \in M$ a *bounded configuration* σ_b.

Define the energy of a ball b for configuration σ by

$$U(\sigma_b) \equiv U(\sigma_b, J) = \frac{1}{2} J_1 \sum_{<x,y>, \; x,y \in b} \sigma(x)\sigma(y) + \tag{10.65}$$

$$J_2 \sum_{x,y \in b : \; d(x,y)=2} \sigma(x)\sigma(y),$$

where $J = (J_1, J_2) \in R^2$.

We shall say that two bounded configurations σ_b and $\sigma'_{b'}$ belong to the same class if $U(\sigma_b) = U(\sigma'_{b'})$ and we write $\sigma'_{b'} \sim \sigma_b$.

Using combinatorial calculations one can prove the following

Lemma 10.18. *1) For any configuration σ_b we have*

$$U(\sigma_b) \in \{U_0, U_1, ..., U_{k+1}\},$$

where

$$U_i = \left(\frac{k+1}{2} - i\right) J_1 + \left(\frac{k(k+1)}{2} + 2i(i-k-1)\right) J_2, \quad i = 0, 1, ..., k+1.$$
$$\tag{10.66}$$

2) Let $\mathbb{C}_i = \Omega_i \cup \Omega_i^-, \quad i = 0, ..., k+1$, where

$$\Omega_i = \left\{\sigma_b : \sigma_b(c_b) = +1, \; |\{x \in b \setminus \{c_b\} : \sigma_b(x) = -1\}| = i\right\},$$

$$\Omega_i^- = \left\{-\sigma_b = \{-\sigma_b(x), x \in b\} : \sigma_b \in \Omega_i\right\},$$

and c_b is the center of the ball b. Then for $\sigma_b \in \mathbb{C}_i$ we have $U(\sigma_b) = U_i$.
3) The class $\mathbb{C}_i$ contains $\frac{2(k+1)!}{i!(k-i+1)!}$ configurations.

Lemma 10.19. *The relative Hamiltonian (10.65) has the form*

$$H(\sigma, \varphi) = \sum_{b \in M} (U(\sigma_b) - U(\varphi_b)). \tag{10.67}$$

Proof. Note that for any two vertices x and y such that $<x, y>$ there exist exactly 2 unit balls $b, b' \in M$ such that $x, y \in b \cap b'$. Also, for any two vertices u and v such that $d(u, v) = 2$ there is a unique ball b such that $u, v \in b$. This completes the proof. $\qquad\qquad\square$

Theorem 10.7. *For any class $\mathbb{C}_i$ and for any bounded configuration $\sigma_b \in \mathbb{C}_i$ there exists a periodic configuration φ with period non-exceeding 2 such that $\varphi_{b'} \in \mathbb{C}_i$ for any $b' \in M$ and $\varphi_b = \sigma_b$.*

Proof. For arbitrary given class $\mathbb{C}_i$ and $\sigma_b \in \mathbb{C}_i$ we shall construct configuration φ as follows. Without loss of generality we can take b as the ball with the center $e \in G_k$ (here e is the identity of G_k), i.e., $b = \{e, a_1, ..., a_{k+1}\}$. Assume $\sigma_b(e) = +1$ (the case $\sigma_b(e) = -1$ is very similar). Denote $F = \{j \in \{1, ..., k+1\} : \sigma_b(a_j) = -1\}$. Note that $|F| = i$ since $\sigma_b(e) = +1$ and $\sigma_b \in \mathbb{C}_i$.

Consider two cases:

Case $i = 0$. In this case we have $\sigma_b(x) = 1$ for any $x \in b$, so configuration φ coincides with translational-invariant one $\varphi^+ = \{\varphi(x) \equiv +1\}$. Thus the period of φ is 1.

Case $i \geq 1$. Consider

$$\mathcal{H}_i = \{x \in G_k : \sum_{j \in F} \omega_j(x) - \text{even}\},$$

where $\omega_j(x)$ is the number of a_j in $x \in G_k$. Note that $\mathcal{H}_i$ is a normal subgroup of index 2 for G_k. By our construction (and assumption $\sigma_b(e) = +1$) we have $\sigma_b(x) = +1$ for any $x \in b \cap \mathcal{H}_i$ and $\sigma_b(u) = -1$ for any $u \in b \cap (G_k \setminus \mathcal{H}_i)$.

We continue the bounded configuration $\sigma_b \in \mathbb{C}_i$ to whole lattice Γ^k (which we denote by φ) by

$$\varphi(x) = \begin{cases} 1 & \text{if} \quad x \in \mathcal{H}_i \\ -1 & \text{if} \quad x \in G_k \setminus \mathcal{H}_i. \end{cases}$$

So we obtain a periodic configuration φ with period 2 (=index of the subgroup); then by the construction $\varphi_b = \sigma_b$. Now we shall prove that all restrictions $\varphi_{b'}, b' \in M$ of the configuration φ belong to $\mathbb{C}_i$. Since $\mathcal{H}_i$ is the subgroup of index 2 in G_k, the quotient group has the form $G_k / \mathcal{H}_i = \{\mathcal{H}^0, \mathcal{H}^1\}$ with the cosets $\mathcal{H}^0 = \mathcal{H}_i, \mathcal{H}^1 = G_k \setminus \mathcal{H}_i$.

Let $q_j(x) = |S_1(x) \cap \mathcal{H}^j|$, $j = 0, 1$; where $S_1(x) = \{y \in G_k : \langle x, y \rangle\}$, the set of all nearest neighbors of $x \in G_k$.

Denote $Q(x) = (q_0(x), q_1(x))$. Clearly, $q_0(x)$ (resp. $q_1(x)$) is the number of points y in $S_1(x)$ such that $\varphi(y) = +1$ (resp. $\varphi(y) = -1$).

We note (see Chapter 1) that for every $x \in G_k$ there is a permutation π_x of the coordinates of the vector $Q(e)$ (where e as before is the identity of G_k) such that

$$\pi_x Q(e) = Q(x).$$

Moreover $Q(x) = Q(e)$ if $x \in \mathcal{H}^0$ and $Q(x) = (q_1(e), q_0(e))$ if $x \in \mathcal{H}^1$. Thus for any $b' \in M$ we have (i) if $c_{b'} \in \mathcal{H}^0$ (where $c_{b'}$ is the center of b') then $\varphi_{b'} = \sigma_b$ up to a rotation; (ii) if $c_{b'} \in \mathcal{H}^1$ then $\varphi_{b'} = -\sigma_b$ up to a rotation. Since both $\sigma_b, -\sigma_b \in \mathbb{C}_i$ we get $\varphi_{b'} \in \mathbb{C}_i$ for any $b' \in M$. $\qquad\square$

Definition 10.4. A configuration φ is called a *ground state* for the relative Hamiltonian H if

$$U(\varphi_b) = \min\{U_0, U_1, ..., U_{k+1}\}, \quad \text{for any } b \in M. \tag{10.68}$$

Remark 10.7.

1. Usually, more simple and interesting ground states are periodic ones. In this section we describe some non-periodic ground states as well (cf. [247], Chapter 2).

2. A periodic ground state can be defined differently (see [247]) as a periodic configuration φ such that for any configuration σ that coincides with φ almost everywhere and $H(\varphi, \sigma) \leq 0$. It is easy to see that if φ is a ground state in the sense of Definition 10.4, then it satisfies $H(\varphi, \sigma) \leq 0$. In [122], [201] it was proved that these two definitions (for periodic ground states) are equivalent for Hamiltonians on $\mathbb{Z}^d$. But there is a problem to prove the equivalence of these definitions for Hamiltonians on the Cayley tree: normally the ratio of the number of boundary sites to the number of interior sites of a lattices becomes small in the thermodynamic limit of a large system. For the Cayley tree it does not, since both numbers grow exponentially like k^n.

We set

$$U_i(J) = U(\sigma_b, J), \quad \text{if } \sigma_b \in \mathbb{C}_i, \quad i = 0, 1, ..., k+1.$$

The quantity $U_i(J)$ is a linear function of the parameter $J \in R^2$. For every fixed $m = 0, 1, ..., k+1$ we denote

$$A_m = \{J \in R^2 : U_m(J) = \min\{U_0(J), U_1(J), ..., U_{k+1}(J)\}\}. \tag{10.69}$$

It is easy to check that

$$A_0 = \{J \in R^2 : J_1 \le 0; \quad J_1 + 2kJ_2 \le 0\};$$

$$A_m = \{J \in R^2 : J_2 \ge 0; \ 2(2m-k-2)J_2 \le J_1 \le 2(2m-k)J_2\}, \ m = 1, ..., k;$$

$$A_{k+1} = \{J \in R^2 : J_1 \ge 0; \quad J_1 - 2kJ_2 \ge 0\}$$

and $R^2 = \cup_{i=0}^{k+1} A_i$.

For any $A_i, A_j, i \ne j$ we have

$$A_i \cap A_j = \begin{cases} \{J : J_1 = 2(2i - k)J_2, \quad J_2 \ge 0\} \text{ if } \ j = i+1, \ \ i = 0, 1, ..., k \\ (0,0) \qquad\qquad\qquad\qquad\quad \text{if } \ \ 1 < |i - j| < k + 1 \\ \{J : J_1 = 0, J_2 \le 0\} \qquad\quad\ \text{if } \ \ |i - j| = k + 1. \end{cases}$$

$$(10.70)$$

Denote

$$B = A_0 \cap A_{k+1}, \quad B_i = A_i \cap A_{i+1}, \quad i = 0, ..., k.$$

$$\tilde{A}_0 = A_0 \setminus (B \cup B_0), \quad \tilde{A}_{k+1} = A_{k+1} \setminus (B \cup B_k),$$

$$\tilde{A}_i = A_i \setminus (B_{i-1} \cup B_i), \quad i = 1, ..., k.$$

Fix $J \in R^2$ and denote

$$N_J(\sigma_b) = |\{j : \sigma_b \in \mathbb{C}_j\}|.$$

Using (10.70) one can prove

Lemma 10.20. *For any $b \in M$ and σ_b we have*

$$N_J(\sigma_b) = \begin{cases} k + 2, \ \ \text{if} \ \ J = (0,0) \\ 2, \ \ \text{if} \ \ J \in \left((\cup_{i=0}^{k} B_i) \cup B \right) \setminus \{(0,0)\}, \\ 1, \ \ \text{otherwise.} \end{cases}$$

Let $GS(H)$ be the set of all ground states of the relative Hamiltonian H (see (10.65)).

For any $\sigma = \{\sigma(x), \ x \in V\} \in \Omega$ denote $\bar{\sigma} = -\sigma = \{-\sigma(x), \ x \in V\}$.

Theorem 10.8.

(i) If $J = (0,0)$ then $GS(H) = \Omega$.
(ii) If $J \in \tilde{A}_i, \ i = 0, ..., k + 1$ then

$$GS(H) = \{\sigma^{(i)}, \bar{\sigma}^{(i)}\}.$$

(iii) If $J \in B_i \setminus \{(0,0)\}$, $i = 0, ..., k$ then

$$GS(H) = \{\sigma^{(i)}, \overline{\sigma}^{(i)}, \sigma^{(i+1)}, \overline{\sigma}^{(i+1)}\} \cup S_i,$$

where S_i contains at least a countable subset of non-periodic ground states.

(iv) If $J \in B \setminus \{(0,0)\}$, then

$$GS(H) = \{\sigma^{(0)}, \overline{\sigma}^{(0)}, \sigma^{(k+1)}, \overline{\sigma}^{(k+1)}\}.$$

Here $\sigma^{(i)}$, $\overline{\sigma}^{(i)}$, $i = 0, ..., k+1$ are periodic ground states such that on any $b \in M$ the bounded configurations $\sigma_b^{(i)}, \overline{\sigma}_b^{(i)} \in \mathbb{C}_i$, i.e., $\sigma^{(0)}, \overline{\sigma}^{(0)}$ are translational-invariant and $\sigma^{(i)}, \overline{\sigma}^{(i)}$, $i = 1, ..., k+1$ are periodic with period two.

Proof. The assertion (i) is trivial. In each case (ii)-(iv) for a given configuration σ_b which makes $U(\sigma_b)$ minimal, by Theorem 10.7 one can construct the periodic ground states $\sigma^{(i)}, \overline{\sigma}^{(i)}$ (with period not exceeding two). For each case the exact number of such ground state coincides with the number of the configurations σ_b which make $U(\sigma_b)$ minimal. Thus it remains to prove the existence of the set S_i defined in case (iii). If $J \in B_i \setminus \{(0,0)\}$ then the minimum points of $U(\sigma_b)$ belong to the classes $\mathbb{C}_i$ and $\mathbb{C}_{i+1}$, i.e., $\sigma_b^{(i)} = \{\sigma_b^{(i)}(x), x \in b\}, \overline{\sigma}_b^{(i)} = \{-\sigma_b^{(i)}(x), x \in b\}$ such that

$$\sigma_b^{(j)}(c_b) = +1, \quad |\{x \in b \setminus \{c_b\} : \sigma_b^{(j)}(x) = -1\}| = j, \quad j = i, i+1, \quad b \in M.$$
$$(10.71)$$

Thus any ground state $\varphi \in \Omega$ must satisfy

$$\varphi_b \in \{\sigma_b^{(i)}, \overline{\sigma}_b^{(i)}, \sigma_b^{(i+1)}, \overline{\sigma}_b^{(i+1)}\}, \quad b \in M. \tag{10.72}$$

Now we shall construct ground states $\varphi \in \Omega$ which satisfy (10.72).

Note that the configurations $\sigma_b^{(i)}$ and $\sigma_{b'}^{(i)}$ $(b, b' \in M)$ are the same up to a motion in G_k so we shall omit b. Thus configuration $\sigma^{(i)}$ is the configuration such that on any unit ball $b \in M$ the condition (10.71) is satisfied.

Suppose two unit balls b and b' are neighbors, i.e., they have a common edge. We shall then say that the two bounded configurations σ_b and $\sigma_{b'}'$ are *compatible* if they coincide on the common edge of the balls b and b'. Denote by $\mathcal{B}(b)$ the set of all neighbor balls of b.

Denote $\tilde{\Omega}_i = \{\sigma^{(i)}, \overline{\sigma}^{(i)}, \sigma^{(i+1)}, \overline{\sigma}^{(i+1)}\}$. For any $\omega, \nu \in \tilde{\Omega}_i$ denote by $n(\omega, \nu) \equiv n_i(\omega, \nu)$ the number of possibilities to set up the configuration ν as a compatible configuration (with ω) around (i.e., on neighboring balls of the ball on which ω is given) the configuration ω. Clearly $n(\omega, \nu) \in \{0, 1, ..., k+1\}$, for any $\omega, \nu \in \tilde{\Omega}_i$ $i = 0, ..., k+1$.

Denote $\mathbf{N}_i \equiv \mathbf{N}_i^{(k)} =$

$$
\begin{pmatrix}
n(\sigma^{(i)}, \sigma^{(i)}) & n(\sigma^{(i)}, \overline{\sigma}^{(i)}) & n(\sigma^{(i)}, \sigma^{(i+1)}) & n(\sigma^{(i)}, \overline{\sigma}^{(i+1)}) \\
n(\overline{\sigma}^{(i)}, \sigma^{(i)}) & n(\overline{\sigma}^{(i)}, \overline{\sigma}^{(i)}) & n(\overline{\sigma}^{(i)}, \sigma^{(i+1)}) & n(\overline{\sigma}^{(i)}, \overline{\sigma}^{(i+1)}) \\
n(\sigma^{(i+1)}, \sigma^{(i)}) & n(\sigma^{(i+1)}, \overline{\sigma}^{(i)}) & n(\sigma^{(i+1)}, \sigma^{(i+1)}) & n(\sigma^{(i+1)}, \overline{\sigma}^{(i+1)}) \\
n(\overline{\sigma}^{(i+1)}, \sigma^{(i)}) & n(\overline{\sigma}^{(i+1)}, \overline{\sigma}^{(i)}) & n(\overline{\sigma}^{(i+1)}, \sigma^{(i+1)}) & n(\overline{\sigma}^{(i+1)}, \overline{\sigma}^{(i+1)})
\end{pmatrix}.
$$

It is easy to see that

$$
\mathbf{N}_0 = \begin{pmatrix}
k+1 & 0 & k+1 & 0 \\
0 & k+1 & 0 & k+1 \\
k & 0 & k & 1 \\
0 & k & 1 & k
\end{pmatrix},
$$

$$
\mathbf{N}_i = \begin{pmatrix}
k-i+1 & i & k-i+1 & i \\
i & k-i+1 & i & k-i+1 \\
k-i & i+1 & k-i & i+1 \\
i+1 & k-i & i+1 & k-i
\end{pmatrix}, \quad i = 1, ..., k-1.
$$

$$
\mathbf{N}_k = \begin{pmatrix}
1 & k & 0 & k \\
k & 1 & k & 0 \\
0 & k+1 & 0 & k+1 \\
k+1 & 0 & k+1 & 0
\end{pmatrix}, \quad
\mathbf{N}_{k+1} = \begin{pmatrix}
k+1 & 0 & 0 & 0 \\
0 & k+1 & 0 & 0 \\
0 & 0 & 0 & k+1 \\
0 & 0 & k+1 & 0
\end{pmatrix}.
$$

Consider $k+1$ sets $\mathbf{Q}_i = \{Q\}$, $i = 0, ..., k$ of matrices $Q = \{q(u,v)\}_{u,v \in \tilde{\Omega}_i}$ such that

$$
q(u,v) \in \{0, 1, ..., n(u,v)\}, \quad \sum_{v \in \tilde{\Omega}_i} q(u,v) = k+1, \quad \forall u \in \tilde{\Omega}_i,
$$

$$
q(u, \sigma^{(i)}) + q(u, \sigma^{(i+1)}) = n(u, \sigma^{(i)}),
$$

$$
q(u, \overline{\sigma}^{(i)}) + q(u, \overline{\sigma}^{(i+1)}) = n(u, \overline{\sigma}^{(i)}), \quad i = 0, ..., k
$$

and $q(u,v) = 0$ if and only if $q(v,u) = 0$, $u, v \in \tilde{\Omega}_i$.

Using matrices $\mathbf{N}_i$ we have

$$\mathbf{Q}_0 = \left\{ Q = \begin{pmatrix} a & 0 & k-a+1 & 0 \\ 0 & b & 0 & k-b+1 \\ c & 0 & k-c & 1 \\ 0 & d & 1 & k-d \end{pmatrix} \right\},$$

where $a, b \in \{0, 1, ..., k+1\}$; $c, d \in \{0, 1, ..., k\}$; $a = k+1$ iff $c = 0$; $b = k+1$ iff $d = 0$.

For $i = 1, ..., k-1$ we have

$$\mathbf{Q}_i = \left\{ Q = \begin{pmatrix} a_1 & b_1 & k-i-a_1+1 & i-b_1 \\ b_2 & a_2 & i-b_2 & k-i-a_2+1 \\ a_3 & b_3 & k-i-a_3 & i-b_3+1 \\ b_4 & a_4 & i-b_4+1 & k-i-a_4 \end{pmatrix} \right\},$$

where $a_1, a_2 \in \{0, 1, ..., k-i+1\}$; $a_3, a_4 \in \{0, 1, ..., k-i\}$; $b_1, b_2 \in \{0, ..., i\}$; $b_3, b_4 \in \{0, ..., i+1\}$; $a_1 = k-i+1$ iff $a_3 = 0$; $a_2 = k-i+1$ iff $a_4 = 0$; $b_1 = 0$ iff $b_2 = 0$; $b_1 = i$ iff $b_4 = 0$; $b_2 = i$ iff $b_3 = 0$; $b_3 = i+1$ iff $b_4 = i+1$.

For $i = k$ we have

$$\mathbf{Q}_k = \left\{ Q = \begin{pmatrix} 1 & a & 0 & k-a \\ b & 1 & k-b & 0 \\ 0 & c & 0 & k-c+1 \\ d & 0 & k-d+1 & 0 \end{pmatrix} \right\},$$

here $a, b \in \{0, 1, ..., k\}$; $c, d \in \{0, 1, ..., k+1\}$; $a = 0$ iff $b = 0$; $a = k$ iff $d = 0$; $b = k$ iff $c = 0$; $c = k+1$ iff $d = k+1$.

For a given $\xi \in \tilde{\Omega}_i$ and $Q = \{q(u, v)\}_{u, v \in \tilde{\Omega}_i} \in \mathbf{Q}_i$ we recurrently construct a ground state $\varphi^{Q, \xi}$ by the following way: fix a ball $b \in M$ and put on b the configuration $\varphi_b^{Q, \xi} := \xi$. On balls taken from $\mathcal{B}(b)$ we set exactly $q(\xi, \omega)$ copies of ω for any $\omega \in \tilde{\Omega}_i$. Thus configurations $\varphi_{b'}^{Q, \xi}, b' \in \mathcal{B}(b)$ are defined. Using these configurations, we define configurations on the balls $\mathcal{B}(b') \setminus \{b\}, (b' \in \mathcal{B}(b))$ putting $q(\varphi_{b'}^{Q, \xi}, \nu)$ copies of $\nu \in \tilde{\Omega}_i \setminus \{\xi\}$ and $q(\varphi_{b'}^{Q, \xi}, \xi) - 1$ copies of ξ which are compatible with $\varphi_{b'}^{Q, \xi}$. Further, on the balls $\mathcal{B}(b'') \setminus \{b'\}, (b'' \in \mathcal{B}(b'), b' \in \mathcal{B}(b))$ we set $q(\varphi_{b''}^{Q, \xi}, \varepsilon)$ copies of $\varepsilon \in \tilde{\Omega}_i \setminus \{\varphi_{b'}^{Q, \xi}\}$ and $q(\varphi_{b''}^{Q, \xi}, \varphi_{b'}^{Q, \xi}) - 1$ copies of $\varphi_{b'}^{Q, \xi}$ which are compatible with $\varphi_{b''}^{Q,\xi}$. Repeating this construction one can obtain a ground state $\varphi^{Q,\xi}$ such that

$$\varphi_b^{Q,\xi} \in \tilde{\Omega}_i, \quad |\{b' \in \mathcal{B}(b) : \varphi_b^{Q,\xi} = \omega, \varphi_{b'}^{Q,\xi} = \nu\}| = q(\omega,\nu),$$

for any $b \in M$ and $\omega, \nu \in \tilde{\Omega}_i$.

In general the ground state $\varphi^{Q,\xi}$ is non-periodic (see example below). It is easy to see that

$$\varphi^{Q^{(i)},\sigma^{(j)}} \equiv \sigma^{(j)}, \quad \varphi^{Q^{(i)},\overline{\sigma}^{(j)}} \equiv \overline{\sigma}^{(j)}, \quad j = i, i+1, \quad i = 0, ..., k,$$

where

$$Q^{(i)} = \begin{pmatrix} k-i+1 & i & 0 & 0 \\ i & k-i+1 & 0 & 0 \\ 0 & 0 & k-i & i+1 \\ 0 & 0 & i+1 & k-i \end{pmatrix}. \tag{10.73}$$

Now using the ground states $\varphi^{Q,\xi}$ we shall construct an infinite set of ground states by the following way: one can choose $\xi \neq \eta$, $\xi, \eta \in \tilde{\Omega}_i$ and $Q_1, Q_2 \in \mathbf{Q}_i$ such that for configurations $\varphi^{Q_1,\xi}, \varphi^{Q_2,\eta}$ there are infinitely many $b \in M$ on which $\varphi_b^{Q_1,\xi}$ and $\varphi_{b'}^{Q_2,\eta}$ are compatible for some $b' \in \mathcal{B}(b)$. Indeed it is sufficient to take $\xi \neq \eta$ such that $q_1(\xi,\eta)q_2(\xi,\eta) \neq 0$ (see example below).

Denote

$$M_1 \equiv M_1^{\xi\eta}(Q_1, Q_2) = \{b \in M : \varphi_b^{Q_1,\xi} \text{ is compatible with}$$
$$\varphi_{b'}^{Q_2,\eta} \text{ for some } b' \in \mathcal{B}(b)\};$$

$$\mathcal{N}_1 = \{n \in \{0, 1, ...\} : \exists b \in M_1 \text{ such that } |c_b| = n\};$$

$$V^{(y)} = \{z \in V : y < z\}.$$

Fix $m \in \mathcal{N}_1$ and denote

$$\tilde{W}_m = \{x \in W_m : \exists b \in M_1 \text{ such that } c_b = x\}.$$

Consider the configuration

$$\varphi_m^{Q_1,Q_2,\xi,\eta}(x) = \begin{cases} \varphi^{Q_1,\xi}(x) & \text{if } x \in V_m \cup \{V^{(y)}, y \in W_m \setminus \tilde{W}_m\} \\ \varphi^{Q_2,\eta}(x) & \text{if } x \in V^{(y)}, y \in \tilde{W}_m. \end{cases}$$

Clearly $\varphi_m^{Q_1,Q_2,\xi,\eta}$, $m \in \mathcal{N}_1$ is a ground state and the number of such ground states is infinite, since $|\mathcal{N}_1| = \infty$. This completes the proof of assertion (iii). $\qquad\square$

Remark 10.8. The proof of (ii) and (iv) can be obtained by using the above matrices. Indeed in case (i) $\tilde{\Omega}_i$ contains just $\sigma^{(i)}$ and $\overline{\sigma}^{(i)}$. Thus $\mathbf{Q}_i$ contains just matrices of type $Q^{(i)}$ (see (10.73)). In case (iv) $\tilde{\Omega}_{k+1} = \{\sigma^{(0)}, \overline{\sigma}^{(0)}, \sigma^{(k+1)}, \overline{\sigma}^{(k+1)}\}$ and $\mathbf{Q}_{k+1}$ contains the unique matrix

$$Q_{k+1} = \begin{pmatrix} k+1 & 0 & 0 & 0 \\ 0 & k+1 & 0 & 0 \\ 0 & 0 & 0 & k+1 \\ 0 & 0 & k+1 & 0 \end{pmatrix}.$$

Consequently,

$$\varphi^{Q_{k+1},\xi} = \begin{cases} \varphi^+ & \text{if} \quad \xi = \sigma^{(0)} \\ \varphi^- & \text{if} \quad \xi = \overline{\sigma}^{(0)} \\ \varphi^\pm & \text{if} \quad \xi = \sigma^{(k+1)} \\ \varphi^\mp & \text{if} \quad \xi = \overline{\sigma}^{(k+1)}. \end{cases}$$

Here $\varphi^\varepsilon = \{\varphi(x) \equiv \varepsilon\}, \varepsilon = +1, -1$ is translational-invariant which coincides with either $\sigma^{(0)}$ or $\overline{\sigma}^{(0)}$. The configuration $\varphi^\pm = -\varphi^\mp$ is periodic with respect to the subgroup $G_k^{(2)} = \{x \in G_k : |x| - \text{even}\} \subset G_k$ (chessboard) and coincides with $\sigma^{(k+1)} = -\overline{\sigma}^{(k+1)}$.

Example 10.2. Consider $k = 2, i = 0, J \in B_0 \setminus \{(0,0)\}$. Take matrices

$$Q_1 = \begin{pmatrix} 0 & 0 & 3 & 0 \\ 0 & 1 & 0 & 2 \\ 1 & 0 & 1 & 1 \\ 0 & 2 & 1 & 0 \end{pmatrix}, \quad Q_2 = \begin{pmatrix} 1 & 0 & 2 & 0 \\ 0 & 1 & 0 & 2 \\ 2 & 0 & 0 & 1 \\ 0 & 1 & 1 & 1 \end{pmatrix}$$

and $\xi = \sigma^{(0)}, \eta = \sigma^{(1)}$. The configurations $\varphi^{Q_1,\xi}, \varphi^{Q_2,\eta}$ and $\varphi_2^{Q_1,Q_2,\xi,\eta}$ are represented in Fig. 10.2-10.4 respectively.

Remark 10.9. Note that the way of the description of an infinite number of ground states used in the proof of (iii) is not unique. One can use $\varphi^{Q,\xi}$ for another way to describe another infinite set of ground states.

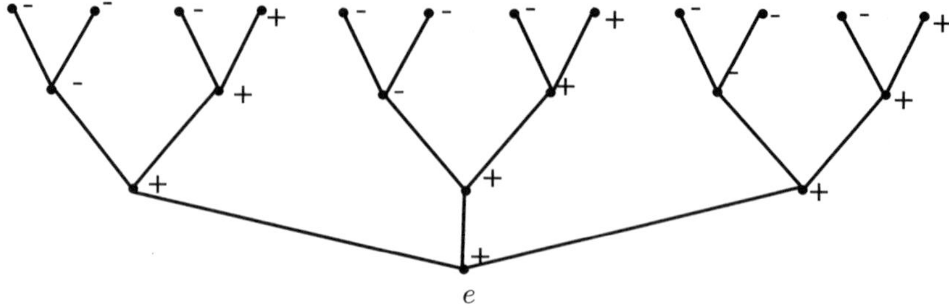

Fig. 10.2   Configuration $\varphi^{Q_1,\xi}$.

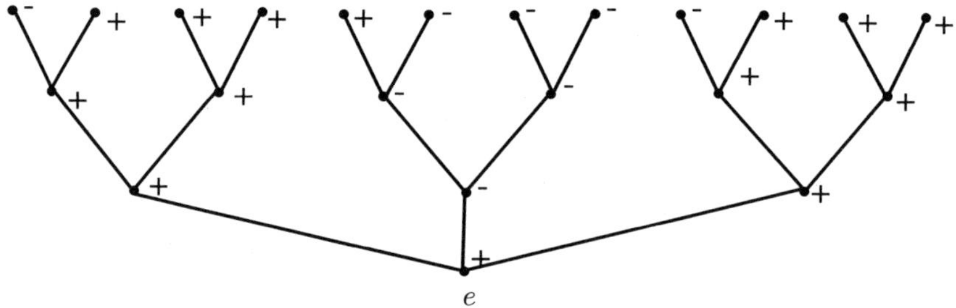

Fig. 10.3   Configuration $\varphi^{Q_2,\eta}$.

10.3.2   *Weakly periodic ground states*

Let $\mathcal{H}$ be a normal subgroup of the group G_k. Let $G_k/\mathcal{H} = \{H_1, ..., H_r\}$ be the quotient (factor) group.

Recall that for any $x \in G_k$, the set $\{y \in G_k : d(x,y) = 1\} \setminus S(x)$ has a unique element; we let $x_\downarrow$ denote it.

A configuration $\sigma(x)$ is called $\mathcal{H}$-weakly periodic if $\sigma(x) = a_{ij}$ for all $x_\downarrow \in H_i$ and $x \in H_j$ where $a_{ij} \in \Phi = \{-1, 1\}$, $i, j = 1, ..., r$.

We set

$$\mathcal{I} = \mathcal{I}(\mathcal{H}) = \left\{ (i,j) \in \{1, ..., r\}^2 : \exists x \in H_i, \exists y \in H_j \text{ such that } d(x,y) = 1 \right\},$$

$$\mathcal{I}_i = \mathcal{I}_i(\mathcal{H}) = \left\{ j \in \{1, ..., r\} : (i,j) \in \mathcal{I} \right\}.$$

We associate each $\mathcal{H}$-weakly periodic configuration σ, i.e.,

$$\sigma(x) = a_{ij}, x_\downarrow \in H_i, x \in H_j, (i,j) \in \mathcal{I}, \tag{10.74}$$

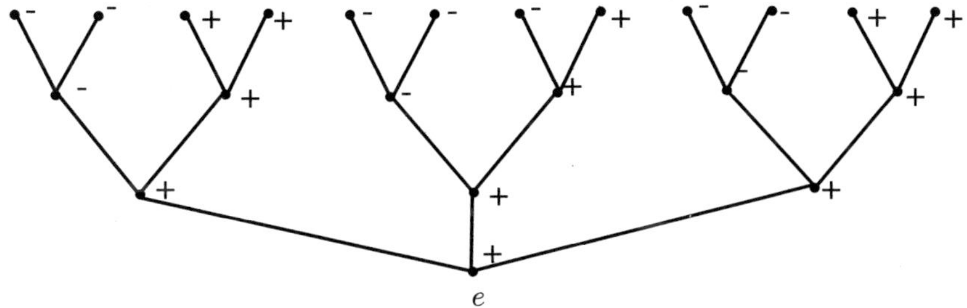

Fig. 10.4   Configuration $\varphi_2^{Q_1,Q_2,\xi,\eta}$.

with a quadratic matrix $\mathcal{B} = \mathcal{B}(\sigma) = \{b_{ij}\}_{i,j=1}^{r}$ with the elements

$$
b_{ij} = \begin{cases} a_{ij}, & (i,j) \in \mathcal{I}; \\ 0, & (i,j) \notin \mathcal{I}. \end{cases}
$$

Let $S_1(x)$ be the set of all nearest neighbors of a point x, and let

$$
q_j(x) = |S_1(x) \cap H_j|, j = 1, ..., r; \quad Q(x) = (q_1(x), ..., q_r(x)).
$$

By results of Chapter 1, for any $x \in G_k$ we have a permutation π_x of the coordinates of the vector $Q(e)$ such that

$$
\pi_x Q(e) = Q(x). \tag{10.75}
$$

The following theorem imposes conditions on the elements of the matrix $\mathcal{B}(\sigma)$ under which σ is a ground state.

Theorem 10.9.

1. *An $\mathcal{H}$-weakly periodic configuration σ is $\mathcal{H}$-periodic if and only if the matrix $\mathcal{B}(\sigma)$ consists of equal rows.*
2. *An $\mathcal{H}$-weakly periodic configuration σ is a ground state of the Hamiltonian H if and only if there exists $j \in \{0, \ldots, k\}$ such that $J_1 = 2(2j - k)J_2, J_2 \geq 0$ and*

$$
a_{pm} - a_{nm} + \sum_{s=1}^{r} q_i(x)a_{ns} = (k - 2j \pm 1)a_{mn}, \tag{10.76}
$$

where $x \in H_n$, $m \in \mathcal{I}_p$ and $n \in \mathcal{I}_m$.

Proof.   1) Assertion 1 follows from (10.74) and the definition of $\mathcal{B}$.

2) The condition $J_1 = 2(2j - k)J_2, J_2 \geq 0$, follows from assertion (iii) of Theorem 10.8 and (10.70). We note that for $J_1 = 2(2j - k)J_2$, the configuration σ is a ground state if there exists $j \in \{0, \ldots, k\}$ such, that $\sigma_b \in \mathbb{C}_j \cup \mathbb{C}_{j+1}$ for all $b \in M$. This means that for any x, and for $\sigma(x) = 1$ or $\sigma(x) = -1$ the configuration $\sigma_{S_1(x)}$ contains either j or $j + 1$ or either $k + 1 - j$ or $k - j$ elements -1. The sum of all values $\sigma(y), y \in S_1(x)$ is therefore $(k - 2j \pm 1)\sigma(x)$, which implies the statement in the theorem. $\square$

Remark 10.10. System (10.76) consists of linear equations, and it is necessary to find $a_{ij} \in \{-1, 1\}$. The best method for finding such solutions is to consider all possible versions $a_{ij} \in \{-1, 1\}$, $j \in \mathcal{I}_i$ (the number of versions does not exceed $2^{|\mathcal{I}|}$) and to check which of them satisfies system (10.76) for some $j \in \{0, \ldots, k\}$. Here, we solve system (10.76) for normal subgroups of indices 2 and 4.

10.3.2.1 *Case of index two*

Let $A \subset \{1, 2, \ldots, k+1\}$, $H_A = \{x \in G_k : \sum_{j \in A} w_j(x) - \text{even}\}$, where $w_j(x)$ is number of letters a_j in the word x.

Let $G_k/H_A = \{H_1, H_2\}$ be the quotient group, where $H_1 = H_A, H_2 = G_k \setminus H_A$.

Then H_A - weakly periodic configuration has the form

$$\varphi(x) = \pm \begin{cases} a_{11}, \ x_{\downarrow} \in H_1, \ x \in H_1 \\ a_{12}, \ x_{\downarrow} \in H_1, \ x \in H_2 \\ a_{21}, \ x_{\downarrow} \in H_2, \ x \in H_1 \\ a_{22}, \ x_{\downarrow} \in H_2, \ x \in H_2. \end{cases} \tag{10.77}$$

Let $|A| = i$. Then system (10.76) has the form:

$$\begin{cases} a_{11} + (k-i)a_{11} + ia_{12} = (k-2j \pm 1)a_{11} \\ a_{21} + (k-i)a_{11} + ia_{12} = (k-2j \pm 1)a_{11} \\ a_{12} + (k-i+1)a_{11} + (i-1)a_{12} = (k-2j \pm 1)a_{21} \\ a_{22} + (k-i+1)a_{11} + (i-1)a_{12} = (k-2j \pm 1)a_{21} \\ a_{11} + (k-i)a_{21} + ia_{22} = (k-2j \pm 1)a_{12} \\ a_{21} + (k-i)a_{21} + ia_{22} = (k-2j \pm 1)a_{12} \\ a_{12} + (k-i+1)a_{21} + (i-1)a_{22} = (k-2j \pm 1)a_{22} \\ a_{22} + (k-i+1)a_{21} + (i-1)a_{11} = (k-2j \pm 1)a_{22}, \end{cases} \tag{10.78}$$

where $i \in \{0, \dots, k+1\}$, and $j \in \{0, \dots, k\}$.

We note that the vector $(a_{11}, a_{12}, a_{21}, a_{22}) \in \{(\pm 1, \pm 1, \pm 1, \pm 1)\}$ takes 16 values. Substituting the coordinates of these vectors in system (10.78), we obtain the following theorem.

Theorem 10.10. *Let* $|A| = i$.

1. *If* $i \neq (k+1)/2$, *then each* H_A-*weakly periodic ground state is* H_A-*periodic or translation invariant, i.e., corresponds to one of the vectors* $\pm(1,1,1,1)$ *and* $\pm(-1,1,-1,1)$.
2. *For* $i = (k+1)/2$ *and* $J_1 = 2J_2$, $J_2 \geq 0$, *there exist two* H_A-*weakly periodic (non-periodic) ground states corresponding to the vectors* $\pm(1,-1,-1,1)$.

10.3.2.2 *Case of index four*

Let $H_A = \{x \in G_k : \sum_{j \in A} w_j(x) - \text{even}\}$, $A \subset \{1, 2, \dots, k+1\}$ and $G_k^{(2)} = \{x \in G_k : |x| - \text{even}\}$, where $|x|$ is the length of the word x. Then $G_k^{(4)} = H_A \cap G_k^{(2)}$ is a normal subgroup of index four. Let $G_k/G_k^{(4)} = \{H_1, H_2, H_3, H_4\}$, be the quotient group.

We note that

$$\mathcal{I}(G_k^{(4)}) = \{(1,3),(3,1),(1,4),(4,1),(2,3),(3,2),(2,4),(4,2)\},$$

and any $G_k^{(4)}$-weakly periodic configuration therefore has the form

$$\varphi(x) = a_{ij}, \quad \text{if } x_\downarrow \in H_i, \ x \in H_j, \ (i,j) \in \mathcal{I}(G_k^{(4)}).$$

Each weakly periodic configuration is thus associated with an eight-dimensional vector $(a_{ij}, (i,j) \in \mathcal{I})$. In this case it is easy to verify the following assertion.

Theorem 10.11. *Let $|A| = i$.*

i) *If $i \neq \frac{k+1}{2}$, then each $G_k^{(4)}$-weakly periodic ground state is periodic or translation-invariant.*

ii) *If $i = \frac{k+1}{2}$ and $J_1 = 2J_2, J_2 \geq 0$, then there are four $G_k^{(4)}$-weakly periodic (non periodic) ground states $\pm\varphi', \pm\varphi''$ corresponding to vectors $\pm(1, 1, -1, -1, -1, -1, 1, 1)$ and $\pm(-1, 1, 1, -1, -1, 1, -1, 1)$:*

$$\pm\varphi'(x) = \pm \begin{cases} +1, \ x_{\downarrow} \in H_2, \ x \in H_4 \\ +1, \ x_{\downarrow} \in H_4, \ x \in H_2 \\ -1, \ x_{\downarrow} \in H_1, \ x \in H_4 \\ -1, \ x_{\downarrow} \in H_4, \ x \in H_1 \\ -1, \ x_{\downarrow} \in H_3, \ x \in H_2 \\ -1, \ x_{\downarrow} \in H_2, \ x \in H_3 \\ +1, \ x_{\downarrow} \in H_1, \ x \in H_3 \\ +1, \ x_{\downarrow} \in H_3, \ x \in H_1, \end{cases}$$

$$\pm\varphi''(x) = \pm \begin{cases} -1, \ x_{\downarrow} \in H_2, \ x \in H_4 \\ +1, \ x_{\downarrow} \in H_4, \ x \in H_2 \\ +1, \ x_{\downarrow} \in H_1, \ x \in H_4 \\ -1, \ x_{\downarrow} \in H_4, \ x \in H_1 \\ -1, \ x_{\downarrow} \in H_3, \ x \in H_2 \\ +1, \ x_{\downarrow} \in H_2, \ x \in H_3 \\ -1, \ x_{\downarrow} \in H_1, \ x \in H_3 \\ +1, \ x_{\downarrow} \in H_3, \ x \in H_1. \end{cases}$$

Corollary 10.2. *If k is an even number, then each weakly periodic ground state is periodic.*

It follows from (10.66) and (10.70) that Theorems 10.10 and 10.11 imply the following assertion.

Corollary 10.3. *For the ground states constructed in Theorems 10.10 and 10.11, the energy of these configurations in any ball of radius 1 is given by $-(k+1)J_2/2, J_2 \geq 0$.*

Remark 10.11. We note that the matrices of the configurations constructed in Theorem 10.10 are symmetric or skew-symmetric. The matrix $\mathcal{B}(\varphi')$ is also symmetric, and the matrix $\mathcal{B}(\varphi'')$ is skew-symmetric.

The following conjecture therefore seems to hold.

Conjecture 1. Let an $\mathcal{H}$-weakly periodic configuration φ be a ground state for Hamiltonian (10.63). Then the matrix $\mathcal{B}(\varphi)$ is either symmetric or skew-symmetric.

The following example shows that it does not follow from the symmetry or skew-symmetry of the matrix $\mathcal{B}(\varphi)$ that φ is a ground state.

Example 10.3. We consider

$$\varphi_0(x) = \begin{cases} 1, & x_\downarrow \in H_2 \; x \in H_4 \\ 1, & x_\downarrow \in H_4 \; x \in H_2 \\ 1, & x_\downarrow \in H_1 \; x \in H_4 \\ 1, & x_\downarrow \in H_4 \; x \in H_1 \\ 1, & x_\downarrow \in H_3 \; x \in H_2 \\ 1, & x_\downarrow \in H_2 \; x \in H_3 \\ -1, & x_\downarrow \in H_1 \; x \in H_3 \\ -1, & x_\downarrow \in H_3 \; x \in H_1, \end{cases} \qquad \mathcal{B}(\varphi_0) = \begin{pmatrix} 0 & 0 & -1 & 1 \\ 0 & 0 & 1 & 1 \\ -1 & 1 & 0 & 0 \\ 1 & 1 & 0 & 0 \end{pmatrix}.$$

The matrix $\mathcal{B}(\varphi_0)$ is symmetric, but it is easy to verify that φ_0 is not a ground state.

Theorems 10.10 and 10.11 also imply the following conjecture.

Conjecture 2. Let $\mathcal{H}$ be an arbitrary normal subgroup of finite index. An $\mathcal{H}$-weakly periodic configuration φ is a ground state of Hamiltonian (10.63), if and only if $\varphi \in \mathbb{C}_{\frac{k+1}{2}} \cup \mathbb{C}_{\frac{k+1}{2}+1}$ and $J_1 = 2J_2$, $J_2 \geq 0$.

10.3.3 The Peierls condition

Definition 10.5. Let $GS(H)$ be the complete set of all ground states of the relative Hamiltonian H. A ball $b \in M$ is said to be an *improper* ball of the configuration σ if $\sigma_b \neq \varphi_b$ for any $\varphi \in GS(H)$. The union of the improper balls of a configuration σ is called the *boundary* of the configuration and denoted by $\partial(\sigma)$.

Definition 10.6. The relative Hamiltonian H with the set of ground states $GS(H)$ satisfies the Peierls condition if for any $\varphi \in GS(H)$ and any configuration σ coinciding almost everywhere with φ,

$$H(\sigma, \varphi) \geq \lambda |\partial(\sigma)|,$$

where λ is a positive constant which does not depend on σ, and $|\partial(\sigma)|$ is the number of unit balls in $\partial(\sigma)$.

Theorem 10.12. *If $J \neq (0,0)$ then the Peierls condition is satisfied.*

Proof. Denote $\mathbf{U} = \{U_0, ..., U_{k+1}\}$ (see (10.66)), $U^{\min} = \min\{U_0, ..., U_{k+1}\}$ and

$$\lambda_0 = \min\left\{\mathbf{U} \setminus \{U_j : U_j = U^{\min}\}\right\} - U^{\min}. \qquad (10.79)$$

Note that $U_0 = ... = U_{k+1}$ if and only if $J = (0,0)$, consequently $\lambda_0 > 0$ if $J \neq (0,0)$.

Suppose σ coincides almost everywhere with a ground state $\varphi \in GS(H)$ then we have $U(\sigma_b) - U(\varphi_b) \geq \lambda_0$ for any $b \in \partial(\sigma)$ since φ is a ground state. Thus

$$H(\sigma, \varphi) = \sum_{b \in M} (U(\sigma_b) - U(\varphi_b)) = \sum_{b \in \triangle(\sigma)} (U(\sigma_b) - U(\varphi_b)) \geq \lambda_0 |\partial(\sigma)|.$$

Therefore, the Peierls condition is satisfied for $\lambda = \lambda_0$. $\qquad \square$

10.3.4 *Contours and Gibbs measures*

Let $\Lambda \subset V$ be a finite set, $\Lambda' = V \setminus \Lambda$ and $\omega_\Lambda = \{\omega(x), x \in \Lambda'\}$, $\sigma_\Lambda = \{\sigma(x), x \in \Lambda\}$ be given configurations. The energy of the configuration σ_Λ has the form

$$H_\Lambda(\sigma|\omega_\Lambda) = J_1 \sum_{\substack{<x,y> \\ x,y \in \Lambda}} \sigma(x)\sigma(y) + J_1 \sum_{\substack{<x,y> \\ x \in \Lambda, y \in \Lambda'}} \sigma(x)\omega(y) + \qquad (10.80)$$

$$J_2 \sum_{\substack{x,y \in \Lambda \\ d(x,y)=2}} \sigma(x)\sigma(y) + J_2 \sum_{\substack{x \in \Lambda, y \in \Lambda' \\ d(x,y)=2}} \sigma(x)\omega(y).$$

Let $\omega_{\Lambda'}^\varepsilon \equiv \varepsilon, \varepsilon = \pm 1$ be a constant configuration outside Λ. For a given ε we extend the configuration σ_Λ inside Λ to the Cayley tree by the constant configuration and denote this configuration by $\sigma_\Lambda^\varepsilon$ and $\Omega_\Lambda^\varepsilon$ the set of all such configurations.

Now we describe a boundary of the configuration $\sigma_\Lambda^\varepsilon$. For the sake of simplicity we consider only case $J \in \tilde{A}_0$. In this case by Theorem 10.8 we have $GS(H) = \{\sigma^{(0)}, \overline{\sigma}^{(0)}\} = \{\sigma^+ \equiv +1, \sigma^- \equiv -1\}$. Fix $+$-boundary condition. Put $\sigma_n = \sigma_{V_n}^+$ and $\sigma_{n,b} = (\sigma_n)_b$. By Definition 10.5 the boundary of the configuration σ_n is

$$\partial \equiv \partial(\sigma_n) = \{b \in M_{n+2} : \sigma_{n,b} \neq \sigma_b^+ \text{ or } \sigma_b^-\},$$

where $M_n = \{b \in M : b \cap V_n \neq \emptyset\}$.

The boundary ∂ contains of $2k+2$ parts

$$\partial_i^+ = \{b \in M_{n+2} : \sigma_{n,b} \in \Omega_i\}, \; i = 1, 2, \ldots, k+1;$$

$$\partial_i^- = \{b \in M_{n+2} : \sigma_{n,b} \in \Omega_i^-\}, \; i = 1, 2, \ldots, k+1,$$

where Ω_i and Ω_i^- are defined in Lemma 10.18.

Consider V_n and for a given configuration σ_n (with "+"-boundary condition) denote

$$V_n^- \equiv V_n^-(\sigma_n) = \{t \in V_n : \sigma_n(t) = -1\}.$$

Let $G^n = (V_n^-, L_n^-)$ be the graph such that

$$L_n^- = \{l = <x,y> \in L : x, y \in V_n^-\}.$$

It is clear, that for a fixed n the graph G^n contains a finite $(= m)$ of maximal connected subgraphs G_r^n, i.e.,

$$G_r^n = \{G_1^n, ..., G_m^n\}, \quad G_r^n = (V_{n,r}^-, L_{n,r}^-), r = 1, ..., m.$$

Here $V_{n,r}^-$ is the set of vertices and $L_{n,r}^-$ the set of edges of G_r^n.

Two edges $l_1, l_2 \in L$ are called nearest neighboring edges if $|i(l_1) \cap i(l_2)| = 1$, and we write $< l_1, l_2 >_1$.

For a given graph G denote by $V(G)$ the set of vertices and by $E(G)$ the set of edges of G.

$$D_{\text{edge}}(K) = \{l_1 \in L \setminus E(K) : \exists l_2 \in E(K) \text{ such that } < l_1, l_2 >_1\}.$$

The (finite) sets $D_{\text{edge}}(G_r^n)$ are called *subcontours* of the boundary $\triangle$.

The set $V_{n,r}^-, r = 1, ..., m$ is called the *interior*, $\text{Int}D_{\text{edge}}(G_r^n)$, of $D_{\text{edge}}(G_r^n)$.

For any two subcontours T_1, T_2 the distance $\text{dist}(T_1, T_2)$ is defined by

$$\text{dist}(T_1, T_2) = \min_{\substack{x \in V(T_1) \\ y \in V(T_2)}} d(x,y),$$

where $d(x,y)$ is the distance between $x, y \in V$.

Definition 10.7. The subcontours T_1, T_2 are called *adjacent* if $\text{dist}(T_1, T_2) \leq 2$. A set of subcontours $\mathcal{A}$ is called connected if for any two subcontours $T_1, T_2 \in \mathcal{A}$ there is a collection of subcontours $T_1 = \tilde{T}_1, \tilde{T}_2, ..., \tilde{T}_l = T_2$ in the set $\mathcal{A}$ such that for each $i = 1, ..., l-1$ the subcontours $\tilde{T}_i$ and $\tilde{T}_{i+1}$ are adjacent.

Definition 10.8. Any maximal connected set (component) of subcontours is called *contour* of the set ∂.

The set of edges from a contour γ is denoted by $\mathrm{supp}\gamma$.

Remark 10.12. Note that Definition 10.8 of contours coincides with the Definition 10.3. For q-component models the quantity $|\mathrm{supp}\gamma|$ plays very important role. But in the present section instead of $|\mathrm{supp}\gamma|$ we will use the number of improper (see Definition 10.5) balls of γ.

For a given contour γ we put

$$\mathrm{imp}_i^\varepsilon \gamma = \{b \in \partial_i^\varepsilon : b \cap \gamma \neq \emptyset\}, \varepsilon = -1, 1; i = 1, \ldots, k+1;$$

$$\mathrm{imp}^\varepsilon \gamma = \cup_{i=1}^{k+1} \mathrm{imp}_i^\varepsilon \gamma, \quad \mathrm{imp}\gamma = \cup_{\varepsilon = \pm 1} \mathrm{imp}^\varepsilon \gamma;$$

$$|\gamma| = |\mathrm{imp}\gamma|, \quad |\gamma_i^\varepsilon| = |\mathrm{imp}_i^\varepsilon \gamma|, \quad |\gamma_i| = |\gamma_i^+| + |\gamma_i^-|.$$

It is easy to see that the collection of contours $\alpha = \{\gamma_r\}$ generated by the boundary σ_n has the following properties

(i) Every contour $\gamma \in \alpha$ lies inside of the set V_{n+1};

(ii) For every two contours $\gamma_1, \gamma_2 \in \alpha$ we have $\mathrm{dist}(\gamma_1, \gamma_2) > 2$, thus their supports $\mathrm{supp}\gamma_1$ and $\mathrm{supp}\gamma_2$ are disjoint.

A collection of contours $\alpha = \{\gamma\}$ that has the properties (i)-(ii) is called a configuration of contours. As we have seen, the configuration σ_n of spin generates the configuration of contours $\alpha = \alpha(\sigma_n)$. The converse assertion is also true. Indeed, for a given collection of contours $\{\gamma_r\}_{r=1}^m$ we put $\sigma_n(x) = -1$ for each $x \in \mathrm{Int}\gamma_r, r = 1, ..., m$ and $\sigma_n(x) = +1$ for each $x \in V_n \setminus \cup_{r=1}^m \mathrm{Int}\gamma_r$.

Let us define a graph structure on M (i.e., on the set of all unit balls of the Cayley tree) as follows. Two balls $b, b' \in M$ are connected by an edge if they are neighbors, i.e., have a common edge. Denote this graph by $G(M)$. Note that the graph $G(M)$ is a Cayley tree of order $k \geq 1$. Here the vertices of this graph are balls of M.

For $A \subset V$ denote

$$B(A) = \{b \in M : b \subset A\};$$

$$D(A) = \{x \in V \setminus A : \exists y \in A, \ \text{such that} \ < x, y >\};$$

$$D_{\mathrm{int}}(A) = \{x \in A : \exists y \in V \setminus A, \ \text{such that} \ < x, y >\}.$$

For $x \in V$ we will write $x \in \gamma$ if $x \in V(\gamma)$.

Denote $N_r(x) = |\{\gamma : x \in \gamma, |\gamma| = r\}|$, where as before $|\gamma| = |\mathrm{imp}\gamma|$.

Lemma 10.21. *If $k = 2$. Then*

$$N_r(x) \leq \mathrm{Const} \cdot (4e)^{2r}. \tag{10.81}$$

Proof. Denote by K_γ the minimal connected subgraph of Γ^2, which contains a contour γ. It is easy to see that if $\gamma = \{\gamma_1, ..., \gamma_m\}, m \geq 1$, (where γ_i is subcontour) then

$$B(V(K_\gamma)) \subset \mathrm{imp}\gamma \cup B(\mathrm{Int}\gamma). \tag{10.82}$$

Note that $D(\mathrm{Int}\gamma)$ as a set contains different points. So we have

$$|\gamma| = |D(\mathrm{Int}\gamma)| + |D_{\mathrm{int}}(\mathrm{Int}\gamma)|;$$

$$|B(\mathrm{Int}\gamma)| = |\mathrm{Int}\gamma \setminus D_{\mathrm{int}}(\mathrm{Int}\gamma)| = |\mathrm{Int}\gamma| - |D_{\mathrm{int}}(\mathrm{Int}\gamma)|.$$

Using Lemma 10.10 we have $|\mathrm{Int}\gamma| = |D(\mathrm{Int}\gamma)| - 2$. Consequently,

$$|B(\mathrm{Int}\gamma)| = |D(\mathrm{Int}\gamma)| - |D_{\mathrm{int}}(\mathrm{Int}\gamma)| - 2 =$$

$$|\gamma| - 2|D_{\mathrm{int}}(\mathrm{Int}\gamma)| - 2.$$

Thus from (10.82) we have

$$|B(V(K_\gamma))| \leq 2(|\gamma| - |D_{\mathrm{int}}(\mathrm{Int}\gamma)| - 1). \tag{10.83}$$

Since γ contains m subcontours we have

$$|D_{\mathrm{int}}(\mathrm{Int}\gamma)| \geq m.$$

Hence we get from (10.83)

$$|B(V(K_\gamma))| \leq 2(|\gamma| - m - 1).$$

Since $\gamma \subset K_\gamma$ we get $|\gamma| \leq |B(K_\gamma)| \leq 2(|\gamma| - m - 1)$. Hence $|\gamma| \geq 2m + 2$ which implies $1 \leq m \leq \frac{|\gamma|-2}{2}$.

By combinatorial calculations we get

$$N_r(x) \leq 4 \sum_{m=1}^{[r/2-1]} \binom{2r - 2m - 2}{r} \tilde{N}_{2r-2m-2,\Gamma^2}(b(x)), \tag{10.84}$$

where $[a]$ is the integer part of a and $b(x)$ is a ball b such that $x \in b$.

Using inequality $\binom{n}{r} \leq 2^{n-1}, r \leq n$ and Lemma 10.11 from (10.84) we get (10.81). $\qquad\square$

The following lemma gives a contour representation of Hamiltonian

Lemma 10.22. *The energy* $H_n(\sigma_n) \equiv H_{V_n}(\sigma_n|\omega_{V_n'} = +1)$ *(see (10.80))* *has the form*

$$H_n(\sigma_n) = \sum_{i=1}^{k+1}(U_i - U_0)|\partial_i| + |M_{n+2}|U_0, \tag{10.85}$$

where $|\partial_i| = |\partial_i^+| + |\partial_i^-|$.

Proof. Using equality $U(\sigma_b) = U(-\sigma_b)$ we have

$$H_n(\sigma_n) = \sum_{b \in M_{n+2}} U(\sigma_{n,b}) = \sum_{i=1}^{k+1} U_i |\partial_i| + (|M_{n+2}| - |\partial|)U_0. \qquad (10.86)$$

Now using equality $|\partial| = \sum_{i=1}^{k+1} |\partial_i|$ from (10.86) we get (10.85). $\square$

Lemma 10.23. *Assume $J \in \tilde{A}_0$. Let γ be a fixed contour and*

$$p_+(\gamma) = \frac{\sum_{\sigma_n : \gamma \in \partial} \exp\{-\beta H_n(\sigma_n)\}}{\sum_{\tilde{\sigma}_n} \exp\{-\beta H_n(\tilde{\sigma}_n)\}}.$$

Then

$$p_+(\gamma) \leq \exp\{-\beta \lambda_0 |\gamma|\}, \qquad (10.87)$$

where λ_0 is defined by formula (10.79) and $\beta = \frac{1}{T}, T > 0-$ temperature.

Proof. Put $\Omega_\gamma = \{\sigma_n : \gamma \subset \partial\}$, $\Omega_\gamma^0 = \{\sigma_n : \gamma \cap \partial = \emptyset\}$ and define a map $\chi_\gamma : \Omega_\gamma \to \Omega_\gamma^0$ by

$$\chi_\gamma(\sigma_n)(x) = \begin{cases} +1 & \text{if} \quad x \in \mathrm{Int}\gamma \\ \sigma_n(x) & \text{if} \quad x \notin \mathrm{Int}\gamma. \end{cases}$$

For a given γ the map χ_γ is one-to-one map. We need the following

Lemma 10.24. *For any $\sigma_n \in \Omega_{V_n}$ and $i = 1, \ldots, k+1$ we have*

$$|\partial_i(\sigma_n)| = |\partial_i(\chi_\gamma(\sigma_n))| + |\gamma_i|.$$

Proof. It is easy to see that the map χ_γ destroys the contour γ and all other contours are invariant with respect to χ_γ. $\square$

Now we shall continue with the proof of Lemma 10.23. By Lemma 10.22 we have

$$p_+(\gamma) = \frac{\sum_{\sigma_n \in \Omega_\gamma} \exp\{-\beta \sum_{i=1}^{k+1}(U_i - U_0)|\partial_i(\sigma_n)|\}}{\sum_{\tilde{\sigma}_n} \exp\{-\beta \sum_{i=1}^{k+1}(U_i - U_0)|\partial_i(\tilde{\sigma}_n)|\}} \leq \qquad (10.88)$$

$$\frac{\sum_{\sigma_n \in \Omega_\gamma} \exp\{-\beta \sum_{i=1}^{k+1}(U_i - U_0)|\partial_i(\sigma_n)|\}}{\sum_{\tilde{\sigma}_n \in \Omega_\gamma^0} \exp\{-\beta \sum_{i=1}^{k+1}(U_i - U_0)|\partial_i(\tilde{\sigma}_n)|\}} =$$

$$\frac{\sum_{\sigma_n \in \Omega_\gamma} \exp\{-\beta \sum_{i=1}^{k+1}(U_i - U_0)|\partial_i(\sigma_n)|\}}{\sum_{\tilde{\sigma}_n \in \Omega_\gamma} \exp\{-\beta \sum_{i=1}^{k+1}(U_i - U_0)|\partial_i(\chi_\gamma(\tilde{\sigma}_n))|\}}.$$

Since $J \in \tilde{A}_0$ by Theorem 10.8 we have $GS(H) = \{\sigma^+, \sigma^-\}$ hence $U_i - U_0 \geq \lambda_0$ for any $i = 1, \ldots, k+1$. Thus using this fact and Lemma 10.24 from (10.88) we get (10.87). $\square$

Using Lemmas 10.21 and 10.23 by very similar argument used for q-component models one can prove

Theorem 10.13. *If $J \in \tilde{A}_0$ then for all sufficiently large β there are at least two Gibbs measures for the model (10.63) on Cayley tree of order two.*

10.4 Finite-range models: general contours

10.4.1 *Configuration space and the model*

Recall for $A \subseteq V$ a spin *configuration* σ_A on A is defined as a function $x \in A \to \sigma_A(x) \in \Phi = \{1, 2, \ldots, q\}$; the set of all configurations coincides with $\Omega_A = \Phi^A$. We denote $\Omega = \Omega_V$ and $\sigma = \sigma_V$. Also we define a *periodic configuration* as a configuration $\sigma \in \Omega$ which is invariant under a subgroup of shifts $G_k^* \subset G_k$ of finite index.

For a given periodic configuration the index of the subgroup is called the *period of the configuration*. A configuration that is invariant with respect to all shifts is called *translational-invariant*.

The energy of the configuration $\sigma \in \Omega$ is given by the formal Hamiltonian

$$H(\sigma) = \sum_{\substack{A \subset V: \\ \mathrm{diam}(A) \leq r}} I(\sigma_A) \tag{10.89}$$

where $r \in N = \{1, 2, \ldots\}$, $\mathrm{diam}(A) = \max_{x,y \in A} d(x,y)$, $I(\sigma_A) : \Omega_A \to R$ is a given translation invariant potential.

Fix $r \in N$ and put $r' = [\frac{r+1}{2}]$, where $[a]$ is the integer part of a. Denote by M_r the set of all balls $b_r(x) = \{y \in V : d(x,y) \leq r'\}$ with radius r', i.e.,

$$M_r = \{b_r(x) : x \in V\}.$$

For $A \subset V$ with $\mathrm{diam}(A) \leq r$ denote

$$n(A) = |\{b \in M_r : A \subset b\}|,$$

where $|A|$ stands for the number of elements of a set A.

The Hamiltonian (10.89) can be written as

$$H(\sigma) = \sum_{b \in M_r} U(\sigma_b), \tag{10.90}$$

where $U(\sigma_b) = \sum_{A \subset b} \frac{I(\sigma_A)}{n(A)}.$

For a finite domain $D \subset V$ with the boundary condition φ_{D^c} given on its complement $D^c = V \setminus D$, the conditional Hamiltonian is

$$H(\sigma_D | \varphi_{D^c}) = \sum_{\substack{b \in M_r: \\ b \cap D \neq \emptyset}} U(\sigma_b), \qquad (10.91)$$

where

$$\sigma_b(x) = \begin{cases} \sigma(x) & \text{if } x \in b \cap D \\ \varphi(x) & \text{if } x \in b \cap D^c. \end{cases}$$

A *ground state* of (10.90) is a configuration φ in Γ^k whose energy cannot be lowered by changing φ in some local region. We assume that (10.90) has a finite number of translation-periodic (i.e., invariant under the action of some subgroup of G_k of finite index) ground states. By a standard trick of partitioning the tree into disjoint sets $Q(x)$ centered at $x \in G_k^*$ (the corresponding subgroup of finite index) and enlarging the spin space from Φ to Φ^Q one can transform the model above into a model with only translation-invariant or non-periodic ground states. Such a transformation was considered in [147] for models on $\mathbb{Z}^d$. Hence, without loss of generality, we assume translation-invariance instead of translational-periodic and we permute the spin so that the set of ground states of the model be $GS = GS(H) = \{\sigma^{(i)}, i = 1, 2, ..., s\}, 1 \leq s \leq q$ with $\sigma^{(i)}(x) = i$ for any $x \in V$.

10.4.2   *The assumptions and Peierls condition*

Denote by $\mathbf{U}$ the set of all possible values of $U(\sigma_b)$ for any configuration σ_b, $b \in M_r$. Since $r < +\infty$ we have $|\mathbf{U}| < +\infty$. Put $U^{\min} = \min\{U : U \in \mathbf{U}\}$ and

$$\lambda_0 = \min\left\{ \mathbf{U} \setminus \{U \in \mathbf{U} : U = U^{\min}\} \right\} - U^{\min}. \qquad (10.92)$$

The important assumptions of this section are the following:

Assumption A1.   The set of all ground states is $GS = \{\sigma^{(i)}, i = 1, 2, ..., s\}, 1 \leq s \leq q$.

Assumption A2. $\lambda_0 > 0$.

Assumption A3. Each $\varphi \in GS$ satisfies

$$U(\varphi_b) = U^{\min} \quad \text{for every } b \in M_r. \qquad (10.93)$$

Remark 10.13. If a configuration σ satisfies (10.93) i.e., $U(\sigma_b) = U^{\min}$ $\forall b \in M_r$ then it is a ground state. Moreover for Hamiltonians on $\mathbb{Z}^d$ it is well known that a configuration is a ground state if and only if the condition (10.93) is satisfied (see e.g. [247]). But such a fact is not clear for Hamiltonians on the Cayley tree, since the tree is a non-amenable graph, i.e., $\inf\{\frac{|\text{boundary of } W|}{|W|} : W \subset V, 0 < |W| < \infty\} > 0$ for $k \geq 2$ (see e.g. [18], [111]).

The *relative Hamiltonian* is defined by

$$H(\sigma, \varphi) = \sum_{b \in M_r} (U(\sigma_b) - U(\varphi_b)).$$

Definition 10.9. Let GS be the complete set of all ground states of the relative Hamiltonian H. A ball $b \in M_r$ is said to be an *improper* ball of the configuration σ if $\sigma_b \neq \varphi_b$ for any $\varphi \in GS$. The union of the improper balls of a configuration σ is called the *boundary* of the configuration and denoted by $\partial(\sigma)$.

Definition 10.10. The relative Hamiltonian H with the set of ground states GS satisfies the Peierls condition if for any $\varphi \in GS$ and any configuration σ coinciding almost everywhere with φ (i.e., $|\{x \in V : \sigma(x) \neq \varphi(x)\}| < \infty$)

$$H(\sigma, \varphi) \geq \lambda |\partial(\sigma)|,$$

where λ is a positive constant which does not depend on σ, and $|\partial(\sigma)|$ is the number of balls in $\partial(\sigma)$.

Theorem 10.14. *If assumptions A1-A3 are satisfied then the Peierls condition holds.*

Proof. Suppose σ coincides almost everywhere with a ground state $\varphi \in GS$ then we have $U(\sigma_b) - U^{\min} \geq \lambda_0$ for any $b \in \partial(\sigma)$ since φ is a ground state. Thus

$$H(\sigma, \varphi) = \sum_{b \in M_r} (U(\sigma_b) - U(\varphi_b)) = \sum_{b \in \partial(\sigma)} (U(\sigma_b) - U^{\min}) \geq \lambda_0 |\partial(\sigma)|.$$

Therefore, the Peierls condition is satisfied for $\lambda = \lambda_0$. $\square$

10.4.3 *Contours*

Let $\Lambda \subset V$ be a finite set. Let $\sigma_{\Lambda^c}^{(i)} \equiv i$, $i = 1, ..., s$ be a constant configuration outside of Λ. For each i we extend the configuration σ_Λ inside Λ to the entire tree by the i-th constant configuration and denote it by $\sigma_\Lambda^{(i)}$. The set of such configurations we denote by $\Omega_\Lambda^{(i)}$.

We use a construction of the subcontours defined from Definition 10.2 and up to Definition 10.3. Note that our definition (see Definition 10.12 below) of a contour depends on r, at $r = 1$ we get a contour defined in Definition 10.3. But the definition of a subcontour does not depend on r.

The set of boundary edges of $\sigma_{V_n}^{(i)}$ is called *edge boundary* $\partial_1(\sigma_{V_n}^{(i)}) \equiv \partial_1$ of the configuration.

Recall $r' = \lceil \frac{r+1}{2} \rceil$.

Definition 10.11. The subcontours T_1, T_2 are called *adjacent* if $\mathrm{dist}(T_1, T_2) \leq 2(r' - 1)$. A set of subcontours $\mathcal{A}$ is called *connected* if for any two subcontours $T_1, T_2 \in \mathcal{A}$ there is a collection of subcontours $T_1 = \tilde{T}_1, \tilde{T}_2, ..., \tilde{T}_l = T_2$ in $\mathcal{A}$ such that for each $i = 1, ..., l - 1$ the subcontours $\tilde{T}_i$ and $\tilde{T}_{i+1}$ are adjacent.

Definition 10.12. Any maximal connected set (component) of subcontours (with given marks) is called a *contour* of the set ∂_1.

For contour $\gamma = \{T_p\}$ denote $\mathrm{Int}\gamma = \cup_p \mathrm{Int}T_p$.

Remark 10.14. We note that

1. Definition 10.12 of contours coincides with Definition 10.3 for $r = 1$. But Definition 10.12 is better than the corresponding Definition 10.8 for $r = 2$. Because, for $r = 2$ from Definition 10.11 we have $\mathrm{dist}(T_1, T_2) = 0$ i.e., the subcontours do not interact if the distance between them is ≥ 1 but in Definition 10.8 the condition was like $\mathrm{dist}(T_1, T_2) \leq 2$.

2. Our definition of a contour is slightly different from the definition of contour of Hamiltonians on $\mathbb{Z}^d$, $d \geq 2$ (see [201], [202], [247]). For any two contours γ, γ' we have $\mathrm{dist}(\gamma, \gamma') > 2(r' - 1)$. Thus our contours do not interact. This means that for any $\sigma \in \Omega$ there is no ball $b \in \partial(\sigma)$ with $b \cap \gamma \neq \emptyset$ and $b \cap \gamma' \neq \emptyset$. Such property allows us to use a *contour-removal* operation. This operation is similar to the one in ordinary Peierls argument [72]: Given a family of contours defining a configuration $\sigma \in \Omega_\Lambda^{(i)}$, the family obtained by omitting one of them is also the family of contours of a (different) configuration in $\Omega_\Lambda^{(i)}$. There is an algorithm of the contour-removal operation to obtain a new configuration as follows. Take the configuration σ and change all the spins in the interior of γ (which must be removed) to value i. This makes γ disappear, but leaves intact the other contours.

For a given (sub)contour γ denote

$$\text{imp}\gamma = \{b \in \triangle : b \cap \gamma \neq \emptyset\}, \quad |\gamma| = |\text{imp}\gamma|.$$

By the construction we have $\text{imp}\gamma \cap \text{imp}\gamma' = \emptyset$ for any contours $\gamma \neq \gamma'$. For $A \subset V$ denote

$$C(A) = \{b \in M_r : b \cap A \neq \emptyset\};$$

$$D(A) = \{x \in V \setminus A : \exists y \in A, \quad \text{such that} \quad <x,y>\}.$$

Lemma 10.25. *Let K be a connected subgraph of the Cayley tree Γ^k of order $k \geq 2$, such that $|V(K)| = n$, then*

(i) $|D(V(K))| = (k-1)n + 2.$
(ii) $|C(V(K))| = k^{r'-1}((k-1)n + 2).$

Proof. (i) We shall use the induction over n. For $n = 1$ and 2 the assertion is trivial. Assume for $n = m$ the lemma is true, i.e., from $|V(K)| = m$ follows that $|D(V(K))| = (k-1)m + 2$. We shall prove the assertion for $n = m + 1$, i.e., for $\tilde{K} = K \cup \{x\}$. Since $\tilde{K}$ is connected graph we have $x \in D(V(K))$ and there is a unique $y \in S_1(x) = \{u \in V : d(x,u) = 1\}$ such that $y \in V(K)$. Thus $D(V(\tilde{K})) = (D(V(K)) \setminus \{x\}) \cup (S_1(x) \setminus \{y\})$. Consequently,

$$|D(V(\tilde{K}))| = |D(V(K))| - 1 + k = (k-1)(m+1) + 2.$$

(ii) Using (i) we obtain $|C(V(K))| = u_{r'}$, where $u_{r'}$ is the last term of the collection $u_1, u_2, ..., u_{r'}$ which is defined by the following recurrent relations

$$u_l = 2 + (k-1) \sum_{i=0}^{l-1} u_i, \quad l = 1, 2, ..., r', \quad u_0 = n. \tag{10.94}$$

Iterating (10.94) we get $u_1 = (k-1)n + 2$, $u_2 = k((k-1)n + 2)$, then using the induction over l we obtain $u_l = k^{l-1}((k-1)n + 2)$. This completes the proof. $\square$

Let us define a graph structure on M_r as follows. Two balls $b, b' \in M_r$ are connected by an edge if their centers are nearest neighbors. Denote this graph by $G(M_r)$. Note that the graph $G(M_r)$ is a Cayley tree of order $k \geq 1$. Here the vertices of this graph are balls of M_r.

For $x \in V$ we will write $x \in \gamma$ if $x \in V(\gamma)$.

Denote $N_l(x) = |\{\gamma : x \in \gamma, |\gamma| = l\}|$, where $|\gamma| = |\text{imp}\gamma|$.

Lemma 10.26. *If $k \geq 2$ then*

$$N_l(x) \leq C_0 \theta^l, \tag{10.95}$$

where $C_0 = 1 + \frac{k+1}{k-1}(k^{r'} - 1)$, $\theta = \theta(k,r) = (2ek)^{2(k+1)(r'-1)k^{r'-1}+2}$.

Proof. Denote by K_γ the minimal connected subgraph of Γ^k, which contains a contour $\gamma = \{\gamma_1, ..., \gamma_m\}, m \geq 1$, where γ_i is a subcontour. Put

$$M_{r,\gamma}^{-} = \{x \in \mathrm{Int}\gamma : \mathrm{dist}(x, V \setminus \mathrm{Int}\gamma) > r'\};$$

$$M_{r,\gamma}^{0} = \{x \in \mathrm{Int}\gamma : \mathrm{dist}(x, V \setminus \mathrm{Int}\gamma) \leq r'\};$$

$$M_{r,\gamma}^{+} = \{x \in V \setminus \mathrm{Int}\gamma : \mathrm{dist}(x, \mathrm{Int}\gamma) \leq r'\};$$

$$Y_\gamma = V(K_\gamma) \setminus \big(\mathrm{Int}\gamma \cup D(\mathrm{Int}\gamma)\big).$$

We have

$$|\gamma| = |M_{r,\gamma}^{0}| + |M_{r,\gamma}^{+}|;$$

$$|C(V(K_\gamma))| \leq |M_{r,\gamma}^{-}| + |\gamma| + |C(Y_\gamma)|. \tag{10.96}$$

For any $k \geq 2$, $r \geq 1$ by Lemma 10.25 we have

$$|M_{r,\gamma}^{-}| = \frac{|D(M_{r,\gamma}^{-})| - 2}{k - 1} < |D(M_{r,\gamma}^{-})| \leq |M_{r,\gamma}^{0}| < |\gamma|. \tag{10.97}$$

Note that $0 \leq |Y_\gamma| \leq 2(m - 1)(r' - 1)$. Thus

$$|C(T_\gamma)| \leq 2(m - 1)(r' - 1)|C(\{y\})|, \tag{10.98}$$

where y is an arbitrary point of Y_γ. By Lemma 10.25 we have $|C(\{y\})| = k^{r'-1}(k + 1)$ since $|V(\{y\})| = 1$. Hence from (10.96)-(10.98) we get

$$|C(V(K_\gamma))| < 2|\gamma| + 2(m - 1)(r' - 1)(k + 1)k^{r'-1}. \tag{10.99}$$

Since γ contains m subcontours we have $m < |\gamma|$. A combinatorial calculations show that

$$N_l(x) \leq C_0 \sum_{m=1}^{l} \binom{2l + 2k^{r'-1}(k + 1)(r' - 1)(m - 1)}{l} \times$$

$$\tilde{N}_{2l+2k^{r'-1}(k+1)(r'-1)(m-1),\Gamma^k}(b_x), \tag{10.100}$$

where $\tilde{N}_{l,\Gamma^k}$ is defined in Lemma 10.11 and b_x is a ball $b \in M_r$ such that $x \in b$. Using inequality $\binom{n}{l} \leq 2^{n-1}, l \leq n$ and Lemma 10.11 from (10.100) we get (10.95). $\qquad\square$

10.4.4 *Non-uniqueness of Gibbs measure*

For $\sigma_n \in \Omega_{V_n}^{(i)}$ the conditional Hamiltonian (10.91) has the form

$$H^{(i)}(\sigma_n) \equiv H(\sigma_n | \sigma_{V_n^c} = i) = \sum_{\substack{b \in M_r: \\ b \cap V_n \neq \emptyset}} U(\sigma_{n,b}) = \qquad (10.101)$$

$$\sum_{b \in \partial(\sigma_n)} (U(\sigma_{n,b}) - U^{\min}) + |C(V_n)| U^{\min},$$

where $\sigma_{n,b} = (\sigma_n)_b$.

The Gibbs measure on the space $\Omega_{V_n}^{(i)}$ with boundary condition $\sigma^{(i)}$ is defined as

$$\mu_{n,\beta}^{(i)}(\sigma_n) = \mathbf{Z}_{n,i}^{-1} \exp(-\beta H^{(i)}(\sigma_n)), \qquad (10.102)$$

where $\mathbf{Z}_{n,i}$ is the normalizing factor.

Let us consider a sequence of balls on Γ^k

$$V_1 \subset V_2 \subset ... \subset V_n \subset ..., \quad \cup V_n = V,$$

and s sequences of boundary conditions outside these balls:

$$\sigma_n^{(i)} \equiv i, n = 1, 2, ..., i = 1, ..., s.$$

Note that each of s sequences of measures $\{\mu_{n,\beta}^{(i)}, n = 1, 2, ...\}, i = 1, ..., s$ contains a convergent subsequence.

We denote the corresponding limits by $\mu_\beta^{(i)}, i = 1, ..., s$.

Our purpose is to show for a sufficiently large β these measures are different.

Lemma 10.27. *Suppose assumptions A1-A3 are satisfied. Let γ be a fixed contour and $p_i(\gamma) = \mu_\beta^{(i)}(\sigma_n : \gamma \in \partial(\sigma_n))$. Then*

$$p_i(\gamma) \leq \exp\{-\beta\lambda_0|\gamma|\}, \qquad (10.103)$$

where λ_0 is defined by formula (10.92).

Proof. Put $\Omega_\gamma = \{\sigma_n \in \Omega_{V_n}^{(i)} : \gamma \subset \partial(\sigma_n)\}$, $\Omega_\gamma^0 = \{\sigma_n : \gamma \cap \partial = \emptyset\}$ and define a (contour-removal) map $\chi_\gamma : \Omega_\gamma \to \Omega_\gamma^0$ by

$$\chi_\gamma(\sigma_n)(x) = \begin{cases} i & \text{if} \quad x \in \text{Int}\gamma \\ \sigma_n(x) & \text{if} \quad x \notin \text{Int}\gamma. \end{cases}$$

When γ is fixed then the configuration on $\text{Int}\gamma$ is also fixed. Therefore the map χ_γ is one-to-one map.

For any $\sigma_n \in \Omega_{V_n}^{(i)}$ we have

$$|\partial(\sigma_n)| = |\partial(\chi_\gamma(\sigma_n))| + |\gamma|.$$

Consequently, using (10.101) one finds

$$p_i(\gamma) = \frac{\sum_{\sigma_n \in \Omega_\gamma} \exp\{-\beta \sum_{b \in \partial(\sigma_n)} (U(\sigma_{n,b}) - U^{\min})\}}{\sum_{\tilde{\sigma}_n} \exp\{-\beta \sum_{b \in \partial(\tilde{\sigma}_n)} (U(\tilde{\sigma}_{n,b}) - U^{\min})\}} \le \qquad (10.104)$$

$$\frac{\sum_{\sigma_n \in \Omega_\gamma} \exp\{-\beta \sum_{b \in \partial(\sigma_n)} (U(\sigma_{n,b}) - U^{\min})\}}{\sum_{\tilde{\sigma}_n \in \Omega_\gamma^0} \exp\{-\beta \sum_{b \in \partial(\tilde{\sigma}_n)} (U(\tilde{\sigma}_{n,b}) - U^{\min})\}} =$$

$$\frac{\sum_{\sigma_n \in \Omega_\gamma} \exp\{-\beta \sum_{b \in \partial(\sigma_n)} (U(\sigma_{n,b}) - U^{\min})\}}{\sum_{\tilde{\sigma}_n \in \Omega_\gamma} \exp\{-\beta \sum_{b \in \partial(\chi_\gamma(\tilde{\sigma}_n))} (U(\chi_\gamma(\tilde{\sigma}_{n,b})) - U^{\min})\}}.$$

Since $\sigma_{n,b} = \chi_\gamma(\sigma_{n,b})$, for any $b \in \partial(\sigma_n) \setminus \mathrm{imp}\gamma$ we have

$$\sum_{b \in \partial(\sigma_n)} (U(\sigma_{n,b}) - U^{\min}) = S_1 + S_2, \qquad (10.105)$$

where

$$S_1 = \sum_{b \in \partial(\chi_\gamma(\sigma_n))} (U(\sigma_{n,b}) - U^{\min}); \quad S_2 = \sum_{b \in \mathrm{imp}\gamma} (U(\sigma_{n,b}) - U^{\min}).$$

By our construction γ is a contour of $\partial(\sigma_n)$ iff $\sigma_n(x) = i$ for any $x \in M_{r,\gamma}^+$. Consequently, $\mathrm{imp}\gamma$ does not depend on $\sigma_n \in \Omega_\gamma$. By assumptions A1-A3 we have $U(\sigma_{n,b}) - U^{\min} \ge \lambda_0 > 0$, for any $b \in \mathrm{imp}\gamma$.

Hence

$$S_2 = \sum_{b \in \mathrm{imp}\gamma} (U(\sigma_{n,b}) - U^{\min}) \ge \lambda_0 |\gamma|, \quad \text{for any } \sigma_n \in \Omega_\gamma. \qquad (10.106)$$

Thus from (10.104)-(10.106) one gets (10.103). $\qquad\qquad\square$

Now using Lemmas 10.99 and 10.101 one can prove the following

Lemma 10.28. *If assumptions A1-A3 are satisfied then for fixed $x \in \Lambda$ uniformly by Λ the following relation holds*

$$\mu_\beta^{(i)}(\sigma_\Lambda : \sigma(x) = j) \to 0, j \ne i \ \ as \ \ \beta \to \infty.$$

This lemma implies the main result, i.e.,

Theorem 10.15. *If A1-A3 are satisfied then for all sufficiently large β there are at least s (= number of ground states) Gibbs measures for the model (10.90) on Cayley tree of order $k \ge 2$.*

10.4.5 *Examples*

In this subsection we shall give several examples with the properties A1-A3.

10.4.5.1 *q-component models.*

Note that under some suitable conditions on the parameters of q-component models (with nearest neighbor interactions) on Cayley tree the assumptions A1-A3 are satisfied. In particular, the *ferromagnetic* Ising, Potts and SOS models have the properties A1-A3.

10.4.5.2 *The Potts model with competing interactions*

Consider the Hamiltonian

$$H(\sigma) = J_1 \sum_{\substack{\langle x,y \rangle, \\ x,y \in V}} \delta_{\sigma(x)\sigma(y)} + J_2 \sum_{\substack{x,y \in V: \\ d(x,y)=2}} \delta_{\sigma(x)\sigma(y)}, \tag{10.107}$$

where $J = (J_1, J_2) \in R^2$, $\sigma(x) \in \Phi = \{1, 2, ..., q\}$ and δ is the Kronecker's symbol.

Note that the Ising model with competing interactions is a particular case of the model (10.107). For the model (10.107) with $k = 2$, $q = 3$ we put

$$U(\sigma_b) \equiv U(\sigma_b, J) = \frac{1}{2} J_1 \sum_{\substack{\langle x,y \rangle, \\ x,y \in b}} \delta_{\sigma(x)\sigma(y)} + J_2 \sum_{\substack{x,y \in b: \\ d(x,y)=2}} \delta_{\sigma(x)\sigma(y)}. \tag{10.108}$$

Simple calculations show that

$$\mathbf{U} = \{U(\sigma_b)\} = \left\{ \frac{3}{2} J_1 + 3J_2, \quad J_1 + J_2, \quad 3J_2, \quad \frac{1}{2} J_1, \quad J_2, \quad \frac{1}{2} J_1 + J_2 \right\}.$$

One can show that for the model (10.107) the assumptions A1-A3 are satisfied if $J \in \{J \in R^2 : J_1 < 0, J_1 + 4J_2 < 0\}$.

10.4.5.3 *A model with the interaction radius $r \geq 1$.*

For $A \subset V$ let us define a generalized Kronecker symbol as the function $U_0(\sigma_A) : \Omega_A \to \{|A| - 1, |A| - 2, ..., |A| - \min\{|A|, |\Phi|\}\}$ by

$$U_0(\sigma_A) = |A| - |\sigma_A \cap \Phi|, \tag{10.109}$$

where as before $\Phi = \{1, 2, ..., q\}$ and $|\sigma_A \cap \Phi|$ is the number of different values of $\sigma_A(x), x \in A$. For instance if σ_A is a constant configuration then $|\sigma_A \cap \Phi| = 1$.

Note that if $|A| = 2$, say, $A = \{x, y\}$, then $U_0(\{\sigma(x), \sigma(y)\}) = \delta_{\sigma(x),\sigma(y)}$.

Now consider the Hamiltonian

$$H(\sigma) = -J \sum_{b \in M_r} U_0(\sigma_b), \tag{10.110}$$

where $J \in R$.

It is easy to see that if $J > 0$ then the assumptions A1-A3 are satisfied for any $r \geq 1$ and $k \geq 2$.

Thus (by Theorem 10.15) these models have q different Gibbs measures for all sufficiently large β.

Commentaries and references. In the usual one-dimensional models the Gibbs measure is unique i.e., there is no phase transition. But it is known that phase transition occurs in the following cases:

a) The set of spin values is $\{-1, 1\}$ and interactions are long range (Dyson's model).

b) The set of spin values is $\{-1, 1\}$ and interactions are nearest neighbors, but they are spatially inhomogeneous (Sullivan's model).

c) The set of spin values is a countably infinite set and interactions are nearest neighbors (Spitzer's model).

For a detailed description of history of phase separation properties of lattice models see [78].

In this chapter we have considered a one-dimensional model of type b). As mentioned above for such kind of model a phase transition occurs (see Theorem 10.1). In such a case of the existence of the phase transition, it would be interesting to know certain properties of a phase separation point (PSP) (curve (membrane) in two (three) dimensional case).

In two (resp. three) dimensional case a phase separation curve (resp. membrane) is defined as an "open" contour [59], [77]. But this construction does not work for one dimensional case with interactions of only nearest neighbors. A notion of PSP, to our knowledge, have not yet been introduced for one-dimensional models. In one-dimensional case the separation "line" is a point. We have introduced here a natural definition of the PSP between two phases in one-dimensional case. We studied asymptotical properties of the PSP. Our definition of PSP is rather natural and properties of the PSP more special than those in two and three dimensional cases.

Results devoted to the one-dimensional models are due to [103], [224], [232].

A theory of phase transitions at low temperatures in general classical lattice (on $\mathbb{Z}^d$) systems was developed by Pirogov and Sinai. This theory is now globally known as Pirogov-Sinai theory or contour arguments. This technique was pioneered by Peierls [193] in his study of the Ising model. The original argument benefited from the particular symmetries of the Ising model. The adaptation of the method to the treatment of non-symmetric models is not trivial, and was developed by Pirogov and Sinai [201], [202], [247], [161], [269]. Later, an alternative version of the argument was set forward in [268].

Note that Pirogov-Sinai theory on Cayley tree was not developed. The method used for the description of Gibbs measures on Cayley tree is the method of Markov random field theory and recurrent equations of this theory. But, if we consider non-symmetric models on Cayley tree, then the description of Gibbs measures by the method becomes a difficult problem: in this situation, a non-linear operator W that maps R^r (for some $r \geq 1$) into itself appears and the problem is then to describe the fixed points of this operator.

The contour methods of this chapter are due to papers [31], [32], [220], [222], [224], [227].

Chapter 11

Other models

This chapter contains several models which were not discussed in previous chapters. In Sections 1-9 we give the definitions of these models and describe some known results for each model. The last section contains a review of remaining models. This last section is divided into three subsections: the first is devoted to classical (real valued) models, the second subsection to quantum models, and the third to models with p-adic values. Moreover, we give a brief description of the differences of behavior between classical (real) models and p-adic models on Cayley trees.

11.1   Inhomogeneous Ising model

The inhomogeneous Ising model is defined by the formal Hamiltonian

$$H(\sigma) = -\sum_{\langle x,y \rangle} J_{xy}\sigma(x)\sigma(y), \tag{11.1}$$

where the summation runs over all pairs of nearest neighbors $\langle x, y \rangle$, J_{xy} is a real-valued function specifying the interaction of the neighbors x and y, and $\sigma(x) \in \{-1, 1\}$ for any $x \in V$.

Theorem 11.1. *There exists one-to-one correspondence between the set of (splitting) Gibbs measures of Hamiltonian (11.1) and the set of functions $h = \{h_x, x \in V\}$ which satisfy the following equation:*

$$h_x = \sum_{y \in S(x)} f(h_y, \theta_{xy}). \tag{11.2}$$

Here, $\theta = \tanh(J_{xy}\beta)$, $f(h, \theta_{xy}) = \operatorname{arctanh}(\theta_{xy} \tanh h)$ and $S(x)$ is the set of direct successors of x on Cayley tree of order k.

Proof.   Similar to the proof of Theorem 2.1. $\square$

321

Let $\mathcal{H}$ be a normal subgroup of G_k of finite index $r \geq 1$, and $G_k/\mathcal{H} = \{\mathcal{H}_0, \ldots, \mathcal{H}_{r-1}\}$ be the quotient group with respect to $\mathcal{H}$.

Note that if there exist $x, y, u, v \in V$ such that $J_{xy} \neq J_{uv}$ then the equation (11.2) has no translation-invariant solutions. Moreover, equation (11.2) has a $\mathcal{H}$-periodic solution iff function J_{xy} is $\mathcal{H}$-periodic, i.e.,

$$J_{xy} = J_{ij}, \quad x \in \mathcal{H}_i, \, y \in \mathcal{H}_j.$$

In [217], under some conditions on J_{xy}, three $\mathcal{H}$-periodic solutions of (11.2) are found. Let μ_0, μ_1, μ_2 be the corresponding $\mathcal{H}$-periodic Gibbs measures.

In [212] we proved that, under the same conditions listed in [217], there is an uncountable set of Gibbs measures, and we constructively described them. In these measures, "1/3" of the Cayley tree is occupied by the measure μ_0, another "1/3" by the measure μ_1, and another "1/3" by the measure μ_2. Measures of this type were first constructed by Dobrushin [60] in the Ising model on $\mathbb{Z}^d$. Recall that Gibbs measures on the Cayley tree in which one "half" of the lattice is occupied by one measure and the other "half" by a different measure were constructed by Bleher and Ganikhodjaev.

In this section we shall describe some countably periodic Gibbs measures of the Hamiltonian (11.1), such measures are invariant with respect to a normal subgroup of infinite index (see Chapter 1).

We consider a normal subgroup $\mathcal{H}_0$ of infinite index constructed as follows. Let the mapping $\pi_0 : \{a_1, \ldots, a_{k+1}\} \longrightarrow \{e, a_1, a_2\}$ be defined by

$$\pi_0(a_i) = \begin{cases} a_i, & \text{if } i = 1, 2 \\ e, & \text{if } i \neq 1, 2. \end{cases}$$

Denote by G_1 the free product of cyclic groups $\{e, a_1\}, \{e, a_2\}$. Consider

$$f_0(x) = f_0(a_{i_1} a_{i_2} \ldots a_{i_m}) = \pi_0(a_{i_1}) \pi_0(a_{i_2}) \ldots \pi_0(a_{i_m}).$$

Then it is easy to see that f_0 is a homomorphism and hence $\mathcal{H}_0 = \{x \in G_k : f_0(x) = e\}$ is a normal subgroup of infinite index.

Now we consider the factor group

$$G_k/\mathcal{H}_0 = \{\mathcal{H}_0, \mathcal{H}_0(a_1), \mathcal{H}_0(a_2), \mathcal{H}_0(a_1 a_2), \ldots\},$$

where $\mathcal{H}_0(y) = \{x \in G_k : f_0(x) = y\}$. We introduce the notations

$$\mathcal{H}_n = \mathcal{H}_0(\underbrace{a_1 a_2 \ldots}_{n}),$$

$$\mathcal{H}_{-n} = \mathcal{H}_0(\underbrace{a_2 a_1 \ldots}_{n}).$$

In this notation, the factor group can be represented as

$$G_k/\mathcal{H}_0 = \{\ldots, \mathcal{H}_{-2}, \mathcal{H}_{-1}, \mathcal{H}_0, \mathcal{H}_1, \mathcal{H}_2, \ldots\}.$$

The partition of the Cayley tree Γ^2 w.r.t. $\mathcal{H}_0$ is shown in Fig. 11.1 (the elements of the class $\mathcal{H}_i$, $i \in \mathbb{Z}$, are merely denoted by i).

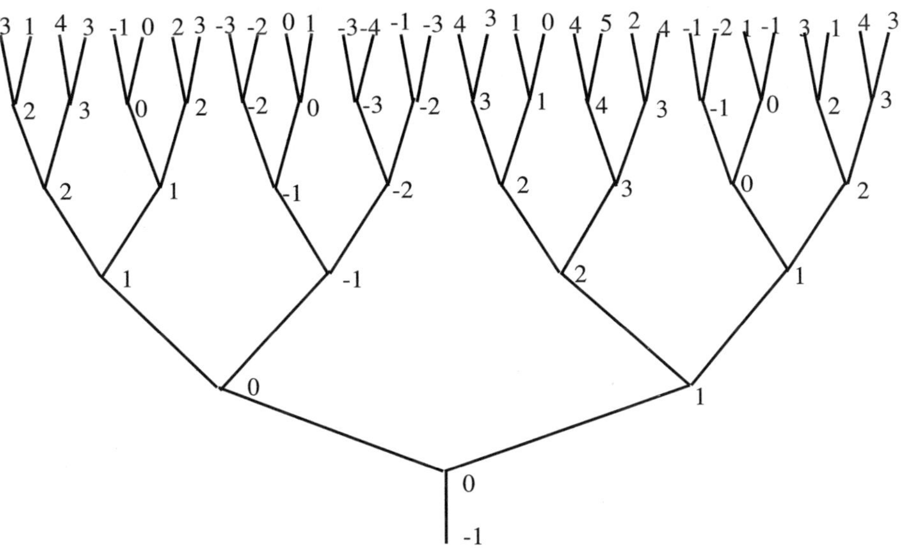

Fig. 11.1 The partition of the Cayley tree Γ^2 w.r.t. $\mathcal{H}_0$, the elements of the class $\mathcal{H}_i$, $i \in \mathbb{Z}$, are denoted by i.

To describe countably periodic Gibbs measures, we must construct $\mathcal{H}_0$-periodic solutions of (11.2) for the normal subgroup $\mathcal{H}_0$ of infinite index. In this case, we can explicitly write the unknown function h.

We assume that

$$h_x = h_i \quad \text{if} \quad x \in \mathcal{H}_i.$$

$$\theta_{xy} = \theta_{i,j} \quad \text{if} \quad x \in \mathcal{H}_i, \, y \in \mathcal{H}_j.$$

Then from equation (11.2) we get (see Fig. 11.1)

$$h_n = f(h_n, \theta_{n,n}) + f(h_{n+1}, \theta_{n,n+1}),$$

$$h_n = f(h_n, \theta_{n,n}) + f(h_{n-1}, \theta_{n,n-1}), \tag{11.3}$$

$$h_n = f(h_{n+1}, \theta_{n,n+1}) + f(h_{n-1}, \theta_{n,n-1}),$$

where $n \in \mathbb{Z}$.

It is easy to see that $h_n \equiv 0$ solves system (11.3) independently of the values of the parameters $\theta_{i,j}$. We also note that if there exists n_0 such that $h_{n_0} = 0$, then we obtain $h_n \equiv 0$.

We now choose parameters $\theta_{i,j}$ such that system (11.3) admits nonzero solutions. It follows from (11.3) that for $n = 0$, we have

$$h_0 = f(h_0, \theta_{0,0}) + f(h_1, \theta_{0,1}),$$
$$h_0 = f(h_0, \theta_{0,0}) + f(h_{-1}, \theta_{0,-1}), \tag{11.4}$$
$$h_0 = f(h_1, \theta_{0,1}) + f(h_{-1}, \theta_{0,-1}).$$

This system implies the equation

$$f(h_{-1}, \theta_{0,-1}) = f(h_0, \theta_{0,0}) = f(h_1, \theta_{0,1}),$$

and hence

$$\theta_{0,-1} \tanh h_{-1} = \theta_{0,0} \tanh h_0 = \theta_{0,1} \tanh h_1.$$

It follows from this that for $\theta_{0,-1} \neq 0$, and $\theta_{0,1} \neq 0$, we have

$$h_0 = 2f(h_0, \theta_{0,0}),$$
$$h_1 = f(h_0, \tfrac{\theta_{0,0}}{\theta_{0,1}}), \tag{11.5}$$
$$h_{-1} = f(h_0, \tfrac{\theta_{0,0}}{\theta_{0,-1}}).$$

The quantity h_0 is therefore determined from the equation

$$h_0 = 2f(h_0, \theta_{0,0}). \tag{11.6}$$

Solving equation (11.6) for h_0 we obtain

$$h_0 = \begin{cases} 0, & \text{if } \theta_{0,0} < 1/2, \\ 0, \pm h_0^*, & \text{if } \theta_{0,0} \geq 1/2, \end{cases}$$

where $h_0^* = \log[(\theta_{0,0} + \sqrt{2\theta_{0,0} - 1})/(1 - \theta_{0,0})]$.

Substituting these quantities into (11.5), we obtain

$$h_0^* = \pm \log \frac{\theta_{0,0} + \sqrt{2\theta_{0,0} - 1}}{1 - \theta_{0,0}},$$
$$h_1^* = f(\pm h_0^*, \tfrac{\theta_{0,0}}{\theta_{0,1}}), \tag{11.7}$$
$$h_{-1}^* = f(\pm h_0^*, \tfrac{\theta_{0,0}}{\theta_{0,-1}}).$$
$$\theta_{0,0} > \tfrac{1}{2}, \ 0 < |\theta_{0,1}| < 1, \ 0 < |\theta_{0,-1}| < 1.$$

We have thus found all possible values of the function h_n for $n = -1, 0, 1$.

We now determine h_2 using the found values h_0^* and h_1^*. For $n = 1$, we obtain from (11.3)

$$
\begin{aligned}
h_1^* &= f(h_1^*, \theta_{1,1}) + f(h_2, \theta_{1,2}), \\
h_1^* &= f(h_0^*, \theta_{1,0}) + f(h_1^*, \theta_{1,1}), \\
h_1^* &= f(h_0^*, \theta_{1,0}) + f(h_2, \theta_{1,2}).
\end{aligned}
\tag{11.8}
$$

Hence

$$
f(h_1^*, \theta_{1,1}) = f(h_0^*, \theta_{1,0}) = f(h_2, \theta_{1,2}).
$$

Simple algebra yields

$$
\begin{aligned}
h_2 &= h_2^* = f(\pm h_1^*, \tfrac{\theta_{1,1}}{\theta_{1,2}}), \\
\theta_{1,0} &= \frac{\tanh(h_1^*/2)}{\tanh h_0^*} \\
\theta_{1,1} &= \frac{\tanh(h_1^*/2)}{\tanh h_1^*} \\
h_0 &\neq 0, \ 0 < |\theta_{1,2}| < 1,
\end{aligned}
\tag{11.9}
$$

and $h_2^* = 0$ for $h_0^* = 0$.

We have thus found three possible values of h_n for $n = 2$.

Using h_1^* and h_2^*, we can now determine h_3, and so on. Induction on $n \geq 1$ yields

$$
\begin{aligned}
h_{n+1} &= h_{n+1}^* = f(\pm h_n^*, \tfrac{\theta_{n,n}}{\theta_{n,n+1}}), \\
\theta_{n,n-1} &= \frac{\tanh(h_n^*/2)}{\tanh h_{n-1}^*} \\
\theta_{n,n} &= \frac{\tanh(h_n^*/2)}{\tanh h_n^*} \\
h_{n-1}^* &\neq 0, \ 0 < |\theta_{n,n+1}| < 1.
\end{aligned}
\tag{11.10}
$$

Similarly, for $n \leq 1$, we have

$$
\begin{aligned}
h_{n-1} &= h_{n+1}^* = f(\pm h_n^*, \tfrac{\theta_{n,n}}{\theta_{n,n-1}}), \\
\theta_{n,n+1} &= \frac{\tanh(h_n^*/2)}{\tanh h_{n+1}^*} \\
\theta_{n,n} &= \frac{\tanh(h_n^*/2)}{\tanh h_n^*} \\
h_{n+1}^* &\neq 0, \ 0 < |\theta_{n,n-1}| < 1.
\end{aligned}
\tag{11.11}
$$

We note that no conditions are imposed on the parameters $\theta_{n,n+1} \neq 0$, $n \geq 0$, and $\theta_{n,n-1} \neq 0$, $n \leq 0$. We can therefore choose these such that

$h_n \neq h_m$ is ensured for any $n \neq m$. We have thus constructed three functions $h_n = 0$ and $\pm h_n^*$, $n \in \mathbb{Z}$, that satisfy functional equation (11.2). We let μ_0 and $\mu_{\pm}$ denote the corresponding limiting Gibbs measures.

We have thus proved the following theorem.

Theorem 11.2. *For an inhomogeneous Ising model, there exists a nonempty subset of the set of parameters J_{xy}, $x, y \in V$, such that there are three countably periodic Gibbs measures for each point of this subset.*

We note that the measure μ_0 is translation-invariant.

Notes. Consider first the one-dimensional model of a linear chain with different interactions and external fields from one site to the next. The recurrence relations for the partition function can only be solved in special cases but one can always calculate the field necessary to produce a given magnetization profile-the "inverse problem" which was solved by Percus. In [235] an inhomogeneous Ising model with site-dependent external field on a Cayley tree is considered. It is shown that Percus's result can be generalized to a Cayley tree of any order k.

In [210], [212], [215] a sufficient conditions for non-uniqueness of Gibbs measure are established. It is shown that for an inhomogeneous Ising model, there exists an uncountable set of limit Gibbs measures. A constructive description of these measures is given.

Results of this section due to [217], where a family of normal subgroups $\mathcal{H}$ of the group G_k of infinite index (i.e., such that $G_k/\mathcal{H}$ is of infinite order) is constructed. By means of such subgroups, a notion of countably-periodic Gibbs measures is introduced and for the inhomogeneous Ising model such measures are constructed. As a consequence, a new type of limiting Gibbs states of the model are obtained.

We note that countably periodic (non-finitely periodic) measures may only exist for inhomogeneous or for countably periodic interactions.

11.2 Random field Ising model

The Hamiltonian of the ferromagnetic random field Ising model (RFIM) is given by

$$H(\sigma) = -\sum_{x,y} J_{xy}\sigma(x)\sigma(y) - \sum_{x} \xi_x\sigma(x), \tag{11.12}$$

where $\sigma(x)$ are spin variables taking values ± 1, ξ_x stands for the random external field, and $J_{xy} > 0$ are the coupling constants.

We shall consider uniform interactions $J_{xy} = 1$, and a dichotomous field, i.e., the ξ_x are real independent random variables taking values $\pm \alpha$ with probability $1/2$. These variables are defined for each site of a Cayley tree Γ^k.

The case $k = 1$ corresponds to the one-dimensional RFIM. In this case it is proven that (see [26]) there is no phase transition for the inverse temperature β (i.e., the Gibbs state $\mu_{\beta,\xi}$, is unique for all $\beta < \infty$ and all configurations $\xi = \{\xi_x, x \in V\}$). The structure of the ground states $\mu_{\infty,\xi} = \lim_{\beta \to \infty} \mu_{\beta,\xi}$ is described in [26]. For the case $k = 2$ Bruinsma [40] proposed some theoretical arguments, to describe the structure of the ground states for RFIM on the Cayley tree, and to estimate the residual entropy at $\beta \to \infty$.

Now following [27] we give some results related to Gibbs measures of the RFIM on the Cayley tree.

The following theorem is used as a main tool:

Theorem 11.3. *There exists one-to-one correspondence between the set of splitting Gibbs measures of Hamiltonian (11.12) and the set of functions* $h = \{h_x, x \in V\}$ *which satisfy the following equation:*

$$h_x = \sum_{y \in S(x)} f_\beta(h_y + \xi_y). \tag{11.13}$$

Here,

$$f_\beta(h) = \frac{1}{\beta} \text{arctanh}[(\tanh \beta)(\tanh \beta h)]. \tag{11.14}$$

Proof. Similar to the proof of Theorem 2.1. $\qquad\qquad\square$

Denote $g_x = f_\beta(h_x + \xi_x)$. Then the recursive equation (11.13) reads

$$h_x = \sum_{y \in S(x)} g_y. \tag{11.15}$$

This implies that g_x satisfies

$$g_x = f_\beta\left(\xi_x + \sum_{y \in S(x)} g_y\right). \tag{11.16}$$

For given $g = \{g_x : x \in V\}$ we consider the following finite dimensional probability distribution (cf. with (2.6)):

$$\mu(\sigma_n|g) = Z_n^{-1} \exp\left\{\beta \sum_{\langle x,y \rangle \in L_n} \sigma(x)\sigma(y) + \beta \sum_{x \in V_n} \xi_x \sigma(x) + \beta \sum_{\substack{x \in W_n, \\ y \in S(x)}} \sigma(x)g_y\right\}. \tag{11.17}$$

We call $g = \{g_x\}$ the *effective* external field, along with $h = \{h_x\}$.

By F.K.G. inequalities [74] one has the following proposition (see [146]).

Proposition 11.1. *Two Gibbs measures $\mu^+_{\beta,\xi}$ and $\mu^-_{\beta,\xi}$, which are limiting Gibbs measures with $+$ and $-$ boundary conditions, exist and they are extreme for all ξ. If $\mu^-_{\beta,\xi} = \mu^+_{\beta,\xi}$ for a given ξ, then the Gibbs measure is unique for this ξ.*

We denote by $g^\pm = \{g^\pm_x,\ x \in V\}$ the configurations that correspond to the Gibbs measures $\mu^\pm_{\beta,\xi}$.

Proposition 11.2. *The function (11.14) has the following properties*

$$f_\beta(-x) = -f_\beta(x), \quad f_\beta(\infty) = 1;$$

$$0 < \frac{d}{dx} f_\beta(x) < \tanh\beta, \quad x \neq 0, \quad \frac{d}{dx} f_\beta(0) = \tanh\beta; \tag{11.18}$$

$$\frac{d}{dx} f_\beta(x) < \frac{1}{2}, \quad x \geq 1; \tag{11.19}$$

$$\frac{d}{dx} f_\beta(x) < \frac{1}{2}(1 - \tanh\beta), \quad x \geq 2; \tag{11.20}$$

$$\frac{d^2}{dx^2} f_\beta(x) < 0, \quad x > 0, \quad \frac{d^2}{dx^2} f_\beta(0) = 0. \tag{11.21}$$

Proof. Follows from the following two equations:

$$\frac{d}{dx} f_\beta(x) = \tfrac{1}{2}[\tanh\beta(x+1) - \tanh\beta(x-1)],$$

$$\frac{d^2}{dx^2} f_\beta(x) = \frac{\beta}{2}\left[\frac{1}{\cosh^2\beta(x+1)} - \frac{1}{\cosh^2\beta(x-1)}\right]. \tag{11.22}$$

$\square$

Recall that by $\mu_{\beta,\xi}$, where $0 < \beta < \infty$ and $\xi = \{\xi_x = \pm\alpha,\ x \in V\}$, we denote a Gibbs measure corresponding to a solution of (11.13) (or equivalently (11.16)).

Theorem 11.4. *If $T > T_c$, (where T_c is the critical temperature of the model in the absence of external field) then $\mu_{\beta,\xi}$ is unique for all ξ.*

Proof. By recursive equations (11.16) we have

$$g_x^+ - g_x^- = f_\beta'(c) \sum_{y \in S(x)} \left(g_y^+ - g_y^- \right),$$

for some $c \in [\xi_x + \sum_{y \in S(x)} g_y^-, \; \xi_x + \sum_{y \in S(x)} g_y^+]$ so that

$$|g_x^+ - g_x^-| \leq k |f_\beta'(c)| \sup_{y \in S(x)} |g_y^+ - g_y^-|. \tag{11.23}$$

When $T > T_c$, one has $\tanh \beta < 1/k$, so that by (11.18), $f_\beta'(c) < 1/k$. By applying recursively the inequality (11.23), we get that $g_x^+ = g_x^-$. Hence $\mu_{\beta,\xi}^- = \mu_{\beta,\xi}^+$. By virtue of Proposition 11.1 this proves Theorem 11.4. $\quad\square$

Theorem 11.5. *Let $k = 2$. If $\alpha > 3$, then*

(a) *$\mu_{\beta,\xi}$ is unique for all $0 < \beta < \infty$ and all ξ.*
(b) *the limit $\mu_{\infty,\xi} \equiv \lim_{\beta \to \infty} \mu_{\beta,\xi}$ exists and it is concentrated on a configuration such that $\sigma(x) = \mathrm{sign}\xi_x$ for all $x \in V$.*
If $2 \leq \alpha \leq 3$, then $\mu_{\beta,\xi}$ is unique for all ξ, and all $0 < \beta < \infty$ such that

$$\tanh[\beta(\alpha - 1)] + \tanh[\beta(3 - \alpha)] \leq 1.$$

Proof. By Proposition 11.2 we have $|f_\beta(x)| < 1$. This implies that $|g_y| < 1 - \delta$ for some $\delta = \delta(\alpha, \beta, k) > 0$ and any y. Hence, when $\alpha \geq 2$ and $k = 2$, we have $|c| > \alpha - 2 + 2\delta$ in (11.23). By (11.22) we obtain

$$f_\beta'(c) = \frac{1}{2}(\tanh \beta(|c| + 1) - \tanh \beta(|c| - 1))$$

consequently, for some $\delta_0 > 0$,

$$0 < f_\beta'(c) < \frac{1}{2} - \delta_0$$

provided $\alpha \geq 3$ or $3 > \alpha \geq 2$ and

$$\tanh \beta(\alpha - 1) + \tanh \beta(3 - \alpha) \leq 1.$$

In both cases, iterating (11.23) we conclude that $g_x^+ = g_x^-$. Hence, the uniqueness part of Theorem 11.5 follows from Proposition 11.1.

To prove part (b) we note that for $n = 1$ one gets by (11.17) that

$$\mu(\sigma(x)|g) = Z^{-1} \exp\left\{ \beta \left[\xi_x + \sum_{y : \langle x,y \rangle} g_y \right] \sigma(x) \right\}.$$

If $|\xi_x| > 3$ then $|\xi_x + \sum_{y:\langle x,y \rangle} g_y| > 0$ and $\mathrm{sign}(\xi_x + \sum_{y:\langle x,y \rangle} g_y) = \mathrm{sign}\xi_x$. Thus taking the limit $\beta \to \infty$ finishes the proof. $\quad\square$

Let $H^{\mathrm{F}}(T)$ be the critical external field in the ferromagnetic Ising model:

$$H^{\mathrm{F}}(T) = \frac{1}{\beta}\left[k\operatorname{artanh}\left(\frac{k\theta-1}{k/\theta-1}\right)^{1/2} - \operatorname{artanh}\left(\frac{k-1/\theta}{k-\theta}\right)^{1/2}\right], \quad (11.24)$$

where $\theta = \tanh\beta$.

Theorem 11.6. *Assume that $0 < T < T_c$ and $\alpha \le H^{\mathrm{F}}(T)$. Then for any $k \ge 2$ and all realizations ξ of the external field,*

(a) there exist two different extreme Gibbs measures $\mu^{+}_{\beta,\xi}$ and $\mu^{-}_{\beta,\xi}$ which are limiting Gibbs measures with $+$ and $-$ boundary conditions;

(b) if $\alpha < 1$ then the limits $\mu^{\pm}_{\infty,\xi} \equiv \lim_{\beta\to\infty}\mu^{\pm}_{\beta,\xi}$ exist and they are concentrated on configurations $\{\sigma(x) = 1,\ x \in V\}$ and $\{\sigma(x) = -1,\ x \in V\}$, respectively.

Proof. Let $\langle\,\cdot\,\rangle^{\pm}_{\beta\xi}$ be the expectation with respect to the measure $\mu^{\pm}_{\beta,\xi}$. Then by the conditions of the theorem and by the F.K.G. inequality one gets that

$$\langle\sigma(x)\rangle^{+}_{\beta\xi} \ge \langle\sigma(x)\rangle^{+}_{\beta\{-\alpha\}} > 0$$

and

$$\langle\sigma(x)\rangle^{-}_{\beta\xi} \le \langle\sigma(x)\rangle^{-}_{\beta\{\alpha\}} < 0.$$

This proves that $\mu^{+}_{\beta,\xi} \ne \mu^{-}_{\beta,\xi}$. Their extremality follows from Proposition 11.1. Since for $\alpha < 1$ $\lim_{\beta\to\infty}\langle\sigma(x)\rangle^{\pm}_{\beta\{\mp\alpha\}} = \pm 1$, we obtain that for all realizations ξ, $\sigma(x) = 1$ a.e. with respect to $\mu^{+}_{\beta,\xi}$ and $\sigma(x) = -1$ a.e. with respect to $\mu^{-}_{\beta,\xi}$. $\qquad\square$

Theorem 11.7. *Let $k = 2$ and assume that $2 \le \alpha \le 3$. Then for all $\beta < \infty$ and for almost all realizations ξ of the external field, there exists a unique Gibbs state $\mu_{\beta,\xi}$.*

Proof. For any x, we denote by y and z, its two direct successors. The recursive equation (11.16) reads

$$g_x = f_\beta(\xi_x + g_y + g_z). \quad (11.25)$$

For a given ξ, let g_x^{+} and g_x^{-} be the g_x corresponding respectively to the measures $\mu^{+}_{\beta,\xi}$ and $\mu^{-}_{\beta,\xi}$. We shall estimate recursively the expectation $\mathbb{E}_\xi|g_x^{+} - g_x^{-}|$. We have

$$g_x^{+} - g_x^{-} = f_\beta(\xi_x + g_y^{+} + g_z^{+}) - f_\beta(\xi_x + g_y^{-} + g_z^{-}) \quad (11.26)$$

$$= f'_\beta(c)[g_y^+ - g_y^- + g_z^+ - g_z^-],$$

where $c \in (\xi_x + g_y^- + g_z^-, \xi_x + g_y^+ + g_z^+)$.

Let us estimate $f'_\beta(c)$. Assume that $\xi_x = \alpha > 0$. Consider different cases for ξ_y and ξ_z.

Case (i) $\xi_y = \xi_z = -\alpha$. Then we use the estimate

$$f'_\beta(c) \le \tanh\beta \tag{11.27}$$

which is valid for all c (see (11.18)).

Case (ii) $\xi_y + \xi_z = 0$. Let for instance $\xi_y = \alpha$ and $\xi_z = -\alpha$. Then

$$g_y^\pm > 0, \quad g_z^\pm > -1.$$

Hence

$$\xi_x + g_y^\pm + g_z^\pm > 1, \quad c > 1.$$

We have (see (11.19))

$$f'_\beta(c) < \frac{1}{2}. \tag{11.28}$$

Case (iii) $\xi_y = \xi_z = \alpha$. Then

$$\xi_x + g_y^\pm + g_z^\pm > 2,$$

hence $c > 2$ and

$$f'_\beta(c) < \frac{1}{2}(1 - \tanh\beta). \tag{11.29}$$

(see (11.20)).

The probabilities of the cases (i), (ii) and (iii) are, respectively, $1/4$, $1/2$, and $1/4$. Thus by (11.26)-(11.29) one gets for $x \in W_n$ that

$$\mathbb{E}_\xi |g_x^+ - g_x^-| < [\frac{1}{4} \cdot \tanh\beta + \frac{1}{2} \cdot \frac{1}{2} + \frac{1}{4} \cdot \frac{1}{2} \cdot (1 - \tanh\beta)] \cdot 2 \cdot E_{n+1},$$

where

$$E_{n+1} = \max_{t \in W_{n+1}} \mathbb{E}_\xi |g_t^+ - g_t^-|.$$

This gives that

$$E_n < \frac{3 + \tanh\beta}{4} \cdot E_{n+1}.$$

Since $(3 + \tanh\beta)/4 < 1$ and $E_n < 2$ for all n, this implies by iterations that $E_n = 0$ for all n. Hence for all x, $g_x^+ = g_x^-$, for almost all configurations ξ, which implies uniqueness by Proposition 11.15. $\qquad\square$

Notes. In [44] a one-dimensional Ising spin system, with ferromagnetic long-range interactions decaying as $1/|i-j|^{2-\alpha}$, is considered in the presence of an external random magnetic field. The field $h = \{h_i, i \in \mathbb{Z}\}$ is given by a family of independent, identically distributed, mean-zero Bernoulli random variables. The authors prove, with probability 1 with respect to the random field, the existence of at least two distinct extremal Gibbs measures for temperature and strength of the randomness small enough.

In [57] exact results for the critical behavior of the random field Ising model on complete graphs and trees is presented, both at equilibrium and away from equilibrium, i.e., models for hysteresis and Barkhausen noise. It is shown that for stretched exponential and power law distributions of random fields the behavior on complete graphs is non-universal, while the behavior on Cayley trees is universal even in the limit of large coordination.

In [66] low temperature series expansions are derived for the random field Ising model with a δ-function distribution on a Cayley tree by two independent methods: (a) the finite-cluster method which uses graph embeddings and appropriate weighting functions; (b) the use of a recursion relation specific to the Cayley tree. Numerical values are evaluated when the order of the tree $k = 3, 4$ and the coefficients are analyzed to assess critical behavior.

The results of this section are due to [26], [27].

11.3 Ashkin-Teller model

The Ashkin-Teller model may be considered as two superposed Ising models which are coupled by a four-spin interaction term. The Hamiltonian is given by

$$H(\sigma,\tau) = -J_2 \sum_{\langle x,y\rangle} (\sigma(x)\sigma(y) + \tau(x)\tau(y)) - J_4 \sum_{\langle x,y\rangle} (\sigma(x)\sigma(y)\tau(x)\tau(y))$$

$$(11.30)$$

where $\sigma(x), \tau(x)$ are classical Ising spins variables taking the values ± 1 on each site of Cayley tree, J_2 (J_4) denotes the strength of the two (four)-spin interactions, and $\langle x, y \rangle$ denotes a summation over all distinct pairs of nearest-neighbors.

In this section following [12] we use an iterative procedure to determine the local properties of the Ashkin-Teller model on a Cayley tree. We are mainly interested in the fixed points of a three-dimensional mapping generated by this scheme. For a Cayley tree of order k these fixed points correspond to the Bethe-Peierls approximation. The stability analysis of these fixed points are then used to obtain the phase diagrams as a function of the order k.

Considering a site on W_n connected to k sites on W_{n+1} we obtain

$$\sum_{\sigma(x),\tau(x)} \exp\left[(K_2\sigma(x) + H_{1,n}) \sum_{y\in S(x)} \sigma(y) + (K_2\tau(x) + H_{2,n}) \sum_{y\in S(x)} \tau(y)+ \right.$$

$$\left. (K_4\sigma(x)\tau(x) + H_{4,n}) \sum_{y\in S(x)} \sigma(y)\tau(y) \right] = \tag{11.31}$$

$$8^k \left[\cosh(K_2\sigma + H_{1,n}) \cosh(K_2\tau + H_{2,n}) \cosh(K_4\sigma\tau + H_{4,n})+ \right.$$

$$\left. \sinh(K_2\sigma + H_{1,n}) \sinh(K_2\tau + H_{2,n}) \sinh(K_4\sigma\tau + H_{4,n})\right]^k$$

where $K_i = J_i\beta$, $i = 2, 4$. The effective fields $H_{1,n}$, $H_{2,n}$ and $H_{4,n}$ are generated by the decimation process.

Equation (11.31) can be written as a single-site effective partition function

$$\mathcal{Z}_{n+1}(\sigma, \tau) = A \exp[H_{1,n+1}\sigma + H_{2,n+1}\tau + H_{4,n+1}\sigma\tau] \tag{11.32}$$

where $H_{1,n+1}$, $H_{2,n+1}$ and $H_{4,n+1}$ are the new effective fields acting on W_{n+1}.

From equations (11.31) and (11.32) it is easy to obtain the recursion relation between successive effective fields. However, it is convenient to introduce new variables defined by

$$m_{1,n} = \frac{\sum_{\{\sigma,\tau=\pm 1\}} \sigma \mathcal{Z}_n(\sigma, \tau)}{\mathcal{Z}_n}, \tag{11.33}$$

$$m_{2,n} = \frac{\sum_{\{\sigma,\tau=\pm 1\}} \tau \mathcal{Z}_n(\sigma, \tau)}{\mathcal{Z}_n}, \tag{11.34}$$

$$q_n = \frac{\sum_{\{\sigma,\tau=\pm 1\}} \sigma\tau \mathcal{Z}_n(\sigma, \tau)}{\mathcal{Z}_n}, \tag{11.35}$$

where

$$\mathcal{Z}_n = \mathcal{Z}_n(+1, +1) + \mathcal{Z}_n(+1, -1)+ \tag{11.36}$$

$$\mathcal{Z}_n(-1, +1) + \mathcal{Z}_n(-1, -1).$$

Variables m_1 and m_2 are considered as local magnetization and variable q as polarization.

The relation between these parameters and the effective fields can be easily obtained. For instance, on the W_{n+1} we have

$$m_{1,n+1} = \frac{\tanh H_{1,n+1} + \tanh H_{2,n+1} \tanh H_{4,n+1}}{D}, \tag{11.37}$$

$$m_{2,n+1} = \frac{\tanh H_{2,n+1} + \tanh H_{1,n+1} \tanh H_{4,n+1}}{D}, \tag{11.38}$$

$$q_{n+1} = \frac{\tanh H_{4,n+1} + \tanh H_{1,n+1} \tanh H_{2,n+1}}{D}, \tag{11.39}$$

where

$$D = 1 + \tanh H_{1,n+1} \tanh H_{2,n+1} \tanh H_{4,n+1}. \tag{11.40}$$

On the other hand the effective fields on W_{n+1} can be expressed in terms of the local magnetization on W_n by

$$H_{1,n+1} = \frac{k}{2} \tanh^{-1} \left[\frac{2\alpha(m_{1,n} - \gamma m_{2,n} q_n)}{1 + \alpha^2 m_{1,n}^2 - \alpha^2 m_{2,n}^2 - \gamma^2 q_n^2} \right], \tag{11.41}$$

$$H_{2,n+1} = \frac{k}{2} \tanh^{-1} \left[\frac{2\alpha(m_{2,n} - \gamma m_{1,n} q_n)}{1 + \alpha^2 m_{1,n}^2 - \alpha^2 m_{2,n}^2 - \gamma^2 q_n^2} \right], \tag{11.42}$$

$$H_{4,n+1} = \frac{k}{2} \tanh^{-1} \left[\frac{2(\gamma q_n - \alpha^2 m_{1,n} m_{2,n})}{1 + \alpha^2 m_{1,n}^2 - \alpha^2 m_{2,n}^2 - \gamma^2 q_n^2} \right], \tag{11.43}$$

where

$$\alpha = \frac{t_2 + t_2 t_4}{1 + t_2^2 t_4}, \quad \gamma = \frac{t_4 + t_2^2}{1 + t_2^2 t_4} \tag{11.44}$$

and

$$t_i = \tanh(J_i \beta), \quad i = 2, 4. \tag{11.45}$$

Equations (11.37) and (11.41) define together a three-dimensional mapping. Given a set of parameters (temperature, coupling constants and order of tree) and a set of boundary conditions (initial conditions), we iterate the mapping and look for its fixed points, i.e., for point such that

$$(m_{1;n+1}, m_{2,n+1}, q_{n+1}) = (m_{1,n}, m_{2,n}, q_n) = (m_1^*, m_2^*, q^*)$$

for some $n \geq L > 0$.

These fixed points, if they do exist, are then associated with the solutions deep within the Cayley tree. It has been shown that these solutions correspond to the so-called Bethe-Peierls approximation on a physical lattice with the same order [18].

Now we shall study the phase diagrams as the coordination number is varied. We will use the stability analysis of the fixed points of the mapping to obtain the phase diagrams in the variables

$$\theta = 1/\beta J_2 k, \quad p = J_4/J_2. \tag{11.46}$$

A fixed point is considered stable if all eigenvalues Λ_i, for $i = 1, 2, 3$, of the Jacobian matrix

$$\mathbf{M} = \begin{pmatrix} \dfrac{\partial m_{1,n+1}}{\partial m_{1,n}} & \dfrac{\partial m_{1,n+1}}{\partial m_{2,n}} & \dfrac{\partial m_{1,n+1}}{\partial q_n} \\[2mm] \dfrac{\partial m_{2,n+1}}{\partial m_{1,n}} & \dfrac{\partial m_{2,n+1}}{\partial m_{2,n}} & \dfrac{\partial m_{2,n+1}}{\partial q_n} \\[2mm] \dfrac{\partial q_{n+1}}{\partial m_{1,n}} & \dfrac{\partial q_{n+1}}{\partial m_{2,n}} & \dfrac{\partial q_{n+1}}{\partial q_n} \end{pmatrix} \tag{11.47}$$

evaluated at the fixed point are less than one in absolute values (see [52]).

The stability boundary of a fixed point is defined by

$$|\Lambda|_{\max} = 1, \tag{11.48}$$

for the largest eigenvalue in absolute value.

Each fixed point is associated with a thermodynamic phase and a corresponding region of stability in the phase diagram.

It may happen that different regions of stability overlap between themselves. In this case the phase which is realized deep within the tree depends on the boundary condition. Physically the overlap among regions of stability indicates the existence of a first-order transition.

It turns out that to determine the loci of the first-order transitions we need a free energy functional associated with the corresponding fixed points. A free energy functional could be obtained by standard methods [18], [253].

11.3.1 *Paramagnetic fixed point*

The paramagnetic (trivial) fixed point is given by $m_1^* = m_2^* = q^* = 0$ and it always exists for any value of k, p, and θ. Their stability boundaries can be obtained analytically. These boundaries are given by the lines L_1, L_2 and L_5. Their equation are given by

$$p = -k\theta \tanh^{-1}\left(\frac{1 + kt_1^2}{k + t_1^2}\right), \quad \text{for} \quad L_1,$$

$$p = k\theta \tanh^{-1}\left(\frac{1 - kt_1}{kt_1 - t_1^2}\right), \quad \text{for} \ \ L_2,$$

$$p = k\theta \tanh^{-1}\left(\frac{1 - kt_1^2}{k - t_1^2}\right), \quad \text{for} \ \ L_5.$$

For $k = 2$ and 3 the lines L_1 and L_2 merge together at the multiphase point $\theta = 0$, $p = -1$ as seen in Figs.11.2 and 11.3. For $k \geq 4$ they still meet at $p = -1$ but with $\theta \neq 0$. We note also that for any value of the coordination number the lines L_2 and L_5 meet together at the four-state Potts point $p = 1$. Note that L_1 is the sole border between the paramagnetic and the $\langle\sigma\tau\rangle$ (see below) phases. Thus we conclude that the transitions between these phases are always continuous, irrespective of the order of the tree.

11.3.2   *Non-trivial fixed points*

Depending on the order k one can find up to five distinct non-trivial fixed points. Each fixed point corresponds to a different phase of the original model (see [49]) and are given by

$$m_1^* = m_2^* = 0, \ q \neq 0 \ \ \text{(Ising phase)},$$

$$m_1^* = m_2^* \neq 0, \ q \neq 0 \ \ \text{(Baxter phase)},$$

$$m_1^* \neq 0, \ m_2^* = 0, \ q = 0 \ \ (\langle\sigma\rangle \ \text{phase}),$$

$$m_1^* \neq m_2^* \neq 0, \ q \neq 0 \ \ \text{(asymmetric Baxter phase)},$$

and a cycle-two fixed point

$$(0, 0, q^*) \longleftrightarrow (0, 0, -q^*) \ \ (\langle\sigma\tau\rangle_{\text{AF}} \ \text{phase}).$$

In [12] the stability boundaries of these fixed points have been determined numerically. Given a point (p, θ) on the phase diagram one chooses an initial condition and iterate the mapping. When the fixed point is attained one then computes the eigenvalues of the stability matrix.

For the $\langle\sigma\tau\rangle_{\text{AF}}$ we must take a product of two matrices, one for $(0, 0, q^*)$ and another for $(0, 0, -q^*)$. The stability criterion (11.48) still holds in this case.

Let us discuss some representative phase diagrams thus obtained.

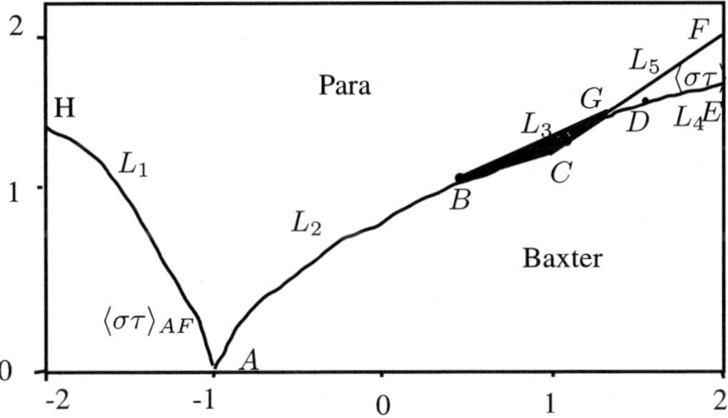

Fig. 11.2   The phase diagram on the Cayley tree with $k = 2$. In the shaded area $BCGB$ there exist two stable fixed points corresponding to the paramagnetic and the Baxter phases, whereas inside $CDGC$ there are also two stable fixed points associated with the $\langle \sigma \tau \rangle$ and the Baxter phases. The points B and D are tricritical points. This general behavior remains valid on the next phase diagrams.

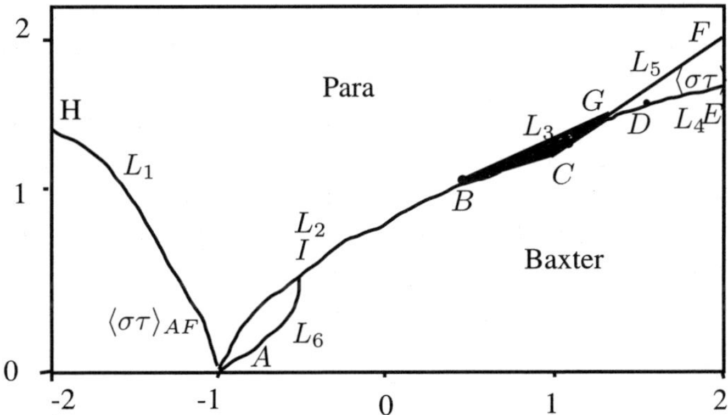

Fig. 11.3   For $k = 3$ there appears an additional phase $\langle \sigma \rangle$ between L_2 and L_6, for $p < 0$. The line L_6 is a first-order transition whereas I is a bicritical point.

Case $k = 2$. Figure 11.2 shows the phase diagram for $k = 2$. We note the existence of three phases besides the already discussed paramagnetic phase. The $\langle \sigma \tau \rangle_{\mathrm{AF}}$ phase shares a common border L_1 with the paramagnetic one. Hence L_1 is a continuous-transition line. The Baxter phase is bordered from above by the lines L_2, L_3 and L_4 along the points A, B, G, D and E. The

$\langle \sigma\tau \rangle$ phase stability region is bordered by L_4 (from point C to E) and L_5. There are two contiguous overlapping regions. Inside the triangular shaped area $BCGB$ there are two kinds of stable fixed points corresponding to the paramagnetic and Baxter phases. A first-order transition line, meeting at point B with the continuous-transition line L_2, is located within this region. Hence B is a tricritical point.

Similarly, within the region $CDGC$ there is a presence of two stable fixed points associated with the Baxter and $\langle \sigma\tau \rangle$ phases. Note that a first-order transition line between these phases exists inside the above mentioned region. Thus D is also a tricritical point. Finally, there is a critical endpoint somewhere along the line GC as two distinct first-order transition lines meets a continuous-transition line (L_5).

From a qualitative point of view, the structure of the phase diagram does not change as we vary the order k of the Cayley tree, as long as we consider $p > 0$. In fact, this picture already reproduce the well-known mean field results (see, for example [49]).

Therefore, we consider the case $p < 0$ on the next diagrams.

Case $k = 3$. In addition to the phases already presented in the previous case, the $\langle \sigma \rangle$ is also found as shown in Fig.11.3. This phase goes continuously into the paramagnetic phase as we cross L_2. However, in crossing the line L_6 it goes to the Baxter phase with jumps on the order parameters m_1, m_2 and q. This situation is different from what we have discussed so far due to the absence of an overlap between the regions of stability. The line L_6 is a first-order transition one and consequently I is a bicritical point.

Case $k = 4$. In this case there appears an overlap between the regions of stability of the phases $\langle \sigma\tau \rangle_{\text{AF}}$ and $\langle \sigma \rangle$.

Case $k = 15$. This phase diagram is qualitatively similar to the infinite-coordination limit [49]. In addition to the phases previously found, the asymmetric Baxter phase is also presented.

Notes. The Ashkin-Teller model was introduced in [13], [49]. This section is based on paper [12]. In [145] using the recursion method, the phase transitions of the Ashkin-Teller model with ferromagnetic interactions on the Cayley tree are studied. The isotropic Ashkin-Teller model and the anisotropic one are respectively investigated, and exact expressions for the free energy and the magnetization are obtained. It is observed that each of the three varieties of phase diagrams for the anisotropic Ashkin-Teller model consists of four phases, i.e., the fully disordered paramagnetic phase Para, the fully ordered ferromagnetic phase Ferro, and two partially ordered ferromagnetic phases $\langle \sigma \rangle$ and $\langle \sigma\tau \rangle$, while the phase diagram for the isotropic

Ashkin-Teller model contains three phases, i.e., the fully disordered para-magnetic phase Para, the fully ordered ferromagnetic phase Baxter Phase, and the partially ordered ferromagnetic phase $\langle \sigma\tau \rangle$.

11.4 Spin glass model

Spin glasses are magnetic systems characterized by randomness and frustration.

The Hamiltonian typically used to describe spin glasses is the Edwards-Anderson Hamiltonian:

$$H(\sigma) = -\sum_{\langle x,y \rangle} J_{xy}\sigma(x)\sigma(y), \tag{11.49}$$

where the bonds J_{xy} are quenched and independently distributed, and the sum is over nearest neighbor pairs on a Cayley tree. Frustration (i.e., the fact that all interaction energies cannot be minimized simultaneously) makes this problem extremely difficult to analyze in finite dimensions. In an effort to develop an understanding of some of the basic properties of this model, many people have turned to mean field theory.

Note that results obtained for the spin glass models on Cayley trees are enough to a separate book, therefore, here we shall give some very brief survey of the results on Cayley trees (see [24], [41]-[43], [48], [109], [117], [159], [252], [254], [255], [256], [260]).

The analysis of the spin glass model was based on the study of the distribution of the single-site magnetization $\mathbf{M} = \langle \sigma(x) \rangle$ at the end point of the half-space Cayley tree.

As an interaction J_{xy} is random and the magnetization $\mathbf{M}$ is a random variable. It satisfies the fixed point equation

$$\mathbf{M} =_d F(\mathbf{M}), \tag{11.50}$$

where $=_d$ means the equality of the distributions and

$$F(\mathbf{M}) = \frac{p_1\mathbf{M}_1 + p_2\mathbf{M}_2}{1 + p_1 p_2 \mathbf{M}_1\mathbf{M}_2}, \tag{11.51}$$

where p_1, p_2, $\mathbf{M}_1$, $\mathbf{M}_2$ are independent real random variables,

$$\mathbf{M}_1 =_d \mathbf{M}_2 =_d \mathbf{M},$$

and $p_1 =_d p_2 =_d \tanh(J_{xy}\beta)$, i.e., $p_i = \pm\tanh(\beta)$ with probability $1/2$, $i = 1, 2$.

The distribution of $\mathbf{M}$ is symmetric and it is stable in the sense that the iterations

$$\mathbf{M}^{(n+1)} =_d F(\mathbf{M}^{(n)})$$

converge to $\mathbf{M}$ in a neighborhood of $\mathbf{M}$.

In [48] the following results are obtained:

1. using a moment analysis of the recursion relations, some global features of the magnetization distribution as a function of temperature is established. The critical temperature corresponds to $p = \tanh(\beta) = p_G \equiv \frac{1}{\sqrt{2}}$. Above this temperature $(p < p_G)$, the single site magnetizations are globally attracted to zero. That is, any initial distribution of magnetizations converges exponentially fast to a point mass at zero. Below this temperature, upper and lower bounds on the Edwards-Anderson order parameter (the second moment of the distribution) are found. Any (nonzero) distribution of boundary magnetizations will iterate into this band within a finite distance of the boundary. As a consequence of the bounds, the critical behavior of the Edwards-Anderson order parameter is given by

$$q \sim |p - p_G|^{\beta_G} \quad \text{as} \quad p \downarrow p_G \tag{11.52}$$

with $\beta_G = 1$.

   By the moment analysis is guaranteed that the low-temperature behavior is nontrivial, it is described by a fixed distribution of magnetizations.

2. Temporarily abandons probabilistic methods, functional analysis and bifurcation theory are used to prove the existence of a positive density as a fixed point to the recursion relation in a neighborhood above p_G. This fixed point function is denoted by ϱ_A, where $A \equiv p - p_G$. By equation (11.52) the second moment $q \approx (\text{const})A$, so there is a natural rescaling of variables, under which the distribution becomes Gaussian as $p \downarrow p_G$.

3. Some of the properties of the spin glass solution ϱ_A are studied. The first important question is whether the distribution is stable (an attractive fixed point). It is shown that the first order approximation to the functional derivative has all negative eigenvalues; this is a natural first step in a complete stability analysis. The second question is the shape of the distribution near the critical point. It is solved for the function ϱ_A to first order in A, and the leading non-Gaussian correction is found. The asymptotic behavior at the high and low ends of the magnetization distribution are considered. Wherever ϱ_A exists, it is shown that

the magnetization falls to zero as $\exp[-1/\zeta^{c(p)}]$, where ζ measures the distance between $\mathbf{M}$ and its maximum (or minimum) allowed value. In particular, upper and lower bounds on $c(p)$ are obtained. These results are consistent with a picture of ϱ as an attractive, symmetric, Gaussian-like distribution with soft singular behavior at the tails.

In [24] the existence of a stable solution $\mathbf{M}$ of a renormalized version of equation (11.50) at zero-temperature is proved.

In [42], [43] an analysis of the Ising spin-glass model on the Cayley tree with fixed uncorrelated boundary conditions is presented. Phase diagrams are derived as a function of temperature vs. concentration of ferromagnetic bonds and, for a symmetric distribution of bonds, external field vs. temperature.

In [43] the bulk ordered phases using bifurcation theory are characterized: the existence of a distribution of single-site magnetizations far inside the lattice is proved which is stable with respect to changes in the boundary conditions.

In [42] magnetized spin-glass phases are characterized by divergence of an appropriate susceptibility: at zero field this signals the existence of an intermediate magnetized spin-glass phase; at nonzero field, this is used to identify the de Almeida-Thouless line.

For more results see above-mentioned papers and references therein. See [33], [61], [157] for the developed theory of random fields and spin glasses.

11.5   Abelian sandpile model

The Abelian sandpile model of self-organized criticality (SOC) was introduced in [14], [15] and attracted a lot of attention after the discovery of the Abelian property by Dhar in [53].

Consider a Cayley tree Γ^k of order $k \geq 1$. Let $\Lambda \subset \Gamma^k$ be a finite subtree. Define configurations as elements of $\mathbb{N}^\Lambda = \{\sigma : \Lambda \to \mathbb{N}\}$. A configuration σ is *stable* if $\sigma(x) \leq k + 1$ for all $x \in \Lambda$. In sandpile terminology, $\sigma(x)$ is the height of a sand column at site x.

The dynamics of the model is defined by two rules.

(i) *Adding a particle*: if the system is in a stable configuration, we choose at random a site x (all sites equally likely), and increase the height at that site by 1. Heights at other sites remain unchanged.

Configurations which are not stable, can be stabilized by means of repeated application of *toppling* operators.

(ii) *Toppling rule*: if at any site the height of the sand column exceeds $k + 1$ then that site topples, its height decreases by $k + 1$ and the sand particles drop on the nearest neighbors. As a result, the height at each of the nearest neighbors increases by 1. Thus, for every $x \in \Lambda$, the toppling operator T_x only acts on configurations σ such that $\sigma(x) > k + 1$. For such configurations define $\sigma' = T_x(\sigma)$ as follows: for every $y \in \Lambda$,

$$\sigma'(y) = \begin{cases} \sigma(x) - k - 1, & \text{if } y = x, \\ \sigma(x) + 1, & \text{if } \langle x, y \rangle, \\ \sigma(y), & \text{otherwise.} \end{cases}$$

For boundary sites – sites in Λ with fewer than $k + 1$ nearest neighbors in Λ, some grains of sand are lost.

If we view configurations as column vectors in $\mathbb{N}^\Lambda$, then

$$\sigma' = T_x(\sigma) = \sigma - \Delta_\Lambda \delta^{(x)}, \tag{11.53}$$

where $\delta_y^{(x)} = \delta_{xy}$, $y \in \Lambda$, and Δ_Λ is square matrix of size $|\Lambda|$ given by

$$\Delta_{xy} = \begin{cases} k + 1, & \text{if } y = x, \\ -1, & \text{if } \langle x, y \rangle, \\ 0, & \text{otherwise.} \end{cases}$$

In fact, Δ_Λ is the so-called graph Laplacian of Λ. From the representation (11.53) of toppling operators one immediately concludes that for all $x, y \in \Lambda$, the toppling operators T_x, T_y commute:

$$T_x \circ T_y = T_y \circ T_x. \tag{11.54}$$

Given an unstable configuration σ, we keep applying the toppling operators at each unstable site, until a stable configuration is reached. In view of (11.54), the order of topplings is irrelevant. The process will eventually stop, because of the dissipativity of topplings on the boundary.

We denote by $\mathcal{S}(\sigma)$ the result of stabilization of σ.

Given two configurations $\sigma, \tau \in \mathbb{N}^\Lambda$, define the addition operation $\sigma \oplus \tau$ as the result of stabilization of $\sigma + \tau$ (added coordinatewise):

$$\sigma \oplus \tau = \mathcal{S}(\sigma + \tau).$$

A stable configuration σ is called *recurrent* if

$$\sigma = \sigma \oplus \tau = \mathcal{S}(\sigma + \tau)$$

for some configuration τ. It turns out that the set of recurrent configuration $\mathcal{R}_\Lambda$ forms a group with a group operation $\oplus$. The group $(\mathcal{R}_\Lambda, \oplus)$ is called the *sandpile group* of Λ.

Adding a particle at a randomly chosen site may cause it to topple, and the toppling may induce toppling at some of the neighboring sites at the next time step, thus causing an *avalanche*.

Assume a system reached a stable configuration (no more topplings). If we keep on adding particles randomly, the system ultimately reaches a statistically stationary state which is *critical* in the sense that the probability distribution of the size of an avalanche by mass or by duration shows a power-law behavior for a very wide range of sizes (the upper cut-off on the size is determined by the number of sites of the lattice, and is strictly infinite in the thermodynamic limit).

A well-recognized problem with the Cayley tree is that most of its sites are very near to the boundary. Hence, calculation of the thermodynamic limit of the bulk properties from the finite $N = |\Lambda|$ calculation requires special care. Consider a site at a distance r from the boundary. If we add a particle at this site, the expected duration of the resulting avalanche in the SOC state tends to infinity only when $r \to \infty$. Thus in order to study the SOC state, we shall only consider the distribution of avalanches caused by adding a particle at a site very far from the boundary. Note that the particles are added at random near the boundary also (else the existence of a unique SOC state is not guaranteed). But in calculating the distribution of avalanche sizes, avalanches caused by additions near the boundary are not included.

In a stable configuration, at any site x the height of the sand column $\sigma(x)$ can take $k+1$ possible values $1, 2, \ldots, k+1$. As the subtree has N sites, the total number of stable configurations is $(k+1)^\Lambda$. But not all of them are allowed in the SOC state. For example, if in a configuration, there are two adjacent sites (say x and y) both having heights 1, then that configuration occurs with zero probability in the SOC state. Indeed, suppose we start with any arbitrary initial configuration with $\sigma(x) \neq 1$ and $\sigma(y) \neq 1$ and add particles randomly at different sites. In order to reach a configuration with $\sigma(x) = 1$ and $\sigma(y) = 1$, at least one toppling must have occurred at both sites. Say y topples last. But any toppling at y increases $\sigma(x)$, by 1. Then just before ys last toppling, $\sigma(x)$, must be 0, but this is not allowed in the model. Therefore, we will never reach a configuration with $\sigma(x) = 1$ and $\sigma(y) = 1$. Thus, any configuration with unit heights at adjacent sites is forbidden in the SOC state.

In [54] exact expressions for various distribution functions including the height distribution at a site and the joint distribution of heights at two sites separated by an arbitrary distance are obtained. Also the probability distribution of the number of distinct sites that topple at least once, the number of topplings at the origin and the total number of topplings in an avalanche are determined. The probability that an avalanche consists of more than n topplings varies as $n^{-1/2}$ for large n. The probability that its duration exceeds t decreases as $1/t$ for large t. These exponents are the same as for the critical percolation clusters in mean-field theory.

In [55], [56] an introduction to the Abelian sandpile and related models of SOC is given. The notes aim to provide a pedagogical introduction to SOC for students and others wishing to learn about the subject in detail. It will be useful reading to anyone interested in theoretical results on SOC, since it summarizes some of the main results in the field, reflecting choices of the author. A lot of information is condensed, and an extensive list of references to original papers is given where the interested reader can find more information. The following topics are discussed, among others:

1. models of SOC: abelian sandpile; loop-erased random walk; Takayasu's model.
2. mathematical structures in the sandpile model, such as: abelian group of addition operators; the steady state; representation of the sandpile group as a product of cyclic groups; the two-point function; recurrent and transient configurations; the burning test.
3. relation to the $q > 0$ limit of the Potts model and spanning trees; decomposition of avalanches into waves.
4. directed sandpiles: computation of critical exponents in all dimensions; equivalence to other models: river networks, the voter model.
5. summary of results for undirected sandpiles on $\mathbb{Z}^d$ and on the Cayley tree.
6. more general abelian models: Eulerian walkers; the Manna model.
7. time-dependent properties.
8. conjectures regarding generic behavior and universality: the effects of stochasticity in toppling rules and 'stickyness'.
9. open problems.

For a detailed mathematical introduction into the sandpile model see [208]. See also [239] for a relation of sandpile models with dynamical systems.

11.6 $Z(M)$ (or clock) models

11.6.1 *The model and equations*

Consider a Cayley tree Γ^k of order k. On each site, there is a unit vector which can take M orientations in the plane; these orientations are numbered by a parameter s taking the values $0, 1, 2, \ldots, M - 1$. The interaction between two of these unit vectors along a bond is taken to be invariant under

(i) a simultaneous rotation of the unit vectors over an allowed angle $p(2\pi M) = p\varphi_M$ and

(ii) the reflection σ of both unit vectors with respect to the $s = 0$ direction.

The (global) symmetry group $S(M)$ of

(1) the interaction is the direct product of the group $Z(M)$ of all plane rotations $R[p\varphi_M]$ over angles $p\varphi_M$; this is an Abelian group:

$$R[p_1\varphi_M]R[p_2\varphi_M] = R[p_2\varphi_M]R[p_1\varphi_M] = R[(p_1 + p_2)\varphi_M], \quad (11.55)$$

and

(2) of the group $\{1, \sigma\}$ consisting of the identity and σ alone ($\sigma^2 = 1$):

$$S(M) = Z(M) \otimes \{1, \sigma\}. \tag{11.56}$$

This group is now not Abelian; in fact, we have

$$\sigma R[p\varphi_M] = R[(M - p)\varphi_M]\sigma, \tag{11.57}$$

as shown explicitly for the case $M = 5$, $p = 2$ in Fig.11.4.

This choice of group, which contains the same elements as the standard Abelian $Z(M) \otimes \{1, \sigma'\}$ with σ' the interchange of the two unit vectors, is better adapted to the problem at hand (see below).

Symmetry under $S(M)$ implies that the interaction energy $E(s, s')$ between two unit vectors in the states s and s', respectively, must be of the form

$$E(s, s') = \sum_{q=0}^{M-1} \delta_{s,s'+q}^{(M)} E_q, \tag{11.58}$$

where $\delta_{s,t}^{(M)}$ is a slightly redefined Kronecker symbol:

$$\delta_{s,t}^{(M)} = \begin{cases} 1, & \text{if } s \equiv t \mod M, \\ 0, & \text{otherwise,} \end{cases} \tag{11.59}$$

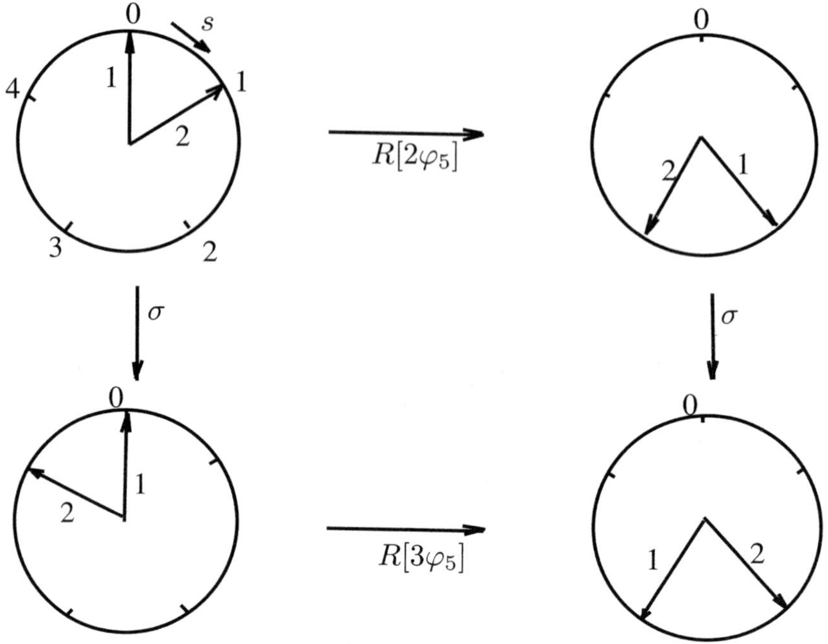

Fig. 11.4   Illustration of the non commutativity of σ and $R[p\varphi_M]$ for $M = 5$, $p = 2$, see (11.57).

and the E_q satisfy

$$E_q = E_{M-q}. \tag{11.60}$$

If the zero of energy is taken such that $E_0 = 0$ in equation (11.58), then the Boltzmann factor corresponding to (11.58) is, with $K_q = -\beta E_q$,

$$\exp \sum_{q=0}^{M-1} K_q \delta^{(M)}_{s,s'+q} = \Omega_{ss'}. \tag{11.61}$$

Defining $w_q = \exp K_q$, so that $w_0 = 1$, it is easily seen that $\boldsymbol{\Omega}$ is a symmetric cyclic matrix:

$$\Omega_{ss'} = w_{|s-s'|}. \tag{11.62}$$

The recursion relation for the partition function $Z^{(n)}(s)$ of an n-generation branch with fixed (state s) unit vector at the root is

$$Z^{(n)}(s) = \left[\sum_{s'} \Omega_{ss'} Z^{(n-1)}(s') \right]^k, \tag{11.63}$$

here and below we use all sums run from 0 to $M - 1$ unless explicitly indicated otherwise.

We now define distribution vectors $\mathbf{x}^{(n)}$ by

$$\mathbf{x}_s^{(n)} = Z^{(n)}(s)/Z^{(n)}, \tag{11.64}$$

with $Z^{(n)}$ the (total) partition function of an n-generation branch,

$$Z^{(n)} = \sum_s Z^{(n)}(s), \tag{11.65}$$

so that $\mathbf{x}^{(n)} \in \ell^1(M)$, i.e., $\sum_s \mathbf{x}_s^{(n)} = 1$ for all n.

For the vectors $\mathbf{x}^{(n)}$ we have

$$\mathbf{x}_s^{(n)} = \left(\mathbf{\Omega}\mathbf{x}^{(n-1)}\right)_s^k \Big/ \sum_{s'} \left(\mathbf{\Omega}\mathbf{x}^{(n-1)}\right)_{s'}^k. \tag{11.66}$$

The non-linear recursion relation (11.66), which can formally be written as

$$\mathbf{x}^{(n)} = A\mathbf{x}^{(n-1)}, \tag{11.67}$$

$$(A\mathbf{x})_s = (\mathbf{\Omega}\mathbf{x})_s^k \Big/ \sum_{s'} (\mathbf{\Omega}\mathbf{x})_{s'}^k, \tag{11.68}$$

now reflects the symmetries of the interaction in the following sense: define $M \times M$ matrices $\mathbf{R}[p\varphi_M]$ and σ by

$$(\mathbf{R}[p\varphi_M]\mathbf{x})_s = \mathbf{x}_{s-p}, \tag{11.69}$$

$$(\sigma\mathbf{x})_s = \mathbf{x}_{-s}, \tag{11.70}$$

where the indices are to be understood mod M. It is easily seen that the matrices $\mathbf{R}[p\varphi_M]$ and σ satisfy (11.55) and (11.57), so that they form a representation of $S(M)$; then the following invariance properties hold:

$$A(\mathbf{R}[p\varphi_M]\mathbf{x}) = \mathbf{R}[p\varphi_M](A\mathbf{x}), \tag{11.71}$$

$$A(\sigma\mathbf{x}) = \sigma(A\mathbf{x}), \tag{11.72}$$

since $\mathbf{\Omega}$ is cyclic and symmetric, we obtain

$$\mathbf{\Omega}\mathbf{R}[p\varphi_M] = \mathbf{R}[p\varphi_M]\mathbf{\Omega}, \quad \sigma\mathbf{\Omega} = \mathbf{\Omega}\sigma. \tag{11.73}$$

By (11.71), all vectors $\mathbf{x}^{(n)}$ will have the same symmetry properties that $\mathbf{x}^{(0)}$ may have. In particular, if $\mathbf{x}^{(0)}$ is fully invariant under the transformations (11.69), (11.70) it follows that

$$\mathbf{x}_s^{(n)} = \mathbf{x}_s^{(0)} = M^{-1}, \quad \text{for all } n \text{ and } s. \tag{11.74}$$

In this case, one says that there is no field on the boundary of the Cayley tree; if, on the contrary, $\mathbf{x}^{(0)}$ is not completely invariant, then a field on the boundary is present. The different types of possible field will be identified below.

Let F be some field parameter; from the recursion relation (11.66) for the $\mathbf{x}^{(n)}$ we can derive the following recursion relation for the partial derivatives of $\mathbf{x}^{(n)}$ with respect to F:

$$\left(\frac{\partial \mathbf{x}^{(n)}}{\partial F}\right)_s = \left[k(\boldsymbol{\Omega}\mathbf{x}^{(n-1)})_s^{k-1} \left(\boldsymbol{\Omega}\frac{\partial \mathbf{x}^{(n-1)}}{\partial F}\right)_s - \right. \tag{11.75}$$

$$\left. k\mathbf{x}_s^{(n)} \sum_{s''} (\boldsymbol{\Omega}\mathbf{x}^{(n-1)})_{s''}^{k-1} \left(\boldsymbol{\Omega}\frac{\partial \mathbf{x}^{(n-1)}}{\partial F}\right)_{s''} \right] \times \left[\sum_{s'} (\boldsymbol{\Omega}\mathbf{x}^{(n-1)})_{s'}^{k} \right]^{-1}.$$

If we evaluate this for all fields zero, so that $\mathbf{x}_s^{(n-1)} = \mathbf{x}_s^{(n)} = M^{-1}$, then we get

$$\left.\frac{\partial \mathbf{x}^{(n)}}{\partial F}\right|_{F=0} = k\lambda_0^{-1}\boldsymbol{\Omega}\left.\frac{\partial \mathbf{x}^{(n-1)}}{\partial F}\right|_{F=0}, \tag{11.76}$$

where

$$\lambda_0 = \sum_{s'} \Omega_{ss'} = \sum_{s'} \Omega_{0s'} = \sum_{q'} w_q, \quad \text{for all} \ \ s. \tag{11.77}$$

Since $\boldsymbol{\Omega}$ is a symmetric cyclic matrix, its eigenvectors and eigenvalues are given by

$$\boldsymbol{\Omega}\mu_p = \lambda_p\mu_p, \tag{11.78}$$

$$(\mu_p)_s = \cos(sp\varphi_M), \quad p = 0, 1, \ldots, [M/2], \tag{11.79}$$

$$(\mu_{M-p})_s = \sin(sp\varphi_M), \quad p = 1, \ldots, M - [M/2] - 1, \tag{11.80}$$

$$\lambda_p = \lambda_{M-p} = \sum_q w_q \cos(sq\varphi_M). \tag{11.81}$$

Here $[x]$ denotes the largest integer smaller than or equal to x.

Note that the λ_0 of (11.77) is actually the zeroth eigenvalue of $\boldsymbol{\Omega}$ as given by (11.81).

The system of M eigenvectors (11.79), (11.80) is complete in M-dimensional vector space; therefore, we can write

$$\left.\frac{\partial \mathbf{x}^{(n)}}{\partial F}\right|_{F=0} = \sum_p \alpha_p^{(n)}\mu_p. \tag{11.82}$$

From (11.76) we get

$$\alpha_p^{(n)} = (k\lambda_p/\lambda_0)\alpha_p^{(n-1)} \tag{11.83}$$

with the solution

$$\alpha_p^{(n)} = (k\lambda_p/\lambda_0)^n \alpha_p^{(0)}. \tag{11.84}$$

We now define fields F_m by

$$\mathbf{x}_s^{(0)} = \frac{\exp\left[\sum_m F_m(\mu_m)_s\right]}{\sum_{s'}\exp\left[\sum_m F_m(\mu_m)_{s'}\right]}, \tag{11.85}$$

so that

$$\left.\frac{\partial \mathbf{x}^{(0)}}{\partial F_m}\right|_{F=0} = M^{-1}\mu_m - M^{-2}\sum_s (\mu_m)_s = \begin{cases} 0, & \text{if } m = 0, \\ M^{-1}\mu_m, & \text{if } m \neq 0. \end{cases} \tag{11.86}$$

Note that only F_m with $m \neq 0$ play the role of field parameters. We have from (11.86), (11.84) and (11.82):

$$\left.\frac{\partial \mathbf{x}^{(0)}}{\partial F_m}\right|_{F=0} = (k\lambda_m/\lambda_0)^n M^{-1}\mu_m, \quad m = 1, 2, \ldots, M-1, \tag{11.87}$$

so that $\mathbf{x}^{(n)}$ may be expanded as

$$\mathbf{x}^{(n)} = \mathbf{x}^{(n)}\Big|_{F=0} + \sum_p \left.\frac{\partial \mathbf{x}^{(n)}}{\partial F_p}\right|_{F=0} F_p + \cdots = \tag{11.88}$$

$$M^{-1}\left(\mu_0 + \sum_{p=1}^{M-1}(k\lambda_p/\lambda_0)^n F_p\mu_p + \ldots\right).$$

By this formula it is clear that the tree cannot sustain an ordering corresponding to a field F_p far from the boundary where it is applied, if $|k\lambda_p/\lambda_0|$ is smaller than one. Such an ordering is obtained, however, for $k\lambda_p/\lambda_0 \geq 1$ (ferromagnetic type ordering) and for $k\lambda_p/\lambda_0 < -1$ (antiferromagnetic type ordering).

Therefore, the phase diagrams of the present models can be obtained by studying the phase boundaries given by

$$k\lambda_p = \lambda_0, \quad k\lambda_p = -\lambda_0. \tag{11.89}$$

The symmetries of the different phases (if any) follow directly from the symmetries of the vectors μ_p which can be sustained by the tree.

Recourse to equations (11.79), (11.80) shows that μ_p and μ_{M-p} are invariant with respect to the group $Z(p)$ of rotations $\mathbf{R}[m\varphi_p]$ if and only if p divides M. Furthermore, only if M is even, we have that $\mu_{M/2}$ is invariant under $S(M/2)$ since all μ_p are eigenfunctions of σ, those of (11.79) with eigenvalue $+1$, those of (11.80) with eigenvalue -1.

Therefore, if the tree can sustain, in a certain phase, order of the types corresponding to the eigenvalues λ_{p_i} for $i = 1, \ldots, n$, $n \le [M/2]$, we have four possibilities:

a) Not all p_i are factors of M: the phase has no symmetry.
b) All p_i are factors of M; then if the largest common factor p of all p_i is $p \ne M/2$, the phase has $Z(p)$ symmetry.
c) M is even and only $\lambda_{M/2}$-type order can be sustained: the phase has $S(M/2)$ symmetry.
d) If the tree cannot sustain any kind of order (disordered phase, $n = 0$), the full $S(M)$ symmetry is, of course, preserved. From this it is clear that most of the $3^{[M/2]}$ phases allowed by the possibilities $|k\lambda_p/\lambda_0| \ne 1$ do not have any invariance properties. This will become evident in the examples below.

11.6.2   *Phases of $Z(M)$ models*

Note that the general structure of phase diagrams is independent of k. In the case $M = 2$ corresponds to the Ising model and the case $M = 3$ corresponds to three-state Potts model.

Case $M = 4$. For the $Z(4)$ model, we have from (11.81) the following eigenvalues of $\boldsymbol{\Omega}$:

$$\lambda_0 = 1 + 2w_1 + 2w_2, \quad \lambda_1 = 1 - w_2, \quad \lambda_2 = 1 - 2w_1 + w_2. \tag{11.90}$$

Thus equation (11.89) yields four phase boundaries:

$$
\begin{aligned}
w_2 &= \tfrac{k-1}{k+2} - \tfrac{2}{k+1}w_1, \quad k\lambda_1 = \lambda_0, \\[4pt]
w_2 &= \tfrac{k+1}{k-1} + \tfrac{2}{k+1}w_1, \quad k\lambda_1 = -\lambda_0, \\[4pt]
w_2 &= 2\tfrac{k+1}{k-1}w_1 - 1, \quad k\lambda_2 = \lambda_0, \\[4pt]
w_2 &= 2\tfrac{k-1}{k+1}w_1 - 1, \quad k\lambda_2 = -\lambda_0.
\end{aligned}
\tag{11.91}
$$

In accordance with the discussion at the end of the previous section, only the ordered phases with λ_2-type ordering and no λ_1-type ordering have symmetries, in this case all $S(2)$ symmetries.

Case $M = 5$. In this case, the eigenvalues of $\boldsymbol{\Omega}$ are

$$\lambda_0 = 1 + 2w_1 + 2w_2,$$
$$\lambda_1 = 1 + 2w_1 \cos(2\pi/5) + 2w_2 \cos(4\pi/5), \qquad (11.92)$$
$$\lambda_2 = 1 + 2w_1 \cos(4\pi/5) + 2w_2 \cos(2\pi/5).$$

Since $M = 5$ is a prime number, all ordered phases lack all symmetries.
Case $M = 6$. In this case, the four eigenvalues of Ω are

$$\lambda_0 = 1 + 2w_1 + 2w_2 + w_3, \quad \lambda_1 = 1 + w_1 - w_2 - w_3,$$
$$\lambda_2 = 1 - w_1 - w_2 + w_3, \quad \lambda_3 = 1 - 2w_1 + 2w_2 - w_3. \qquad (11.93)$$

The phase boundaries as given by (11.89) are now six planes.

Remark 11.1. We note that the general $Z(M)$ model reduces to an M-state Potts model if one takes all $w_p = w$ for $p \neq 0$.

11.7 The planar rotator model

The planar rotator model is obtained from $Z(M)$ model as $M \to \infty$.

In the limit $M \to \infty$, the unit vector giving the state of a spin on a site is not restricted to a finite number of orientations any more, but becomes a plane rotator. Accordingly, the interaction energy (11.58) takes the form

$$E(\varphi, \varphi') = E(|\varphi - \varphi'|), \qquad (11.94)$$

where φ and φ' are the angles defining the directions of the unit vectors with respect to a fixed direction. The recursion relation (11.63) takes the form

$$Z^{(n)}(\varphi) = \left[\int_0^{2\pi} \Omega(\varphi, \varphi') Z^{(n-1)}(\varphi') d\varphi' \right]^k, \qquad (11.95)$$

where the kernel given by

$$\Omega(\varphi, \varphi') = \exp[K(|\varphi - \varphi'|)] = \exp[-\beta E(|\varphi - \varphi'|)]. \qquad (11.96)$$

Distribution functions $\rho^{(n)}(\varphi)$ are now defined by

$$\rho^{(n)}(\varphi) = Z^{(n)}(\varphi)/Z^{(n)}, \qquad (11.97)$$

$$Z^{(n)} = \int_0^{2\pi} Z^{(n)}(\varphi) d\varphi. \qquad (11.98)$$

The eigenfunctions and eigenvalues of $\Omega(\varphi, \varphi')$ are analogous to (11.78)-(11.81):

$$\int_0^{2\pi} \Omega(\varphi, \varphi')\mu_p(\varphi')d\varphi' = \lambda_p\mu_p(\varphi), \tag{11.99}$$

$$\mu_p(\varphi) = \cos(p\varphi), \quad p = 0, 1, \ldots, \tag{11.100}$$

$$\mu_{-p}(\varphi) = \sin(p\varphi), \quad p = 1, 2, \ldots, \tag{11.101}$$

$$\lambda_p = \lambda_{-p} = \int_0^{2\pi} \Omega(|\varphi'|)\cos(p\varphi')d\varphi'. \tag{11.102}$$

From a study of the behavior of the distribution functions far from the boundary, where a small field of λ_p-type is applied, we find (see [167]) that the tree can sustain a perturbation corresponding to λ_p if

$$k\lambda_p > \lambda_0 \quad \text{or} \quad k\lambda_p < -\lambda_0. \tag{11.103}$$

Thus the phase boundaries are given as

$$k\lambda_p = \lambda_0, \quad k\lambda_p = -\lambda_0. \tag{11.104}$$

To be more specific, let the interaction be of the cosine type,

$$E(\varphi, \varphi') = E\cos(\varphi - \varphi'). \tag{11.105}$$

With $K = -\beta E$ we have from (11.102)

$$\lambda_p = \int_0^{2\pi} \exp(K\cos\varphi')\cos(p\varphi')d\varphi' = 2\pi I_p(K), \tag{11.106}$$

where $I_p(x)$ is the pth modified Bessel function. For this case, the phase transitions given by (11.104) occur at the points given as

$$I_p(K)/I_0(K) = \pm k^{-1}. \tag{11.107}$$

The modified Bessel functions satisfy

$$I_p(-K) = (-1)^p I_p(K). \tag{11.108}$$

Furthermore, $I_p(K)/I_0(K)$ is a function which monotonically increases from the value 0 for $K = 0$ to the value 1 in the limit $K \to \infty$, and we have $I_p(K) < I_{p'}(K)$ for $K > 0$ and $p > p'$. From this we have the following picture of the phase transitions for the planar rotator model with cosine interaction (or classical XY model):

(i) For $K > 0$, the ferromagnetic case, equation (11.107) has one solution K_p for all values of p; this solution corresponds to the $+$ sign. Moreover, $K_p > K_{p'}$, for $p > p'$ and $\lim_{p \to \infty} K_p = +\infty$. As is seen from the eigenfunctions (11.100), (11.101), only the disordered state ($K < K_1$) has the full invariance with respect to the group of planar rotations; for $K < K_p$ ordering with $Z(p)$ symmetry is possible; the state for $K_1 < K < K_2$ has no invariances at all.

(ii) For $K < 0$, the anti-ferromagnetic case, equation (11.107) has one solution $-K_p$ for all values of p; for p even, it corresponds to the $+$ sign (ferromagnetic ordering), whereas for p odd, it corresponds to the $-$ sign (anti-ferromagnetic ordering). As before, $Z(p)$ symmetry is possible for $K < -K_p$ there is no symmetry for $-K_2 < K < -K_1$ and full rotation symmetry for $K > -K_1$.

In both cases (i) and (ii), the possibility of $Z(p)$ symmetry does not mean that the phase has this symmetry; on the contrary, since λ_1-type ordering is possible for all phases for which $Z(p)$ symmetry is possible, the phases as such do not show any symmetries properties (except, of course, the disordered phase). The fact that the first phase transition at K_1 immediately destroys all symmetry is typical of the cosine interaction. The same occurs for a $Z(M)$ model with cosine interaction.

Notes. The results for $Z(M)$ models and the planar rotator model are taken from [166]. See [164]-[167] for results concerned with the symmetry groups of spin models, permissible groups and many other results. See also [110].

11.8 $O(n, 1)$-model

Consider on Cayley tree Γ^k of order k the Hamiltonian

$$\mathcal{H} = J \sum_{\langle x,y \rangle} \mathbf{n}_x \cdot \mathbf{n}_y + H \sum_x \sigma_x, \tag{11.109}$$

where $\mathbf{n} = (\sigma, \tau)$ is an $(n+1)$-component vector sweeping the hyperboloid $H^{n,1}$, defined by the equation $\mathbf{n}^2 = \sigma^2 - \tau^2 = 1$. This hyperboloid is the symmetric space, associated with $O(n, 1)$ group: $H^{n,1} = O(n, 1)/O(n)$. We parametrize $\mathbf{n}$ as follows $\sigma = \sqrt{1 + \tau^2} = \cosh\theta$, $0 \le \theta < \infty$,

$$\tau = \begin{pmatrix} \sinh\theta \cos\phi_1 \\ \sinh\theta \sin\phi_1 \cos\phi_2 \\ \vdots \\ \sinh\theta \sin\phi_1 \sin\phi_2 \ldots \sin\phi_{n-1} \end{pmatrix}, \quad \phi_1, \ldots, \phi_{n-2} \in [0, \pi], \quad \phi_{n-1} \in [0, 2\pi).$$

With this parametrization the scalar product is $\mathbf{n}_x \cdot \mathbf{n}_y = \sigma_x \sigma_y - \tau_x \cdot \tau_y \ge \cosh(\theta_x - \theta_y) \ge 1$ and, therefore, the Hamiltonian $\mathcal{H}$ is bounded from below only for $J, H \ge 0$.

The $O(n, 1)$-invariant measure on $H^{n,1}$ is

$$d\mathbf{n} = a d\theta \, \sinh^{n-1}\theta \, d\phi_1 \sin^{n-2}\phi_1 \, d\phi_2 \sin^{n-3}\phi_2 \ldots d\phi_{n-1},$$

where a is a normalization constant to be fixed later.

Now we introduce distribution function of the local order parameter $P(\mathbf{n})$ in the usual manner. Namely, we cut one of the $k+1$ branches coming from site $\mathbf{n}$ and integrate the part of the Boltzmann weight $exp(-\mathcal{H})$ over this branch. The resulting function $P(\mathbf{n})$ satisfies the integral equation

$$P(\mathbf{n}) = \int d\mathbf{n}' L(\mathbf{n}, \mathbf{n}') D(\mathbf{n}') P^k(\mathbf{n}'), \qquad (11.110)$$

where $L(\mathbf{n}, \mathbf{n}') = e^{-J\mathbf{n} \cdot \mathbf{n}'}$ and $D(\mathbf{n}') = e^{-H\sigma'}$.

Introduce the following notations:

$$Z = \int d\mathbf{n} D(\mathbf{n}) P^{k+1}(\mathbf{n}), \quad \text{(the partition function)}$$

$$\langle A(\mathbf{n}) \rangle = Z^{-1} \int d\mathbf{n} A(\mathbf{n}) D(\mathbf{n}) P^{k+1}(\mathbf{n}), \quad \text{(one-site averages)}$$

and weighted correlations:

$$\langle A(\mathbf{n}_0) B(\mathbf{n}_r) \rangle_w = \frac{N(r)}{Z} \int d\mathbf{n}_0 A(\mathbf{n}_0) D(\mathbf{n}_0) P^k(\mathbf{n}_0) \times \qquad (11.111)$$

$$\left(\prod_{i=1}^r \int d\mathbf{n}_i M(\mathbf{n}_{i-1}, \mathbf{n}_i) \right) P(\mathbf{n}_r) B(\mathbf{n}_r),$$

$$M(\mathbf{n}, \mathbf{n}') = L(\mathbf{n}, \mathbf{n}') D(\mathbf{n}') P^{k-1}(\mathbf{n}'),$$

where the factor $N(r) = (k+1)k^{r-1}$ counts the number of sites located at the distance r from a given site (without this factor all the correlators exponentially decay because of the geometry of the Cayley tree).

The constant a in the definition of the measure $d\mathbf{n}$ can be chosen arbitrarily.

It is easy to see that rescaling of the measure changes overall normalization of $P(\mathbf{n})$ and Z but does not affect either one-site averages or correlators. This allows us to choose a convenient normalization for $P(\mathbf{n})$ as follows. Note that when $H = 0$ the equation (11.110) admits constant solution. Then we require that this solution be simply $P(\mathbf{n}) = 1$ or, equivalently, that

$$\int d\mathbf{n}' L(\mathbf{n}, \mathbf{n}') = 1. \tag{11.112}$$

This fixes

$$a = \left(\frac{J}{2\pi}\right)^{(n-1)/2} \left(2K_{(n-1)/2}(J)\right)^{-1}.$$

Magnetic field H breaks the $O(n,1)$ symmetry down to $O(n)$. Then the function P may depend only on σ or, equivalently, on θ. This allows us to perform angular integrations in (11.110), yielding

$$P(\theta) = \int_0^\infty d\theta' L_L(\theta, \theta') D(\theta') P^k(\theta'), \tag{11.113}$$

$$L_L(\theta, \theta') = \left(2K_{(n-1)/2}(J)\right)^{-1} \left(\frac{\sinh \theta'}{\sinh \theta}\right)^{(n-1)/2} e^{-J \cosh \theta \cosh \theta'} \times$$

$$(2\pi J \sinh \theta \sinh \theta')^{1/2} I_{n/2-1}(J \sinh \theta \sinh \theta'),$$

$$D(\theta) = e^{-H \cosh \theta}.$$

Similar integration may be done in expressions for partition function

$$Z = aS_{n-1} \int_0^\infty d\theta \sinh^{n-1} \theta D(\theta) P^{k+1}(\theta),$$

where $S_{n-1} = 2\pi^{n/2}/\Gamma(n/2)$ is the volume of the sphere S^{n-1}, and correlators.

In particular, upon averaging of $\mathbf{n}$ only σ-component survives giving the order parameter

$$\langle \sigma \rangle \equiv \langle \cosh \theta \rangle = \frac{aS_{n-1}}{Z} \int_0^\infty d\theta \sinh^{n-1} \theta \cosh \theta D(\theta) P^{k+1}(\theta).$$

For invariant correlator $\langle \mathbf{n}_0 \cdot \mathbf{n}_r \rangle = \langle \sigma_0 \sigma_r \rangle - \langle \tau_0 \cdot \tau_r \rangle$ the angular integrals give different kernels for longitudinal $G^L(r) \equiv \langle \cosh \theta_0 \cosh \theta_r \rangle_w$ and transverse $G_{ij}^T(r) \equiv \langle \tau_{0i} \tau_{rj} \rangle_w$ parts:

$$G^L(r) = \frac{a S_{n-1} N(r)}{Z} \int_0^\infty d\theta_0 \sinh^{n-1} \theta_0 \cosh \theta_0 D(\theta_0) P^k(\theta_0) \times \quad (11.114)$$

$$\left(\prod_{i=1}^r \int_0^\infty d\theta_i M_L(\theta_{i-1}, \theta_i) \right) P(\theta_r) \cosh \theta_r.$$

$$G_{ij}^T(r) = \delta_{ij} \frac{a S_{n-1} N(r)}{nZ} \int_0^\infty d\theta_0 \sinh^{n-1} \theta_0 D(\theta_0) P^k(\theta_0) \times \quad (11.115)$$

$$\left(\prod_{i=1}^r \int_0^\infty d\theta_i M_T(\theta_{i-1}, \theta_i) \right) P(\theta_r) \sinh \theta_r,$$

where

$$M_L(\theta, \theta') = L_L(\theta, \theta') D(\theta') P^{k-1}(\theta'), \quad M_T(\theta, \theta') = L_T(\theta, \theta') D(\theta') P^{k-1}(\theta'),$$

$$L_T(\theta, \theta') = \left(2K_{(n-1)/2}(J) \right)^{-1} \left(\frac{\sinh \theta'}{\sinh \theta} \right)^{(n-1)/2} e^{-J \cosh \theta \cosh \theta'} \times$$

$$(2\pi J \sinh \theta \sinh \theta')^{1/2} I_{n/2}(J \sinh \theta \sinh \theta').$$

11.9   Supersymmetric $O(n,1)$ model

In this section we consider the supersymmetric version of $O(n,1)$ model, namely, a non-linear model with field taking values on the so-called hyperbolic superplane. This object is constructed as follows. We consider a set of 5-component vectors

$$\bar{\psi} = (\sigma, \tau_1, \tau_2, \bar{\xi}, -\xi),$$

where the first three components are commuting, whereas the last two are Grassmannians (we use the adjoint of the second kind, see [20] for a review of super-analysis).

Next we consider the group G of linear transformations in the space of vectors ψ which preserves the length

$$\|\psi\| = \sigma^2 - \tau_1^2 - \tau_2^2 - 2\bar{\xi}\xi.$$

Let K be the subgroup of G which separately preserves σ^2 and $\tau_1^2 + \tau_2^2 + 2\bar{\xi}\xi$. Then the coset space G/K is isomorphic to the space of vectors ψ of unit length $\|\psi\| = 1$. This is the hyperbolic superplane.

We will use the following parametrization of G/K:

$$\tau_1 = \sinh\theta\cos\phi, \quad \tau_2 = \sinh\theta\sin\phi, \quad \sigma = \cosh\theta + \frac{\bar{\xi}\xi}{\cosh\theta}.$$

In this parametrization the G-invariant measure on G/K is

$$d\psi = \frac{a}{\sigma}d\tau_1 d\tau_2 d\bar{\xi}d\xi = a\left(1 - \frac{\bar{\xi}\xi}{\cosh^2\theta}\right)d\theta\sinh\theta d\phi d\bar{\xi}d\xi.$$

The Hamiltonian in this case is

$$\mathcal{H} = J\sum_{\langle i,j\rangle}\bar{\psi}_i\psi_j + H\sum_i\sigma_i.$$

We again choose the constant a in the definition of $d\psi$ such that $\int d\psi'\exp(-J\bar{\psi}\psi') = 1$. This gives $a = e^J/2\pi$.

Proceeding like in the previous section we introduce function $P(\psi)$ (by symmetry it actually depends only on σ) which satisfies the equation

$$P(\sigma) = \int d\psi'\exp(-J\bar{\psi}\psi' - H\sigma')P^k(\sigma').$$

Expanding both the left-hand side and the right-hand side in powers of Grassmann variables and integrating them out we get from the last equation

$$P(\theta) = e^{J(1-\cosh\theta)}e^{-H}P^k(0) + \int_0^\infty d\theta' L_{L0}(\theta,\theta')D(\theta')P^k(\theta'), \quad (11.116)$$

$$L_{L0}(\theta,\theta') = e^J J\sinh\theta e^{-J\cosh\theta\cosh\theta'}I_1(J\sinh\theta\sinh\theta'). \quad (11.117)$$

The first term in (11.116) is the boundary term resulting from integration by parts.

For $\theta = 0$ from (11.116) we get $P(0) = e^{-H}P^k(0)$ which means that $P(0) = \exp(\frac{H}{k-1})$ or $P(0) = 0$. To have $P(\theta) = 1$ as a solution $H = 0$ we have to choose $P(0) = \exp(\frac{H}{k-1})$.

We can also perform Grassmann integrations in formula for partition function, one-site averages and longitudinal correlators:

$$Z = \exp(J + \frac{2H}{k-1}), \quad \langle A(\sigma)\rangle = A(0), \quad \langle A(\sigma_0)B(\sigma_r)\rangle_w = N(r)A(0)B(0).$$

In particular

$$\langle\cosh\theta\rangle = 1, \quad G^L(r) \equiv \langle\sigma_0\sigma_r\rangle_w = N(r), \quad G_c^L(r) = 0, \quad (11.118)$$

where subscript c refers to connected correlator. For transverse correlator we have

$$G_{ij}^T(r) = \delta_{ij}\frac{e^J N(r)}{Z}\int_0^\infty d\theta_0 D(\theta_0)P^k(\theta_0)\times \qquad (11.119)$$

$$\left(\prod_{i=1}^r \int_0^\infty d\theta_i L_{T0}(\theta_{i-1},\theta_i)D(\theta_i)P^{k-1}(\theta_i)\right)P(\theta_r)\sinh\theta_r,$$

where

$$L_{T0}(\theta,\theta') = e^J J\sinh\theta e^{-J\cosh\theta\cosh\theta'}I_0(J\sinh\theta\sinh\theta').$$

Notes. The $O(n,1)$ model and its supersymmetric version is taken from the paper [112], this paper also discusses the replica limit $n\to 0$ and shows that if one takes it carefully all the results for the $O(n,1)$ model exactly reproduce results of the supersymmetric treatment. The analysis is very close to that of papers [63], [162] devoted to the problem of Anderson localization on Cayley tree. In [112] it is proved that $O(n,1)$ model exhibits two phases with a phase transition between them for any $0\le n<1$. The critical behavior of different correlators near this transition is obtained and showed that it is exactly the same as exhibited by the supersymmetric model.

11.10   The review of remaining models

This section contains reviews and references of several models defined on Cayley trees. Note that most of them were only studied in physical point of view. This review section will be useful for mathematicians who want to develop the theory of Gibbs measures for these models.

11.10.1   *Real values*

In this subsection we shall give classical models.

Asymmetric clock model. Consider q-state asymmetric clock model on a Cayley tree of order k. The Hamiltonian is

$$\mathcal{H} = -J\sum_{\langle x,y\rangle}\cos\left(\frac{2\pi}{q}(\sigma(x)-\sigma(y)-\alpha)\right), \qquad (11.120)$$

where $\sigma(x) \in \{1, 2, \ldots, q\}$, $x \in V$ and $J, \alpha \in R$.

In [73] using similar approach of [257], a non-linear recursion relation is obtained. The recursion map is iterated numerically and different types of attractors are found which characterize different phases.

XY chiral model.

The XY chiral model arises as a generalization of q-state clock model. We define plane spins at each site (vertex) of the tree with ferromagnetic interactions between spins in the plane and competing interactions in the modulated direction (the Z direction). The XY chiral model in the presence of a magnetic field can be defined by the Hamiltonian

$$\mathcal{H} = -J_1 \sum_{\langle x,y \rangle} \mathbf{S}_x \cdot \mathbf{S}_y - J_2 \sum_{\langle x,y \rangle} (\mathbf{S}_x \times \mathbf{S}_y) \cdot \hat{z} - \sum_x \mathbf{H} \cdot \mathbf{S}_y, \qquad (11.121)$$

where the spins $\mathbf{S}_x$ have components in the XY plane only, and $\hat{z}$ represents the direction of modulation. On a Cayley tree $\hat{z}$ points toward the center of the tree and the phase modulation takes place with respect to generations of the tree.

The components of the spin $\mathbf{S}_x$ can be written as $S_{x_X} = \cos \varphi_x$ and $S_{x_X} = \sin \varphi_x$, where $\varphi_x = 2\pi\sigma(x)/q$ represents the angle between the direction of the plan spin $\mathbf{S}_x$ and the X axis and $\sigma(x) = 0, 1, \ldots, q-1$. The Hamiltonian (11.121) can be written in the more usual form

$$\mathcal{H} = -J \sum_{\langle x,y \rangle} \cos \left(\frac{2\pi}{q} (\sigma(x) - \sigma(y) - \alpha) \right) - H \sum_x \cos \left[\frac{2\pi\sigma(x)}{q} \right], \quad (11.122)$$

where $J_1 = J \cos[2\phi\alpha/q]$, $J_2 = J \sin[2\phi\alpha/q]$ and α is the 'chiral field'.

In [21] the phase diagram at zero temperature from a corresponding 1D mapping is calculated. Ferromagnetic, commensurate and incommensurate modulated phases as well as chaotic structures are present in the zero temperature limit. For finite temperatures, the phase diagrams are obtained from a 2D mapping. The chaotic behavior is present only at low temperatures.

Heisenberg model. Let S^d denote the d-dimensional unit sphere. The angle between two points w and z on S^d is denoted by $\angle(w, z)$.

The Heisenberg model is defined by the Hamiltonian

$$H(\sigma) = J \sum_{\langle x,y \rangle} \cos \angle(\sigma(x), \sigma(y)),$$

where $J \in R$.

See [190] for the ferromagnetic ($J > 0$) Heisenberg model, there the spectrum and the wave functions of the single-magnon states are obtained.

In [194] the Heisenberg model on a general tree is studied. The phenomenon of interest was the classification of phase transition (non-uniqueness of the Gibbs measure) according to whether it is robust. The critical value for robust phase transition is computed for the Heisenberg and Potts models on a general tree in terms of the branching number (Hausdorff dimension) of the tree. In some cases, such as the $q \geq 3$ Potts model, it is shown that robust phase transition and usual phase transition do not coincide, while in other cases, such as the Heisenberg models, it was conjectured that robust phase transition and usual phase transition are equivalent.

Spherical model. In [270] the mean spherical model (MSM) on a Cayley tree is studied.

The MSM is made up of N scalar spin variables $-\infty < \sigma(x) < \infty$, with the restriction that the thermal average of the spin length $1/N \langle \sum_x \sigma^2(x) \rangle = 1$; this is called the spherical condition.

Therefore the MSM is described by the following effective Hamiltonian

$$-\beta \mathcal{H}(\sigma) = K \sum_{\langle x, y \rangle} \sigma(x)\sigma(y) - S \sum_x \sigma^2(x) + b \sum_x \sigma(x), \qquad (11.123)$$

where the first term is the exchange interaction, $K = J\beta$.

The second term is a Lagrange term for the total length of the spins, where S must be determined in such a way that $\langle \sum_x \sigma^2(x) \rangle = N$.

The third term is the Zeeman term with b proportional to the magnetic field.

In [270] the phase transition of the MSM on the Cayley tree is investigated by a Lagrange multiplier technique. There is one critical temperature where the specific heat has a finite discontinuity, while the susceptibility has a cusp. There is no spontaneous magnetization. See also [45].

Anderson model.

Let $\Gamma^k = (V, L)$ be a Cayley tree.

The Anderson Hamiltonian, H, on the Hilbert space

$$\ell^2(\Gamma^k) = \{\varphi : V \to \mathbb{C} : \sum_{x \in V} |\varphi(x)|^2 < \infty\}$$

is the operator of the form

$$H = \Delta + \epsilon q$$

where

1. The free Laplacian Δ is defined by

$$(\Delta\varphi)(x) = \sum_{y:\, d(x,y)=1} (\varphi(x) - \varphi(y)), \quad \text{for all} \ \ \varphi \in \ell^2(\Gamma^k),$$

with the usual distance d denoting the number of edges in the shortest (only) path between sites on the Cayley tree.

2. The operator q is a random potential,

$$(q\varphi)(x) = q(x)\varphi(x),$$

where $\{q(x)\}_{x \in V}$ is a family of independent, identically distributed real random variables with common probability distribution ν. The coupling constant ϵ measures the disorder.

The goal is to study spectrum for this operator. In [135] Klein proved the existence of purely absolutely continuous spectrum, under weak disorder, on the Cayley tree.

A new proof of a version of Kleins theorem is given in [75].

In [116] the absolutely continuous spectra for the Anderson model on different types of trees are studied.

The paper [160] reviews the main aspects and problems in the Anderson model on the Cayley tree. It shows how the question as to whether wave functions are extended or localized is related to the existence of complex solutions of a certain non-linear equation. See [51] for more results on the model.

Vertex model. The vertex model on the Cayley tree $\Gamma^k = (V, L)$ of order $k \geq 1$ is defined as follows. Each edge of the tree can be in one of the states $\{0, 1, \ldots, q-1\}$, $q \geq 1$ and vertex weight $w_x(s_1, s_2, \ldots, s_{k+1}) > 0$ of the vertex x with incident edges in the states $\{s_1, \ldots, s_{k+1}\}$ is independent of the permutations of these states.

The Hamiltonian of the model is given by

$$\mathcal{H} = \sum_{x \in V} \ln w_x(s_1, \ldots, s_{k+1}).$$

In [137] the symmetric two-state vertex model on the Cayley tree is studied. Two types of first-order phase transitions are distinguished according to the behavior of correlation functions in the high-temperature phase. A manifold in the model parameter space on which the correlations vanish is shown to be the same as for the honeycomb lattice. See also [138], [236], [237].

Remark 11.2. We note that there are many papers devoted to non-Gibbsian measures of the lattice models on trees (see [65] and the references therein). In [65] the Gibbsian properties of homogeneous low-temperature Ising Gibbs measures on trees are studied, subjected to an infinite-temperature Glauber evolution.

11.10.2   *Quantum case*

First attempts to investigate quantum Markov chains over trees was done in [4]. In [2] a hierarchy of notions of Markovianity for states on discrete infinite tensor products of C^*-algebras is introduced and for each of these notions some explicit examples are constructed. It was showed that the construction of [1] can be generalized to trees.

Note that a noncommutative extension of classical Markov fields, associated with Ising and Potts models on a Cayley tree, were investigated in [179], [181], [185].

We note that phase transitions in a quantum setting play an important role to understand quantum spin systems (see for example [11], [22], [76]).

Let us give some examples of quantum models:

Valence-bond-solid models. To each site $x \in V$ of a Cayley tree we assign a quantum spin variable with spin $s = k/2$. Let $\langle x, y \rangle$ denote a pair of nearest neighbors in the tree and $P_{x,y}^{(k)}$ the orthogonal projection onto the subspace in $\mathbb{C}^{k+1} \otimes \mathbb{C}^{k+1}$, located at the sites x and y, which corresponds to maximal total spin i.e., $k = k/2 + k/2$.

The formal Hamiltonian, the valence-bond-solid (VBS) model, is then defined as

$$\mathcal{H} = \sum_{\langle x,y \rangle} P_{x,y}^{(k)}.$$

This Hamiltonian is a positive operator (being the sum of positive terms), and it has the peculiar property of possessing a ground state with vanishing energy.

In [70] the thermodynamic limit of the ground states of VBS models on a Cayley tree is studied. The uniqueness for $k \leq 4$ and the occurrence of Néel order for $k \geq 5$ are proved. The main technical tool was a transfer matrix description of VBS states. See also [71], [124].

Quantum XY model.   Consider the Pauli spin operators

$\sigma_x^{(u)}, \sigma_y^{(u)}, \sigma_z^{(u)}$ at site $u \in V$:

$$\sigma_x^{(u)} = \begin{pmatrix} 0 & 1 \\ 1 & 0 \end{pmatrix}, \quad \sigma_y^{(u)} = \begin{pmatrix} 0 & -i \\ i & 0 \end{pmatrix}, \quad \sigma_z^{(u)} = \begin{pmatrix} 1 & 0 \\ 0 & -1 \end{pmatrix}.$$

For every edge $\langle u, v \rangle \in L$ put

$$K_{\langle u,v \rangle} = \exp\{\beta H_{\langle u,v \rangle}\},$$

where

$$H_{\langle u,v \rangle} = \frac{1}{2} \left(\sigma_x^{(u)} \sigma_x^{(v)} + \sigma_y^{(u)} \sigma_y^{(v)} \right).$$

Such kind of Hamiltonian is called quantum XY model per edge $\langle u, v \rangle \in L$.

In [3] (see also [178]) forward quantum Markov chains (QMC) defined on a Cayley tree are studied. Using the tree structure, a construction of quantum Markov chains on a Cayley tree is given. By means of such constructions, the existence of a phase transition for the XY model on a Cayley tree of order three in QMC scheme is shown. By the phase transition one means the existence of two distinct QMC for the given family of interaction operators $K_{\langle u,v \rangle}$.

For more information about models of statistical mechanics, condensed matters physics, quantum chaos, quantum field theory see [33], [61], [64], [107], [165], [206], [240], [246], [252], [264].

11.10.3   *p-adic values*

Let us first give some notions with p-adic setting.

p-adic numbers and measures.   Let $\mathbb{Q}$ be the field of rational numbers. For a fixed prime number p, every rational number $x \neq 0$ can be represented in the form $x = p^r \frac{n}{m}$, where $r, n \in \mathbb{Z}$, m is a positive integer, and n and m are relatively prime with p: $(p, n) = 1$, $(p, m) = 1$.

The p-adic norm of x is given by

$$|x|_p = \begin{cases} p^{-r} & \text{for } x \neq 0 \\ 0 & \text{for } x = 0. \end{cases}$$

This norm is non-Archimedean and satisfies the so-called strong triangle inequality

$$|x + y|_p \leq \max\{|x|_p, |y|_p\}.$$

The completion of $\mathbb{Q}$ with respect to the p-adic norm defines the p-adic field $\mathbb{Q}_p$. Any p-adic number $x \neq 0$ can be uniquely represented in the canonical form

$$x = p^{\gamma(x)}(x_0 + x_1 p + x_2 p^2 + \dots), \tag{11.124}$$

where $\gamma(x) \in \mathbb{Z}$ and the integers x_j satisfy: $x_0 > 0$, $0 \leq x_j \leq p-1$ (see [136], [258]). In this case $|x|_p = p^{-\gamma(x)}$.

Given $a \in \mathbb{Q}_p$ and $r > 0$ put

$$B(a,r) = \{x \in \mathbb{Q}_p : |x - a|_p < r\}.$$

The p-adic *logarithm* is defined by the series

$$\log_p(x) = \log_p(1 + (x-1)) = \sum_{n=1}^{\infty}(-1)^{n+1}\frac{(x-1)^n}{n},$$

which converges for $x \in B(1,1)$; the p-adic exponential is defined by

$$\exp_p(x) = \sum_{n=0}^{\infty}\frac{x^n}{n!},$$

which converges for $x \in B(0, p^{-1/(p-1)})$.

Lemma 11.1. [136], [238]. *Let $x \in B(0, p^{-1/(p.1)})$, then*

$$|\exp_p(x)|_p = 1, \quad |\exp_p(x) - 1|_p = |x|_p, \quad |\log_p(1 + x)|_p = |x|_p,$$

$$\log_p(\exp_p(x)) = x, \quad \exp_p(\log_p(1 + x)) = 1 + x.$$

We refer the reader to [136], [238], [258] for the basics of p-adic analysis and p-adic mathematical physics.

Let $(X, \mathcal{B})$ be a measurable space, where $\mathcal{B}$ is an algebra of subsets of X.

A function $\mu : \mathcal{B} \to \mathbb{Q}_p$ is said to be a p-adic measure if for any $A_1, ..., A_n \in \mathcal{B}$ such that $A_i \cap A_j = \emptyset$, $i \neq j$, the following holds:

$$\mu(\bigcup_{j=1}^{n} A_j) = \sum_{j=1}^{n} \mu(A_j).$$

A p-adic measure is called a probability measure if $\mu(X) = 1$, see, e.g. [131], [209].

Note that various models described in the language of p-adic analysis have been actively studied, see e.g. [8], [9], [10] and numerous applications of p-adic analysis to mathematical physics have been proposed in [258].

It is also known that a number of p-adic models in physics cannot be described using ordinary Kolmogorov's probability theory. In [132] an abstract p-adic probability theory was developed by means of the theory of non-Archimedean measures [209].

A non-Archimedean analogue of the Kolmogorov theorem was proved in [91]. Such a result allows to construct wide classes of stochastic processes and the possibility to develop statistical mechanics in the context of p-adic theory.

Note that to define a p-adic Hamiltonian one can consider spin and interactions values as p-adic number.

The p-adic Gibbs measure can be defined by the usual formulas, replacing the real $\exp(x)$ by p-adic exponent $\exp_p(x)$.

We refer the reader to [130], [133], [182]-[184] where various models of statistical physics in the context of p-adic fields are studied.

In [82] p-adic Gibbs measures of a hard core model on the Cayley tree over the p-adic field are studied.

To conclude, we will give a brief description of the differences of behavior between classical (real) models and p-adic models on Cayley trees.

Hard core models. *Real case:* In this model (see [251], Chapter 7), for all $\lambda > 0$ and $k \geq 1$, there exists a unique translational invariant splitting Gibbs measure μ_0. Let $\lambda_c = \frac{1}{(k-1)} \left(\frac{k}{k-1} \right)^k$, then:

(i) for $\lambda \leq \lambda_c$, the Gibbs measure is unique (and coincides with the above measure μ_0),

(ii) for $\lambda > \lambda_c$, in addition to μ_0, there exist two distinct extreme periodic measures, μ^+ and μ^-. In addition, there is a continuum set of distinct, extreme, non-translational-invariant, Gibbs measures.

For $\lambda > \frac{1}{(\sqrt{k}-1)} \left(\frac{\sqrt{k}}{\sqrt{k}-1} \right)^k$, the measure μ_0 is not extreme.

p-adic case: In [82] it is shown that the p-adic HC model is completely different from real HC model. For a fixed k, the p-adic HC model may have a splitting Gibbs measure only if p divides $2^k - 1$. Moreover, if p divides $2^k - 1$ and does not divide $k + 2$ then there exists unique translation invariant p-adic Gibbs measure. The HC model admits only translation invariant and periodic with period two (chess-board) Gibbs measures. For $p \geq 7$, a periodic p-adic Gibbs measure exists iff p divides both $2^k - 1$ and $k - 2$. For $k = 2$, a p-adic splitting Gibbs measure exists if and only if $p = 3$; in this case it is shown that if λ belongs to a p-adic ball of radius $1/27$ then there are precisely two periodic (non translation invariant) p-adic Gibbs measures. Finally, we have proven that a p-adic Gibbs measure is bounded if and only if $p \neq 3$.

Potts model. *Real case:* The ferromagnetic q states Potts model for any $q \geq 2$ exhibits possibly $q + 1$ distinct translation invariant Gibbs measures. Namely, there exist two critical temperatures $0 < T_c' < T_c$ such that:
(i) for $T \in (T_c', T_c]$ there are $q + 1$ extreme Gibbs measures, one of them is

called unordered one;

(ii) for $T \leq T'_c$, q extreme Gibbs measures coexist: there is the unordered which is not extreme;

(iii) for $T > T_c$ there is one Gibbs measure (Chapter 5).

p-adic case: The model exhibits a phase transition whenever $k = 2, q \in p\mathbb{N}$ and $p \geq 3$ (resp. $q \in 2^2\mathbb{N}, p = 2$) [182]. Whenever $k \geq 3$ a phase transition may occur only at $q \in p\mathbb{N}$ if $p \geq 3$ and $q \in 2^2\mathbb{N}$ if $p = 2$. Moreover for the p-adic Ising model ($q = 2$) there is no phase transition. This is one interesting difference between real and p-adic Ising model on trees.

λ-model. *Real case:* A nearest-neighbor λ-model with two spin values on the Cayley tree is considered in [211]. There, it was proven that this model has similar properties like Ising model.

p-adic case: (see [130]) For the p-adic non-homogeneous λ-model there is no phase transition and as well as being unique, the p-adic Gibbs measure is bounded if and only if $p \geq 3$. If $p = 2$, a condition asserting the non-existence of a phase transition was given.

This result shows that, in p-adic case, even non-homogeneous interactions do not lead to the occurrence of a phase transition.

From the above given results it follows that the set of p-adic Gibbs measures is sparse with respect to the set of real Gibbs measures. The main reasons for this could be explained by the following:

(i) The set of values of real norm $|x|$ is continuous $[0, +\infty)$, but the set of values of p-adic norm is discrete $\{p^m : m \in \mathbb{Z}\}$.

(ii) The real function e^x is defined for any $x \in R$ but p-adic function $\exp_p(x)$ is defined only for $x \in \mathbb{Q}_p$ with $|x|_p \leq \frac{1}{p}$.

(iii) The set of values of real function e^x and its norm $|e^x|$ is continuous $(0, +\infty)$, but the set of values of p-adic function $\exp_p(x)$ is $\{x : |x-1|_p \leq \frac{1}{p}\}$ and the set of values of its norm $|\exp_p(x)|_p$ contains unique point 1, i.e., $|\exp_p(x)|_p = 1$ for all x with $|x|_p \leq \frac{1}{p}$.

To avoid these difficulties in [233] the notion of p-adic Markov random field is defined. It is known that in the real case a Gibbs measure is equivalent to a Markov random field [204], [218]. In [233] it is shown that the p-adic Gibbs measure in not equivalent to p-adic Markov random field. In fact, the set of p-adic Gibbs measures is a subset of the set of p-adic Markov random fields.

Nevertheless, we believe that p-adic Gibbs measures might have interesting applications.

Bibliography

[1] L. Accardi, A. Frigerio, *Markovian cocycles*, Proc. Royal Irish Acad. **83** (1983), 251–263.

[2] L. Accardi, H. Ohno, F.M. Mukhamedov, *Quantum Markov fields on graphs*, Inf. Dimens. Anal. Quantum Probab. Relat. Top. **13** (2010), 165–189.

[3] L. Accardi, F. Mukhamedov, M. Saburov, *On quantum Markov chains on Cayley tree II: Phase transitions for the associated chain with XY-model on the Cayley tree of order three*, Ann. Henri Poincare **12**(6) (2011), 1109–1144.

[4] L. Affleck, E. Kennedy, E.H. Lieb, H. Tasaki, *Valence bond ground states in isortopic quantum anti-ferromagnetic*, Commun. Math. Phys. **115**, (1988), 477–528.

[5] F.S. de Aguiar, F.A. Bosco, A.S. Martinez, R.S. Goulart, Jr., *Phase diagram of the one-state Potts model on the Cayley tree*, J. Statist. Phys. **58**(5-6) (1990), 1231–1238.

[6] F.S. de Aguiar, L.B. Bernardes, R.S. Goulart, Jr., *Metastability in the Potts model on the Cayley tree*, J. Statist. Phys. **64**(3-4) (1991), 673–682.

[7] H. Akin, U.A. Rozikov, S. Temir, *A new set of limiting Gibbs measures for the Ising model on a Cayley tree*, J. Statist. Phys. **142**(2) (2011), 314–321.

[8] S. Albeverio, W. Karwowski, *A random walk on p-adics–the generator and its spectrum*, Stoch. Process. Appl. **53** (1994), 1–22.

[9] S. Albeverio, X. Zhao, *Measure-valued branching processes associated with random walks on p-adics*, Ann. Probab. **28** (2000), 1680–1710.

[10] S.Albeverio, W. Karwowski, K.Yasuda, *Trace formula for p-adics*, Acta Appl. Math. **71** (2002), 31–48.

[11] S. Albeverio, Yu. Kondratiev, Yu. Kozitsky, M. Röckner, *The statistical mechanics of quantum lattice systems. A path integral approch*, EMS, Zürich, 2009.

[12] J. M. de Araújo, F. A. da Costa, *Phase diagrams of the Ashkin-Teller model on the Cayley Tree*, Brazilian J. Phys. **27**(2) (1997), 89–95.

[13] J. Ashkin, E.Teller, *Statistics of two-dimensional lattices with four components*, Phys. Rev. **64** (1943), 178–184.

[14] P. Bak, C. Tang, K. Wiesenfeld, *Self-organized criticality: An explanation of the 1/f noise*, Phys. Rev. Lett. **59**(4) (1987), 381–384.

[15] P. Bak, C. Tang, K. Wiesenfeld, *Self-organized criticality*, Phys. Rev. A **38**(1) (1988), 364–374.

[16] A.V. Bakaev, A.N. Ermilov, A.M. Kurbatov, *A frozen Potts model on a Cayley tree*, Sov. Phys. Dokl. **33**(3) (1988), 190–191.

[17] J. C. A. Barata, D.H.U. Marchetti, *Griffiths' singularities in diluted Ising models on the Cayley tree*, J. Statist. Phys. **88**(1-2) (1997), 231–268.

[18] R.J. Baxter, *Exactly Solved Models in Statistical Mechanics*, Academic, London, 1982.

[19] N. Berger, C. Kenyon, E. Mossel, Y. Peres, *Glauber dynamics on trees and hyperbolic graphs*, Probab. Theory Relat. Fields. **131**(3) (2005), 311–340.

[20] F. A. Berezin, *Introduction to Superanalysis*, Reidel, Dodrecht, 1987.

[21] A.T. Bernardes, M.J. de Oliveira, *Field behaviour of the XY chiral model on a Cayley tree*, J. Phys. A **25** (1992), 1405–1415.

[22] M. Biskup, L. Chayes, Sh. Starr, *Quantum spin systems at positive temperature.* Commun. Math. Phys. **269** (2007), 611–657.

[23] P.M. Bleher, N.N. Ganikhodjaev (Ganikhodzhaev), *On pure phases of the Ising model on the Bethe lattice*, Theor. Probab. Appl. **35** (1990), 216–227.

[24] P.M. Bleher, *The Bethe lattice spin glass at zero temperature*, Ann. Inst. H. Poincare Phys. Theor. **54**(1) (1991), 89–113.

[25] P.M. Bleher, J. Ruiz, V.A. Zagrebnov, *On the purity of the limiting Gibbs state for the Ising model on the Bethe lattice*, J. Statist. Phys. **79** (1995), 473–482.

[26] P.M. Bleher, J. Ruiz, V.A. Zagrebnov, *One-dimensional random-field Ising model: Gibbs states and structure of Ground states*, J. Statist. Phys. **84**(5/6) (1996), 1077–1093.

[27] P.M. Bleher, J. Ruiz, V.A. Zagrebnov, *On the phase diagram of the random field Ising model on the Bethe lattice*, J. Statist. Phys. **93**(1/2) (1998), 33–78.

[28] P.M. Bleher, J. Ruiz, R.H. Schonmann, S. Shlosman, V.A. Zagrebnov, *Rigidity of the critical phases on a Cayley tree*, Mosc. Math. J. **1**(3) (2001), 345–363.

[29] B. Bollobás, *Modern Graph Theory*, Graduate Texts in Math., vol. 184, Springer-Verlag, New York, 1998.

[30] C. Borgs, *Statistical physics expansion methods in combinatorics and computer science*, http://research.microsfort.com/ borgs/CBMS.pdf, 2004.

[31] G.I. Botirov, U.A. Rozikov, *On q-component models on Cayley tree: the general case*, J. Statist. Mech.: Theory and Exper., P10006, (2006), 8 pages.

[32] G.I. Botirov, U.A. Rozikov, *Potts model with competing interactions on the Cayley tree: The contour method*, Theor. Math. Phys. **153**(1) (2007), 1423-1433.

[33] A. Bovier, *Statistical Mechanics of Disordered Systems. A Mathematical Perspective.* Cambridge Series in Statistical and Probabilistic Mathemat-

ics. Cambridge Univ. Press, Cambridge, 2006.

[34] J. Bricmont, J. Lebowitz, C. Pfister, *On the local structure of the phase separation line in the two-dimensional Ising system*, J. Statist. Phys. **26** (1981), 313–332.

[35] G.R. Brightwell, P. Winkler, *Graph homomorphisms and phase transitions*, J. Combin. Theory Ser. B **77**(2) (1999), 221–262.

[36] G. R. Brightwell, O. Häggström, P. Winkler, *Nonmonotonic behavior in hard-core and Widom-Rowlinson models*, J. Statist. Phys. **94**(3-4) (1999), 415–435.

[37] G. R. Brightwell, P. Winkler, *Random colorings of a Cayley tree*, Contemporary Combinatorics, B. Bollobas, ed., Bolyai Society Mathematical Studies, 2001.

[38] G. R. Brightwell, P. Winkler, *Hard constraints and the Bethe lattice: adventures at the interface of combinatorics and statistical physics.* Proc. ICM 2002, Higher Education Press, Beijing, IIIi:605–624, 2002.

[39] G. R. Brightwell, P. Winkler, *A second threshold for the hard-core model on a Bethe lattice.* Random Structures Algorithms **24**(3) (2004), 303–314.

[40] R. Bruinsma, *Random field Ising model on a Bethe lattice*, Phys. Rev. B **30** (1984), 289–299.

[41] J.M. Carlson, J.T. Chayes, L.Chayes, J.P. Sethna, D.J. Thouless, *Critical behavior of the Bethe lattice spin glass*, Europhys. Lett. **5**(4) (1988), 355–360.

[42] J.M. Carlson, J.T. Chayes, J.P. Sethna, D.J. Thouless, *Bethe lattice spin glass: the effects of a ferromagnetic bias and external fields. II. Magnetized spin-glass phase and the de Almeida-Thouless line*, J. Statist. Phys. **61**(5-6) (1990), 1069–1084.

[43] J.M. Carlson, J.T. Chayes, L. Chayes, J.P. Sethna, D.J. Thouless, *Bethe lattice spin glass: the effects of a ferromagnetic bias and external fields, I. Bifurcation analysis*, J. Statist. Phys. **61** (5-6) (1990), 987–1067.

[44] M. Cassandro, E. Orlandi, P. Picco, *Phase transition in the 1d random field Ising model with long range interaction*, Comm. Math. Phys. **288**(2) (2009), 731–744.

[45] D. Cassi, A. Pimpinelli, *Spherical model on the Bethe lattice via the statistics of walks*, Internat. J. Mod. Phys. B **4**(11-12) (1990), 1913–1921.

[46] J. Cavender, *Taxonomy with confidence*, Math. BioSci. **40** (1978), 271–280.

[47] P. Caputo, F. Martinelli, *Phase ordering after a deep quench: the stochastic Ising and hard core gas models on a tree*, Probab. Theory Relat. Fields **136**(1) (2006), 37–80.

[48] J.T. Chayes, L. Chayes, J.P. Sethna, D.J. Thouless, *A mean field spin glass with short range interactions*, Comm. Math. Phys. **106** (1986), 41–89.

[49] F. A. da Costa, M. J. de Oliveira, S. R. Salinas, *Asymmetric Baxter phase in the symmetric Ashkin-Teller model*, Phys. Rev. B **36**(13) (1987), 7163–7165.

[50] T.M. Cover, J.A. Thomas, *Elements of Information Theory*, John Wiley

and Sons, 1991.

[51] B. Derrida, G.J. Rodgers, *Anderson model on a Cayley tree: the density of states*, J. Phys. A **26**(9) (1993), L457–L463.

[52] R. L. Devaney, *An Introduction to Chaotic Dynamical System*, Westview Press, 2003.

[53] D. Dhar, *Self-organized critical state of sandpile automaton models*, Phys. Rev. Lett. **64**(14) (1990), 1613–1616.

[54] D. Dhar, S.N. Majumdar, *Abelian sandpile model on the Bethe lattice*, J. Phys. A: Math. Gen. **23** (1990), 4333–4350.

[55] D. Dhar, *Studying self-organized criticality with exactly solved models*, (1999), arXiv:cond-mat/9909009v1 [cond-mat.stat-mech]

[56] D. Dhar, *Theoretical studies of self-organized criticality*, Phys. A **369**(1) (2006), 29–70.

[57] R. Dobrin, J.H. Meinke, P.M. Duxbury, *Random field Ising model on complete graphs and trees*, J. Phys. A: Math. Gen. **35** (2002), L247–L254.

[58] R.L. Dobrushin, *The description of a random field by means of conditional probabilities and conditions of its regularity*, Theory Probab. Appl. **13** (1968), 197–224.

[59] R.L. Dobrushin, *Gibbs state describing coexistence of phases for a three-dimensional Ising model*, Theory Probab. Appl. **17** (1972), 582–600.

[60] R.L. Dobrushin, *Investigation of Gibbsian states for three-dimensional lattice systems*, Theory Probab. Appl. **18**(2) (1973), 253–271.

[61] C. De Dominicis, I. Giardina, *Random Fields and Spin Glasses. A Field Theory Approach*. Cambridge University Press, Cambridge, 2006.

[62] J. Dugundji, *Topology*, Allyn, Bacon, Boston, 1966.

[63] K.B. Efetov, *Anderson metal-insulator transition in a system of metal granules: existence of a minimum metallic conductivity and a maximum dielectric constant*, Sov. Phys. JETP **61**(3) (1985), 606–617.

[64] K. Efetov, *Supersymmetry in Disorder and Chaos*, Cambridge Univ. Press, Cambridge, 1997.

[65] A.C.D. van Enter, V.N. Ermolaev, G. Iacobelli, C. Külske, *Gibbs-non-Gibbs properties for evolving Ising models on trees*, (2010), arXiv.1009.2952v1.

[66] O. Entin-Wohlman, C. Domb, *Random field Ising model on the Bethe lattice*, J. Phys. A **17**(11) (1984), 2247–2256.

[67] Yu. Kh. Eshkabilov, F. H. Haydarov, U. A. Rozikov, *Non-uniqueness of Gibbs measure for models with uncountable set of spin values on a Cayley Tree*, J. Statist. Phys. **147**(4) (2012), 779–794.

[68] Yu. Kh. Eshkabilov, F. H. Haydarov, U. A. Rozikov, *Uniqueness of Gibbs measure for models with uncountable set of spin values on a Cayley tree*, Math. Phys. Anal. Geom. **16**(1) (2013), 1–17.

[69] W. Evans, C. Kenyon, Y. Peres, L.J. Schulman, *Broadcasting on trees and the Ising model*, Ann. Appl. Probab. **10**(2) (2000), 410–433.

[70] M. Fannes, B. Nachtergaele, R.F. Werner, *Ground states of VBS models on Cayley trees*, J. Statist. Phys. **66**(3-4) (1992), 939–973.

[71] M. Fannes, B. Nachtergaele, R.F. Werner, *Finitely correlated states on

quantum spin chains, Comm. Math. Phys. **144**(3) (1992), 443–490.

[72] R. Fernandez, *Contour ensembles and the description of Gibbsian probability distributions at low temperature*, http://www.univ-rouen.fr/LMRS/Persopage/Fernandez/resucont.html (1998).

[73] K. Fesser, H.J. Herrmann, *The asymmetric clock model on a Cayley tree*, J. Phys. A: Math. Gen. **17** (1984), 1493–1507.

[74] C. M. Fortuin, P. W. Kasteleyn, J. Ginibre, *Correlation inequalities on some partially ordered sets*, Commun. Math. Phys. **22** (1971), 89–103.

[75] R. Froese, D. Hasler, W. Spitzer, *Absolutely continuous spectrum for the Anderson model on a tree: a geometric proof of Klein's theorem*, Comm. Math. Phys. **269**(1) (2007), 239–257.

[76] J. Fröhlich, R. Israel, E. Lieb, B. Simon, *Phase transitions and reflection positivity. I. General theory and long range lattice models*, Commun. Math. Phys. **62** (1978), 1–34.

[77] G. Gallavotti, *The phase separation line in the two-dimensional Ising model*, Commun. Math. Phys. **27** (1972), 103–136.

[78] G. Gallavotti, *Statistical Mechanics, A Short Treatise*. Text and Monographs in Physics, Springer, 1999.

[79] D. Galvin, J. Kahn. *On phase transition in the hard-core model on Z^d*, Comb. Prob. Comp. **13** (2004), 137–164.

[80] D. Galvin, P. Tetali, *Slow mixing of Glauber dynamics for the hard-core model on regular bipartite graphs*, Random Structures Algorithms **28**(4) (2006), 427–443.

[81] D. Galvin, F. Martinelli, K. Ramanan, P. Tetali, *The multistate hard core model on a regular tree*, SIAM J. Discrete Math. **25**(2) (2011), 894–915.

[82] D. Gandolfo, U.A. Rozikov, J. Ruiz, *On p-adic Gibbs measures for hard core model on a Cayley tree*, Markov Proc. Relat. Fields **18** (4) (2012), 701–720.

[83] D. Gandolfo, M.M. Rakhmatullaev, U.A. Rozikov, J. Ruiz, *On free energies of the Ising model on the Cayley tree*, J. Statist. Phys. **150**(6) (2013), 1201–1217.

[84] D. Gandolfo, J. Ruiz, S. Shlosman, *A manifold of pure Gibbs states of the Ising model on a Cayley tree*, J. Statist. Phys. **148** (2012), 999–1005.

[85] N.N. Ganikhodzhaev (Ganikhodjaev), *On pure phases of the three-state ferromagnetic Potts model on the second-order Bethe lattice*. Theor. Math. Phys. **85**(2) (1990), 1125–1134.

[86] N.N. Ganikhodzhaev, *On pure phases of the ferromagnetic Potts model Bethe lattices*, Dokl. AN Uzbekistan. No. 6-7 (1992), 4–7.

[87] N.N. Ganikhodjaev, *Group presentations and automorphisms of the Cayley tree*, Dokl. AN Uzbekistan. No. 5 (1994), 3–5.

[88] N.N. Ganikhodjaev, U.A. Rozikov, *On pure Gibbs distributions of the Ising model on the Bethe lattice with competing interactions*, Uzbek Math. J. No. 2 (1995), 36–47 (Russian).

[89] N.N. Ganikhodjaev, U.A. Rozikov, *Classes of normal subgroups of finite index of the group representation of the Cayley tree*, Uzbek. Mat. J. No. 4 (1997), 31–39 (Russian).

[90] N.N. Ganikhodzhaev, U.A. Rozikov, *Description of periodic extreme Gibbs measures of some lattice models on a Cayley tree*, Theor. Math. Phys. **111**(1) (1997), 480–486.

[91] N.N. Ganikhodjaev, F.M. Mukhamedov, U.A. Rozikov, *Phase transitions in the Ising model on Z over the p-adic numbers*, Uzbek Math. J. No. 4 (1998), 23–29.

[92] N.N. Ganikhodjaev, U.A. Rozikov, *On disordered phase in the ferromagnetic Potts model on the Bethe lattice*, Osaka Jour. Math. **37**(2) (2000), 373–383.

[93] N.N. Ganikhodjaev, F.M. Mukhamedov, U.A. Rozikov, *Existence of phase transition for the Potts p-adic model on the set Z*, Theor. Math. Phys. **130**(2) (2002), 500–507.

[94] N.N. Ganikhodjaev, *Exact solution of an Ising model on the Cayley tree with competing ternary and binary interactions*, Theor. Math. Phys. **130**(3) (2002), 419–424.

[95] N.N. Ganikhodjaev, C.H. Pah, M.R.B. Wahiddin, *Exact solution of an Ising model with competing interactions on a Cayley tree*, J. Phys. A **36** (2003), 4283–4289.

[96] N.N. Ganikhodjaev, C.H. Pah, M.R.B. Wahiddin, *An Ising model with three competing interactions on a Cayley tree*, J. Math. Phys. **45** (2004), 3645–3658.

[97] N.N. Ganikhodjaev, *The Potts model on Z^d with countable set of spin values*, J. Math. Phys. **45**(3) (2004), 1121–1127.

[98] N.N. Ganikhodjaev, U.A. Rozikov, *The Potts model with countable set of spin values on a Cayley tree*, Lett. Math. Phys. **75**(2) (2006), 99–109.

[99] N.N. Ganikhodjaev, H. Akin, S. Temir, *Potts model with two competing binary interactions*, Turk. J. Math. **31** (2007), 229–238.

[100] N.N. Ganikhodjaev, *Limiting Gibbs measures of Potts model with countable set of spin values*, J. Math. Anal. Appl. **336**(1) (2007), 693–703.

[101] N.N. Ganikhodjaev, F.M. Mukhamedov, C.H. Pah, *Phase diagram of the three states Potts model with next nearest neighbour interactions on the Bethe lattice*, Phys. Lett. A **373** (2008), 33–38.

[102] N.N. Ganikhodjaev, U.A. Rozikov, *On Ising model with four competing interactions on Cayley tree*, Math. Phys. Anal. Geom. **12**(2) (2009), 141–156.

[103] N.N. Ganikhodjaev, U.A. Rozikov, *On a phase separation point for one-dimensional models*. Siberian Adv. Math. **19**(2) (2009), 75–84.

[104] N.N. Ganikhodjaev, S. Temir, H. Akin, *Modulated phase of a Potts model with competing binary interactions on a Cayley tree*, J. Statist. Phys. **137** (2009), 701–715.

[105] N.N. Ganikhodjaev, *Strange attractor in the Potts model on a Cayley tree in the presence of competing interactions*, J. Concr. Appl. Math. **9**(1) (2011), 47–54.

[106] N.N. Ganikhodjaev, H. Akin, S. Uguz, S. Temir, *Phase diagrams of an Ising system with competing binary, prolonged ternary and next-nearest interactions on a Cayley tree*, J. Concr. Appl. Math. **9**(1) (2011), 26–34.

[107] H.O. Georgii, *Gibbs Measures and Phase Transitions*, Second edition. de Gruyter Studies in Mathematics, 9. Walter de Gruyter, Berlin, 2011.

[108] C. Glaffig, E. Waymire, *Infinite divisibility of a Bethe lattice Ising model*, J. Statist. Phys. **47**(1-2) (1987), 185–192.

[109] Y.Y. Goldschmidt, *Mean-field theory of the Potts spin-glass with finite connectivity*, Nucl. Phys. B **295**(3) (1988), 409–421.

[110] E. Goles, S. Martinez, *The one-site distributions of Gibbs states on Bethe lattice are probability vectors of period ≤ 2 for a nonlinear transformation*, J. Statist. Phys. **52**(1-2) (1988), 267–285.

[111] G. Grimmett, *The Random-Cluster Model*, Springer, Berlin, 2006.

[112] I.A. Gruzberg, A.D. Mirlin, *Phase transition in a model with non-compact symmetry on Bethe lattice and the replica limit*, J. Phys. A **29**(17) (1996), 5333–5345.

[113] O. Häggström, *The random-cluster model on a homogeneous tree*, Probab. Theory Relat. Fields **104**(2) (1996), 231–253.

[114] O. Häggström, *A monotonicity result for hard-core and Widom-Rowlinson models on certain d-dimensional lattices*, Electron. Comm. Probab. **7** (2002), 67–78 (electronic).

[115] Y. Higuchi, *Remarks on the limiting Gibbs states on a tree*, Publ. RIMS Kyoto Univ. **13** (1977), 335–348.

[116] F. Halasan. *Absolutely continuous spectrum for the Anderson model on a Cayley tree*. PhD thesis, Univ. of British Columbia, 2009.

[117] D.A. Huse, D.S. Fisher, *Pure states in spin glasses*, J. Phys. A **20**(15) (1987), L997–L1003.

[118] S. Inawashiro, C.J. Thompson, *Competing Ising interactions and chaotic glass-like behaviour on a Cayley tree*, Phys. Lett. **97A** (1983), 245–248.

[119] D. Ioffe, *On the extremality of the disordered state for the Ising model on the Bethe lattice*, Lett. Math. Phys. **37** (1996), 137–143.

[120] J. Kahn, *An entropy approach to the hard-core model on bipartite graphs*, Combin. Probab. Comput. **10**(3) (2001), 219–237.

[121] M. I. Kargapolov, Yu. I. Merzlyakov, *Fondations of Group Theory* [in Russian], Nauka. Moscow, 1982.

[122] I.A. Kashapov, *Structure of ground states in three-dimensional Ising model with tree-step interaction*, Theor. Math. Phys. **33** (1977), 912–918.

[123] S. Katsura, M. Takizawa, *Bethe lattice and the Bethe approximation*, Prog. Theor. Phys. **51** (1974) 82–98.

[124] T. Kennedy, E. H. Lieb, H. Tasaki, *A two-dimensional isotropic quantum antiferromagnet with unique disordered ground state*, J. Statist. Phys. **53** (1988), 383–415.

[125] H. Kesten, B.P. Stigum, *Additional limit theorem for indecomposable multi-dimensional Galton-Watson processes*, Ann. Math. Statist. **37** (1966), 1463–1481.

[126] H. Kesten, *Quadratic transformations: a model for population growth. I*. Adv. Appl. Probab. **2** (1970), 1–82.

[127] F.P. Kelly, *Stochastic models of computer communication systems. With discussion*, J. Roy. Statist. Soc. Ser. B **47** (1985), 379–395; 415–428.

[128] F. Kelly. *Loss networks*, Ann. Appl. Probab. **1**(3) (1991), 319–378.

[129] R.M. Khakimov. *Uniqueness of wekly periodic Gibbs measure for HC models*. To appear in Math. Notes. 2013.

[130] M. Khamraev, F.M. Mukhamedov, U.A. Rozikov, *On the uniqueness of Gibbs measures for p-adic non homogeneous λ-model on the Cayley tree*, Lett. Math. Phys. **70** (2004), 17–28.

[131] A. Yu. Khrennikov, *p-adic valued probability measures*, Indag. Math., New Ser. **7** (1996), 311–330.

[132] A.Yu. Khrennikov, S. Yamada, A. van Rooij, *The measure-theoretical approach to p-adic probability theory*, Ann. Math. Blaise Pascal **6** (1999), 21–32.

[133] A.Yu. Khrennikov, F.M. Mukhamedov, J.F.F. Mendes, *On p-adic Gibbs measures of the countable state Potts model on the Cayley tree*, Nonlinearity **20** (2007), 2923–2937.

[134] R. Kindermann, J.L. Snell, *Markov random fields and their applications*, Contemporary Mathematics **1**, 1980.

[135] A. Klein. *Extended states in the Anderson model on the Bethe lattice*, Adv. Math. **133** (1998), 163–184.

[136] N. Koblitz, *p-adic Numbers, p-adic Analysis, and zeta-Functions.* Springer, Berlin, 1977.

[137] M. Kolesik, *Correlation functions of the two-state vertex model on the Cayley tree*, Internat. J. Modern Phys. B **6**(21) (1992), 3469–3477.

[138] M. Kolesik, *No-free-ends method for lattice animals and vertex models with arbitrary number of states.* Phys. A **202**(3-4) (1994), 529–539.

[139] R. Kotecky, S.B. Shlosman, *First-order phase transition in large entropy lattice models*, Commun. Math. Phys. **83** (1982), 493–515.

[140] M.A. Krasnosel'ski, *Positive solutions of opertor equations*, Gos. Izd. Moscow, 1969 (Russian).

[141] M. A. Krasnosel'ski, P.P. Zabrejko, *Geometric Methods of Nonlinear Analysis*, Nauka. Moscow, 1975 (Russian).

[142] S. H. Kung, *Sums of integer powers via the Stolz-Cesáro theorem*, Math. Assoc. Amer. **40** (2009), 42–44.

[143] A.G. Kurosh, *Group Theory*, Akademic Verlag, Berlin, 1953.

[144] O.E. Lanford, D. Ruelle, *Observables at infinity and states with short range correlations in statistical mechanics*, Comm. Math. Phys. **13** (1969), 194–215.

[145] J.-X. Le, Z.-R. Yang, *Phase diagram for Ashkin-Teller model on Bethe lattice*, Commun. Theor. Phys., (Beijing) **43**(5) (2005), 841–846.

[146] J. L. Lebowitz, A. Martin-Löf, *On the uniqueness of the equilibrium state for Ising spin systems*, Commun. Math. Phys. **25** (1972), 276–282.

[147] J. L. Lebowitz, A.E. Mazel, *On the uniqueness of Gibbs states in the Pirogov-Sinai theory*, Commun. Math. Phys. **189** (1997), 311–321.

[148] J.-J. Loeffel, *About Gibbs states on Bethe lattices.* Stochastic processes, physics and geometry (Ascona and Locarno, 1988), 497–515, World Scientific Publ., Teaneck, NJ, 1990.

[149] G. Louth, *Stochastic networks: complexity, dependence and routing*, Cam-

bridge University (thesis), 1990.

[150] B. Luen, K. Ramanan, I. Ziedins, *Nonmonotonicity of phase transitions in a loss network with controls*, Ann. Appl. Probab. **16**(3) (2006), 1528–1562.

[151] A. Maritan, A.L. Stella, *Phase transition in a gauge model on a tree-like lattice*, J. Phys. A **16**(5) (1983), L157–L162.

[152] A. Mariz, C. Tsallis, E.L. Albuquerque, *Phase diagram of the Ising model on a Cayley tree in the presence of competing interactions and magnetic field*, J. Statist. Phys. **40** (1985) 577–592.

[153] J.B. Martin, *Reconstruction thresholds on regular trees*. Discrete random walks (Paris, 2003), 191–204 (electronic), Discrete Math. Theor. Comput. Sci. Proc., AC, Assoc. Discrete Math. Theor. Comput. Sci., Nancy, 2003.

[154] J.B. Martin, U.A. Rozikov, Y.M. Suhov, *A three-state hard-core model on a Cayley tree*, J. Nonlin. Math. Phys. **12**(3) (2005), 432–448.

[155] F. Martinelli, A. Sinclair, D. Weitz, *Fast mixing for independent sets, colorings, and other models on trees*. Random Structures Algorithms, **31**(2) (2007), 134–172.

[156] A.E. Mazel, Yu.M. Suhov, *Random surfaces with two-sided constraints: an application of the theory of dominant ground states*, J. Statist. Phys. **64** (1991), 111–134.

[157] D.S. McKenzie, M.A. Saqi, *The random bond Ising model on the Bethe lattice*, J. Phys. A **19**(18) (1986), 3883–3890.

[158] J. Meier, *Groups, Graphs and Trees. An Introduction to the Geometry of Infinite Groups*. London Mathematical Society Student Texts, 73. Cambridge Univ. Press, Cambridge, 2008.

[159] M. Mézard, G. Parisi, *The Bethe lattice spin glass revisited*, Eur. Phys. J. B **20** (2001) 217–233.

[160] J. Miller, B. Derrida, *Weak-disorder expansion for the Anderson model on a tree*, J. Statist. Phys. **75**(3-4) (1994), 357–388.

[161] R.A. Minlos, *Introduction to Mathematical Statistical Physics*, University lecture series, AMS, v.19, 2000.

[162] A.D. Mirlin, Y.V. Fyodorov, *Localization transition in the Anderson model on the Bethe lattice: spontaneous symmetry breaking and correlation functions*, Nucl. Phys. B **366**(3) (1991), 507–532.

[163] P. Mitra, K. Ramanan, A. Sengupta, I. Ziedins, *Markov random field models of multicasting in tree networks*, Adv. Appl. Probab. **34**(1) (2002), 1–27.

[164] H. Moraal, *Symmetries of spin models on bipartite lattices*, Phys. A **291**(1-4) (2001), 410–422.

[165] H. Moraal, *Classical, Discrete Spin Models. Symmetry, Duality and Renormalization*. Lecture Notes in Physics, 214. Springer-Verlag, Berlin, 1984.

[166] H. Moraal, *Statistical mechanics of $Z(M)$-models on Cayley trees*, Phys. A **105**(3) (1981), 472–492.

[167] H. Moraal, *Statistical mechanics of quasi-one-dimensional systems*, Phys. A **85**(3) (1976), 457–484.

[168] T. Morita, *The Ising system with an interaction of finite range on the Cayley tree*, Phys. A **83**(2) (1975), 411–418.

[169] E. Mossel, *Reconstruction on trees: beating the second eigenvalue*, Ann. Appl. Probab. **11**(1) (2001), 285–300.

[170] E. Mossel, Y. Peres, *Information flow on trees*, Ann. Appl. Probab. **13**(3) (2003), 817–844.

[171] E. Mossel, *Survey: Information Flow on Trees*. Graphs, morphisms and statistical physics, 155–170, DIMACS Ser. Discrete Math. Theoret. Comput. Sci., 63, Amer. Math. Soc., Providence, RI, 2004.

[172] E. Mossel, *Phase transitions in phylogeny*, Trans. Amer. Math. Soc. **356**(6) (2004), 2379–2404.

[173] E. Mossel, *On the impossibility of reconstructing ancestral data and phylogenies*, J. Comput. Biol. **10**(5) (2003), 669–678.

[174] E. Mossel, M. Steel, *A phase transition for a random cluster model on phylogenetic trees*, Math. Biosci. **187**(2) (2004), 189–203.

[175] J.L. Monroe, *Phase diagrams of Ising models on Husime trees II*, J. Statist. Phys. **67** (1992), 1185–2000.

[176] J.L. Monroe, *A new criterion for the location of phase transitions for spin system on a recursive lattice*, Phys. Lett. A **188** (1994), 80–84.

[177] E. Müller-Hartmann, *Theory of the Ising Model on a Cayley Tree*, Z. Phys. B **27** (1977), 161–168.

[178] F.M. Mukhamedov, M. Saburov, *Phase transitions for XY-model on the Cayley tree of order three in quantum Markov chain scheme*, C. R. Math. Acad. Sci. Paris **349**(7-8) (2011), 425–428.

[179] F.M. Mukhamedov, *On factor associated with the unordered phase of λ-model on a Cayley tree*, Rep. Math. Phys. **53** (2004), 1–18.

[180] F.M. Mukhamedov, U.A. Rozikov, *On the von Neumann algebra corresponding to one phase of the inhomogeneous Potts model on a Cayley tree*, Theor. Math. Phys. **126**(2) (2001), 169–174.

[181] F.M. Mukhamedov, U.A. Rozikov, *On Gibbs measures of models with competing ternary and binary interactions and corresponding von Neumann algebras*, J. Statist. Phys. **114** (2004), 825–848.

[182] F.M. Mukhamedov, U.A. Rozikov, *On Gibbs measures of p-adic Potts model on the Cayley tree*, Indag. Math., New Ser. **15** (2004), 85–100.

[183] F.M. Mukhamedov, U.A. Rozikov, *On inhomogeneous p-adic Potts model on a Cayley tree*, Infin. Dimens. Anal. Quantum Probab. Relat. Top. **8** (2005), 277–290.

[184] F.M. Mukhamedov, U.A. Rozikov, J.F.F. Mendes, *On phase transitions for p-adic Potts model with competing interactions on a Cayley tree*, in p-Adic Mathematical Physics: Proc. 2nd Int. Conf., Belgrade, 2005 (Am. Inst. Phys., Melville, NY, 2006), AIP Conf. Proc. 826, pp. 140–150.

[185] F.M. Mukhamedov, U.A. Rozikov, *On Gibbs measures of models with competing ternary and binary interactions and corresponding von Neumann algebras. II*, J. Statist. Phys. **119**(1/2) (2005), 427–446.

[186] F.M. Mukhamedov, U.A. Rozikov, J.F.F. Mendes, *On contour arguments for the three state Potts model with competing interactions on a semi-

infinite Cayley tree, J. Math. Phys. **48**(1) (2007), 013301, 14 pp.

[187] L. Nirenberg, *Topics in nonlinear functional analysis* AMS, Courant Lec. Notes in Math, 6, N.Y. 2001.

[188] I. Niven, *Mathematics of Choice*, The Math. Assoc. America. V.15. 1965.

[189] E.P. Normatov, U.A. Rozikov, *Description of harmonic functions using group representations of the Cayley tree*, Math. Notes. **79**(3-4) (2006), 399–407.

[190] T. Ogawa, *The Heisenberg model on the Cayley tree.* International Symposium on Mathematical Problems in Theoretical Physics (Kyoto Univ., Kyoto, 1975), pp. 339-341. Lecture Notes in Phys., 39. Springer, Berlin, 1975.

[191] V.R. Ohanyan, L.N. Ananikyan, N.S. Ananikian, *An exact solution on the ferromagnetic face-cubic spin model on a Bethe lattice*, Physica A **377** (2007), 501–513.

[192] M. Ostilli, *Cayley trees and Bethe lattices: A concise analysis for mathematicians and physicists*, Physica A, 2012, http://dx.doi.org/10.1016/j.physa.2012.01.038.

[193] R. Peierls, *On Ising model of ferro magnetism*, Proc. Cambridge Phil. Soc. **32** (1936), 477–481.

[194] R. Pemantle, J.E. Steif, *Robust phase transitions for Heisenberg and other models on general trees*, Ann. Probab. **27**(2) (1999), 876–912.

[195] F. Peruggi, F. di Liberto, G. Monroy, *Phase diagrams of the q-state Potts model on Bethe lattices*, Phys. A **141**(1) (1987), 151–186.

[196] F. Peruggi, F. di Liberto, G. Monroy, *Critical behaviour in three-state Potts antiferromagnets on a Bethe lattice*, Phys. A **131**(1) (1985), 300–310.

[197] F. Peruggi, *Probability measures and Hamiltonian models on Bethe lattices. II. The solution of thermal and configurational problems*, J. Math. Phys. **25**(11) (1984), 3316–3323.

[198] F. Peruggi, *Probability measures and Hamiltonian models on Bethe lattices. I. Properties and construction of MRT probability measures*, J. Math. Phys. **25**(11) (1984), 3303–3315.

[199] F. Peruggi, *The termodynamic limit on Bethe lattices*, J. Phys. A.: Math. Gen. **16**(18) (1983), L713–L719.

[200] F. Peruggi, F. di Liberto, G. Monroy, *The Potts model on Bethe lattices*, I, *General results*, J. Phys. A **16**(4) (1983), 811–827.

[201] S.A. Pirogov, Ya.G. Sinai, *Phase diagrams of classical lattice systems*, I, Theor. Math. Phys. **25** (1975), 1185–1192.

[202] S.A. Pirogov, Ya.G. Sinai, *Phase diagrams of classical lattice systems*, II, Theor. Math. Phys. **26** (1976), 39–49.

[203] V.V. Prasolov, *Polynomials.* Springer, Berlin, 2004.

[204] C. Preston, *Gibbs States on Countable Sets.* Cambridge Univ. Press, London, 1974.

[205] G.B. Price, *Multivariate Analysis*, Springer-Verlag, New York, Berlin, 1984.

[206] B. Prum, J.C. Fort, *Stochastic Processes on Lattices and Gibbs Measures*,

Kluwer, 1991.

[207] K. Ramanan, A. Sengupta, I. Ziedins, P. Mitra, *Markov random field models of multicasting in tree networks*, Adv. Appl. Probab. **34** (2002), 58–84.

[208] F. Redig, *Mathematical aspects of the abelian sandpile model*, in Mathematical Statistical Physics, Lecture Notes of the Les Houches Summer School 2005 (Les Houches), Eds.: Bovier, A. et al., Amsterdam: Elsevier, 2006, pp. 657–728.

[209] A.C.M. van Rooij, *Non-Archimedean Functional Analysis*, M. Dekker, New York, 1978.

[210] U.A. Rozikov, *Structures of partitions of the group representation of the Cayley tree into cosets by finite-index normal subgroups, and their applications to the description of periodic Gibbs distributions*, Theor. Math. Phys. **112**(1) (1997), 929–933.

[211] U.A. Rozikov, *Description of limit Gibbs measures for λ-models on the Bethe lattice*, Siberan Math. J. **39**(2) (1998), 373–380.

[212] U.A. Rozikov, *Construction of an uncountable number of limiting Gibbs measures in the inhomogeneous Ising model*, Theor. Math. Phys. **118**(1) (1999), 77–84.

[213] U.A. Rozikov, *On pure phase of the anti ferromagnetic Potts model on the Cayley tree*, Uzbek Math. J. No. 1 (1999), 73–77 (Russian).

[214] U.A. Rozikov, *Investigation of Gibbs measures of lattice systems and random walks in random environments on Cayley trees*, Doctor of Science Dissertation, Institute of Mathematics, Tashkent, 2000 (Russian).

[215] U.A. Rozikov, *Description of periodic Gibbs measures of the Ising model on the Cayley tree*, Russ. Math. Surv. **56**(1) (2001), 172–173.

[216] U.A. Rozikov, *Representability of trees, and some of their applications*, Math. Notes **72**(3-4) (2002), 479–488.

[217] U.A. Rozikov, *Countably periodic Gibbs measures of the Ising model on the Cayley tree*, Theor. Math. Phys. **130**(1) (2002), 92–100.

[218] U.A. Rozikov, A.M. Rahmatullaev, *Gibbs measures and Markov random fields with association I*, Math. Notes, **72**(1) (2002), 83–89.

[219] U.A. Rozikov, Kh.A. Nazarov, *Periodic Gibbs measures for the Ising model with competing interactions*, Theor. Math. Phys. **135**(3) (2003), 881–888.

[220] U.A. Rozikov, *On q-component models on Cayley tree: contour method*, Lett. Math. Phys. **71**(1) (2005), 27–38.

[221] U.A. Rozikov, Sh.A. Shoyusupov, *Gibbs measures for the SOS model with four states on a Cayley tree*, Theor. Math. Phys. **149**(1) (2006), 1312–1323.

[222] U.A. Rozikov, *A constructive description of ground states and Gibbs measures for Ising model with two-step interactions on Cayley tree*, J. Stat. Phys. **122**(2) (2006), 217–235.

[223] U.A. Rozikov, Yu.M. Suhov, *Gibbs measures for SOS model on a Cayley tree*, Infin. Dimens. Anal. Quantum Probab. Relat. Top. **9**(3) (2006), 471–488.

[224] U.A. Rozikov, *An example of one-dimensional phase transition*, Siberian Adv. Math. **16**(2) (2006), 121–125.

[225] U.A. Rozikov, Sh.A. Shoyusupov, *Fertile three state HC models on Cayley tree*, Theor. Math. Phys. **156**(3) (2008), 1319–1330.

[226] U.A. Rozikov, M.M. Rakhmatullaev, *On weak periodic Gibbs measures of Ising model on Cayley trees*, Theor. Math. Phys. **156**(2) (2008), 1218–1227.

[227] U.A. Rozikov, *A contour method on Cayley trees*, J. Statist. Phys. **130**(4) (2008), 801–813.

[228] U.A. Rozikov, M.M. Rakhmatullaev, *Weakly periodic ground states and Gibbs measures for the Ising model with competing interactions on the Cayley tree*, Theor. Math. Phys. **160**(3) (2009), 1291–1299.

[229] U.A. Rozikov, Yu.Kh. Eshkobilov, *On models with uncountable set of spin values on a Cayley tree: Integral equations*, Math. Phys. Anal. Geom. **13**(3) (2010), 275–286.

[230] U.A. Rozikov, *Ground subgroups.* TWMS J. Pure Appl. Math. **2**(2) (2011), 271–278.

[231] U.A. Rozikov, G. T. Madg'oziev, *Non-uniqueness of Gibbs measure for a model on a Cayley tree*, Theor. Math. Phys. **167**(2) (2011), 669–680.

[232] U.A. Rozikov, *System of recursive equations for the partition functions of 1D models*, Lobachevskii J. Math. **32**(2) (2011), 109–113.

[233] U.A. Rozikov, F.F. Turaev, *p-adic Gibbs states and Markov random fields*, Dokl. AN Uzbekistan. No. 2 (2011), 3-6.

[234] U.A. Rozikov, R.M. Khakimov, *A condition of uniqueness of weakly periodic Gibbs measure for hard core model*, Theor. Math. Phys. **173** (1) (2012), 1377–1386.

[235] L. Šamaj, *Inhomogeneous Ising model on the Bethe lattice.* Phys. A **153**(3) (1988), 517–529.

[236] L. Šamaj, M. Kolesik, *Mapping of the symmetric vertex model onto the Ising model for an arbitrary lattice coordination*, Phys. A **182**(3) (1992), 455–466.

[237] L. Šamaj, M. Kolesik, *Self-duality of the O(2) gauge transformation and the phase structure of vertex models*, Phys. A **193**(1) (1993), 157–168.

[238] W.H. Schikhof, *Ultrametric Calculus*, Cambridge Univ. Press, Cambridge, 1984.

[239] K. Schmidt, E. Verbitskiy, *New directions in algebraic dynamical systems*, Regular and Chaotic Dynamics **16**(1-2) (2011), 79–89.

[240] R.H. Schonmann, *Multiplicity of phase transitions and mean-field criticality on highly non-amenable graphs*, Commun. Math. Phys. **219**(2) (2001), 271–322.

[241] E. Seneta, *Non-Negative Matrices*, Allen, Unwin, London, 1973.

[242] J.-P. Serre, *Trees.* Translated from the French original by John Stillwell. Springer Monographs in Mathematics. Springer-Verlag, Berlin, 2003.

[243] A. N. Shiryaev, *Probability*, 2nd Ed. (Springer, 1996).

[244] Sh. A. Shoyusupov, *Description of Gibbs measures of non symmetrical models on Cayley tree* [in Russian], PhD Dissertation, Institute of Math-

ematics, Tashkent, 2008.

[245] S. Sidki, *Regular trees and their automorphisms*. Monografias de Matematica [Mathematical Monographs], 56. Instituto de Matematica Pura e Aplicada (IMPA), Rio de Janeiro, 1998.

[246] B. Simon, *The Statistical Mechanics of Lattice Gases*, Princeton University Press, 1993.

[247] Ya.G. Sinai, *Theory of Phase Transitions: Rigorous Results*, Pergamon, Oxford, 1982.

[248] M. Steel, *Distributions in bicolored evolutionary trees*. Ph.D. Thesis, Massey University, Palmerston North, New Zealand, 1989.

[249] M. Steel, M. Charleston, *Five surprising properties of parsimoniously colored trees*, Bull. Math. Biol. **57** (1995), 367–375.

[250] F. Spitzer, *Markov random field on infinite tree*, Ann. Probab. **3** (1975), 387–398.

[251] Yu.M. Suhov, U.A. Rozikov, *A hard-core model on a Cayley tree: an example of a loss network*, Queueing Syst. **46**(1/2) (2004), 197–212.

[252] M. Talagrand, *Mean field models for spin glasses*. Vol. I. *Basic examples.* Ergebnisse der Mathematik und ihrer Grenzgebiete. 3. Folge. A Series of Modern Surveys in Mathematics [Results in Mathematics and Related Areas. 3rd Series. A Series of Modern Surveys in Mathematics], 54. Springer-Verlag, Berlin, 2011.

[253] C. J. Thompson, *Local properties of an Ising model on a Cayley tree*, J. Statist. Phys. **27** (1982), 441–456.

[254] C. J. Thompson, *On mean field equations for spin glasses*, J. Statist. Phys. **27** (1982), 457–472.

[255] D.J. Thouless, *Spin-glass on a Bathe latitice*, Phys. Rev. Lett. **56** (1986), 1082–1085.

[256] D.J. Thouless, C. Kwon, *Ising spin glass at zero temperature on the Bethe lattice*, Phys. Rev. B **37**(13) (1988), 7649–7654.

[257] J. Vannimenus, *Modulated phase of an Ising system with competing interactions on a Cayley tree*, Z. Phys. B **43** (1981), 141–148.

[258] V.S. Vladimirov, I.V. Volovich, E.V. Zelenov, *p-Adic Analysis and Mathematical Physics*, World Scientific, Singapore, 1994.

[259] F. Wagner, D. Grensing, J. Heide, *New order parameters in the Potts model on a Cayley tree*, J. Phys. A **34**(50) (2001), 11261–11272.

[260] J.R. Wedagedera, T.C. Dorlas, *The phase diagram of a spin glass model*, J. Statist. Phys., **103**(5-6) (2001), 697–716.

[261] Y.K. Wang, F.Y. Wu, *Multi-component spin model on a Cayley tree*, J. Phys. A: Math. Gen. **9**(4) (1976), 593–604.

[262] J. C. Wheeler, B. Widom, *Phase equilibrium and critical behavior in a two-component Bethe-lattice gas or three-component Bethe-lattice solution*, J. Chem. Phys. **52** (1970), 5334–5343.

[263] B. Widom, J. S. Rowlinson, *New model for the study of liquid-vapor phase transition*, J. Chem. Phys. **52** (1970), 1670–1684.

[264] W.F. Wreszinski, S.R.A. Salinas, *Disorder and competition in soluble lattice models*. Series on Advances in Stat. Mech., 9. World Scientific Publ.

Co., Inc., River Edge, NJ, 1993.

[265] K. Yosida, *Functional Analysis*, Springer-Verlag, 1965.

[266] S. Zachary, *Countable state space Markov random fields and Markov chains on trees*, Ann. Probab. **11** (1983), 894–903.

[267] S. Zachary, *Bounded, attractive and repulsive Markov specifications on trees and on the one-dimensional lattice*, Stoch. Process. Appl. **20** (1985) 247–256.

[268] M. Zahradnik, *An alternate version of Pirogov-Sinai theory*, Comm. Math. Phys. **93** (1984), 559–581.

[269] M. Zahradnik, *A short course on the Pirogov-Sinai theory*, Rendiconti Math. Serie VII **18** (1998), 411–486.

[270] W.M.X. Zimmer, G.M. Obermair, *Spherical model on the Cayley tree*, J. Phys. A **11**(6) (1978), 1119–1129.

Index